AF763777

Handbook of
Biodegradation, Biodetrioration, and Biostabilization

Michalina Falkiewicz-Dulik
Katarzyna Janda
George Wypych

ChemTec Publishing
Toronto 2010

Published by ChemTec Publishing
38 Earswick Drive, Toronto, Ontario M1E 1C6, Canada

ISBN 978-1-895198-44-7

Cover design: Anita Wypych

Library and Archives Canada Cataloguing in Publication

Falkiewicz-Dulik, Michalina
Handbook of material biodegradation, biodeterioration and biostabilization / Michalina Falkiewicz-Dulik, Katarzyna Janda, George Wypych.

Includes bibliographical references and index.
ISBN 978-1-895198-44-7

1. Materials--Biodegradation. 2. Biodegradation. 3. Stabilizing agents. I. Wypych, George II. Janda, Katarzyna III. Title.

QR135.F35 2010 620.1'1223 C2010-902141-X

Printed in United States and United Kingdom

Table of Contents

1	**Introduction**	**1**
1.1	Short historical note	1
1.2	Classification	2
1.3	Definitions	4
	References	*6*
2	**Microorganism Involved in Biodegradation of Materials**	**7**
2.1	General classification of living things	7
2.2	Bacteria	8
2.2.1	Actinobacteria	9
2.2.2	Bacteroidetes/Chlorobi	11
2.2.3	Chlamydiae/Verrucomicrobiae	12
2.2.4	Chloroflexi	13
2.2.5	Cyanobacteria	13
2.2.6	Fibrobacteres/Acidobacteria	14
2.2.7	Firmicutes	14
2.2.8	Fusobacteria	17
2.2.9	Nitrospirae	17
2.2.10	Planctomycetes	17
2.2.11	Proteobacteria	17
2.2.12	Thermodesulfobacteria	23
2.2.13	Thermotogae	23
2.3	Fungi	23
2.3.1	Ascomycota	24
2.3.2	Basidiomycota	27
2.3.3	Blastocladiomycota	28
2.3.4	Chytridiomycota	28
2.3.5	Glomeromycota	29
2.3.6	Microsporidia	29
2.3.7	Neocallimastigomycota	29
2.4	Protozoa	29

2.5 Biodegradation & biodeterioration mechanisms 30
References *39*

3 Industrial Biocides 41
3.1 General mechanism of biostabilization 41
3.2 Chemical types of biostabilizers 47
3.2.1 Acetal aldehyde-releasing compounds 48
3.2.2 Acid esters 49
3.2.3 Acids 50
3.2.4 Active halogen products 51
3.2.5 Alcohols 53
3.2.6 Aldehydes 54
3.2.7 Amides 55
3.2.8 Azoles 57
3.2.9 Carbamates 58
3.2.10 Formaldehyde-releasing compounds 60
3.2.11 Haloalkylthio compounds 61
3.2.12 Heterocyclic N,S compounds 63
3.2.13 Metal-containing products 64
3.2.14 Oxidizing agents 66
3.2.15 Phenolics 67
3.2.16 Polymeric materials 68
3.2.17 Pyridine derivatives 69
3.2.18 Quaternary ammonium compounds and other surface active agents 70
3.2.19 Other (not included) products 72
3.3 Principles of selection of biostabilizers 73
3.4 Longevity of biostabilized materials 76
References *78*

4 Biodegradation, Biodeterioration, and Biostabilization of Industrial Products 81
4.1 Building materials 82
4.2 Coatings and paints 87
4.3 Cultural heritage 97
4.4 Dental materials 101
4.5 Electrical and electronic products 104

4.6 Fibers and textiles 106
4.7 Leather and leather products 110
4.7.1 General information 110
4.7.2 Rawhide and skin 113
4.7.2.1 Structure 113
4.7.2.2 Composition of raw leather 113
4.7.2.3 Damage caused to the animal 114
4.7.2.4 Damage to a hide by the autolysis 117
4.7.2.5 Bacterial deterioration of fresh skins 118
4.7.2.6 Bacterial deterioration of salted hides 120
4.7.2.7 Microorganisms in the skin processing 124
4.7.3 Leather 132
4.7.3.1 General data 132
4.7.3.2 The microbiological decomposition of tanned leather 133
4.7.4 Contamination of shoe materials 134
4.7.4.1 General data 134
4.7.4.2 Microbiological aspects of the use of footwear 135
4.7.5 Leather and parchment in library materials 142
4.7.5.1 General data 142
4.7.5.2 Archaeological leather 143
4.7.5.3 Hard leather bindings 144
4.7.5.4 Parchment 147
4.7.6 Protection of skin and leather from microflora 153
4.7.6.1 General data 153
4.7.6.2 Preservation of raw hides 153
4.7.6.3 Chemical preservatives 159
4.7.6.4 Skin protection against microorganisms in the tanning processes 164
4.7.6.5 The use of biocides in the leather industry, and legislation, health and environment 170
4.7.6.6 Antimicrobial shoe materials and footwear 173
4.7.6.7 Protection of heritage leather products against germs 175
4.8 Marine transport and installations 184
4.9 Medical applications 194
4.10 Metals 197
4.11 Mineral dispersions 200
4.12 Petroleum products 202

4.12.1. General data 202
4.12.2. Microorganisms in fuels 203
4.12.2.1 Microbes in the aviation fuels and jet fuels 206
4.12.2.2 Microbes in the automotive fuel 210
4.12.2.3 Microbes in marine fuels 211
4.12.2.4 Microorganisms in gasoline 212
4.12.2.5 Microorganisms in heating oil 212
4.12.3 Microorganisms in lubricating oils 214
4.12.4 Microbes in the hydraulic oils 216
4.12.5 Microorganisms in turbine oils 217
4.12.6 Microorganisms in transformer oils 217
4.12.7 Microorganisms in oil emulsions 218
4.12.8. Microbes in the oil-water emulsions used in hydraulic systems in coal mines 224
4.12.9 Microbes in asphalt and insulating materials 226
4.12.10 Biocides in the protection of petroleum products 227
4.13 Pharmaceuticals, cosmetics, and toiletries 235
4.14 Polymers 238
4.14.1 Acrylics 238
4.14.2 Acrylonitrile-butadiene-styrene copolymer 241
4.14.3 Alkyd resins 241
4.14.4 Epoxy resin 243
4.14.5 Ethylene propylene copolymers 244
4.14.6 Ionomer 244
4.14.7 Phenolic resins 245
4.14.8 Polyamides 246
4.14.9 Polycarbonate 248
4.14.10 Polyetheretherketone 250
4.14.11 Polyethylene 250
4.14.12 Poly(ethylene oxide) 252
4.14.13 Poly(ethylene terephthalate) 253
4.14.14 Polyimide 254
4.14.15 Polymethylmethacrylate 255
4.14.16 Polyoxymethylene 255
4.14.17 Polypropylene 256
4.14.18 Polystyrene 257
4.14.19 Polysulfone 258

4.14.20 Polytetrafluoroethyelene 259
4.14.21 Polyurethanes 259
4.14.22 Poly(vinyl alcohol) 262
4.14.23 Poly(vinyl chloride) 263
4.14.24 Silicone 265
4.15 Pulp and paper 266
4.16 Roofing materials 269
4.17 Rubber 271
4.18 Sealants and adhesives 273
4.19 Stones including cultural heritage 275
4.20 Swimming pools 279
4.21 Water 281
4.22 Wood 283

5 Analytical methods in biodegradation, biodeterioration, and biostabilization 291
5.1 Standards 291
5.1.1 Adhesives and sealants 291
5.1.2 Antifouling coatings 291
5.1.3 Antiseptic drugs and handwash 292
5.1.4 Chemical materials in general 293
5.1.5 Coatings and paints 293
5.1.6 Cooling water systems 294
5.1.7 Detergents 294
5.1.8 Fuels and fuel systems 294
5.1.9 Geomembranes and geotextiles 295
5.1.10 Hydraulic fluids 295
5.1.11 Lubricants 295
5.1.12 Lumber, pallets, and wood boxes 296
5.1.13 Metalworking fluids 296
5.1.14 Oilfield and refinery 296
5.1.15 Oil spill response 296
5.1.16 Packaging 297
5.1.17 Paper 297
5.1.18 Plastics and polymers 297
5.1.19 Stone consolidants 298
5.1.20 Surgical implants and medical devices 298

5.1.21 Water systems 299
5.2 Other analytical methods 299
References *302*

6 Biostabilizers - Health & Safety **309**
6.1 Toxic substance control 309
6.2 Carcinogenic effect 311
6.3 Workplace exposure limits 312
6.4 Food regulatory acts 315
References *316*

7 Environmental Fate of Biostabilizers **317**
7.1 Concentration 317
7.2 Toxicity 320
7.3 Decay 322
References *326*

8 Legislation **329**
8.1 European Union 329
8.2 International 331
8.3 USA 332
References *333*

9 Personal Protection **335**
9.1 Clothing 335
9.2 Gloves 336
9.3 Eye protection 338
9.4 Respiratory protection 339
References *342*

Index **345**

Acknowlegment **367**

1

INTRODUCTION

1.1 SHORT HISTORICAL NOTE

A large percentage of the world's monuments is made from stone. It has been estimated that, in the case of limestone, 1.5–3 mm of its surface layer may erode away in 100 years in temperate climates.[1] This may cause the disappearance of inscriptions on tombstones in Europe within 300 years.[1] On one hand, this is an alarming prospect, calling for protective action, and, on the other, the transformation of stone into sand and soil is a natural recycling process, essential to sustain life on earth.

Stone is considered as one of the most durable materials. It is therefore not surprising that protection of products against deterioration is such an important matter and an essential part of production know-how.

It is not possible to establish who was the first to come up with an idea of product preservation because this must have been one of the earlier tasks undertaken by humans, but usually preservation of food and human remains is considered one of the first applications of chemical treatment to preserve material. In mummification processes used in Egypt, palm oil, spices, and natron salt were the first chemical materials used for the purpose of protection against deterioration.[2]

In the area of food preservation, many chemical methods have been used, such as pickling, smoking, salting, etc. most likely for thousands years. But water cleaning may serve here as a very good example since it is also relevant today but was designed thousands of years ago. Romans used silver coins for their bactericide properties to protect water from contamination.[3] Interestingly, a similar method was used aboard Apollo spacecraft, MIR station, and the space shuttle.[3] Alexander the Great was advised by Aristotle to store water in silver vessels during his campaigns.[4] In 1884, a German obstetrician (Crede) used a silver nitrate solution for medical purposes, but silver is now commonly used in bactericidal bandages.[3]

Antifouling boats and ships were used by ancient Greeks and Romans.[5] Their vessels had lead sheeting secured by copper nails. This was only replaced by tar paints in the XV-XVI centuries (Columbus and the English Navy).[5] Many toxic chemicals such as arsenic and mercury have been added to resinous binders in the XIX and the first half of the XX century to replace tar paints. After the Second World War, tributyltin was introduced to antifouling paints which was later found to affect aquatic life.[5] The so-called self-polishing copolymer paints containing tributyltin were used on 70-80% of world's fleets during the late 1990s.[6]

Other materials were initially produced on a small scale, which did not warrant a search for chemical methods of product preservation. Only later, the development of industrial scale production made durability and life span of products to became an important part of quality and new methods of treatment were being developed. Below are a few examples.

Before 1926, zinc chloride was the most frequently used fungicide for textiles, along with salicylic acid, phenol, cresylic acid, and formaldehyde.[7]

The modern cosmetics industry was developed in the 1930s. The major technical problem then was to prevent visible mold growth. This prompted the use of bactericidal preservatives.[8]

Metal corrosion was first linked to microorganisms in 1922 by von Wolzogen Kuhr who postulated that anaerobic microorganisms contribute to iron corrosion through removal of hydrogen accumulated at the cathodic site on surface.[9]

Wood preservation was targeted by ancient Greeks and Romans but with little success. Effective wood preservation began in the mid-1830s with the patenting of the Bethell process and the development of creosote.[10]

1.2 CLASSIFICATION

The Biocidal Products Directive 98/8/EC developed by the European Union is the most convenient method of classification of products involved in prevention of biodegradation and biodeterioration of products and materials. Table 1.1 gives details of this classification.

Table 1.1. Description of biocidal products based on Directive 98/8/EC

#	Product type	Description
	Main Group 1: Disinfectants and general biocidal products	
1	human hygiene biocidal products	used for human hygiene purposes
2	private and public health area disinfectants and other biocidal products	used for the disinfection of air, surfaces, materials, equipment and furniture which are not used for direct food or feed contact in private, public or industrial areas, including hospitals, as well as products used as algaecides. Usage areas include swimming pools, aquariums, bathing and other waters, air-conditioning units, walls and floors in health and other institutions, chemical toilets, waste water, hospital waste, soil and other substrates (in playgrounds)
3	veterinary hygiene biocidal products	includes products used in areas in which animals are housed, kept, or transported
4	food and feed area disinfectants	used for the disinfection of equipment, containers, consumption utensils, surfaces or pipework associated with the production, transport, storage, or consumption of food, feed or drink (including drinking water) for humans and animals
5	drinking water disinfectants	for both humans and animals
	Main Group 2: Preservatives	
6	in-can preservatives	used for the preservation of manufactured products, other than foodstuffs or feeding stuffs, in containers by the control of microbial deterioration to ensure their shelf-life

Table 1.1. Description of biocidal products based on Directive 98/8/EC

#	Product type	Description
7	film preservatives	used for the preservation of films or coatings by the control of microbial deterioration in order to protect the initial properties of the surface of materials or objects such as paints, plastics, sealants, wall adhesives, binders, papers, art works etc.
8	wood preservatives	for wood from and including saw-mill stage, and wood products (including preventative and curative products)
9	fibre, leather, rubber and polymerized materials preservatives	includes the preservation of fibrous materials, such as paper or textile products
10	masonry preservatives	used for the preservation and remedial treatment of masonry or other construction materials other than wood by the control of microbiological algal attack
11	preservatives for liquid-cooling and processing systems	use for the preservation of water and other liquids used in cooling and processing systems by the control of harmful organisms such as microbes, algae and mussels (not drinking water preservation products)
12	slimicides	used for the prevention or control of slime growth on materials, equipment and structures, used in industrial processes, e.g., on wood and paper pulp, and porous sand strata in oil extraction
13	metalworking-fluids preservatives	products used for the preservation of metalworking fluids by the control of microbial deterioration
Main Group 3: Pest control		
14	rodenticides	control of mice, rats or other rodents
15	avicides	control of birds
16	molluscicides	control of molluscs, e.g., snails that may clog pipes
17	piscicides	control of fish; excludes products for the treatment of fish diseases
18	insecticides, acaricides and to control other arthropods	for example insects, arachnids and crustaceans
19	repellents or attractants	used to control harmful organisms (invertebrates such as fleas, vertebrates such as birds), by repelling or attracting, including those that are used for human or veterinary hygiene either directly or indirectly
Main Group 4: Other biocidal products		
20	preservatives for food and feedstocks	used for the preservation of food or feedstuffs by the control of harmful organisms
21	antifouling products	used to control growth and settlement of fouling organisms (microbes and higher forms of plant and animal species) on vessels, aquaculture equipment or other structures used in water
22	embalming or taxidermist fluids	used for the disinfection and preservation of human or animal corpses, or parts of them

Table 1.1. Description of biocidal products based on Directive 98/8/EC

#	Product type	Description
23	control of vertebrates	used to control vermin

Table 1.1 shows that, according to this classification, there are 4 main groups and 23 product groups involved in protection of various materials. For the purpose of this book, the second main group (Main Group 2: Preservatives) is the most important but some representatives of other groups will also be included in the discussion.

1.3 DEFINITIONS

Some terms used in this book are defined for the purpose of clarity (Table 1.2), and some selected terms are further discussed at the end of this section.

Table 1.2. Definition of some important terms.[9-11]

Term	Definition
alteration	change in materials that does not necessarily involve a worsening in their conservative characteristics
antimicrobial	compounds that kill or inhibit the growth of microbes – used as disinfectants or preservatives
antimicrobial pesticides	compounds intended to disinfect, sanitize, reduce, or mitigate growth or development of microbiological organisms, or to protect inanimate objects, industrial processes or systems, surfaces, water, or other chemical substances from contamination, fouling, or deterioration caused by bacteria, viruses, fungi, protozoa, algae, or slime
antiseptic	a substance which destroys or inhibits microorganisms on living tissues, thereby limiting or preventing the harmful results of infection
biodegradation	the harnessing, by man, of the decay abilities of organisms to render a waste material more useful or acceptable
biodegradation	the metabolic breakdown of materials into simpler components by living organisms
biodegradation	the destruction of organic compounds by microorganisms
biodeterioration	every form of irreversible alteration that causes changes/modifications in the properties of the materials due to metabolic activity and the development of one or more microbial population
biodeterioration	any undesirable change in the properties of a material caused by the vital activities of organism
biodeterioration (general)	a process in the environment that recycles complex organic matter (and materials), and is an integral component of the life cycle
bioreceptivity	tendency of a material to be colonized by one or more organisms not always linked to a biodeterioration process
cathodic polarization	consumption of hydrogen accumulated on the surface of metals by microorganisms attached on the surface
decay	change in materials that always involves a worsening in their conservative characteristics

Table 1.2. Definition of some important terms.[9-11]

Term	Definition
disinfection	killing microbes on surfaces to levels that are not harmful to health or the quality of the product
efflorescence	presence of a deposit of inorganic substances on stone or masonry surfaces that was not originally there in the sound material
foxing	deterioration of paper occurring as brownish, reddish stains because of the activity of microorganisms (majority) or the oxidation and/or formation of heavy metals deposits
hydrogen embrittlement	permeation of molecular hydrogen into the metal matrix, resulting in crack and structure deformation
microbial-induced corrosion	an electrochemical process catalyzed by microorganisms through formation of electrochemical potential between the cathode and the anode, cathode depolarization, and attack on metals by generated corrosive products
patina	originally defined by Filippo Baldinucci (Dictionary of Art, 1681) as the time-dependent darkening of frescos and oil painting. Today, the term denotes the sum of all changes that the surface layers of objects of art are submitted to and inherit from environmental, biological, and substrate-related factors
pest	any insect, rodent, nematode, fungus, weed, or any other form of terrestial, aquatic plant or animal life or virus, bacteria, or other microorganisms
pesticide	any substance or mixture of substances intended for preventing, destroying, repelling, or mitigating any pest
preservative	a chemical agent used to prevent microbial growth in finished products. It prevents their multiplication or kills them to prevent spoilage or contamination of the product with pathogens
sanitizer	a chemical agent used to disinfect equipment

There is a major difference in definitions in Europe and in the USA. In Europe, antimicrobial compounds are called biocides, whereas in the USA they are called pesticides or closer antimicrobial pesticides.

In some sections of the scientific community, a strong distinction is being proposed between the term of biodegradation and biodeterioration.[12] According to such a distinction, biodegradation is a modification to useful products whereas biodeterioration converts useful products to wastes. Although there are some merits in attempts to distinguish between good processes (bioremediation or conversion of biodegradable materials) and bad processes (damage to products by microorganisms), this does not justify changes to the meaning of "degradation", which means "changing to a lower state" or "a decline to a lower condition, quality, or level", and addition of "bio" only means that it is being done with the participation of living organisms. Deterioration is defined as "a symptom of reduced quality or strength" or "diminishing or impairing in quality, character, or value". There is no general difference between meanings of degradation and deterioration, and, if there is, it is more related to the extent of processes which occurred than the quality of changes (bad or good). It could be derived from the last sentence that degradation is more of a microscopic phenomenon and deterioration has a visible, macroscopic character which disqualifies a product from further use or limits its use.

Considering that all these differences are about feelings rather than merits, we will use both terms in the book interchangeably.

REFERENCES

1 Scheerer, S; Ortega-Morales, O; Gaylarde, C, *Adv. Appl. Microbiology*, **66**, 97-139, 2009.
2 Lewis, B; Lewis, R, **Tutankhamun and Golden Treasures of Ancient Egypt**. *Star Fire*, London, 2007.
3 Jain, J; Arora, S; Rajwade, J; Khandelwal, S; Paknikar, K M, *Mol. Pharmaceutics*, 1-56, 2009.
4 Russell, A D; Hugo, W B, *Progress in Medical Chemistry*, **31**, 351-70, 1994.
5 Readman, J W, **Antifouling Paint Biocides**, Konstantinou, I K, Ed., *Springer*, Berlin, 2006.
6 Yebra, D M; Kill, S; Weinell, C E; Dam-Johansen, K, *Biofouling*, **22**, 1, 33-41, 2006.
7 Montegut, D; Indictor, N; Koestler, R J, *Int. Biodeterioration*, **28**, 209-226, 1991.
8 Brannan, D K; Geis, P A, Eds., **Cosmetic Microbiology**, *Elsevier*, 2009.
9 Ranalli, G; Zanardini, E; Sorlini, C, Eds., **Encyclopedia of Microbiology**, *Elsevier*, 2009.
10 Morrell, J J, **Encyclopedia of Materials: Science and Technology**, *Elsevier*, 2008.
11 Federal Insecticide, Fungicide, and Rodenticide Act, 7 U.S.C. § 136.
12 Allsopp, D; Seal, K J; Gaylarde, C C, **Introduction to Biodeterioration**, 2nd Ed., *Cambridge University Press,* 2004.

2

Microorganism Involved in Biodegradation of Materials

2.1 GENERAL CLASSIFICATION OF LIVING THINGS

The modern classification system (taxonomic system) of living things was developed by the Swedish botanist Carl Linneaeus in the XVIII century. It is a hierarchical system explained by Figure 2.1.

Domain	Archea	Eubacteria	Eukaryote			
Kingdom	Archeabacteria	Monera	Fungi	Protista	Plantae	Animalia
Phylum						
Class						
Order						
Family						
Genus						
Species						

Figure 2.1. Taxonomic system of classification of living things (six kingdoms classification).

The open sides of cells symbolize that any level of hierarchical system is likely to be further divided until species are reached.

Out of six kingdoms, plants (Plantae) and animals (Animalia) are not the subject of this book. Archeabacteria live in extreme environments, such as salt lakes or hot, acidic springs. They have unique properties and features which distinguish them from other bacteria. Their unique properties permit them to live in these unusual environments.

The Protista kingdom includes unicellular organisms such as amoeba, algae, and slime moulds. They are of interest in studies on material biodegradation. The Monera kingdom includes unicellular bacteria which are known to cause deterioration of materials along with fungi which are unicellular and monocellular organisms such mushrooms, moulds, and yeast.

It should be noted that the attempts of classification seem to be far from over and many new ideas are being proposed. One more recent classification was proposed by Cavalier-Smith.[1,2] This classification is given in Figure 2.2. The classification given in Figure 2.2 has some followers, including those who work on processes of biodegradation of materials.

Empires	Kingdoms				
Prokaryota	Bacteria				
Eukaryota	Animalia	Plantae	Fungi	Chromista	Protozoa

Figure 2.2. Classification of living things.[1,2]

Two Empires are proposed in place of three Domains. Two Domains: Archea and Eubacteria were combined into one Empire, Prokaryota, which has only one Kingdom: Bacteria. In addition, Kingdom Protista has been divided into two Kingdoms: Chromista and Protozoa.

Prokaryota do not have cell nuclei or any other membrane-bound organelles. They differ from the Eukaryota, which have a cell nuclei, containing DNA. The DNA of Prokaryota is stored in nucleoids whereas Eukaryota have DNA organized in chromosomes. One important distinction between the two Empires is that Prokaryota have a larger surface area to volume ratio, which makes them to have a higher growth rate and higher metabolic rate than Eukaryota.

Chromista is the phylogenetic (composed of many races) group of golden-brown algae that includes algae having chloroplasts containing chlorophylls a and c and various colorless forms that are closely related to them. The name Chromista means colored. It includes such diverse organisms as microscopic diatoms, having delicate silica skeletons only 40 μm long, giant kelps, which may grow as long as fifty meters, or mildew.

Protozoa is the kingdom of protozoans. The protozoa are one-celled species considered to be the smallest animals. They can breathe, move, and reproduce like larger animals. The group includes the amoeba, the paramecium (larger than amoeba), and the euglena.

2.2 BACTERIA

Table 2.3 gives information on classification of Kingdom Bacteria.

Table 2.3. Classification of Bacteria Kingdom.

Phylum name	Classes	Applicability
Actinobacteria	*Acidimicrobidae, Actinobacteridae, Coriobacteridae, Rubrobacteridae*	yes
Aquificae	*Aquificae*	no
Bacteroidetes/Chlorobi	*Bacteroidetes, Chlorobia, Flavobacteria, Sphingobacteria*	yes
Chlamydiae/ Verrucomicrobiae	*Chlamydiae, Lentisphaerae, Verrucomicrobiae*	yes
Chloroflexi	*Anaerolineae, Caldilineae, Chloroflexi, Thermomicrobia*	yes
Chrysiogenetes	*Chrysiogenetes*	no
Cyanobacteria	*Gleobacteria,* unclassified *Cyanobacteria*	yes

Table 2.3. Classification of Bacteria Kingdom.

Phylum name	Classes	Applicability
Deferribacteres	*Deferribacteres*	no
Deinococcus-Thermus	*Deinococci*	no
Dictyoglomi	*Dictyoglomi*	no
Fibrobacteres/ Acidobacteria	*Acidobacteria, Fibrobacteres, Holophagae, Solibacteres*	yes
Firmicutes	*Bacilli, Clostridia, Erysipelotrichi, Thermolithobacteria*	yes
Fusobacteria	*Fusobacteria*	yes
Gemmatimonadetes	*Gemmatimonadetes*	no
Nitrospirae	*Nitrospira*	yes
Planctomycetes	*Planctomycetia*	yes
Proteobacteria	*Alphaproteobacteria, Betaproteobacteria, Deltaproteobacteria, Epsilonproteobacteria, Gammaproteobacteria*	yes
Spirochaetes	*Spirochetes*	no
Synergistetes	*unclassified Synergistetes*	no
Tenericutes	*Mollicutes*	no
Thermodesulfobacteria	*Thermodesulfobacteria*	yes
Thermotogae	*Thermotogae*	yes

Table 2.3 shows that there are 22 distinct groups with taxonomic Phylum status included in classification of Kingdom Bacteria. Some of these Phyla contain only one class but some Phyla have diversity of species, for which they need several Classes to include classified species.

Applicability in Table 2.3 determines whether a particular group (Phylum) participates in deterioration of materials. The groups important to our subject are discussed in more detail below.

2.2.1 ACTINOBACTERIA

Actinobacteria is a Gram-positive bacteria which contains a large number of species, which are antibiotic producers, soil bacteria, medically important patogens, and very resilient species which can grow in a hostile polluted environment. Table 2.4 shows the names of families and genuses included in *Actinobacteria.*

Table 2.4. Structure of Actinobacteria

Phylum	Family	Genera
Actinobacteria	*Acidimicrobiaceae*	*Acidimicrobium*
	Acidothermaceae	*Acidothermus*
	Actinosynnemataceae	*Actinokineospora, Actinosynnema, Lechevalieria, Lentzea, Saccharothrix, Saccharothrixopsis*
	Beutenbergiaceae	*Beutenbergia, Georgenia, Salana*

Table 2.4. Structure of Actinobacteria

Phylum	Family	Genera
Actinobacteria	*Bifidobacteriaceae*	*Aeriscardovia, Alloscardovia, Bifidobacterium, Falcivibrio, Gardnerella, Metascardovia, Parascardovia, Scardovia*
	Bogoriellaceae	*Bogoriella*
	Brevibacteriaceae	*Brevibacterium*
	Cellulomonadaceae	*Actinotalea, Cellulomonas, Oerskovia, Tropheryma*
	Corynebacteriaceae	*Corynebacterium*
	Dermabacteraceae	*Brachybacterium, Dermabacter, Devriesea*
	Dermacoccaceae	*Demetria, Dermacoccus, Kytococcus*
	Dermatophilaceae	*Dermatophilus, Kineosphaera*
	Dietziaceae	*Dietzia*
	Frankiaceae	*Frankia*
	Geodermatophilaceae	*Actinotelluria, Blastococcus, Geodermatophilus, Modestobacter*
	Glycomycetaceae	*Glycomyces, Stackebrandtia*
	Gordoniaceae	*Gordonia*
	Intrasporangiaceae	*Arsenicicoccus, Humihabitans, Intrasporangium, Janibacter, Knoellia, Kribbia, Ornithinicoccus, Ornithinimicrobium, Oryzihumus, Serinicoccus, Terrabacter, Terracoccus, Tetrasphaera*
	Jonesiaceae	*Jonesia*
	Kineosporiaceae	*Cryptosporangium, Kineococcus, Kineosporia*
	Microbacteriaceae	*Agreia, Agrococcus, Agromyces, Aureobacterium, Clavibacter, Crustibacterium, Cryobacterium, Cryocola, Curtobacterium, Frigoribacterium, Frondicola, Gulosibacter, Humibacter, Labedella, Leifsonia, Leucobacter, Microbacterium, Microcella, Mycetocola, Okibacterium, Plantibacter, Pseudoclavibacter, Rathayibacter, Rhodoglobus, Salinibacterium, Subtercola, Yonghaparkia, Zimmermannella*
	Micrococcaceae	*Acaricomes, Arthrobacter, Citricoccus, Kocuria, Micrococcus, Nesterenkonia, Renibacterium, Rothia, Sinomonas, Zhihengliuella*
	Micromonosporaceae	*Actinoplanes, Catellatospora, Catenuloplanes, Couchioplanes, Dactylosporangium, Microbispora, Micromonospora, Micropolyspora, Pilimelia, Salinispora, Spirilliplanes, Thermoactinomyces, Thermomonospora, Verrucosispora*
	Mycobacteriaceae	*Mycobacterium, Nocardia, Rhodococcus*
	Nakamurellaceae	*Nakamurella*
	Nocardiaceae	*Nocardia, Rhodococcus, Smaragdicoccus, Micropolyspora*
	Nocardioidaceae	*Aeromicrobium, Friedmanniella, Hongia, Kribbella, Marmoricola, Micropruina, Nocardioides*
	Nocardiopsaceae	*Nocardiopsis, Streptomonospora, Thermobifida*
	Promicromonosporaceae	*Cellulosimicrobium, Isoptericola, Myceligenerans, Promicromonospora, Xylanibacterium, Xylanimicrobium, Xylanimonas*
	Propionibacteriaceae	*Aestuariimicrobium, Friedmanniela, Luteococcus, Microlunatus, Micropruina, Propionibacterium, Propionimicrobium, Propioniferax, Tessaracoccus*

Table 2.4. Structure of Actinobacteria

Phylum	Family	Genera
Actinobacteria	*Pseudonocardiaceae*	*Amycolatopsis, Pseudonocardia, Saccharomonospora, Saccharopolyspora*
	Rarobacteraceae	*Rarobacter*
	Sanguibacteraceae	*Sanguibacter*
	Segniliparaceae	*Segniliparus*
	Sporichthyaceae	*Sporichthya*
	Streptomycetaceae	*Kitasatospora, Streptomyces, Streptoverticillium*
	Streptosporangiaceae	*Acrocarpospora, Astrosporangium, Clavisporangium, Herbidospora, Microbispora, Microtetraspora, Nonomuraea, Planobispora, Planomonospora, Planotetraspora, Sphaerisporangium, Streptosporangium, Thermopolyspora*
	Thermomonosporaceae	*Actinocorallia, Actinomadura, Spirillospora, Thermomonospora*
	Tsukamurellaceae	*Tsukamurella*
	Williamsiaceae	*Williamsia*
	Yaniellaceae	*Yaniella*

Actinobacteria is one of the largest phyla in the bacterial kingdom. It is very diverse from the point of view of its morphology, including branching mycelium, cocoids, and rods. These bacteria are found in different environments, such as medieval wall paintings, desert soil, butter, marine sponges, and radon-containing hot springs.[3] *Actinobacteria* is known to produce metabolites with antibacterial and antifungal properties.

2.2.2 BACTEROIDETES/CHLOROBI

Bacteroidetes are Gram-negative bacteria mostly associated with gastrointestinal colonization and marine sediments. *Chlorobi* is the one of five phyla (*Firmicutes, Chloroflexi, Chlorobi, Proteobacteria, and Cyanobacteria*) with members capable of chlorophyll-based phototrophy. *Bacteroidetes* are found on medieval paintings and stone monuments. They also cause fouling on painted and unpainted concrete structures. *Chlorobi* are green sulfur bacteria mostly found in a marine environment. Table 2.5 gives the structure of this phylum.

Table 2.5. Bacteroidetes/Chlorobi

Phylum	Family	Genera
Bacteroidetes	*Bacteroidaceae*	*Acetofilamentum, Acetomicrobium, Acetothermus, Anaerorhabdus, Bacteroides, Fusobacterium, Leptotrichia, Megamonas, Porphyromonas, Selenomonas*
	Blattabacteriaceae	*Blattabacterium*
	Cryomorphaceae	*Cryomorpha*
	Flammeovirgaceae	*Fabibacter, Flammeovirga, Flexithrix, Fulvivirga, Limibacter, Perexilibacter, Persicobacter, Rapidithrix, Reichenbachiella, Roseivirga, Sediminitomix, Thermonema*

Table 2.5. Bacteroidetes/Chlorobi

Phylum	Family	Genera
Bacteroidetes	*Flexibacteraceae*	*Adhaeribacter, Arcicella, Candidatus Magnospira, Cytophaga, Dyadobacter, Effluviibacter, Ekhidna, Emticicia, Flectobacillus, Flexibacter, Hymenobacter, Larkinella, Leadbetterella, Marinoscillum, Microscilla, Persicitalea, Pontibacter, Rhodocytophaga, Rhodonellum, Rudanella, Runella, Spirosoma, Sporocytophaga*
	Flavobacteriaceae	*Actibacter, Aequorivita, Aestuariicola, Algibacter, Antarcticimonas, Aquimarina, Arenibacter, Ascidianibacter, Aureimarina, Aureimonas, Bergeyella, Bizionia, Candidatus Amoebinatus, Capnocytophaga, Cellulophaga, Chryseobacterium, Cloacibacterium, Coccinimonas, Coccinistipes, Coenonia, Coralibacter, Costertonia, Croceibacter, Croceimarina, Croceitalea, Dokdonia, Elizabethkingia, Empedobacter, Epilithonimonas, Eudoraea, Flagellimonas, Flaviramulus, Flavobacterium, Flavococcus, Formosa, Fulvibacter, Gaetbulibacter, Galbibacter, Gelidibacter, Gillisia, Gilvibacter, Gramella, Haloanella, Joostella, Kaistella, Kordia, Kriegella, Krokinobacter, Lacinutrix, Leeuwenhoekiella, Leptobacterium, Lutibacter, Lutimonas, Maribacter, Mariniflexile, Marixanthomonas, Mesoflavibacter, Mesonia, Muricauda, Nonlabens, Oceanistipes, Olleya, Orabacterium, Ornithobacterium, Persicivirga, Pibocella, Planobacterium, Polaribacter, Poriferibacter, Pseudozobellia, Psychroflexus, Psychroserpens, Riemerella, Robiginitalea, Salegentibacter, Salibacter, Salinimicrobium, Sandarakinotalea, Sediminibacter, Sediminicola, Sediminitalea, Sejongia, Stenothermobacter, Subsaxibacter, Subsaximicrobium, Sufflavibacter, Tamlana, Tenacibaculum, Ulvibacter, Vitellibacter, Wautersiella, Weeksella, Winogradskyella, Yeosuana, Zeaxanthinibacter, Zhouia, Zobellia, Zunongwangia*
	Marinilabiaceae	*Alkaliflexus, Anaerophaga, Marinilabilia*
	Myroidaceae	*Myroides, Psychromonas*
	Porphyromonadaceae	*Barnesiella, Butyricimonas, Candidatus Vestibaculum, Dysgonomonas, Odoribacter, Paludibacter, Parabacteroides Petrimonas, Porphyromonas, Proteiniphilum, Tannerella*
	Prevotellaceae	*Hallella, Paraprevotella, Prevotella, Xylanibacter*
	Rikenellaceae	*Alistipes, Alkaliflexus, Marinilabilia, Petrimonas, Rikenella, Ruminofilibacter*
	Saprospiraceae	*Aureispira, Candidatus Aquirestis, Haliscomenobacter, Lewinella, Saprospira*
	Sphingobacteriaceae	*Mucilaginibacter, Nubsella, Olivibacter, Parapedobacter, Pedobacter, Pseudosphingobacterium, Sphingobacterium, Sphingoterrabacterium*
Chlorobi	*Chlorobiaceae*	*Ancalochloris, Chlorobaculum, Chlorobium, Chloroherpeton, Clathrochloris, Pelodictyon, Prosthecochloris*

2.2.3 CHLAMYDIAE/VERRUCOMICROBIAE

Chlamydiae are Gram-negative bacteria that are pathogenic for humans and animals. They are as small as, or smaller than, many viruses. *Chlamydiae* were identified as participating in biodeterioration of the Lions Fountain at the Alhambra Palace. They are also present in decaying wood. Table 2.6 gives the structure of this phylum.

Table 2.6. Chlamydiae/Verrucomicrobiae.

Phylum	Family	Genera
Chlamydiae	*Chlamydiaceae*	*Chlamydia, Chlamydophila*
	Criblamydiaceae	*Criblamydia, Estrella*
	Lentisphaceae	
	Parachlamydiaceae	*Candidatus Protochlamydia, Neochlamydia, Parachlamydia*
	Rhabdochlamydiaceae	*Rhabdochlamydia*
	Simkaniaceae	*Candidatus Fritschea, Simkania*
	Victivallales	*Victivallaceae*
	Waddliaceae	*Waddlia*
Verrucomicrobiae	*Opitutaceae*	*Alterococcus, Opitutus*
	Puniceicoccaceae	*Cerasicoccus, Coraliomargarita, Pelagicoccus, Puniceicoccus*
	Verrucomicrobiaceae	*Akkermansia, Fucophilus, Haloferula, Luteolibacter, Persicirhabdus, Prosthecobacter, Roseibacillus, Rubritalea, Verrucomicrobium*

2.2.4 CHLOROFLEXI

Chloroflexi produce energy from light. They have an elongated filamentous shape. They are abundant in waste water treatment. They were found in ancient caves, including paintings in these caves (Lascaux Cave). They participate in the wood decay process and can be used for biodegradation of transformer oil.

Table 2.7. Chloroflexi.

Family	Genera
Anaerolinaceae	*Anaerolinea, Bellilinea, Leptolinea, Levilinea, Longilinea*
Caldilineaceae	*Caldilinea*
Chloroflexaceae	*Chloroflexus, Chloronema, Heliothrix, Roseiflexus*
Herpetosiphonaceae	*Herpetosiphon*
Oscillochloridaceae	*Oscillochloris*
Sphaerobacteraceae	*Sphaerobacter*
Thermomicrobiaceae	*Thermomicrobium*

2.2.5 CYANOBACTERIA

Cyanobacteria are morphologically diverse bacteria that carry out water-oxidizing, oxygen-evolving, plant-like photosynthesis. They grow using CO_2 as the primary source of carbon. Some cyanobacteria have aerobic, nitrogen-fixing capacity. They can live in many diverse environments, including many extreme environments, hot and cold deserts, hot springs, and hypersaline environments. They have a very diverse cellular structure from single (small and large) cells to lateral branching, sheaths, and helical coils. They are known to participate in deterioration of limestone walls of Jerusalem, marble monuments in Rome, Great Jaguar Pyramid in Guatemala, various historical buildings in Brazil, and many other monuments. They are known to degrade synthetic polymer materials and wood. They cause fouling of painted and unpainted concrete structures. Table 2.8 gives the structure of this phylum.

Table 2.8. Cyanobacteria.

Family	Genera
Chroococcales	*Aphanocapsa, Aphanothece, Chamaesiphon, Chondrocystis, Chroococcus, Chroogloeocystis, Coelosphaerium, Crocosphaera, Cyanobacterium, Cyanobium, Cyanodictyon, Cyanosarcina, Cyanothece, Dactylococcopsis, Gloeocapsa, Gloeothece, Halothece cluster, Johannesbaptistia, Merismopedia, Microcystis, Radiocystis, Rhabdoderma, Rubidibacter, Snowella, Sphaerocavum, Synechococcus, Synechocystis, Thermosynechococcus, Woronichinia*
Gloeobacteria	*Gloeobacterales*
Nostocales	*Microchaetaceae, Nostocaceae, Rivulariaceae, Scytonemataceae*
Oscillatoriales	*Arthronema, Arthrospira, Blennothrix, Crinalium, Geitlerinema, Halomicronema, Halospirulina, Hydrocoleum, Jaaginema, Katagnymene, Komvophoron, Leptolyngbya, Limnothrix, Lyngbya, Microcoleus, Oscillatoria, Phormidium, Planktolyngbya, Planktothricoides, Planktothrix, Plectonema, Pseudanabaena, Pseudophormidium, Schizothrix, Spirulina, Starria, Symploca, Trichocoleus, Trichodesmium, Tychonema*
Pleurocapsales	*Chroococcidiopsis, Dermocarpa, Dermocarpella, Myxosarcina, Pleurocapsa, Solentia, Stanieria, Xenococcus*
Prochlorophytes	*Prochloraceae, Prochlorococcaceae, Prochlorotrichaceae*
Stigonematales	*Capsosira, Chlorogloeopsis, Fischerella, Hapalosiphon, Mastigocladopsis, Mastigocladus, Nostochopsis, Stigonema, Symphyonema, Symphyonemopsis, Umezakia, Westiellopsis*

2.2.6 FIBROBACTERES/ACIDOBACTERIA

Fibrobacteres include rumen bacteria, which allow ruminant animals to degrade cellulose. This group does not seem to have relevance to biodeterioration. *Acidobacteria* is a new phylum (recently created). Their representatives are very popular in soil. Table 2.9 characterizes the structure of this phylum.

Table 2.9. Fibrobacteres/Acidobacteria

Phylum	Family	Genera
Acidobacteria	*Acanthopleuribacteraceae*	*Acanthopleuribacter*
	Holophagaceae	*Geothrix, Holophaga*
	Solibacteraceae	*Candidatus Solibacter*
Fibrobacteres	*Fibrobacteraceae*	*Fibrobacter*

2.2.7 FIRMICUTES

This is one of the two largest phyla of Bacteria Kingdom. Approximately, there are 2475 species in 255 genera, 40% of which are in six genera *(Lactobacillus, Mycoplasma, Bacillus, Clostridium, and Streptomyces*). The majority of *Firmicutes* is Gram-positive, some (*Mollicutes or Mycoplasmas*) do not have cell walls and do not respond to Gram staining, others (*Megasphaera, Pectinatus, Selenomonas, and Zymophilus*) are Gram-negative. Many *Firmicutes* are resistant to desiccation and can survive extreme conditions. *Firmicutes* are found on medieval paintings, caves, the USS Arizona at Pearl Harbour, Hawaii, a stained glass window in Florence Cathedral, numerous stone monuments, decaying wood, petroleum samples, wastewater systems, olive fermentation brine, small intestine, corroded sewer system, saline water, and many other substrates.

Table 2.9. Firmicutes.

Class	Family	Genera
Bacilli	*Aerococcaceae*	*Abiotrophia, Aerococcus, Dolosicoccus, Eremococcus, Facklamia, Globicatella, Ignavigranum*
	Alicyclobacillaceae	*Alicyclobacillus, Caldibacillus*
	Bacillaceae	*Alkalibacillus, Amphibacillus, Amylobacillus, Anoxybacillus, Aquisalibacillus, Bacillus, Caldalkalibacillus, Caldaterra, Cerasibacillus, Filobacillus, Geobacillus, Gracilibacillus, Halalkalibacillus, Halobacillus, Halolactibacillus, Lentibacillus, Marinococcus, Natronobacillus, Oceanobacillus, Ornithinibacillus, Paraliobacillus, Paucisalibacillus, Piscibacillus, Pontibacillus, Saccharococcus, Salimicrobium, Salinibacillus, Salirhabdus, Salsuginibacillus, Sediminibacillus, Tenuibacillus, Terribacillus, Thalassobacillus, Tumebacillus, Virgibacillus, Vulcanibacillus*
	Carnobacteriaceae	*Alkalibacterium, Allofustis, Alloiococcus, Atopobacter, Atopococcus, Atopostipes, Carnobacterium, Desemzia, Dolosigranulum, Granulicatella, Isobaculum, Lacticigenium, Marinilactibacillus, Trichococcus*
	Caryophanaceae	*Caryophanon*
	Enterococcaceae	*Catellicoccus, Enterococcus, Melissococcus, Pilibacter, Tetragenococcus, Vagococcus*
	Lactobacillaceae	*Lactobacillus, Paralactobacillus, Pediococcus, Sharpea*
	Leuconostocaceae	*Fructobacillus, Leuconostoc, Oenococcus, Weissella*
	Listeriaceae	*Brochothrix, Listeria*
	Paenibacillaceae	*Aneurinibacillus, Brevibacillus, Cohnella, Paenibacillus, Saccharibacillus, Thermobacillus*
	Pasteuriaceae	*Pasteuria*
	Planococcaceae	*Crocinobacterium, Filibacter, Jeotgalibacillus, Kurthia, Lysinibacillus, Marinibacillus, Paenisporosarcina, Planococcus, Planomicrobium, Rummeliibacillus, Solibacillus, Sporosarcina, Ureibacillus, Viridibacillus*
	Sporolactobacillaceae	*Sinobaca, Sporolactobacillus, Tuberibacillus*
	Staphylococcaceae	*Jeotgalicoccus, Macrococcus, Nosocomiicoccus, Salinicoccus, Staphylococcus*
	Streptococcaceae	*Lactococcus, Lactovum, Streptococcus*
	Thermoactinomycetaceae	*Desmospora, Laceyella, Mechercharimyces, Planifilum, Seinonella, Shimazuella, Thermoactinomyces, Thermoflavimicrobium*
	Turicibacteraceae	*Turicibacter*
Clostridia	*Catabacteriaceae*	*Catabacter*

Class	Family	Genera
Clostridia	*Clostridiaceae*	*Alkaliphilus, Anaerobacter, Anaerosporobacter, Anoxynatronum, Butyricicoccus, Caldanaerocella, Caloramator, Caloranaerobacter, Caminicella, Candidatus Arthromitus, Clostridiisalibacter, Clostridium, Ethanoligenens, Geosporobacter, Lactonifactor, Linmingia, Natronincola, Oxobacter, Sarcina, Sporacetigenium, Tepidimicrobium, Thermobrachium, Thermohalobacter, Thermotalea, Tindallia*
	Eubacteriaceae	*Acetobacterium, Alkalibacter, Anaerofustis, Eubacterium, Garciella, Pseudoramibacter*
	Gracilibacteraceae	*Gracilibacter*
	Halanaerobiaceae	*Halanaerobium, Halarsenatibacter, Halocella, Halothermothrix*
	Halobacteroidaceae	*Acetohalobium, Halanaerobacter, Halanaerobaculum, Halobacteroides, Halonatronum, Natroniella, Orenia, Selenihalanaerobacter, Sporohalobacter*
	Heliobacteriaceae	*Candidatus Helioclostridium, Heliobacillus, Heliobacterium, Heliophilum, Heliorestis*
	Lachnospiraceae	*Acetitomaculum, Anaerostipes, Bryantella, Butyrivibrio, Catonella, Coprococcus, Dorea, Hespellia, Johnsonella, Lachnobacterium Lachnospira, Moryella, Oribacterium, Parasporobacterium, Pseudobutyrivibrio, Robinsoniella, Roseburia, Shuttleworthia, Sporobacterium, Syntrophococcus*
	Natranaerobiaceae	*Natranaerobius*
	Oscillospiraceae	*Oscillibacter*
	Peptococcaceae	*Candidatus Desulforudis, Cryptanaerobacter, Dehalobacter, Desulfitobacterium, Desulfonispora, Desulfosporosinus, Desulfotomaculum, Desulfurispora, Pelotomaculum, Peptococcus, Sporotomaculum, Syntrophobotulus, Thermincola*
	Peptostreptococcaceae	*Filifactor, Peptostreptococcus, Tepidibacter*
	Ruminococcaceae	*Acetanaerobacterium, Acetivibrio, Anaerofilum, Anaerotruncus, Faecalibacterium, Fastidiosipila, Hydrogenoanaerobacterium, Oscillospira, Papillibacter, Ruminococcus, Sporobacter, Subdoligranulum*
	Syntrophomonadaceae	*Candidatus Contubernalis, Dethiobacter, Pelospora, Syntrophomonas, Syntrophothermus, Thermosyntropha*
	Thermoanaerobacteri-aceae	*Caldanaerobacter, Caldanaerobium, Carboxydibrachium, Carboxydothermus, Desulfovirgula, Fervidicola, Gelria, Moorella, Tepidanaerobacter, Thermacetogenium, Thermanaeromonas, Thermoanaerobacter*
	Thermodesulfobiaceae	*Coprothermobacter, Thermodesulfobium*
	Veillonellaceae	*Acetonema, Acidaminococcus, Allisonella, Anaeroarcus, Anaeroglobus, Anaeromusa, Anaerosinus, Anaerospora, Anaerovibrio, Centipeda, Dendrosporobacter, Desulfosporomusa, Dialister, Megamonas, Megasphaera, Mitsuokella, Pectinatus, Pelosinus, Phascolarctobacterium, Propionispira, Propionispora, Psychrosinus, Quinella, Schwartzia, Selenomonas, Sporomusa, Sporotalea, Succiniclasticum, Succinispira, Thermosinus, Veillonella, Zymophilus*

Class	Family	Genera
Erysipelotrichi	*Erysipelotrichaceae*	*Allobaculum, Bulleidia, Catenibacterium, Coprobacillus, Erysipelothrix, Holdemania, Solobacterium, Turicibacter*
Thermolithobacteria	*Thermolithobacteraceae*	*Thermolithobacter*

2.2.8 FUSOBACTERIA

It is an anaerobic, Gram-negative bacteria. Some species are pathogenic to humans. It was found on ancient paintings.

Table 2.10. Fusobacteria

Phylum	Family	Genera
Fusobacteria	*Fusobacteriaceae*	*Cetobacterium, Fusobacterium, Ilyobacter, Leptotrichia, Propionigenium, Psychrilyobacter, Sebaldella, Sneathia, Streptobacillus,*

2.2.9 NITROSPIRAE

These are single-cell microorganisms, some of which may cause infections and disease in animals and humans. Gram-negative, nitrogen oxidizing microorganisms were first found on a corroded pipe in a Moscow heating system. They were also found in caves having Paleolithic paintings.

Table 2.11. Nitrospirae.

Phylum	Family	Genera
Nitrospirae	*Nitrospiraceae*	*Candidatus Magnetobacterium, Leptospirillum, Nitrospira, Thermodesulfovibrio*

2.2.10 PLANCTOMYCETES

These are aquatic bacteria, which are found in marine and fresh water samples. They are also available in soil. They are capable of anaerobic ammonium oxidation. They were found to deteriorate building materials of Roman catacombs. They were present in caves having Paleolithic paintings. They participate in water treatment systems.

Table 2.12. Planctomycetes.

Phylum	Family	Genera
Planctomycetes	*Planctomycetaceae*	*Blastopirellula, Candidatus Jettenia, Gemmata, Isosphaera, Pirellula, Planctomyces, Rhodopirellula, Schlesneria, Singulisphaera, Zavarzinella*

2.2.11 PROTEOBACTERIA

Proteobacteria form one of the largest phyla. They are Gram-negative. They include a wide variety of pathogens, including *Escherichia, Salmonella, Vibrio,* and *Helicobacter.* Some of the bacteria in this phylum are responsible for nitrogen fixation. Most members of the group are anaerobic. They have been involved in deterioration of Milan Cathedral marble, Roman catacombs, the USS Arizona, and the Lions Fountain at Alhambra Palace in Granada. They were found in Altamira Cave in Spain, archeological species of wood, and several European museums. They were involved in alteration of glass, volcanic rocks, crude oil, stones, polymers, copper pipes, metal working fluids, and pulp and paper.

Table 2.13. Proteobacteria.

Class	Family	Genera
Alphaproteobacteria	*Acetobacteraceae*	*Acetobacter, Acidicaldus, Acidiphilium, Acidisoma, Acidisphaera, Acidocella, Acidomonas, Asaia, Belnapia, Craurococcus, Gluconacetobacter, Gluconobacter, Granulibacter, Kozakia, Leahibacter, Neoasaia, Oleomonas, Paracraurococcus, Rhodopila, Rhodovastum, Roseococcus, Roseomonas, Rubritepida, Saccharibacter, Stella, Swaminathania, Tanticharoenia,*
	Anaplasmataceae	*Aegyptianella, Anaplasma, Candidatus Neoehrlichia, Ehrlichia, Neorickettsia*
	Aurantimonadaceae	*Aurantimonas, Fulvimarina, Martelella*
	Bartonellaceae	*Bartonella*
	Beijerinckiaceae	*Beijerinckia, Chelatococcus, Methylocapsa, Methylocella*
	Bradyrhizobiaceae	*Afipia, Agromonas, Balneimonas, Blastobacter, Bosea, Bradyrhizobium, Nitrobacter, Oligotropha, Photorhizobium, Rhodoblastus, Rhodopseudomonas*
	Brucellaceae	*Brucella, Daegunia, Mycoplana, Ochrobactrum, Pseudochrobactrum*
	Caulobacteraceae	*Asticcacaulis, Brevundimonas, Caulobacter, Nitrobacteria, Phenylobacterium, Woodsholea*
	Cohaesibacteraceae	*Cohaesibacter*
	Erythrobacteraceae	*Altererythrobacter, Erythrobacter, Erythromicrobium, Porphyrobacter*
	Holosporaceae	*Holospora*
	Hyphomicrobiaceae	*Ancalomicrobium, Angulomicrobium, Blastochloris, Devosia, Filomicrobium, Hyphomicrobium, Maritalea, Methylorhabdus, Pedomicrobium, Polyprosthecobacterium, Prosthecomicrobium, Rhodomicrobium, Rhodoplanes*
	Hyphomonadaceae	*Delongia, Hellea, Henriciella, Hirschia, Hyphomonas, Maricaulis, Oceanicaulis, Robiginitomaculum*
	Kiloniellaceae	*Kiloniella*
	Kopriimonadaceae	*Kopriimonas*
	Kordiimonadales	*Kopriimonas*
	Methylobacteriaceae	*Meganema, Methylobacterium, Microvirga*
	Methylocystaceae	*Albibacter, Hansschlegelia, Methylocystis, Methylopila, Methylosinus, Methylosporovibrio, Pleomorphomonas, Terasakiella*
	Parvularculaceae	*Parvularcula*
	Phyllobacteriaceae	*Aliihoeflea, Aminobacter, Aquamicrobium, Chelatovorus, Defluvibacter, Hoeflea, Mesorhizobium, Nitratireductor, Parvibaculum, Phyllobacterium, Pseudaminobacter*
	Rhizobiaceae	*Amorphomonas, Candidatus Liberibacter, Carbophilus, Kaistia, Rhizobium/Agrobacterium* group, *Shinella, Sinorhizobium/Ensifer* group

Class	Family	Genera
Alphaproteobacteria	*Rhodobacteraceae*	*Ahrensia, Albidovulum, Amaricoccus, Antarcticicola, Antarctobacter, Catellibacterium, Citreicella, Citreimonas, Dinoroseobacter, Donghicola, Haematobacter, Jannaschia, Ketogulonicigenium, Labrenzia, Leisingera, Loktanella, Lutimaribacter, Maribius, Marinosulfonomonas, Marinovum, Maritimibacter, Marivita, Methylarcula, Nautella, Nereida, Nesiotobacter, Oceanibulbus, Oceanicola, Octadecabacter, Palleronia, Pannonibacter, Paracoccus, Pelagibaca, Phaeobacter, Pseudorhodobacter, Pseudoroseobacter, Pseudoruegeria, Pseudovibrio, Rhodobaca, Rhodobacter, Rhodothalassium, Rhodovulum, Roseibacterium, Roseibium, Roseicyclus, Roseinatronobacter, Roseisalinus, Roseivivax, Roseobacter, Roseovarius, Rubellimicrobium, Rubrimonas, Ruegeria, Sagittula, Salipiger, Sediminimonas, Seohicola, Shimia, Stappia, Sulfitobacter, Tateyamaria, Tetracoccus, Thalassobacter, Thalassobium, Thalassobius, Thalassococcus, Thioclava, Tropicibacter, Tropicimonas, Wenxinia, Yangia*
	Rhodobiaceae	*Afifella, Anderseniella, Rhodobium, Roseospirillum*
	Rhodospirillaceae	*Azospirillum, Caenispirillum, Dechlorospirillum, Defluviicoccus, Inquilinus, Magnetospirillum, Marispirillum, Nisaea, Oceanibaculum, Pelagibius, Phaeospirillum, Rhodocista, Rhodospira, Rhodospirillum, Rhodovibrio, Roseospira, Skermanella, Telmatospirillum, Thalassobaculum, Thalassospira, Tistrella*
	Rickettsiaceae	*Rickettsieae, Wolbachieae*
	Sneathiellaceae	*Sneathiella*
	Sphingomonadaceae	*Blastomonas, Citromicrobium, Kaistobacter, Lutibacterium, Novosphingobium, Sandaracinobacter, Sandarakinorhabdus, Sphingobium, Sphingomonas, Sphingopyxis, Sphingosinicella, Zymomonas*
	Xanthobacteraceae	*Ancylobacter, Azorhizobium, Labrys, Pseudolabrys, Pseudoxanthobacter, Starkeya, Xanthobacter*
Betaproteobacteria	*Alcaligenaceae*	*Achromobacter, Advenella, Alcaligenes, Azohydromonas, Bordetella, Brackiella, Castellaniella, Derxia, Kerstersia, Oligella, Parasutterella, Pelistega, Pigmentiphaga, Pusillimonas, Sutterella, Taylorella, Tetrathiobacter*
	Burkholderiaceae	*Burkholderia, Candidatus Glomeribacter, Chitinimonas, Cupriavidus, Lautropia, Limnobacter, Pandoraea, Paucimonas, Polynucleobacter, Ralstonia, Thermothrix*
	Comamonadaceae	*Acidovorax, Alicycliphilus, Brachymonas, Caldimonas, Comamonas, Curvibacter, Delftia, Diaphorobacter, Doohwaniella, Giesbergeria, Hydrogenophaga, Hylemonella, Lampropedia, Macromonas, Malikia, Ottowia, Pelomonas, Polaromonas, Pseudacidovorax, Ramlibacter, Rhodoferax, Schlegelella, Simplicispira, Variovorax, Verminephrobacter, Xenophilus*
	Gallionellaceae	*Gallionella*
	Hydrogenophilaceae	*Hydrogenophilus, Tepidiphilus, Thiobacillus*
	Methylophilaceae	*Methylobacillus, Methylophilus, Methylotenera, Methylovorus*

Class	Family	Genera
Betaproteobacteria	*Neisseriaceae*	*Alysiella, Andreprevotia, Aquaspirillum, Aquitalea, Bergeriella, Chitinibacter, Chitinilyticum, Chitiniphilus, Chromobacterium group, Conchiformibius, Deefgea, Eikenella, Formivibrio, Gulbenkiania, Kingella, Laribacter, Leeia, Microvirgula, Neisseria, Paludimonas, Pseudogulbenkiania, Silvimonas, Simonsiella, Stenoxybacter, Uruburuella, Vogesella*
	Nitrosomonadaceae	*Nitrosomonas, Nitrosospira, Nitrosovibrio*
	Oxalobacteraceae	*Collimonas, Duganella, Herbaspirillum, Herminiimonas, Janthinobacterium, Massilia, Naxibacter, Oxalicibacterium, Oxalobacter, Telluria, Undibacterium*
	Procabacteriaceae	*Procabacter*
	Rhodocyclaceae	*Aromatoleum, Azoarcus, Azonexus, Azospira, Azovibrio, Dechloromonas, Denitratisoma, Ferribacterium, Garrityella, Methyloversatilis, Petrobacter, Propionivibrio, Quatrionicoccus, Rhodocyclus, Sterolibacterium, Thauera, Zoogloea*
	Spirillaceae	*Spirillum*
Deltaproteobacteria	*Bacteriovoracaceae*	*Bacteriovora, Peredibacter*
	Bdellovibrionaceae	*Bdellovibrio*
	Cystobacterineae	*Cystobacteraceae, Myxococcaceae*
	Desulfarculaceae	*Desulfarculus*
	Desulfoarculaceae	*Desulfoarculus*
	Desulfobacteraceae	*Algorimarina, Desulfatibacillum, Desulfatiferula, Desulfatirhabdium, Desulfobacter, Desulfobacterium, Desulfobacula, Desulfobotulus, Desulfocella, Desulfococcus, Desulfodehalobacter, Desulfofaba, Desulfofrigus, Desulfoluna, Desulfonema, Desulforegula, Desulfosalina, Desulfosarcina, Desulfospira, Desulfotignum*
	Desulfobulbaceae	*Desulfobulbus, Desulfocapsa, Desulfofustis, Desulfopila, Desulforhopalus, Desulfotalea, Desulfurivibrio*
	Desulfohalobiaceae	*Desulfohalobium, Desulfonatronospira, Desulfonatronovibrio, Desulfonauticus, Desulfothermus, Desulfovermiculus*
	Desulfomicrobiaceae	*Desulfomicrobium*
	Desulfonatronumaceae	*Desulfonatronum*
	Desulfovibrionaceae	*Bilophila, Desulfovibrio, Lawsonia*
	Desulfurellaceae	*Desulfurella, Hippea*
	Desulfuromonadaceae	*Desulfuromonas, Desulfuromusa*
	Geobacteraceae	*Geoalkalibacter, Geobacter, Geopsychrobacter, Geothermobacter*
	Nannocystineae	*Haliangiaceae, Nannocystaceae*
	Nitrospinaceae	*Nitrospina*
	Pelobacteraceae	*Malonomonas, Pelobacter*
	Sorangiineae	*Phaselicystidaceae, Polyangiaceae*
	Syntrophaceae	*Desulfobacca, Desulfomonile, Smithella, Syntrophus*
Deltaproteobacteria	*Syntrophobacteraceae*	*Desulfacinum, Desulfatimicrobium, Desulfoglaeba, Desulforhabdus, Desulfovirga, Syntrophobacter, Thermodesulforhabdus*

Class	Family	Genera
Epsilonproteobacteria	*Campylobacteraceae*	*Arcobacter, Campylobacter, Candidatus Thioturbo, Sulfurospirillum*
	Helicobacteraceae	*Flexispira, Helicobacter, Sulfuricurvum, Sulfurimonas, Thiovulum, Wolinella*
	Hydrogenimonaceae	*Hydrogenimonas*
	Nautiliaceae	*Caminibacter, Lebetimonas, Nautilia*
Gammaproteobacteria	*Acidithiobacillaceae*	*Acidithiobacillus*
	Aeromonadaceae	*Aeromonas, Oceanimonas, Oceanisphaera, Tolumonas, Zobellella*
	Alcanivoracaceae	*Alcanivorax*
	Alteromonadaceae	*Aestuariibacter, Agarivorans, Alishewanella, Alteromonas, Bowmanella, Glaciecola, Haliea, Marinimicrobium, Marinobacter, Marinobacterium, Melitea, Microbulbifer, Saccharophagus, Salinimonas*
	Cardiobacteriaceae	*Cardiobacterium, Dichelobacter, Suttonella*
	Chromatiaceae	*Allochromatium, Arsukibacterium, Chromatium, Halochromatium, Isochromatium, Lamprocystis, Marichromatium, Nitrosococcus, Rhabdochromatium, Rheinheimera, Thermochromatium, Thioalkalicoccus, Thiobaca, Thiocapsa, Thiococcus, Thiocystis, Thiodictyon, Thioflavicoccus, Thiohalocapsa, Thiolamprovum, Thiophaeococcus, Thiorhodococcus, Thiorhodovibrio*
	Colwelliaceae	*Colwellia, Thalassomonas*
	Coxiellaceae	*Aquicella, Coxiella, Rickettsiella*
	Crenotrichaceae	*Crenothrix*
	Ectothiorhodospiraceae	*Alkalilimnicola, Alkalispirillum, Aquisalimonas, Arhodomonas, Ectothiorhodosinus, Ectothiorhodospira, Halorhodospira, Natronocella, Nitrococcus, Thioalkalispira, Thioalkalivibrio, Thiohalospira, Thiorhodospira*
	Enterobacteriaceae	*Aranicola, Arsenophonus, Averyella, Biostraticola, Brenneria, Buchnera, Budvicia, Buttiauxella, Candidatus Ishikawaella, Candidatus Phlomobacter, Candidatus Riesia, Candidatus Stammerula, Cedecea, Citrobacter, Cronobacter, Dickeya, Edwardsiella, Enterobacter, Erwinia, Escherichia, Ewingella, Grimontella, Hafnia, Klebsiella, Kluyvera, Leclercia, Leminorella, Margalefia, Moellerella, Morganella, Obesumbacterium, Pantoea, Pectobacterium, Photorhabdus, Phytobacter, Plesiomonas, Pragia, Proteus, Providencia, Rahnella, Raoultella, Salmonella, Samsonia, Serratia, Shigella, Sodalis, Tatumella, Thorsellia, Tiedjeia, Trabulsiella, Wigglesworthia, Xenorhabdus, Yersinia, Yokenella*
	Ferrimonadaceae	*Ferrimonas, Paraferrimonas*
	Francisellaceae	*Francisella*
	Granulosicoccaceae	*Granulosicoccus*

Class	Family	Genera
Gammaproteobacteria	*Hahellaceae*	*Endozoicomonas, Hahella, Zooshikella*
	Halomonadaceae	*Carnimonas, Chromohalobacter, Cobetia, Halomonas, Halotalea, Halovibrio, Kushneria, Modicisalibacter, Salicola, Salinicola, Zymobacter group*
	Halothiobacillaceae	*Halothiobacillus, Thioalkalibacter, Thiofaba, Thiovirga*
	Idiomarinaceae	*Idiomarina, Pseudidiomarina*
	Legionellaceae	*Fluoribacter, Legionella, Tatlockia*
	Litoricolaceae	*Litoricola*
	Methylococcaceae	*Clonothrix, Methylobacter, Methylocaldum, Methylococcus, Methylohalobius, Methylomicrobium, Methylomonas, Methylosarcina, Methylosoma, Methylosphaera, Methylothermus*
	Moraxellaceae	*Acinetobacter, Alkanindiges, Enhydrobacter, Moraxella, Psychrobacter*
	Moritellaceae	*Moritella, Paramoritella*
	Oceanospirillaceae	*Amphritea, Balneatrix, Bermanella, Marinomonas, Marinospirillum, Neptuniibacter, Neptunomonas, Nitrincola, Oceanobacter, Oceanospirillum, Oleispira, Pseudospirillum, Thalassolituus*
	Oleiphilaceae	*Oleiphilus*
	Pasteurellaceae	*Actinobacillus, Aggregatibacter, Avibacterium, Bibersteinia, Bisgaard taxa, Bisgaardia, Chelonobacter, Gallibacterium, Haemophilus, Histophilus, Lonepinella, Mannheimia, Nicoletella, Pasteurella, Phocoenobacter, Terrahaemophilus, Volucribacter*
	Piscirickettsiaceae	*Cycloclasticus, Hydrogenovibrio, Methylophaga, Piscirickettsia, Sulfurivirga, Thioalkalimicrobium, Thiomicrospira*
	Pseudoaltero-monadaceae	*Algicola, Pseudoalteromonas*
	Pseudomonadaceae	*Azotobacter group, Cellvibrio, Pseudomonas, Rhizobacter*
	Psychromonadaceae	*Psychromonas*
	Saccharospirillaceae	*Saccharospirillum*
	Shewanellaceae	*Shewanella*
	Sinobacteraceae	*Alkanibacter, Hydrocarboniphaga, Nevskia, Singularimonas, Sinobacter, Steroidobacter*
	Succinivibrionaceae	*Anaerobiospirillum, Ruminobacter, Succinimonas, Succinivibrio*
	Thermithiobacillaceae	*Thermithiobacillus*
	Thiotrichaceae	*Achromatium, Beggiatoa, Leucothrix, Thiomargarita, Thioploca, Thiothrix*
	Vibrionaceae	*Aliivibrio, Allomonas, Catenococcus, Enterovibrio, Grimontia, Listonella, Photobacterium, Photococcus, Salinivibrio, Vibrio*
Gammaproteobacteria	*Xanthomonadaceae*	*Aquimonas, Arenimonas, Aspromonas, Dokdonella, Dyella, Ferritrophicum, Frateuria, Fulvimonas, Ignatzschineria, Kaistibacter, Luteibacter, Luteimonas, Lysobacter, Panacagrimonas, Pseudoxanthomonas, Rhodanobacter, Sideroxydans, Silanimonas, Stenotrophomonas, Thermomonas, Wohlfahrtiimonas, Xanthomonas, Xylella*

Class	Family	Genera
Zetaproteobacteria	*Mariprofundaceae*	*Mariprofundus*

2.2.12 THERMODESULFOBACTERIA

Thermodesulfobacteria are thermophilic sulfate-reducing bacteria. They are found in oil-fields.

Table 2.14. Thermodesulfobacteria.

Phylum	Family	Genera
Thermodesulfobacteria	*Thermodesulfobacteriaceae*	*Caldimicrobium, Geothermobacterium, Thermodesulfatator, Thermodesulfobacterium*

2.2.13 THERMOTOGAE

They inhabit hot springs and deep ocean vents, and have a specialized outer covering called a toga. The optimal temperature for their growth is 70-80°C. They are found in oil-fields.

Table 2.15 Thermotogae.

Phylum	Family	Genera
Thermotogae	*Thermotogaceae*	*Fervidobacterium, Geotoga, Kosmotoga, Marinitoga, Petrotoga, Thermopallium, Thermosipho, Thermotoga*

2.3 FUNGI

One of the two Empires is divided into 5 Kingdoms and one of them is the Kingdom of Fungi. Table 2.16 shows the general division of Kingdom Fungi.

Table 2.16. Kingdom of Fungi.

Subkingdom	Phylum	Subphyla
Dikarya	*Ascomycota*	*Pezizomycotina, Saccharomycotina, Taphrinomycotina*
	Basidiomycota	*Agaricomycotina, Pucciniomycotina, Ustilaginomycotina*
Do not belong to any Subkingdom	*Blastocladiomycota*	They are not divided into Subphyla but their next taxonomic level is Class
	Chytridiomycota	
	Glomeromycota	
	Microsporidia	
	Neocallimastigomycota	

Fungi are divided into several phyla which are not specific having species such as fungi, molds, and yeast which can be found in different phyla. *Ascomycota* is the largest phylum containing above 60,000 classified species. It is expected that there is about 1.5 million species which belong to this kingdom but only about 5% have been classified so far. It is also the most important phylum from the point of view of biodegradation and biodeterioration since most species active in these processes can be found in *Ascomycota*. Sections below give further details on the breakdown of Kingdom Fungi. Fungi are usually associated with plants but they have more similarity to animals. For example, their cell walls are composed of chitin, unlike in the case of plants, where cells are composed of cellulose. Also, genetic studies show that fungi are related to animals. Present use of DNA for classification purposes will provide many interesting data for both classification and understanding similarities and differences between various kingdoms.

Here are the most important characteristics of fungi:

- they do not contain chlorophyll
- digest food first and then absorb it into their bodies
- release digestive enzymes to break down organic material of their host
- they are decomposers and recyclers of nutrients in the environment
- most live on dead organisms
- they are mostly multicellular, but some are unicellular, such as yeast
- grow best in warm, moist environments preferring shade

2.3.1 ASCOMYCOTA

Reproduction in *Ascomycota* always involves the formation of an "ascus", a sac-like structure that distinguishes the *Ascomycota* from all other fungi. This is probably the only factor which unites *Ascomycota* since the group is very large and diverse, containing mushrooms, fungi, molds, and yeasts (the last three groups are involved in the subject of this book). They are mostly terrestial but some have adapted to marine and freshwater environments.

Their cell walls are, like in the case of all other fungi, built of chitin but they contain galactose and mannose (unlike xylose and mannose as in *Basidiomycota*).

This phylum contains species of edible mushrooms, yeasts used for baking and fermentation, fungi used for production of antibiotics, variety of microorganisms which spoil food, and very large number of microorganisms which cause biodeterioration of product groups discussed in Chapter 4.

Tables 2.17 and 2.18 show the classification of *Ascomycota.*

Table 2.17. Classification of Ascomycota.

Subphylum	Class	Order/Subclass
Mitrosporic *Ascomycota*		*Acremonium* (Genus), *Stachybotrys* (Genus)
Pezizomycotina	*Arthoniomycetes*	*Arthoniales*
	Dothideomycetes	*Dothideomycetes incertae sedis, Dothideomycetidae, mitosporic Dothideomycetes, Pleosporomycetidae*
	Eurotiomycetes	*Chaetothyriomycetidae, Eurotiomycetes incertae sedis, Eurotiomycetidae, Mycocaliciomycetidae*
	Laboulbeniomycetes	*Laboulbeniales, Laboulbeniomycetes incertae sedis, Pyxidiophorales*

Table 2.17. Classification of Ascomycota.

Subphylum	Class	Order/Subclass
Pezizomycotina	*Lecanoromycetes*	*Acarosporomycetidae, Lecanoromycetes incertae sedis, Lecanoromycetidae, Ostropomycetidae*
	Leotiomycetes	*Cyttariales, Erysiphales, Helotiales, Leotiomycetes incertae sedis, mitosporic Leotiomycetes, Rhytismatales, Thelebolales*
	Lichinomycetes	*Eremithallales, Lichinales*
	Orbiliomycetes	*Orbiliales*
	Pezizomycetes	*Pezizales*
	Pezizomycotina incertae sedis	*Alysidiella, Batistiaceae, Everhartia, Geoglossales, Medeolariales, Phacidiella, Sarcosagium, Sphaeriothyrium, Thelocarpaceae*
	Sordariomycetes	*Hypocreomycetidae, mitosporic Sordariomycetes, Sordariomycetes incertae sedis, Sordariomycetidae, Xylariomycetidae*
Saccharomycotina	*Saccharomycetes*	*Saccharomycetales, Saccharomycetes incertae sedis*
Taphrinomycotina	*Neolectomycetes*	*Neolectales*
	Pneumocystidomycetes	*Pneumocystidales*
	Schizosaccharomycetes	*Schizosaccharomycetales*
	Taphrinomycetes	*Taphrinales*
	Taphrinomycotina incertae sedis	*Saitoella*

Table 2.18. Families of Ascomycota.

Order/Subclass	Family/Genus
Acarosporomycetidae	*Acarosporaceae*
Arthoniales	*Arthoniaceae, Roccellaceae*
Chaetothyriomycetidae	*Chaetothyriaceae, Chaetothyriales incertae sedis, Herpotrichiellaceae, mitosporic Chaetothyriales*
Cyttariales	*Cyttariaceae*
Dothideomycetes incertae sedis	*Acidomyces, Acrospermaceae, Argynnaceae, Asterinaceae, Botryosphaeriales, Crivellia, Dangeardiella, Eremomycetaceae, Farlowiella, Fusculina, Helicascus, Jahnulales, Julella, Leptospora, Meliolinaceae, Micropeltidaceae, Mycopepon, Otthia, Paraliomyces, Parodiopsidaceae, Patellariales, Planistromellaceae, Thyridaria, Tubeufiaceae, Tyrannosorus, Zopfiaceae*
Dothideomycetidae	*Capnodiales (Cladosporium), Dothideales, Myriangiales*
Eremithallales	*Eremithallaceae*
Erysiphales	*Erysiphaceae*
Eurotiales	*Elaphomycetaceae, Trichocomaceae (Aspergillus, Penicillium)*
Eurotiomycetes incertae sedis	*Calyptrozyma*

Table 2.18. Families of Ascomycota.

Order/Subclass	Family/Genus
Helotiales	*Bulgariaceae, Dermateaceae, Helotiaceae, Helotiales incertae sedis, Hemiphacidiaceae, Hyaloscyphaceae, Leotiaceae, Loramycetaceae, mitosporic Helotiales, Phacidiaceae, Rutstroemiaceae, Sclerotiniaceae, Vibrisseaceae*
Hypocreomycetidae	*Ascocodinaea, Bertiaceae, Bionectriaceae, Ceratostomataceae, Chadefaudiellaceae, Chaetosphaerellaceae, Clavicipitaceae (Paecilomyces), Conioscyphascus, Cordycipitaceae, Etheirophora, Halosphaeriaceae, Hypocreaceae, Hypocreales incertae sedis, Melanosporaceae, Microascaceae, Microascales incertae sedis, mitosporic Hypocreales (Fusarium, Trichoderma), mitosporic Microascales, Nectriaceae, Niessliaceae, Nitschkiaceae, Ophiocordycipitaceae, Porosphaerellopsis, Scortechiniaceae, Swampomyces, Torpedospora*
Laboulbeniales	*Ceratomycetaceae, Laboulbeniaceae*
Laboulbeniomycetes incertae sedis	*Laboulbeniopsis*
Lecanoromycetes incertae sedis	*Candelariaceae, Umbilicariaceae*
Lecanoromycetidae	*Anziaceae, Arthrorhaphidaceae, Biatoridium, Bilimbia, Caliciaceae, Catillariaceae, Cetradoniaceae, Cladoniaceae, Coccocarpiaceae, Collemataceae, Crocyniaceae, Dactylosporaceae, Ectolechiaceae, Elixiaceae, Furcaspora, Fuscideaceae, Gypsoplacaceae, Haematommataceae, Lecanoraceae, Lecideaceae, Leprocaulon, Letrouitiaceae, Lobariaceae, Lopezaria, Loxosporaceae, Megalariaceae, Megalosporaceae, Micareaceae, Mycoblastaceae, Nephromataceae, Notolecidea, Ophioparmaceae, Pannariaceae, Parmeliaceae, Peltigeraceae, Peltigerineae incertae sedis, Phlyctidaceae, Physciaceae, Pilocarpaceae, Placynthiaceae, Porpidiaceae, Psoraceae, Ramalinaceae, Rhizocarpaceae, Sphaerophoraceae, Stenhammarella, Stereocaulaceae, Teloschistaceae,*
Leotiomycetes	*Bulgariaceae, Cudoniaceae, Cyttariaceae, Erysiphaceae, Dermateaceae, Helotiaceae, Helotiales incertae sedis, Hemiphacidiaceae, Hyaloscyphaceae, Leotiaceae, Loramycetaceae, mitosporic Helotiales, Phacidiaceae, Rhytismataceae, Rutstroemiaceae, Sclerotiniaceae, Thelebolaceae, Vibrisseaceae*
Leotiomycetes incertae sedis	*Darkera, Microglossum, Myxotrichaceae, Pseudeurotiaceae, Thuemenidium*
Lichinales	*Heppiaceae, Lichinaceae, Peltulaceae*
mitosporic Dothideomycetes	*Asteromella, Botryomyces, Cenococcum, Cyclothyrium, Cystocoleus, Racodium, Sclerostagonospora, Seifertia*
mitosporic Leotiomycetes	*Lauriomyces, Leohumicola, Rhexocercosporidium*
mitosporic Sordariomycetes	*Asteroma, Myrmecridium, Pleurothecium, Polyscytalum, Rhodoveronaea*
Mycocaliciomycetidae	*Mycocaliciaceae, Sphinctrinaceae*
Neolectales	*Neolectaceae*
Orbiliales	*Orbiliaceae*
Ostropomycetidae	*Agyriaceae, Anamylopsoraceae, Arctomiaceae, Asterothyriaceae, Baeomycetaceae, Coenogoniaceae, Gomphillaceae, Graphidaceae, Gyalectaceae, Hymeneliaceae, Icmadophilaceae, Megasporaceae, Ochrolechiaceae, Odontotremataceae, Pertusariaceae, Porinaceae, Stictidaceae, Thelotremataceae*

Table 2.18. Families of Ascomycota.

Order/Subclass	Family/Genus
Pezizales	*Ascobolaceae, Ascodesmidaceae, Caloscyphaceae, Discinaceae, Glaziellaceae, Helvellaceae, Karstenellaceae, Morchellaceae, Pezizaceae, Pezizales incertae sedis, Pyronemataceae, Rhizinaceae, Sarcoscyphaceae, Sarcosomataceae, Tuberaceae*
Pleosporomycetidae	*Gloniaceae, Hysteriales, Mytilinidiales, Pleosporales*
Pneumocystidales	*Pneumocystidaceae*
Pyxidiophorales	*Pyxidiophoraceae*
Rhytismatales	*Ascodichaenaceae, Cudoniaceae, Rhytismataceae*
Saccharomycetales	*Ascoideaceae, Cephaloascaceae, Dipodascaceae, Endomycetaceae, Lipomycetaceae, Metschnikowiaceae, mitosporic Saccharomycetales (Candida), Phaffomycetaceae, Saccharomycetaceae (Saccharomyces), Saccharomycetales incertae sedis, Saccharomycodaceae, Saccharomycopsidaceae, Trichomonascaceae, Wickerhamomycetaceae*
Saccharomycetes incertae sedis	*Cyniclomyces, Starmera, Starmerella*
Schizosaccharomycetales	*Schizosaccharomycetaceae*
Sordariomycetes incertae sedis	*Acremonium, Hilberina, Koralionastetales, Lulworthiales, Meliolales, Papulosaceae, Phlogicylindrium, Phomatospora, Phyllachorales, Stereosphaeria, Teracosphaeria, Thyridiaceae, Trichosphaeriales, Wallrothiella*
Sordariomycetidae	*Annulatascaceae, Apiosporaceae, Boliniaceae, Calosphaeriaceae, Catabotrydaceae, Caudatispora, Cephalothecaceae, Chaetomiaceae, Chaetosphaeriaceae, Coniochaetaceae, Cryphonectriaceae, Diaporthales incertae sedis, Erythromada, Gnomoniaceae, Helminthosphaeriaceae, Kathistaceae, Lasiosphaeriaceae, Lasiosphaeriella, Lentomitella, Leptosporella, Magnaporthaceae, Melanconidaceae, mitosporic Diaporthales, mitosporic Sordariales, Ophiostomataceae, Pleurotremataceae, Valsaceae, Sordariaceae, Sordariales incertae sedis, Xylomelasma*
Taphrinales	*Protomycetaceae, Taphrinaceae*
Taphrinomycotina incertae sedis	*Saitoella*
Thelebolales	*Thelebolaceae*
Xylariomycetidae	*Amphisphaeriaceae Clypeosphaeriaceae, Diatrypaceae, Graphostromataceae, Hyponectriaceae, mitosporic Xylariales, Xylariaceae, Xylariales incertae sedis*

2.3.2 BASIDIOMYCOTA

Dead wood habitats support a wide variety of specialized fungi, which may exceed 100,000 species. Of these, a large number are *Basidiomycota*. There are at least 30,000 species of *Basidiomycota* known today. The process of multiplication involves the production of haploid basidiospores. *Basidiomycota* are involved in biodeterioration of wood; especially in biodegradation of lignin. Table 2.19 gives the classification breakdown of *Basidiomycota*.

Table 2.19. Families of Basidiomycota

Subphylum	Family
Agaricomycotina	*Agaricaceae, Agaricales incertae sedis, Albatrellaceae, Amanitaceae, Aporpiaceae, Atheliaceae, Auriscalpiaceae, Auriculariaceae, Auriculariales incertae sedis, Bankeraceae, Battarreaceae, Bolbitiaceae, Boletales incertae sedis, Boletineae, Bondarzewiaceae, Botryobasidiaceae, Cantharellaceae, Cantharellales incertae sedis, Ceratobasidiaceae, Cerinomycetaceae, Christianseniaceae, Clathraceae, Clavariaceae, Clavulinaceae, Coniophorineae, Coriolaceae, Corticiaceae, Cortinariaceae, Crepidotaceae, Cryptotrichosporon, Cyphellaceae, Cystofilobasidiaceae, Dacrymycetaceae, Dictyonemataceae, Echinodontiaceae, Entolomataceae, Epitheliaceae, Exidiaceae, Favolaschiaceae, Filobasidiaceae, Fistulinaceae, Ganodermataceae, Gautieriaceae, Geastraceae, Gastrosporiaceae, Gloeocystidiellaceae, Gloeophyllaceae, Gomphaceae, Hapalopilaceae, Hericiaceae, Hyaloriaceae, Hydnaceae, Hydnangiaceae, Hygrophoraceae, Hymenochaetaceae, Hymenochaetales incertae sedis, Hymenogastraceae, Hyphodermataceae, Hysterangiaceae, Lachnocladiaceae, Lentinaceae, Lepidostromataceae, Lycoperdaceae, Marasmiaceae, Melanogastraceae, Meripilaceae, Meruliaceae, Mesophelliaceae, Mycenastraceae, Niaceae, Nidulariaceae, Omphalotaceae, Paxilineae, Phallaceae, Phanerochaetaceae, Phelloriniaceae, Pleurotaceae, Pluteaceae, Podaxaceae, Podoscyphaceae, Polyporaceae, Polyporales incertae sedis, Psathyrellaceae, Pterulaceae, Russulaceae, Schizophyllaceae, Schizoporaceae, Sclerodermatineae, Sebacinaceae, Sirobasidiaceae, Sparassidaceae, Sphaerobolaceae, Stephanosporaceae, Stereaceae, Strophariaceae, Suillineae, Syzygosporaceae, Tapinellineae, Thelephoraceae, Trechisporaceae, Tremellaceae, Tricholomataceae, Tulasnellaceae, Typhulaceae, Xenasmataceae*
Basidiomycota incertae sedis	*Entorrhizaceae*
Pucciniomycotina	*Agaricostilbaceae, Chaconiaceae, Chionosphaeraceae, Coleosporiaceae, Cronartiaceae, Cryptomycocolacaceae, Cystobasidiaceae, Eocronartiaceae, Heterogastridiaceae, Hoehnelomycetaceae, Melampsoraceae, Microbotryaceae, Mikronegeriaceae, Mixiaceae, Pachnocybaceae, Phakopsoraceae, Phragmidiaceae, Phleogenaceae, Pileolariaceae, Platygloeaceae, Pucciniaceae, Pucciniastraceae, Pucciniosiraceae, Raveneliaceae, Saccoblastiaceae, Septobasidiaceae, Sphaerophragmiaceae, Spiculogloeaceae, Uropyxidaceae, Ustilentylomataceae*
Ustilaginomycotina	*Anthracoideaceae, Brachybasidiaceae, Cintractiaceae, Cryptobasidiaceae, Doassansiaceae, Doassansiopsaceae, Eballistraceae, Entylomataceae, Exobasidiaceae, Georgefischeriaceae, Gjaerumiaceae, Glomosporiaceae, Graphiolaceae, Melaniellaceae, Melanotaeniaceae, Microstromataceae, Quambalariaceae, Rhamphosporaceae, Tilletiaceae, Tilletiariaceae, Urocystaceae, Ustilaginaceae, Volvocisporiaceae*

2.3.3 BLASTOCLADIOMYCOTA

This phylum includes the following families: *Blastocladiaceae, Catenariaceae, Coelomomycetaceae*, and *Physodermataceae*. They are parasitic fungi on plants and animals. No references exist regarding their participation in biodegradation and biodeterioration.

2.3.4 CHYTRIDIOMYCOTA

Chytridiomycota are the smallest and simplest fungi, which are ancestors to all Fungi. They are known to be able to utilize keratin (e.g., wool). Table 2.20 gives the breakdown of their classes and families.

Table 2.20. Families of Chytridiomycota.

Class	Family
Chytridiomycetes	*Chytridiaceae, Cladochytriaceae, Endochytriaceae, Kappamycetaceae, Lobulomycetaceae, Megachytriaceae, Olpidiaceae, Rhizophydiaceae, Spizellomycetaceae, Synchytriaceae, Terramycetaceae*
Monoblepharidomycetes	*Gonapodyaceae, Harpochytriaceae, Monoblepharidaceae, Oedogoniomycetaceae*

2.3.5 GLOMEROMYCOTA

It is a small phylum containing about 200 species. There is no reference regarding their participation in biodegradation and biodeterioration but their presence in soil may be important for some degradative processes. They include the following families contained in four Orders: *Acaulosporaceae, Ambisporaceae, Archaeosporaceae, Diversisporaceae, Geosiphonaceae, Gigasporaceae, Glomeraceae, Pacisporaceae, Paraglomeraceae*, and *Scutellosporaceae*.

2.3.6 MICROSPORIDIA

Microsporidia is a phylum of spore-forming unicellular parasites. This phylum includes the following families: *Cougourdellidae, Culicosporidae, Endoreticulatus, Gurleyidae, Nadelsporidae, Ordosporidae, Pereziidae, Pseudonosematidae*, and *Tetramicriidae*. Representatives of this phylum were not found to biodegrade materials, they mostly multiply inside living organisms, causing numerous diseases.

2.3.7 NEOCALLIMASTIGOMYCOTA

This phylum has only one family containing anaerobic fungi, found mainly in the stomachs of ruminants. Representatives of this phylum are not known to participate in degradation of materials, which are the subject of this book.

2.4 PROTOZOA

Kingdom Protozoa includes unicellular microorganisms. They move using whip-like tails called flagella. Over 30,000 different types have been found. Protozoa participate in biodegradation and biodeterioration of many plastics and polymers. They are found on walls of churches in Brazil, cause biodeterioration of stone monuments in Turkey, cause biodegradation of crude oil and related products, participate in bioconversion of lignocellulosic materials, cause fouling of paints, participate in biodegradation of solid municipal wastes, are found on railway tracks in European countries, and participate in many other biodegradation reactions of common products. It is important that humid conditions or water immersion are the part of the conditions of degradation. Table 2.21 gives breakdown of their phyla and families.

Table 2.21. Phyla and families of Kingdom Protozoa.

Phylum	Family (Order)
Apicomplexa	*Cryptosporidiidae*
Ciliophora	*Acinetidae, Amphileptidae, Amphisiellidae, Apostomatida, Aspidiscidae, Astomatida, Astylozoonidae, Bursariidae, Amphisiellidae, Chonotrichida, Colpodidae, Cyrtolophosididae, Cyrtophorida, Dendrosomatidae, Didiniidae, Enchelyidae, Epalxellidae, Ephelotidae, Epistylidae, Euplotidae, Halteriidae, Holophryidae, Holostichidae, Hymenostomatida, Keronidae, Lacrymariidae, Lagenophryidae, Leptopharyngidae, Loxodidae, Microthoracidae, Metopidae, Mobilina, Nassulidae, Opercullariidae, Ophrydiidae, Oxytrichidae, Podophryidae, Prorodontidae, Rhynchodida, Scaphidiodontidae, Scyphidiidae, Spirofilidae, Spirostomidae, Stentoridae, Strobilididae, Strombidiidae, Tintinnina, Tracheliidae, Trachelophyllidae, Trachelostylidae, Trichostomatida, Urnulidae, Urotrichidae, Urostylidae, Vaginicolidae, Vorticellidae*
Protozoa	*Acanthocistidae, Acanthometridae, Acanthoniidae, Acrasea, Actinocephalidae, Actinophyridae, Actinosphaeridae, Actinopodea, Acytosteliaceae, Anurosporidiidae, Archigregarinida, Aulacanthidae, Cavosteliidae, Ciliophryidae, Clathrellidae, Eucoccidiida, Eucyrtidiidae, Haplosporidiidae, Labyrinthulida, Neogregarinida, Plasmodiophorida, Porosporidae, Proteomyxida, Protococcida, Stethopiliidae, Toxoplasmida*
Sarcomastigophora	*Chloromonadida, Choanoflagellida, Chrysomonadida, Coccolithophorida, Cryptomonadida, Dinoflagellida, Diplomonadida, Ebriida, Euglenida, Heterochlorida, Hypermastigida, Kinetoplastida, Opalinida, Oxymonadida, Retortamonadida, Rhizomastigida, Silicoflagellida, Trichomonadida, Volvocida*

2.5 BIODEGRADATION & BIODETERIORATION MECHANISMS

Biodegradation of materials consists of three main stages: biodeterioration, biofragmentation, and assimilation.[4] Figure 2.1 shows these stages in a more complex way for polymeric materials. The first step, generally called biodeterioration, differs in meaning, depending on the technology used for production and design of material. There are two groups of modern materials designed to minimize the impact of polymers on the environment: durable materials and biodegradeable materials.

Durable polymeric materials are used for production of various goods which are expected to have long service life. In the case of these materials, biodeterioration will frequently affect only a material's surface in the initial step of degradation. Surface deterioration may frequently be sufficient for material replacement or it may initiate further changes caused by this initial modification.

Biodegradable materials are produced to have much shorter lifespan. They are expected to deteriorate rapidly after serving the purpose for which they were designed. In the case of these materials, biodeterioration means fragmenting material to smaller pieces, which can be efficiently converted within the framework described by Figure 2.1 and assimilated back into the environment.

It has to be stressed that biodeterioration of both groups of materials does not occur only because of biodegradative processes but by the combined action of biotic and abiotic processes. Abiotic processes include mechanical degradation, thermal degradation, degradation by availability of moisture and water vapor, oxygen, ultraviolet light, and environ-

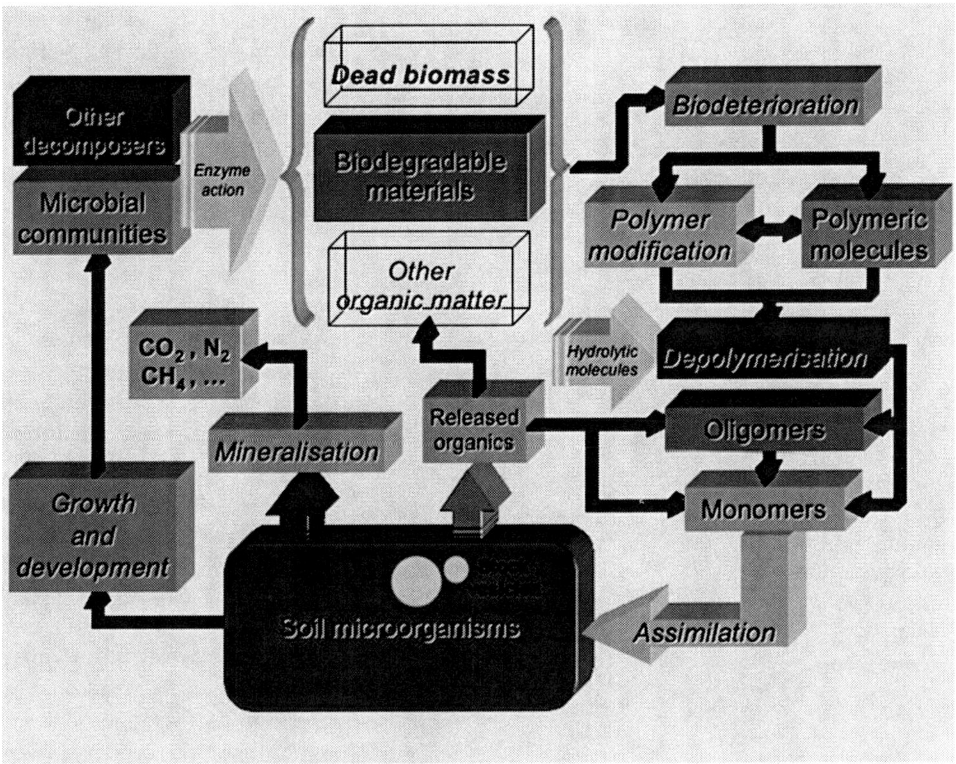

Figure 2.1. Polymer biodegradation scheme. [Adapted, by permission, from Lucas, N; Bienaime, C; Belloy, C; Queneudec, M; Silvestre, F; Nava-Saucedo, J-E, *Chemosphere*, **73**, 429-42, 2008.]

mental pollutants. All these abiotic processes are frequently more pronounced in the initial stages of material degradation (deterioration) than biotic processes, although they are frequently mediated by the action of microorganisms.

Abiotic processes have a different impact on durable goods and biodegradable materials. In the case of durable goods they usually cause limited damage to the surface, making it accessible to moisture and microorganisms penetration. In the case of biodegradable materials, abiotic processes are intensified by compositional design and they are expected to produce fragmentation of material, since materials in smaller fragments are easier to biodegrade. It is clear that there is essential difference between time scales and extent of biodeterioration of both groups of materials.

The process of biofragmentation follows biodeterioration. Here, in addition to the action of abiotic factors, microorganisms may secrete catalytic agents (enzymes and free radicals) which cleave macromolecules to oligomers, dimers, and monomers.[4]

The final steps include assimilation (formation of energy, new biomass, and various metabolites used by microorganisms) and mineralization (simple gaseous molecules and mineral salts are released into environment).[4]

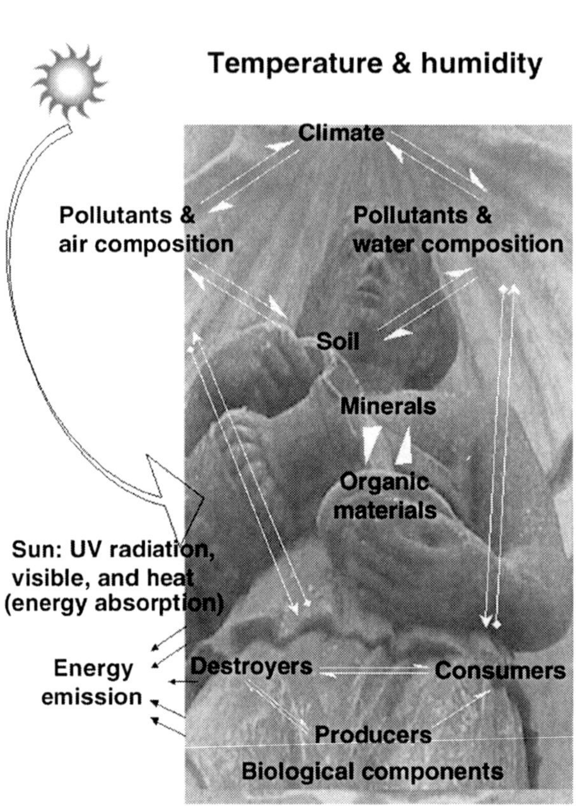

Figure 2.2. Biodeterioration of stone monument. [Adapted, by permission, from Ranalli, G; Zanardini, E; Sorlini, C, **Biodeterioration – Including Cultural Heritage, in Encyclopedia of Microbiology,** *Elsevier*, 2009, pages 191-205.]

Figure 2.2 reaffirms the cooperation of biotic and abiotic factors based on the example of stone monuments.[5] In the background (around the figure), climatic factors are displayed. They include effects of temperature, humidity, rain and its composition, wind, air composition (presence of various pollutants), UV radiation, energy of visible light (important in various photobleaching processes), and infrared radiation (transfer of heat from sun radiation).

Three interconnected cycles were included on the body of the stone monument. These include: climate cycle, deposition cycle, and biotic cycle.

The effect of the climate cycle depends on the material type involved in degradation and the climatic condition in a particular site during the entire duration of exposure. The effect of climatic conditions on various materials is not the subject of this book and relevant information can be found in the specialized monograph.[6] It also should be noted that the properties of materials are affected by its processing conditions. There are many monographs devoted to this subject (e.g., see ref. 7). The appropriate monograph discussing thermal degradation of material of interest should be consulted to find details. Finally, Chapter 4 contains discussion of mechanisms of biodegradation developed for specific purposes of different products.

Microorganisms deposition is facilitated by rain, air circulation, and wind, which move around microorganisms from soil to air and to the surfaces of materials in surrounding. It has been confirmed by many researchers that the outdoor microbial population in the air is similar to the communities of organisms found in soil.[8] (This prompted the idea that soil burial tests will provide similar conditions to outdoor exposures but in accelerated pace.)

The biotic cycle contains three groups of microorganisms: destroyers, producers and consumers. These organisms cooperate in biodeterioration processes. Producers, also called autotrophic microorganisms, are equipped to gain energy from synthesis. They produce complex molecules from simple organic compounds taking energy from sunlight (photosynthesis) or from oxidation of inorganic compounds. They are most likely bacteria or algae. Consumers, or heterotrophs, are most likely bacteria, fungi, or protozoa. They break down more complex molecules, produced by autotrophs, and they gain energy required to support their life-functions. Finally, destroyers damage accessible surfaces of material by lowering molecular weight of building blocks, changing the structure of building blocks, or depositing corrosive substances. Many microorganisms are capable of hydrolysis, redox reactions, denitrification, radical formation, and many other reactions performed by their enzymatic apparatus. These reactions cause changes in structure, crystallinity, and molecular weight of building blocks of various materials which decreases their mechanical performance characteristics. Various microorganisms also produce acids, nitroxides, sulfates, and other substances which may react with surface of product causing irreversible damage.

Microorganisms colonize the surface of material in two steps: adhesion and aggregation. Adhesion of microorganism depends on its properties as well as the properties of the surface in contact. It is well know that sticky surfaces (e.g., surfaces which exude some low molecular compounds or surfaces covered with sticky oligomers formed by degradation) are more likely to retain microorganisms which come into contact with them. Also, van der Waals and Debye forces of repulsion and attraction play an essential role in this first step. In some cases, enzymatic processes occur directly in contact with polymeric materials. This happens in medical materials which are in direct contact with living tissues equipped with an enzymatic apparatus.[9]

Materials are frequently a combination of different compounds. Cellulosic materials (e.g., wood, or fibers) are a good example here. They are composed of pectin, hemicellulose, and lignin.[8] Pectins are a food source and they promote microbiological attack. Lignins and naturally occurring waxes are deterrents which lower the probability of a successful attack.

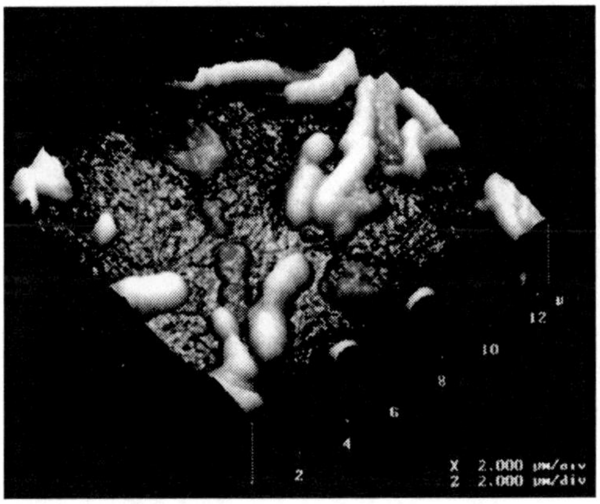

Figure 2.3. AFM micrograph of stainless steel surface colonized by one-year-old biofilm formed by *Pseudomonas aeruginosa*. [Adapted, by permission, from Steele, A; Goddard, D T; Beech, I B, *Intl. Biodeterioration Biodegrad.*, 35-46, 1994.]

Aggregation is the second step in colonization which leads to formation of biofilm.[3] Microbial biofilms can exist as a single layer mat or a three-dimensional architecture with channels permitting liquid and gas flow and dispersion of nutrients and waste components.[10]

In another source,[11] three steps are itemized: the primary attachment of bacteria to surfaces, followed by proliferation of the attached bacterial cells, which leads to the accumulation of multilayered clusters of cells and glycocalyx formation (covering of cell surface with a polysaccharide or glycoprotein).

Still another step is sometimes being:[12,13]

- the transport of organic molecules and cells to the surface

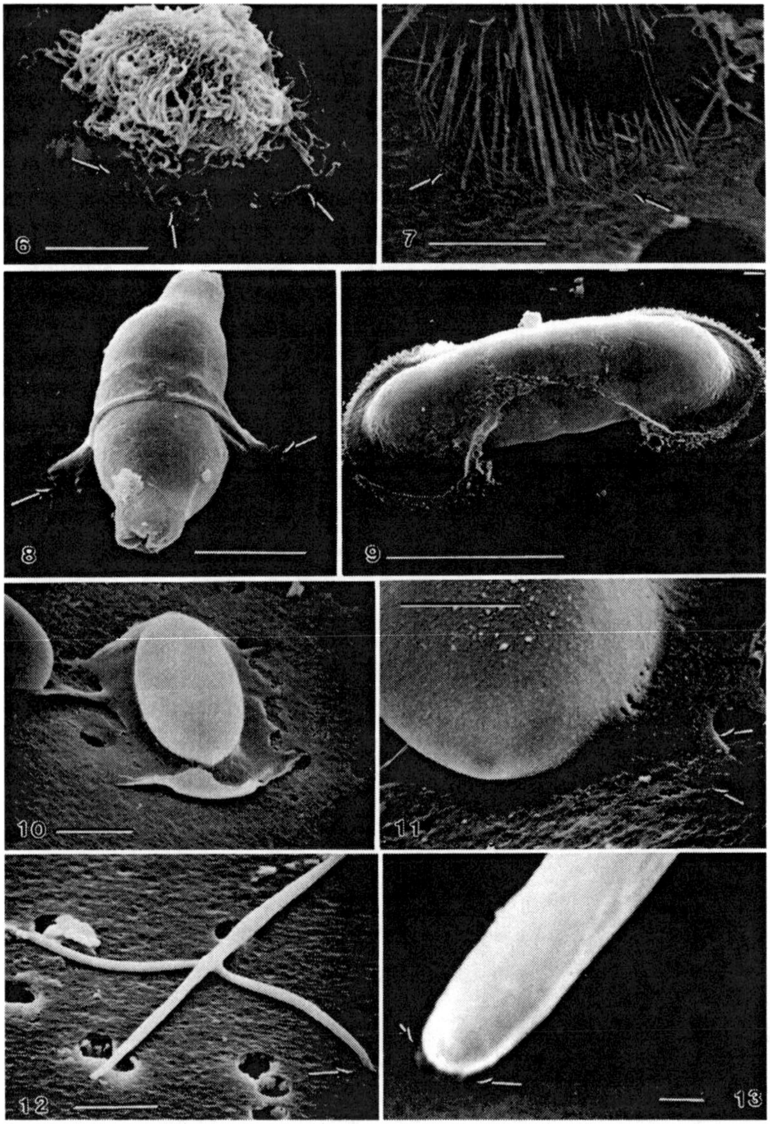

Figure 2.4. Scanning electron micrographs. (6) *Appendichordella amicta*. Ascospore with loop-like appendages attaching (arrowed) spore to substratum; (7) *Nereiospora comata*. Hair-like appendages attaching spore to surface (arrowed); (8) *Ceriosporopsis tubulifera*. Ascospore with annulus-like equatorial appendage forming attachment pad (arrowed) to substratum; (9) Attachment of *Remispora galerita* ascospore by its polar appendages to the surface; (10) and (11) *Sordaria fimicola*. Ascospore enveloped by a mucilaginous sheath (10) and showing attachment to the substratum (11); (12) and (13) *Alatospora acuminata*. Conidia attached to a surface by secreted mucilage (arrowed) at the tip of the conidial arms. Bar lines: (6), (8), (9), (10), (12) are 10 μm, (7), (11) are 5 μm; and (13) is 1 μm. [Adapted, by permission, from Jones, E B G, *Mycol. Res.*, **98**, 9, 961-81, 1994.]

- the adsorption of organic molecules to give a "conditioned" surface
- the adsorption of cells to the conditioned surface
- the growth of adsorbed cells with associated synthesis of expolymeric substances.

In spite of general understanding of the steps leading to biofilm formation, it is generally accepted[14] that biofilm formation is poorly understood. The cell attachment to the material surface and subsequent stabilization of the cell association with the surface are considered crucial stages in biofilm formation.[14] The attached cells produce adhesive extracellular polymeric substances that improve microorganism binding to the surface and thereby enhance the survival and robustness of the biofilm microorganisms.[14] Microbial adhesion is the first critical step in biofilm formation.[15] If the surface properties of the material prevent cell adhesion or if the attached cells are not able to survive on the surface, biofilm formation is retarded.[15] The propensity of a solid surface to attach cellular particles depends on both the surface characteristics (surface charge, roughness, etc.) and environmental conditions (ionic strength, pH, etc.).[15] These are some important directions for the design of biocide mechanisms of action. It should be noted that a biofilm can form an effective barrier against antimicrobial agents.[12] Figure 2.3 shows an AFM micrograph of the surface of material colonized by one-week-old biofilm formed by *Pseudomonas aeruginosa*. The surface of the substrate and some bacteria are encased in an extracellular polymeric substance, EPS, other cells are only attached to the surface of EPS. Figure 2.4 shows different methods of attachment, which enhances fungal adhesion.[16]

Biofilms are composed of microbial cells, extracellular polymers produced by them, and water (95%).[17] This shows that they have a very porous structure. The distribution of microorganisms is not uniform. In multi-species biofilms, voids, connecting channels between these voids, and microbial clusters or layers are found.[18]

The composition of different species depends on conditions and substrate. It is clear from studies under limited access to light (catacombs) that the susceptibility of a material to be colonized by different groups of living organisms (bioreceptivity) depends on environmental factors, such as water availability, pH, climatic exposure, light, nutrient sources, and on mineral composition, type of cement as well as porosity and permeability of the rock material.[19] Dim light habitats like the catacombs are characterized by the development of terrestrial epilithic cyanobacteria and a few eukaryotic algae developing along light gradients.[19] Figure 2.5 shows SEM micrographs of different details of biofilms formed under such conditions.

Figure 2.6 shows some microorganisms identified in biofilms on cinematographic films. Biofilms contained bacteria, yeast, and fungi. The composition of microorganisms identified in different films depended on location in which films were stored. Films from Barcelona had 15 identified species, from Grand Canaria 10 species, and from Madrid only 7 species.[20] Most microorganisms that were isolated can resist environmental conditions adverse to microbial life, e.g., desiccation or nutrient starvation, and they can adhere to the substrates by hyphae or exopolymers.[20] This may give rise to biofilm development, increasing the corrosive action of microbial metabolic products when suitable conditions become available.[20]

Table 2.1 gives data for biofilm thickness on different substrates and in different environments

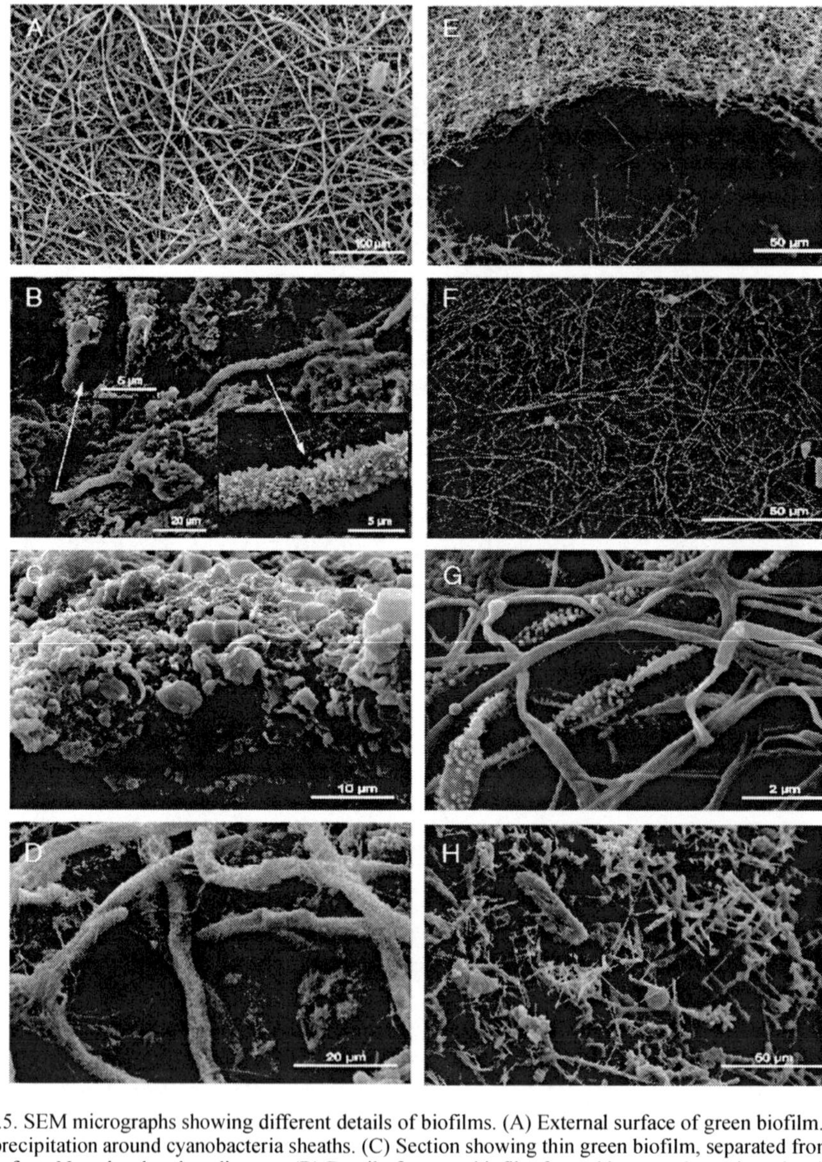

Figure 2.5. SEM micrographs showing different details of biofilms. (A) External surface of green biofilm. (B) Calcite precipitation around cyanobacteria sheaths. (C) Section showing thin green biofilm, separated from the stucco surface. Note the abundant diatoms. (D) Detail of a green biofilm formed by cyanobacteria and actinobacteria. (E) Section showing white biofilm with a dense net of needle-like calcite crystals. (F) Detail of the external surface of white biofilm. (G) Details of filaments in the internal area of white biofilm close to the surface. Note the calcium carbonate nucleation points. (H) Detail of the inner part of white biofilm showing spherical particles associated with needle-fiber calcite crystals and beaded filaments. [Adapted, by permission, from Sanchez-Morala, S; Luquea, L; Cuezvaa, S; Solerb, V; Benaventec, D; Laizd, L; Gonzalezd, J M; Saiz-Jimenez, C, *Sci. Total Environment*, **349**, 260-76, 2005.]

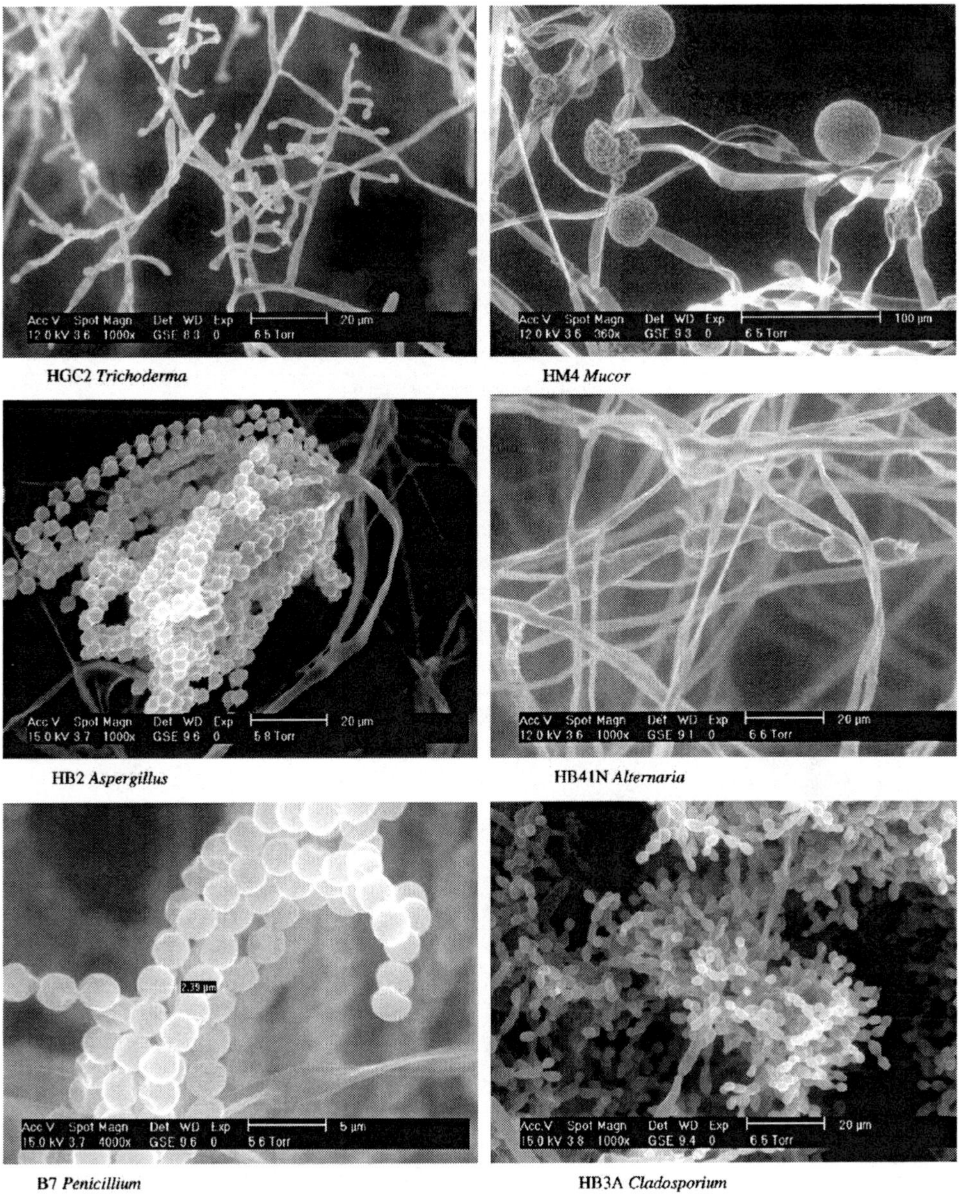

Figure 2.6. SEM micrographs of different genera of fungi isolated from cinematographic films showing their characteristic hyphae and conidia. [Adapted, by permission, from Abrusci, C; Martin-Gonzalez, A; Del Amo, A; Catalina, F; Collado, J; Platas, G, *Intl. Biodeteriration Biodeg.*, **56**, 58-68, 2005.]

Table 2.22. Biofilm thickness.

Thickness, μm	Substrate	Reference
0-12	surface of antifouling paint in marine env. (plateau after 10 days)	26
3-5	polymer	21
3-18	heat exchanger	22
10-60	historical glass window panels (various churches in Germany)	24
27	reverse osmosis membrane	27
30-125	heat exchanger (depends on water velocity; max at 1 ms^{-1})	23
50-150	polymer	21
65-270	Plexiglas panels in an Indian estuary (accumulated in 7 days)	25
a few 10,000	paper manufacture and waste water treatment	21

Processes of biodeterioration can be classified according to processes occurring on material exposure, as follows:[12]

- mechanical deterioration (the material is damaged as a direct result of the activity of an organism, such as its movement or growth)
- assimilatory biodeterioration (a material is degraded for its nutritive value, e.g., the breakdown of cellulosic materials by cellulolytic microorganism)
- dissimilatory biodeterioration (metabolic products damage a material by causing corrosion, pigmentation, or by the release of toxic metabolites into a substance, e.g., poisoning of grain by mycotoxins or dissolution of stones by acids produced by microorganisms)
- soiling/biofouling (the presence of an organism or its excrement renders the product unacceptable, e.g., biofouling of ships' hulls or the formation of slime in fuel lines)

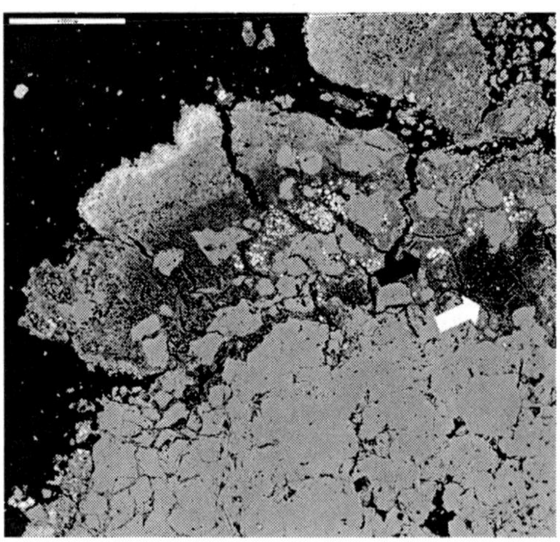

Figure 2.7. SEM image of the surface of a stone figure. Black arrow indicates green algal colony; white arrow indicates fungal cells. Bar = 200 μm. [Adapted, by permission, from de los Rios, A; Galvan, V; Ascasoa, C, *Intl. Biodeterioration Biodeg.*, **54**, 113-20, 2004.]

The above biodeterioration processes are product and microorganism composition specific and they are broadly discussed in Chapter 4.

Many processes of biodegradation occur on material surface but there are also numerous exceptions. For example, the presence of an extensive pore surface (>5 m^2/g) resulting from a high porosity or the presence of clay minerals, facilitates the spreading of microflora within the pore system.[28] Also, high-porosity values (from around 14 vol% with an average pore radius between 1 and 10 μm), permit deep penetration of moisture into the material and thus microbial penetration to a depth of up to 3-5 cm.[28] Figure 2.7 shows penetration of algae and fungi into the body of a figure in Santa Cruz la Real, Segovia, Spain.[29]

The depth of penetration depends on the stone substrate and the type of lichen; lichens can sometimes be found at least 3 cm below the stone surface.[30] Lichens cause mechanical damage due to penetration of their rhizines, composed of fungal filaments, and the expansion/contraction of the thallus on wetting/drying.[30] As a result of hyphal penetration, surface water can reach a considerable depth, where the so-called ice-wedging action caused by freezing expansion of the thallus takes place when temperatures fall below zero.[31]

The examples of the above general mechanisms are further discussed in more detail in Chapter 4 for groups of products which are affected by biodegradative processes.

REFERENCES

1 Cavalier-Smith, T, *Biol. Rev. Camb. Philos. Soc.*, **73**, 3, 203-66, 1998.
2 Cavalier-Smith, T, *Microbiology Today*, 166-69, 2006.
3 Ul-Hassan, A; Wellington, E M, **Encyclopedia of Microbiology**, *Elsevier*, 2009.
4 Lucas, N; Bienaime, C; Belloy, C; Queneudec, M; Silvestre, F; Nava-Saucedo, J-E, *Chemosphere*, **73**, 429-42, 2008.
5 Ranalli, G; Zanardini, E; Sorlini, C, **Biodeterioration – Including Cultural Heritage, in Encyclopedia of Microbiology,** *Elsevier*, 2009, Pages 191-205.
6 Wypych, G, **Handbook of Material Weathering**, 4th Ed., *ChemTec Publishing*, Toronto, 2008.
7 Wypych, G, **PVC Degradation and Stabilization**, *ChemTec Publishing*, Toronto, 2008.
8 Montegut, D; Indictor, N; Koestler, R J, *Intl. Biodeterioration,* **28**, 209-26, 1991.
9 Christenson, E M; Anderson, J M; Hiltner, A, *Corrosion Eng. Sci. Technol.*, **42**, 4, 312-23, 2007.
10 Chavant, P; Gaillard-Martinie, B; Talon, R; Hébraud, M; Bernardi, T, *J. Microbiol. Methods*, **68**, 605–12, 2007.
11 Campanac, C; Pineau, L; Payard, A; Baziard-Mouysset, G; Roques, C, *Antimicrobial Agents Chemotherapy*, **46**, 5, 1469-74, 2002.
12 Morton, L H G; Surman, S B, *Intl. Biodeterioration Biodegradation*, 203-21, 1994.
13 Kim, D; Jung, S; Sohn, J; Kim, H; Lee, S, *Desalination*, **238**, 43–52, 2009.
14 Nigmatullin, R; Konovalova, V; Pobigay, G, *J. Appl. Polym. Sci.*, **111**, 1697-705, 2009.
15 Guillemot, G; Despax, B; Raynaud, P; Zanna, S, Marcus, P; Schmitz, P; Mercier-Bonin, M, *Plasma Process. Polym.*, **5**, 228-38, 2008.
16 Jones, E B G, *Myco/. Res.*, **98**, 9, 961-81, 1994.
17 Herbert-Guillou, D; Tribollet, B; Festy, D, *Bioelectrochemistry*, **53**, 119-25, 2000.
18 Steele, A; Goddard, D T; Beech, I B, *Intl. Biodeterioration Biodegrad.*, 35-46, 1994.
19 Sanchez-Morala, S; Luquea, L; Cuezvaa, S; Solerb, V; Benaventec, D; Laizd, L; Gonzalezd, J M; Saiz-Jimenez, C, *Sci. Total Environment*, **349**, 260-76, 2005.
20 Abrusci, C; Martin-Gonzalez, A; Del Amo, A; Catalina, F; Collado, J; Platas, G, *Intl. Biodeteriration Biodeg.*, **56**, 58-68, 2005.
21 Flemming, H-C, *Polym. Deg. Stab.*, **99**, 309-15, 1998.
22 Sriyutha Murthy, P; Venkatesan, R; Nair, K V K; Ravindran, M, *Intl. Biodeterioration Biodeg.*, **53**, 133-40, 2004.
23 Bott, T R, Biological growth on heat exchanger surfaces in **Fouling Heat Exchangers**, *Elsevier*, 1995, pages 223-67.
24 Müller, E; Drewello, U; Drewello, R; Weißmann, R; Wuertz, S, *J. Cultural Heritage*, **2**, 31-42, 2001.
25 Rao, T S, *Estaurine, Coastal Shelf Sci.*, **58**, 67-75, 2003.
26 Meseguer Yebra, D; Kiil, S; Weinell, C E; Dam-Johansen, K, *Prog. Org. Coat.*, **57**, 56-66, 2006.
27 Flemming, H-C, *Experimental Thermal Fluid Sci.*, **14**, 382-91, 1997.
28 Warscheid, T; Braams, J, *Intl. Biodeterioration Biodeg.*, **46**, 343-68, 2000.
29 de los Rios, A; Galvan, V; Ascasoa, C, *Intl. Biodeterioration Biodeg.*, **54**, 113-20, 2004.
30 Scheerer, S; Ortega-Morales, O; Gaylarde, C, in Microbial Deterioration of Stone Monuments—An Updated Overview, **Advances in Applied Microbiology**, Volume 66, *Elsevier*, 2009, Pages 97-139.

31 Chen, J; Blume, H-P; Beyer, L, *Catena*, **39**, 121-46, 2000.

3

INDUSTRIAL BIOCIDES

3.1 GENERAL MECHANISM OF BIOSTABILIZATION

Biocide and microorganism interaction is controlled by diffusion that determines different levels of concentration, as follows:[1]

- uptake of biocide by cell
- partition/passage of biocide to target(s) of its action
- concentration of biocide at target(s)
- damage to target(s).

The initiating step is the migration of biocide through an interface. This process is regulated by the physicochemical characteristics of both the cell and material containing biocide. A biocide encounters structures which account, to varying degrees, for the different sensitivities of individual species of microorganisms. Figure 3.1 presents differences between Gram-negative and Gram-positive bacteria.[1] Gram-negative cells have an additional barrier of the lipopolysaccharide (LPS) layer to biocide penetration which Gram-positive cells do not possess.[1] This structure has a significant moderating influence on the penetration of both hydrophilic and hydrophobic molecules, establishing a molecular weight cut-off for the passage of the water-filled pores (porins), and requiring optimal lipophilic properties for the progress of hydrophobic biocides.[1]

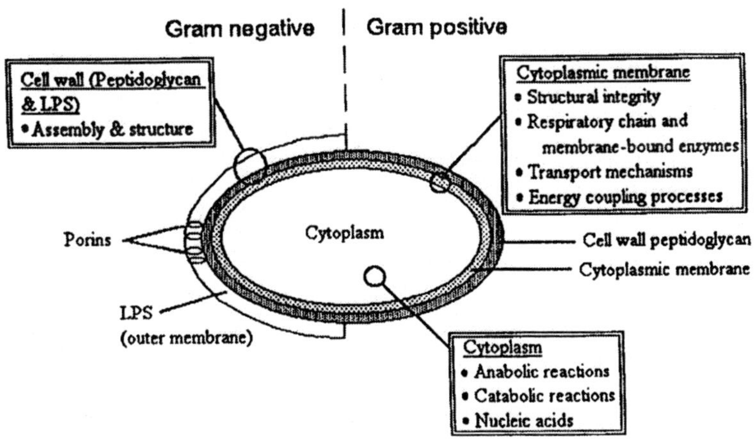

Figure 3.1. Diffusion barriers for biocide on its pathways to reach potential targets in bacterial cells. [Adapted, by permission, from Denyer, S P, *Intl. Biodeterioration Biodeg.*, 227-45, 1995.]

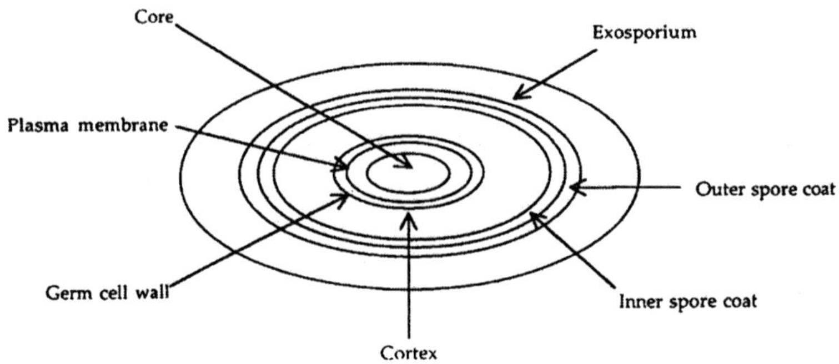

Figure 3.2. A structure of bacterial spore (not drawn to scale). [Adapted, by permission, from Russell, A D, *Intl. Biodeterioration Biodeg.*, 247-65, 1995.]

This barrier effect can be moderated by the introduction of agents such as ethylenediaminetetraacetic acid, which increases the permeability of the Gram-negative outer membrane.[1]

Intrinsic resistance is frequently a chromosome-controlled property of a bacterial cell that enables it to circumvent the action of a biocide. It is most commonly found in Gram-negative bacteria, mycobacteria, and bacterial spores.[2] Many biocides will kill (at low concentrations) non-sporulating bacteria but not bacterial spores; others are also sporicidal (kill spores), although high concentrations may be needed to achieve this effect.[2]

The reasons for high resistance of spores is immediately understood from Figure 3.2. The goal of biocide diffusion is to reach the core which contains proteins, DNA, and RNA which are targets to be destroyed by a biocide to perform its role. But on its way it has numerous barriers, including[2]

- outer spore coat (alkali resistant; removed by disulfide bond-reducing agents)
- inner spore coat (alkali soluble, composed mostly of proteins)
- cortex (composed of peptidoglycan)
- plasma cell membrane
- germ cell wall

Similar differences exist between fungus and its spores, which are also more resistant.[2] The order of microbial resistance is as follows: enveloped virus (least resistant) < Gram-positive bacteria < protozoa < vegetative fungi and algae < Gram negative bacteria < fungal spores < protozoal cysts < small non-enveloed viruses < mycobacteria < protozoal oocysts < bacterial spores (most resistant).[3]

In the above analysis, we did not mention, so far, the effect of biofilm formation on the diffusion process. Biofilms are complex three-dimensional structures in which microorganisms are embedded in a matrix of extracellular polymeric substances, EPS. The EPS matrix provides biofilm with mechanical stability by filling and forming the space between the bacterial cells to keep them together.[4]

There are four main groups of extracellular microbial polymers, including exopolysaccharides, hydrophobins (a group of small fungal proteins), glomalin (a glycoproteinaceous substance produced by arbuscular mycorrhizal fungi), and extracellular lipid compounds (for example, glycolipids such as rhamnolipids).[7] The production of these substances is a common feature of Archaea, Eubacteria, and eukaryotic microorganisms (e.g., algae, fungi).[5]

It is well established that bacterial biofilms exhibit an increased resistance to antimicrobial treatments. This resistance is attributed to reduced diffusion, physiological changes because of lower oxygen level and fewer nutrients, and the production of enzymes degrading antimicrobial substances.[6]

Possible explanations for the increased resistance when biofilm is present include[7]

- limited diffusion of antimicrobial agents through the biofilm matrix (the glycocalyx creates a diffusion barrier to the antimicrobial agent). Diffusion through a biofilm may be affected by charge (ionic) interactions between the glycocalyx and the antimicrobial agent, by an increase in the distance the agent must diffuse, by molecular sieving (size exclusion), and by the viscosity of the glycocalyx
- interaction and neutralization of the antimicrobial agents with/by the biofilm matrix (all components of the cell envelope except peptidoglycan play a role in the barrier mechanisms because peptidoglycan is spongy and therefore permeable; it is also possible that the glycocalyx matrix in a biofilm reacts with and neutralizes the antimicrobial agent, for example, it has been suggested that iodine reacts with glycocalyx compounds; also, it was demonstrated that chemical interactions between chlorine and biofilm glycocalyx also take place)
- enzyme mediated resistance (known examples include heavy metal and formaldehyde resistance)
- level of metabolic activity within the biofilm (biofilm reduced susceptibility requires that at least some of the cells within a biofilm experience a nutrient limitation that causes them to enter a slow-growing or starved state)
- genetic adaptation (a variety of bacteria, mostly fermentative, exhibit oxidizing stress response by producing oxidant-degrading and repair enzymes; this mechanism implies that reduced susceptibility of biofilm is genetically programmed)
- outer membrane structure (hydrophilic antibacterial agents are prevented from entering through the outer membrane by the lipopolysaccharide layer and the underlying phospholipids, whereas hydrophobic agents are excluded by outer membrane proteins)

There are no quantitative data which can rate the importance of various influences, listed above, on resistance of microorganisms to biocide action caused by the biofilm formation.

Figure 3.3 shows the targets in various microorganisms, methods of their inactivation, and examples of biocides which are capable to proceed according to these mechansims.

Biocide mechanisms of action can be divided into four broad categories:[9]

- oxidants (rapid speed of kill agents, such as chlorine and peroxides that work directly *via* radical-mediated reactions to oxidize organic material)
- electrophilic agents (inorganic ions such as silver, copper, and mercury, and organic biocides such as formaldehyde and isothiazolones which react covalently

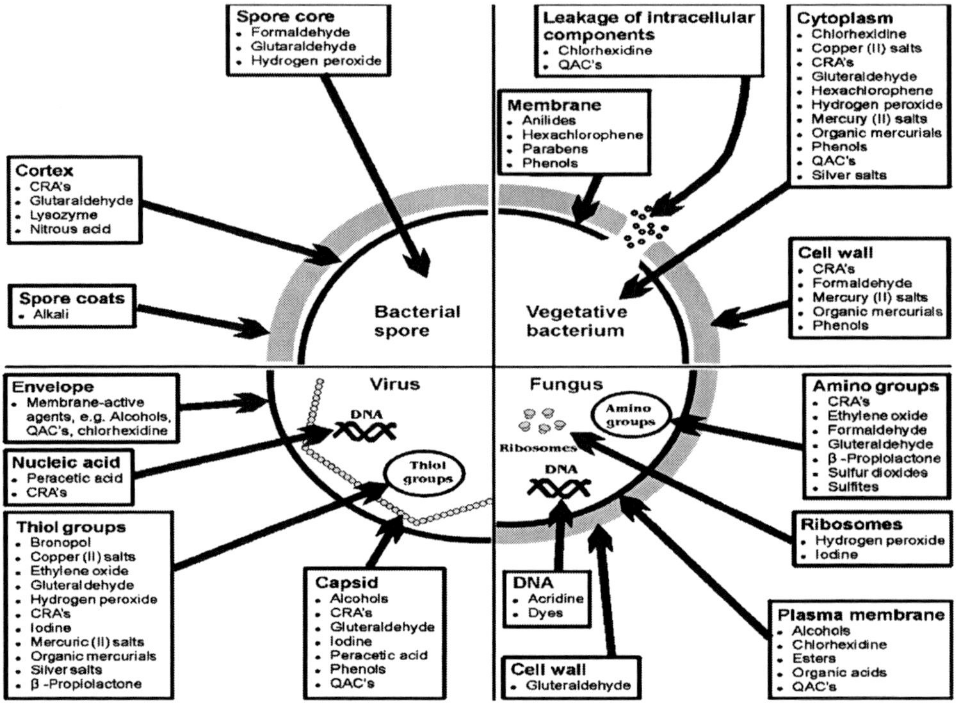

Figure 3.3. Mechanism of microorganism inactivation by biocides. CRAs = chlorine-releasing agents, QACs = quaternary ammonium compounds. [Adapted, by permission, from Russell, A D; Furr, R J; Maillard, J Y, *ASM News*, **63**, 481–487, 1997.]

with cellular nucleophiles to inactivate enzymes and initiate the formation of intracellular free radicals which contribute to their lethal action)

- cationic membrane active biocides (chlorhexidine, quaternary ammonium compounds, phenols, and alcohols (e.g., phenoxyethanol) destabilize membranes leading to rapid cell lysis
- protonophores (parabens, weak acids, such as sorbic and benzoic acids, and pyrithione interfere with the ability of the cell membrane to maintain a proper pH balance, resulting in acidification of the cell interior and widespread disruption of metabolism).

Reactive biocides have multiple targets in the cell and the number of targets attacked by any one biocide may depend to some extent on the reactivity of the biocide.[9] Powerful oxidants, such as the hypohalous acids, react with all organic molecules while less reactive electrophiles such as isothiazolones can react only with strong nucleophiles such as sulfhydryls.[9] Compounds such as aldehydes react with hydroxyls, amines, and sulfhydryls. The multiplicity of targets makes it difficult if not impossible to make biocide ineffective.

The activity of biocides depends on three main factors:[10]

- the nature of the physical environment (time of contact, temperature, concentration, pH, and the presence of organic matter)
- the nature and condition of the microorganism (vegetative, spores; see above)
- the ability of the organism to render the biocide inactive (mechanisms of resistance, see below).

The nature of the physical environment has to be tailored to a biocide and microorganism types but there are some general principles. Contact time and concentration increase rises the probability of successful action of the biocide. Temperature increase usually increases lethal action of biocides but it frequently depends on pH (for example, at low pH, temperature may increase lethal action but, at high pH, temperature increase reduces effectiveness of biocide). The presence of organic matter is the potential source of nutrient, and as such, it facilitates the growth of microorganisms.

Presence of biocide may affect many different processes which are vital for microorganism functions and survival. Here are examples of such processes and the molecular target site or mechanism (in parentheses), based on the effect of different biocides on algae:[11]

- porphirin biosynthesis (protoporphyrinogene oxidase)
- RNA synthesis and function (nucleotide analogue disturbing codon-anticodon recognition or RNA polymerase I)
- nucleotide biosynthesis (inhibition of purine biosynthesis)
- proton translocation and ATP synthesis (protonophoric uncoupling)
- protein biosynthesis (50S subunit of 70S ribosomes)
- membrane functions (membrane disintegration)
- respiratory electron transport (succinate ubiquinone oxido-reductase complex II or cytochrome bc complex III)
- lipid biosynthesis (elongase catalyzing the formation of very-long-chain fatty acids)
- aminoacid biosynthesis (acetolactate synthase)
- DNA synthesis and function (DNA gyrase)
- carotenoid biosynthesis (phytoene desaturase)
- photosynthetic electron transport (redox catalyst at photosystem I or D1 protein of photosystem II)
- steroid biosynthesis (cytochrome P450 steroid 14a demethylases)

The above list is given as an example of an usually large number of target options which can be considered in the design and selection of a system of biocide(s). In real conditions, the affected process and target depend on a pair of biocide and microorganism (and its condition). The above list, developed based on the action of a mixture of 16 different test substances, also suggests that there is potential for synergism of action of biocides, considering that it is unlikely that one biocide will be able to attack so many different targets and it is highly unlikely that one biocide is able to provide lethal action to a combination of wide spectrum of microorganisms present in biofilms.

Bacterial resistance to biocides is of two types:[12]

- intrinsic resistance (intrinsic insusceptibility), a natural property of an organism

- acquired resistance, which results from mutation or acquisition of generic material (such as a plasmid or transposon)

Resistance, which is a genetically determined phenomenon, is distinguished from phenotypic adaptation processes, which are not hereditary. Adaptation can be avoided by rigorous cleaning and disinfection (and/or using proper levels of biocide), avoiding concentrations of biocides below the microbiocidal concentration.[13]

Regarding intrinsic resistance, resistant cells usually adsorb less biocide than sensitive cells.[12] This is either because the resistant cells have more protective layers (see for example, Figure 3.2) through which a biocide must diffuse or because the biocide (which is reactive) undergoes either chemical or physical interaction with a cell fragment which is not a target.

The following intrinsic resistance mechanisms are established:[12]

- Gram-negative bacteria – impermeability of the outer membrane, (glycocalyx)
- Mycobacteria – impermeability (waxy cell walls)
- bacterial spores (spore coat and cortex)
- Gram-positive bacteria (glycocalyx)
- inactivation (chromosome-mediated)

Unlike intrinsic resistance (usually expressed by chromosomal genes), the acquired resistance is a consequence of mutations in chromosomal genes or the acquisition of plasmids or transposons.[12] Chromosomal mutations are changes in the base sequence of DNA. The following acquired resistance mechanisms have been found:[12]

- enzymatic detoxification or modification (modification of enzymes, such as acetyltransferase, adenyllase, phosphotransferase; inactivation of β-lactamase, acetyltransferase, hydrolase, lyase, and dehydrogenase)
- enhanced efflux (by ATPase pump and in MRSA strains)
- enzymatic trapping (trapping via β-lactamase in periplasm)
- impaired uptake (plasmid-borne mechanism, mutational alterations to porins, plasmid-encoded porin modification)
- modification of regulation (mutation in chromosomal gene involved in uptake)
- pathway compensation (overproduction of dihydrofolate reductase)
- ribosomal protection (transposon tetracycline-resistant determinants or cytoplasmic proteins acting in conjuction with ribosomes)

In addition to intrinsic and acquired resistance, the effect of biofilm formation discussed above should also be considered.

It should be pointed out that resistance has been reported to many preservatives and biocides, including benzalkonium chloride, benzisothiazolone, benzoic acid, chloroallyltriazine-azoniadamantane, chloramine, chlorhexidine, chlorophenol, dibromodicyanobutane, dimethyloxazolidine, dimethyldithiocarbamate, dimethoxydimethylhydantoin, formaldehyde, glutaraldehyde, hexahydrotriethyltriazine, hydrogen peroxide, imidazolidinyl urea, iodine, mercuric salts, methylenebischlorophenol, methylchloro/methylisothiazolone, methyl paraben, phenoxyethanol, phenylethyl alcohol, phenylmercuric acetate, propyl paraben, povidone iodine, quaternary ammonium compounds, sorbic acid, tetrahydrothiadiazinthione, and trifluoromethyl dichlorocarbanilide.[14]

Three ways of delaying the build-up of resistance against fungicides have been suggested:[15]

- the use of molecules showing negatively correlated cross-resistance
- alternation of fungicide applications
- the use of mixtures of fungicides

The use of mixtures of fungicides with different modes of action was found to be the best strategy.[15] Three fungicides with different modes of action – oxadixyl (phenylamide; systemic), mancozeb (dithiocarbamate; protectant), and cymoxanil (penetrant) – were tested alone and in different combinations.[15] The level of synergistic interaction of fungicide mixtures was found to be dependent on the proportion of the single components in the mixture as well as on the activity of the individual compounds against the fungal strains.[15]

Chlorine dioxide is an effective disinfectant, which exhibits synergism with monochloramine and free chlorine.[16] Combined effects of diuron with its metabolites resulted in synergism of antifouling biocides tested using phytoplankton bioassay.[17] Oxine and copper oxinate were resistant to inactivation by organic materials, interacted synergistically *in vitro,* and did not elicit resistance on long-term exposure.[18]

3.2 CHEMICAL TYPES OF BIOSTABILIZERS

Biostabilizer can be classified according to the European Union classification given in Chapter 1. European Union classification is based on application. Applications of biostabilizers to different groups of products will be discussed in Chapter 4. In the present chapter, information on biostabilizers will be organized according to their chemical composition. The reasons are as follows:

- it is easier to search products by their chemical composition
- it is easier to compare compounds of similar structure and composition
- groups of products having similar composition can be compared with other groups to find similarities and differences
- it is possible to generalize properties of materials within a group.

This review of biostabilizers will include:

- chemical and commercial names of stabilizers
- their identifiers (e.g., CAS number and EC number)
- typical chemical composition
- chemical and physical properties of main component
- optical and electrical properties
- toxicity data
- environmental impact data
- antimicrobial effectiveness
- typical applications
- suggested concentrations
- manufacturers.

It is apparent from this list that the goal of selected information is to show the expected performance for different groups of biostabilizers, the economy of their use, their impact on health safety and environment, how to apply them, and where to find them.

Care is taken to include only products which are approved for use in either the European Union or the United States. Exceptions are made for recent patented inventions, which are also included in the tables below. The relevant patents are listed in the references.

Historical products which were eliminated from use for various reasons will not be included. The chemical groups are given in alphabetical order. The information is provided in tabular form having the same structure of data and presentation to assist comparison.

3.2.1 ACETAL ALDEHYDE-RELEASING COMPOUNDS

Table 3.1. Properties of acetal aldehyde-releasing compounds.

Chemical name (CAS and EC #s): dimethoxane, 2,6-dimethyl-l,3-dioxan-4-acetate (828-00-2, 212-579-9)		
Commercial name (manufacturer): Bioban DXN (DOW)		
Chemical formula: $C_8H_{14}O_4$	**Molecular weight**: 174.22	
Other components in composition (%): acetic acid – 1.1, vinyl acetate – 5, crotonaldehyde – 2, acetalaldehyde – 2		
Main product content, %: 85-95		
CHEMICAL PROPERTIES		
Water solubility, %: miscible		
Chemical reactivity: hydrolyzed by water		
PHYSICAL PROPERTIES		
Melting point, °C: below 25	**Boiling point**, °C: 185	**Density**, g/cm^3: 1.06-1.08
Refractive index: 1.43-1.437	**Vapor pressure**, Pa (20°C): 29	**pH**: 5.2
Viscosity, cps: 114_{25}	**Odor**: mustard-like	
OPTICAL & ELECTRICAL PROPERTIES		
Color: yellow liquid		
HEALTH & SAFETY		
Flash point, °C: 87	**LD50, oral, rat**, mg/kg: 1930	
ENVIRONMENTAL IMPACT		
Rainbow trout, 96-h LC50, mg/l: 370	**Daphnia magna**, 48-h LC50, mg/l: 24	
Biodegradability: Over 50% is hydrolyzed in two hours. It reacts with water to form acetic acid and dioxins; the latter breaks down into acetaldehyde and aldol in 14 hours.		
ANTIMICROBIAL ACTIVITY		
Product type: group 2, # 6	**Typical concentration**, %: 0.1-0.2	
Minimum inhibition concentration, mg/l: *bacteria*: *Bacillus subtilis* 625, *Brevibacterium ammoniagenes* 625, *Enterobacter aerogenes* 625, *Escherichia coli* 625, *Pseudomonas aeruginosa* 625, *Pseudomonas fluorescens* 625, *Salmonella choleraesuis* 312, *Salmonella typhosa* 625, *Shigella sonnei* 625, *Staphylococcus aureus* 1250, *fungi*: *Aspergillus flavus* 1250, *Aspergillus niger* 1250, *Aspergillus oryzae* 1250, *Aspergillus terreus* 1250, *Candida albicans* 1250, *Penicillium piscarium* 625, *Penicillium sp.* 1250, *Pityrosporum ovale* 625, *Saccharomyces cerevisiae* 2500		
pH application range: 3-8.5		

Table 3.1. Properties of acetal aldehyde-releasing compounds.

APPLICATIONS
water-based paints, adhesives, latex, ink and fountain, dyestuffs, fabric softeners, sizing and spinning emulsions
Remark: to be phased out in European Union in 2009

Chemical formula:

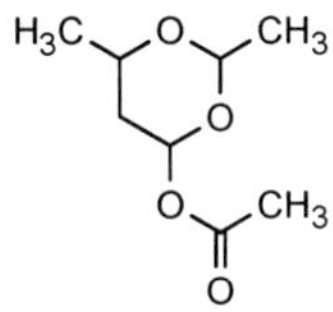

3.2.2 ACID ESTERS

Table 3.2. Properties of acid esters.

Chemical name (CAS and EC #s): 1,4-bis(bromoacetoxy)2-butene (20679-58-7); 2,3-dihydroxypropyl dodecanoate, glyceryl monolaurate (27215-38-9); silicic acid esters (Henkel)[19]		
Commercial name (manufacturer): BBAB (Dead Sea Bromine Group, Buckman Laboratories), Monolaurin (Colonial)		
Chemical formula: $C_8H_{10}Br_2O_4$, $C_{15}H_{30}O_4$	**Molecular weight**: 329.97, 274,4	
Main product content, %: 90; 90		
CHEMICAL PROPERTIES		
Water solubility, %: ins.; ins.	**Chemical reactivity**: hydrolysis	
PHYSICAL PROPERTIES		
Melting point, °C: -20; 55	**Boiling point**, °C: –; $186_{1\ mmHg}$	**Density**, g/cm^3: 1.69-1.75; 0.96
pH: 2-3; 4-5	**Odor**: irritant; faint	
OPTICAL & ELECTRICAL PROPERTIES		
Color: dark-brown liquid; cream		
HEALTH & SAFETY		
LD50, oral, rat, mg/kg: 191; 53,000		
ENVIRONMENTAL IMPACT		
Daphnia magna, 24-h LC50, mg/l: 0.024	**Biodegradability**: biodegradable	
ANTIMICROBIAL ACTIVITY		
Product types: group 2, # 12	**Typical concentration**, %: –; 0.5-1	
Minimum inhibition concentration, mg/l: 20-200; 5,000		
APPLICATIONS		
slime-forming bacteria, yeasts and fungi in paper-mill process water; emulsifier with bactericide properties		
Remark: BBAB and glyceryl monolaureate are not on the list of European Union biocides		

Chemical formulas:

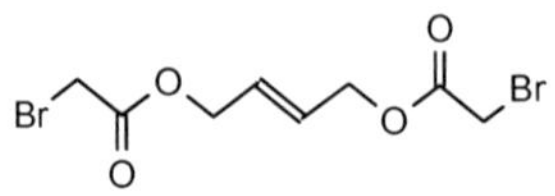

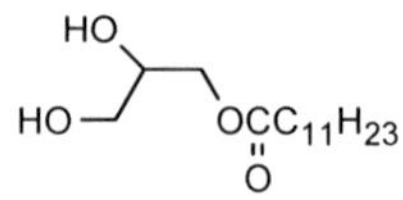

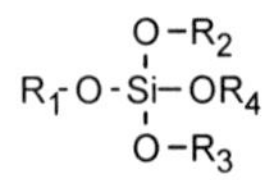

3.2.3 ACIDS

Table 3.3. Properties of acids.

Chemical name (CAS and EC #s): benzoic (65-85-0, 200-618-2), boric (11113-50-1, 233-139-2), bromoacetic (79-08-3, 201-175-8), citric (77-92-9, 201-069-1), decanoic (334-48-5, 206-376-4), formic (64-18-6, 200-579-1), hydrochloric (7647-01-0, 231-595-7), lactic (79-33-4, 201-196-2), lauric (143-07-7, 205-582-1), nonanoic (112-05-0, 203-931-2), octanoic (124-07-2, 204-677-5), peracetic (79-21-0, 201,186,8), peroxyoctanoic (33734-57-5, –), salicylic (69-72-7, 200-712-3), sorbic (110-44-1, 203-768-7)		
Commercial name (manufacturer): commercial names the same (Albermale, BASF, DuPont, Huels, Kimberley Clark, Wacker, and many other companies)		
Main product content, %: 98-99	**Molecular weight**: 36.5-192.12	
CHEMICAL PROPERTIES		
Water solubility, %: 0.2-misc.	**Chemical reactivity**: corrosive, skin burns	
PHYSICAL PROPERTIES		
Melting point, °C: -114-171	**Boiling point**, °C: -85-300	**Density**, g/cm^3: 0.89-1.934
Refractive index: 1.37-1.49	**Vapor pressure**, kPa (20°C): traces-4103	
Thermal expansion coef., 10^{-4} °C^{-1}: 9.9-10.01	**pH**: 1-5.1	
Surface tension, mN/m: 37.58-49.7	**Odor**: odorless to pungent	
Heat of combustion, MJ/kg: 3.23-15.61		
Henry's law constant, atm m^3/mole: 4.3E-14 -2.04E+6		
OPTICAL & ELECTRICAL PROPERTIES		
Color: colorless to white	**Dielectric constant**: 4.12-58	
HEALTH & SAFETY		
Autoignition temperature, °C: 485-601	**Flash point**, °C: from non-flammable to 11-157	
LD50, oral, rat, mg/kg: 730-7,360	**LD50, dermal, rabbit**, mg/kg: 1,006-10,000	
Threshold limiting value, ACGIH, ppm: 5-10	**Max. concentration**, IDHL, ppm: 30-50	
ENVIRONMENTAL IMPACT		
Daphnia magna, 48-h EC50, mg/l: 34-133	**log Kow**: 0.41-4.90	
ANTIMICROBIAL ACTIVITY		
Product types: group 1, #1-4, group 2, #6, 8, 10-13, group 3, #15-19,	**Typical concentration**, %: 0.05-0.6	
Minimum inhibition concentration, mg/l: 100-1000	**Use at pH**: 3-6.5	

Table 3.3. Properties of acids.

APPLICATIONS
the most frequently used as disinfectants (except for drinking water), but also as in-can preservatives, wood, masonry, and cooling system preservatives, and slimicides

Chemical formulas:

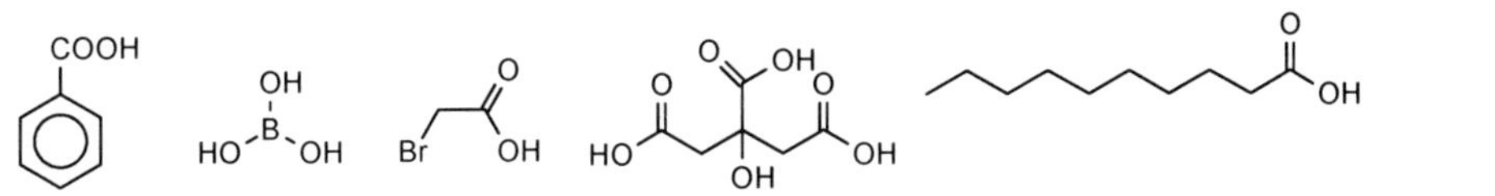

benzoic boric bromoacetic citric decanoic

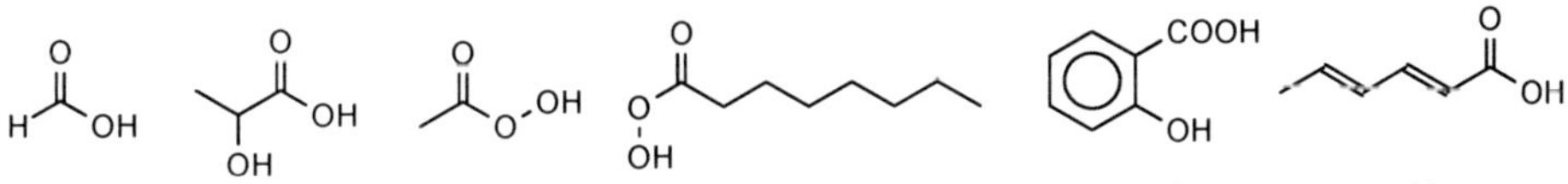

formic lactic peracetic peroxyoctanoic salicylic sorbic

3.2.4 ACTIVE HALOGEN PRODUCTS

Table 3.4. Properties of active halogen products.

Chemical name (CAS and EC #s): bis(trichloromethyl) sulfone (3064-70-8, 221-310-4), 1-bromo-1-(bromomethyl)-1,3-propanedicarbonitrile (bromothalonil) (35691-65-7, 252-681-0), 2-bromo-2-nitropropane-1,3-diol (Bronopol) (52-51-7, 200-143-0), bromochlorodimethylhydantoin (32718-18-6, 251-171-5), 2-bromo-4'-hydroxyacetophenone (2491-38-5, 219-655-0), calcium hypochlorite (7778-54-3, 231-908-7), chlorine (7782-50-5, 231-959-5), chlorine dioxide (10049-04-4, 233-162-8), chlorocresol (59-50-7, 200-431-6), chloramin (127-65-1, 204-854-7), chlorfenapyr (122453-73-0, -), chlorophene (120-32-1, 204-385-8), cypermethrin (52315-07-8, 257-842-9), diiodomethyl p-tolyl sulfone (20018-09-1, 243-468-3), didecylammonium chloride (7173-51-5, 230-525-2), dodecylguanidine monohydrochloride (13590-97-1, 237-030-0), flufenoxuron (101463-69-8, 417-680-3), iodine (7553-56-2, 231-442-4), octenidine dihydrochloride (70775-75-6, 274-861-8), permethrin (52645-53-1, 258-067-9), sodium bromide (7647-15-6, 231-599-9), sodium dichloroisocyanurate dihydrate (51580-86-0, 220-767-7), sodium hypochlorite (7681-52-9, 231-668-3), sulfuryl fluoride (2699-79-8, 220-281-5), tetrachloroisophthalonitrile (1897-45-6, 217-588-1), triclosan (3380-34-5, 222-182-2)		
Commercial name (manufacturer): chemical name, abbreviation, or brand name: Amical, Bioban, Bronopol, Chloramin T, DDAC, Triclosan (BASF, Buckman Laboratories, Clearon, DOW, Eka Chemicals, Euro Chlor, Lanxess, Wacker and many other companies)		
Main product content, %: 2-100	**Molecular weight**: 70.905-422.03	
CHEMICAL PROPERTIES		
Water solubility, %: traces to 29.3		
Chemical reactivity: decomposition to chlorine in water; oxidizing material, corrosivity		
PHYSICAL PROPERTIES		
Melting point, °C: -135.8-250	**Boiling point**, °C: -55.4 to 350	**Density**, g/cm^3: 0.92-4.93
Refractive index: 1.0008-3.34	**Vapor pressure**, hPa (20°C): 0.0000167-7772	
Surface tension, mN/m: 18.4	**Viscosity**, mPas: 14-340	

Table 3.4. Properties of active halogen products.

Henry law constant, atm m^3/mol: 1.E-11 to 0.0117	
pH: 4.5-7.2	**Odor**: odorless to pungent to suffocating
OPTICAL & ELECTRICAL PROPERTIES	
Color: colorless to yellowish; bluish	**Dielectric constant**: 1.0-1.45
HEALTH & SAFETY	
Autoignition temperature, °C: none to 590	**Flash point**, °C: none to 192
LD50, oral, rat, mg/kg: 94-14,000	**LD50, dermal, rabbit**, mg/kg: 2,000-10,000
Threshold limiting value, NIOSH, ppm: 0.1-5	**Max. concentration**, IDHL, ppm: 2-200
ENVIRONMENTAL IMPACT	
Bluegill sunfish, 96-h LC50, mg/l: 0.000051-36.1	**Daphnia magna**, 48-h LC50, mg/l: 0.017-209
Fathead minnow, 96-h LC50, mg/l: 0.02-59.6	**Rainbow trout**, 96-h LC50, mg/l: 0.0054-41.6
log Kow: -0.64-6.60	
ANTIMICROBIAL ACTIVITY	
Product types: groups 1 and 2, # from 1 to 13	**Typical concentration**, %: 0.02-1.6
Minimum inhibition concentration, mg/l: 0.2-6,400	**pH of use**: 2-11
APPLICATIONS	
all disinfectant and preservative groups use some active chlorine compounds with DDAC and Triclosan being the most common	

Chemical formulas:

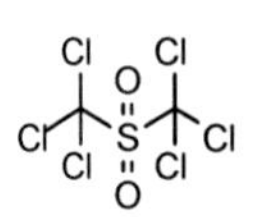

bis(trichloromethyl)sulfone

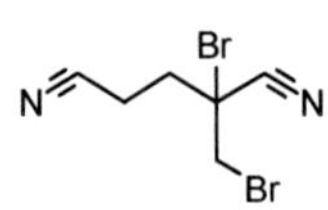

bromothalonil

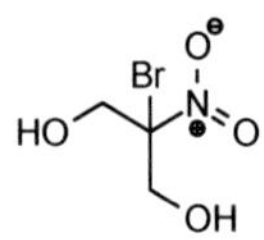

Bronopol

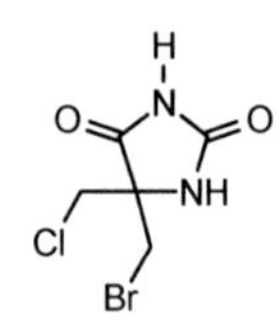

bromochlorodimethylhydantoin

$Ca(OCl)_2$

Ca hypochlorite

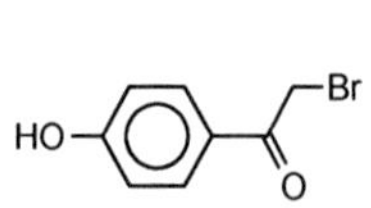

2-bromo-4'-hydroxyacetophenone

Cl-Cl

chlorine

O·Cl-O•

chlorine dioxide

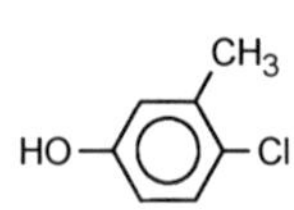

chlorocresol

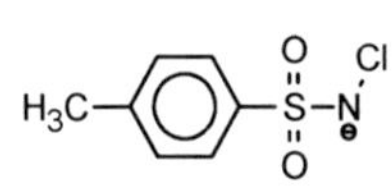

chloramin

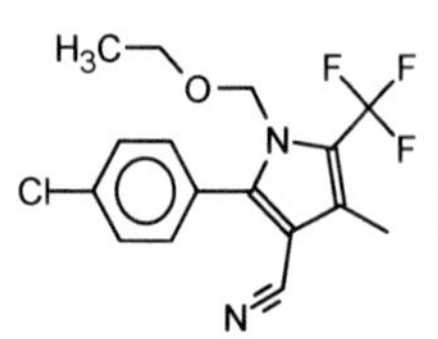

chlorfenapyr

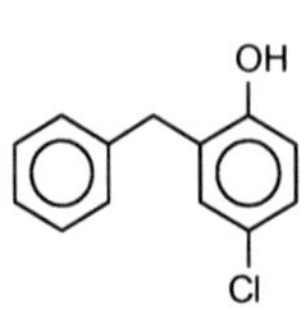

chlorophene

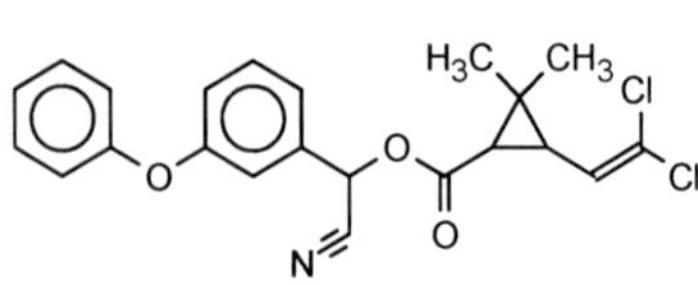

cypermethrin

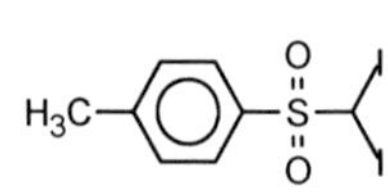

diiodomethyl p-tolyl sulfone

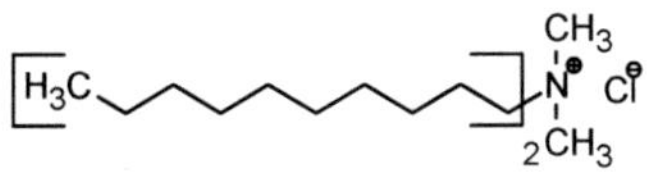

didecylammonium chloride

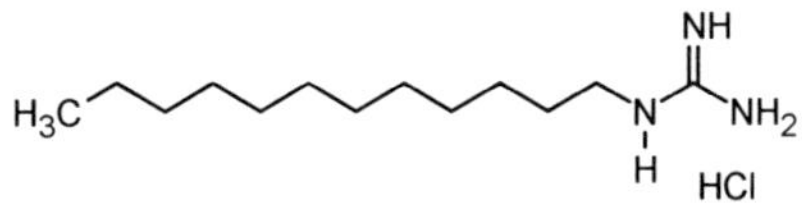

dodecylguanidine monohydrochloride

iodine

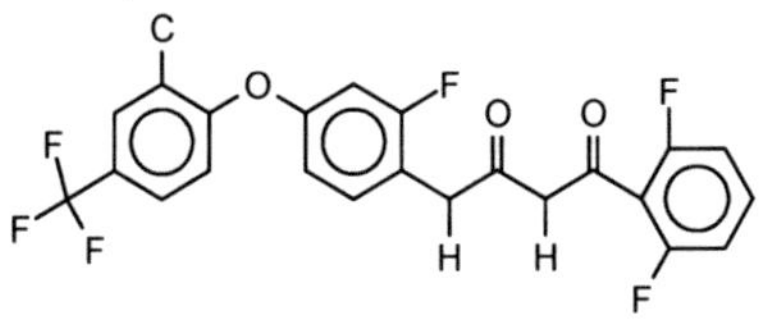

flufenoxuron

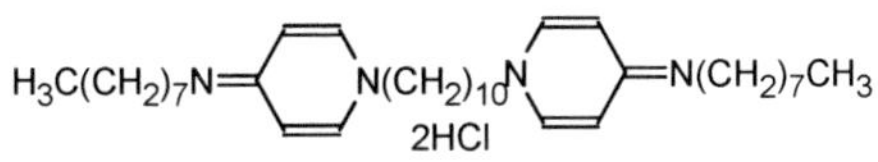

octenidine dihydrochloride

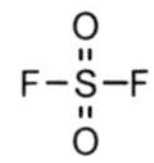

sulfuryl fluoride

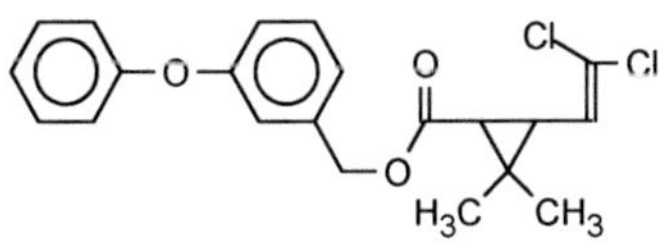

permethrin

NaOCl

sodium hypochlorite

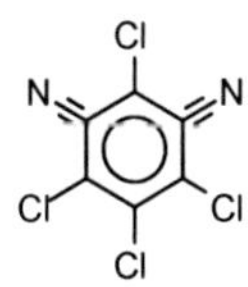

tetrachloroisophthalonitrile

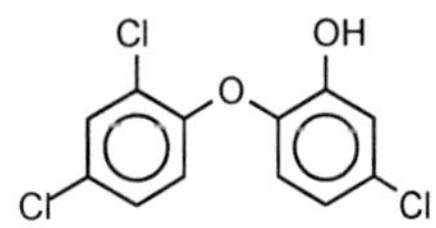

triclosan

3.2.5 ALCOHOLS

Table 3.5. Properties of alcohols.

Chemical name (CAS and EC #s): biphenyl-2-ol (90-43-7, 201-993-5), ethanol (64-17-5, 200-578-6), 2-phenoxyethanol (122-99-6, 204-589-7), propan-1-ol (71-23-8, 200-746-9), propan-2-ol (67-63-0, 200-661-7)		
Commercial name (manufacturer): general purpose chemicals use chemical name (BASF, DOW, Ecolab, and many other manufacturers)		
Main product content, %: 98-100	**Molecular weight**: 46.07-170.21	
CHEMICAL PROPERTIES		
Water solubility, %: misc-0.7	**Chemical reactivity**: esterification	
PHYSICAL PROPERTIES		
Melting point, °C: -127-59	**Boiling point**, °C: 78.5-286	**Density**, g/cm^3: 0.789-1.213
Refractive index: 1.36-1.53	**Vapor pressure**, kPa (20°C): 2.66E-4 to 7.19	
Thermal expansion coef., 10^{-7} °C^{-1}: 9.56	**Surface tension**, mN/m: 20.93-23.7	
pH: 11.2-11.6	**Viscosity**, cP: 1.17-2.25	**Odor**: mild to faint
Heat of combustion, MJ/kg: 29.7-33.6		
Henry law constant, atm m^3/mol: 4.3E-5 to 7.4E-6		
OPTICAL & ELECTRICAL PROPERTIES		
Color: colorless to white to pinkish	**Dielectric constant**: 18.3-25	
HEALTH & SAFETY		
Autoignition temperature, °C: 363-520	**Flash point**, °C: 11-138	

Table 3.5. Properties of alcohols.

LD50, oral, rat, mg/kg: 1260-7060	
Threshold limiting value, NIOSH, ppm: 200-1,000	**Max. concentration**, IDLH, ppm: 800-3,300
ENVIRONMENTAL IMPACT	
Daphnia magna, 48-h LC50, mg/l: 110-11,850	**Rainbow trout**, 96-h LC50, mg/l: 13,000
Fathead minnow, 96-h LC50, mg/l: 6.24-15300	**log Kow**: -0.31-3.09
ANTIMICROBIAL ACTIVITY	
Product types: group 1, #1-4 and group 2, #10	**Typical concentration**, %: 1-80
Minimum inhibition concentration, mg/l: 60-11,000	
APPLICATIONS	
disinfectant used in public hygiene, private and public health area disinfection, veterinary hygiene, food and feed area disinfection, and masonry (biphenyl-2-ol)	

Chemical formulas:

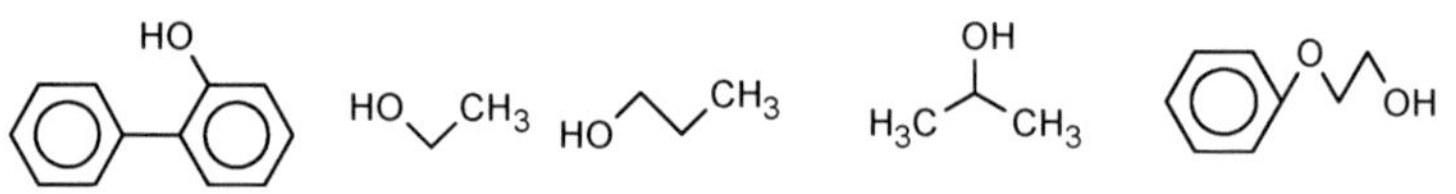

biphenyl-2-ol ethanol propan-1-ol propan-2-ol phenoxyethanol

3.2.6 ALDEHYDES

Table 3.6. Properties of aldehydes.

Chemical name (CAS and EC #s): acrylic aldehyde (107-02-8, 203-453-4), cinnamic aldehyde (104-55-2, 203-213-9), formaldehyde (50-00-0, 200-001-8), glutaraldehyde (111-30-8, 203-856-5), glyoxal (107-22-2, 203-474-9)		
Commercial name (manufacturer): general purpose chemicals used by chemical name (Baker Petrolite, BASF, Biotal Technologies, DOW)		
Main product content, %: 20-100	**Molecular weight**: 30.03-132.15	
CHEMICAL PROPERTIES		
Water solubility, %: 0.1-misc.	**Chemical reactivity**: flammable, polymerization	
PHYSICAL PROPERTIES		
Melting point, °C: -92 to 15	**Boiling point**, °C: -19.5 to 246	**Density**, g/cm^3: 0.72-1.14
Refractive index: 1.375-1.618	**Vapor pressure**, kPa (20°C): 0.225-518	
Heat of combustion, MJ/kg: 17.1	**Odor**: mild-pungent	
Henry law constant, atm m^3/mol: 3.33E-9 to 3.37E-7	**pH**: 2.1-4	
OPTICAL & ELECTRICAL PROPERTIES		
Color: colorless to yellow	**Dielectric constant**: 16.9-84	
HEALTH & SAFETY		
Autoignition temperature, °C: 225-430	**Flash point**, °C: -29 to 100	

Table 3.6. Properties of aldehydes.

LD50, oral, rat, mg/kg: 46-7,070	**LD50, dermal, rabbit**, mg/kg: 562-10,000
Threshold limiting value, NIOSH, ppm: 0.05-0.3	**Max. concentration**, IDHL, ppm: 2-20
ENVIRONMENTAL IMPACT	
Bluegill sunfish, 96-h LC50, µg/l: 79-100	**Daphnia magna**, 48-h LC50, mg/l: 0.057-11.5
Fathead minnow, 96-h LC50, mg/l: 0.0226-550	**Rainbow trout**, 96-h LC50, µg/l: 16-118
Biological oxygen demand, 5 days: 97.4	
Biodegradability: readily biodegradable	**log Pow**: -2.54 to 2.22
ANTIMICROBIAL ACTIVITY	
Product types: group 1, #1-5; group 2, #6, 11-13	**Typical concentration**, %: 0.1-4
Minimum inhibition concentration, mg/l: 15-1,400	**pH of use**: 3-10
APPLICATIONS	
disinfectant, in-can preservative, preservative for liquid cooling system, slimicide, metal-working fluids	

Chemical formulas:

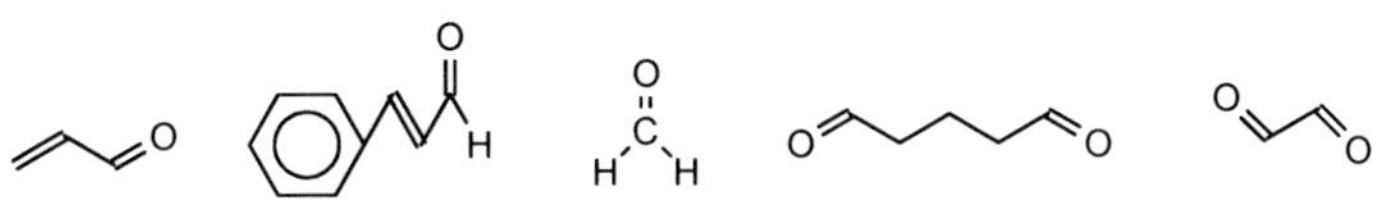

acrolein cinnamic formaldehyde glutaraldehyde glyoxal

3.2.7 AMIDES

Table 3.7. Properties of amides.

Chemical name (CAS and EC #s): benzamide, 2,2'-dithiobis(N-methyl- (2527-58-4, 219-768-5), 2-chloroacetamide (79-07-2, 201-174-2), clothianidin (210880-92-5, 433-460-1), cyanamide (420-04-2, 206-992-3), 2,2-dibromo-3-nitrilo-propanamide (10222-01-2, 233-539-7), diuron (330-54-1, 206-354-4), 3-(4-isopropylphenyl)-1,1-dimethylurea (34123-59-6, 251-835-4), tetramethylolacetylene diurea (5395-50-6, 226-408-0), thioperoxydicarbonic diamide (137-26-8, 205-286-2)		
Commercial name (manufacturer): DBNPA, Diurex, Dynex, Isopruton, Thiram, TMAD (BASF, Bayer, Chemtura, DOW, DuPont, Global Technologies, Rohm & Haas, Troy)		
Main product content, %: 20-100	**Molecular weight**: 42.04-249.7	
CHEMICAL PROPERTIES		
Water solubility, %: 0.0036-9	**Odor**: odorless to mild	
PHYSICAL PROPERTIES		
Melting point, °C: 45-176.8	**Boiling point**, °C: 225	**Density**, g/cm^3: 1.20-1.61
Refractive index: 1.442	**Vapor pressure**, hPa (20°C): 0.000000024-0.0069	
Henry law constant, atm m^3/mol: 2.6E-10 to 1.9E-8		**pH**: 6.2-6.6

Table 3.7. Properties of amides.

OPTICAL PROPERTIES	
Color: colorless to pale yellow	
HEALTH & SAFETY	
Flash point, °C: 89 to noncombustible solid	
LD50, oral, rat, mg/kg: 70 to >5,000	**LD50, dermal, rabbit**, mg/kg: 590-2,000
Threshold limiting value, NIOSH, mg/m^3: 1-10	**Max. concentration**, IDLH, mg/m^3: 50-100
ENVIRONMENTAL IMPACT	
Bluegill sunfish, 96-h LC50, mg/l: 8.2-117	**Daphnia magna**, 48-h LC50, mg/l: 0.21-5.3
Rainbow trout, 96-h LC50, mg/l: 16-105	**log Kow**: -0.53 to 2.68
Biodegradability: biodegradable	
ANTIMICROBIAL ACTIVITY	
Product types: group 1, #2, 3, 4; group 2, #6, 7, 9, 12, 13	**Typical concentration**, %: 0.001-10
Minimum inhibition concentration, mg/l: 15-500	
APPLICATIONS	
disinfectants in private and public health areas, veterinary disinfectants, and food and feed disinfectants; in-can preservatives, film preservatives, fiber, leather, rubber, and polymer biocides, liquid-cooling system preservatives, slimicides, and metal-working fluid preservatives	

Chemical formulas:

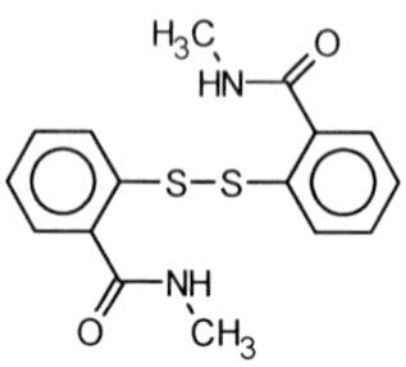

benzamide, 2,2'-dithiobis(N-methyl-

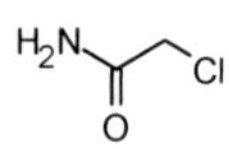

2-chloroacetamide

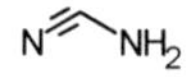

cyanamide

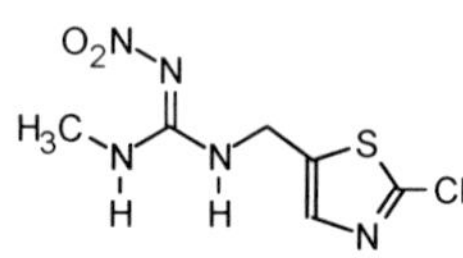

clothianidin

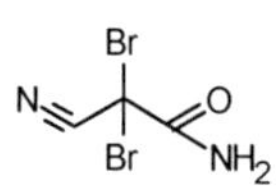

2,2-dibromo-3-nitrilo-propanamide

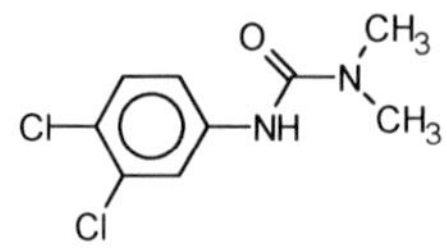

diuron

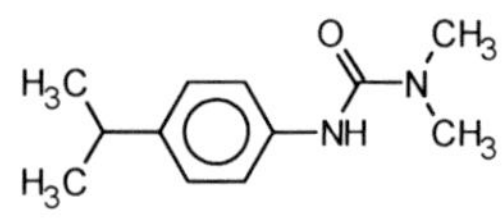

3-(4-isopropylphenyl)-1,1-dimethylurea

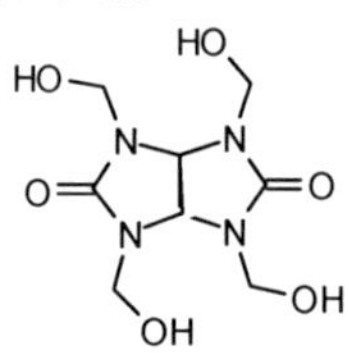

tetramethylolacetylene diurea

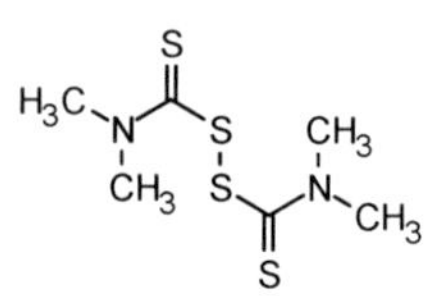

thioperoxydicarbonic diamide

3.2.8 AZOLES

Table 3.8. Properties of azoles.

Chemical name (CAS and EC #s): cyproconazole (94361-06-5, –), imazalil (35554-44-0, 252-615-0), propiconazole (60207-90-1, 262-104-4), tebuconazole (107534-96-3, 403-640-2)		
Commercial name (manufacturer): Cyproconazole, Enilconazole, Imazalil, Propiconazole, Tebuconazole (Janssen, Lanxess, Makhteshim, Syngenta)		
Main product content, %: 94-98	**Molecular weight**: 291.78-342.224	
CHEMICAL PROPERTIES		
Water solubility, %: 0.00014-0.14		
PHYSICAL PROPERTIES		
Melting point, °C: 53-106	**Boiling point**, °C: 347-450	**Density**, g/cm^3: 1.25-1.348
Refractive index: 1.5643	**Vapor pressure**, hPa (20°C): 0.0000000017-0.03	
Surface tension, mN/m: 65.2	**Viscosity**, mPas: 0.000093	**Odor**: odorless
Henry law constant, atm m^3/mol: 1.45E-10-4.1E-9		**pH**: 6-7
OPTICAL & ELECTRICAL PROPERTIES		
Color: white, yellow, brown		
HEALTH & SAFETY		
Autoignition temperature, °C: 468	**Flash point**, °C: 192	
LD50, oral, rat, mg/kg: 227-5,000	**LD50, dermal, rabbit**, mg/kg: 2,000-4,200	
ENVIRONMENTAL IMPACT		
Bluegill sunfish, 96-h LC50, mg/l: 1.3-23	**Daphnia magna**, 48-h EC50, ppm: 4-86	
Rainbow trout, 96-h LC50, mg/l: 0.83-19	**log Kow**: 3.70-3.82	
ANTIMICROBIAL ACTIVITY		
Product types: group 1, #3; group 2, #7 & 8	**Typical concentration**, %: 0.1-1.5	
Minimum inhibition concentration, mg/l: 5-1,000		
APPLICATIONS		
disinfectant in veterinary hygiene and film and wood preservative. In a new inventions as fungicide in pharmaceutical products[20] and antifreeze concentrate[21]		

Chemical formulas:

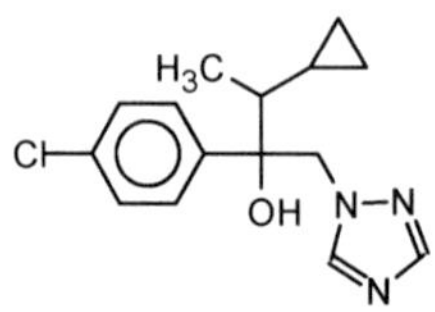

cyproconazole

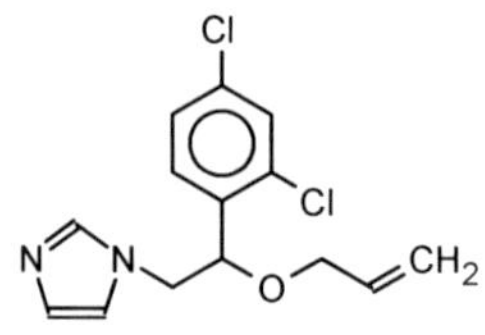

imazalil

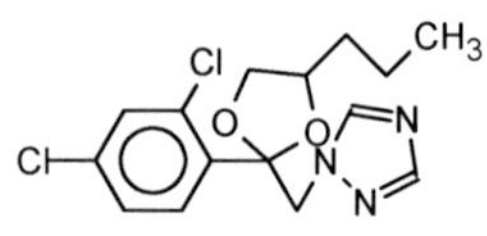

propiconazole

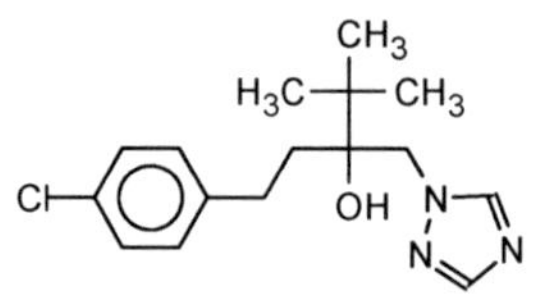

tebuconazole

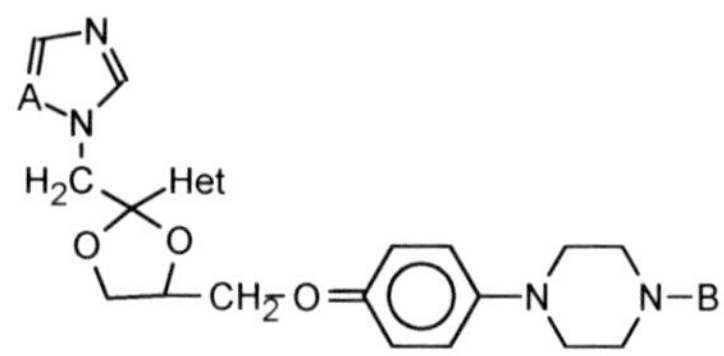

N-oxide of azole[20]

3.2.9 CARBAMATES

Table 3.9. Properties of carbamates.

Chemical name (CAS and EC #s): disodium ethylenebis(dithiocarbamate) (142-59-6, 205-547-0), fenoxycarb (72490-01-8, 276-696-7), iodopropynyl butylcarbamate (55406-53-6, 259-627-5), Metam-sodium (137-42-8, 205-293-0), methyl 2-benzimidazolecarbamate (10605-21-7, 234-232-0), potassium dimethyldithiocarbamate (128-03-0, 204-875-1), sodium dimethyldithiocarbamate (128-04-1, 204-876-7), Thiram (137-26-8, 205-286-2), zinc dimethyldithiocarbamate (137-30-4, 205-288-3), zinc ethylenebis(dithiocarbamate) (12122-67-7, 235-180-1)		
Commercial name (manufacturer): Bioban, BMC, Carbendazim, Dithane, MBC, Metam, Nabam, Nalco, Sistan, Vancide, Vapam, Zineb, Ziram (AKZO, BASF, Buckman Laboratories, DOW, Flexsys, Intace, Lanxess, Rohm & Haas, Taminco, Troy, Vanderbilt)		
Main product content, %: 40-100	**Molecular weight**: 129.18-305.83	
CHEMICAL PROPERTIES		
Water solubility, %: 0.000168 to misc.		
Chemical reactivity: with acids		
PHYSICAL PROPERTIES		
Melting point, °C: -15-300	**Boiling point**, °C: 108	**Density**, g/cm^3: 1.02-1.77
Viscosity, mPas: 4.8-1,500	**Vapor pressure**, hPa (20°C): 0.0000017-1.17	
Odor: odorless, sulfurous, pungent	**Henry law constant**, atm m^3/mol: 1.2E-7	
pH: 6.5-12		
OPTICAL & ELECTRICAL PROPERTIES		
Color: white to yellow		
HEALTH & SAFETY		
Autoignition temperature, °C: 385-500	**Flash point**, °C: 89-100	

Table 3.9. Properties of carbamates.

LD50, oral, rat, mg/kg: 267-6,400	**LD50, dermal, rabbit**, mg/kg: 2,000-3,074
Threshold limiting value, NIOSH, mg/m^3: 5	**Max. concentration**, IDLH, mg/m^3: 100
ENVIRONMENTAL IMPACT	
Bluegill sunfish, 96-h LC50, mg/l: 0.042-38.5	**Daphnia magna**, 48-h LC50, mg/l: 0.04-1.5
Fathead minnow, 96-h LC50, mg/l: 0.06-0.2	**Rainbow trout**, 96-h LC50, mg/l: 0.067-6.69
log Kow: 1-1.23	
ANTIMICROBIAL ACTIVITY	
Product types: group 2, #7, 9-13; group 4, # 21	**Typical concentration**, %: 1-50
Minimum inhibition concentration, mg/l: 0.5-1000	
APPLICATIONS	
film, fiber, leather, rubber and polymer preservatives; masonry preservatives; preservatives for liquid-cooling systems; slimicides; preservatives for metal-working fluids; antifouling products	

Chemical formulas:

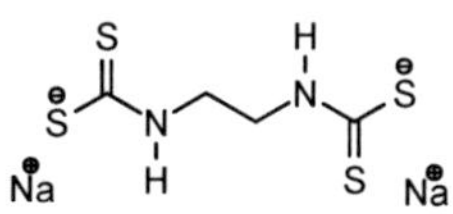
disodium ethylenebis(dithiocarbamate)

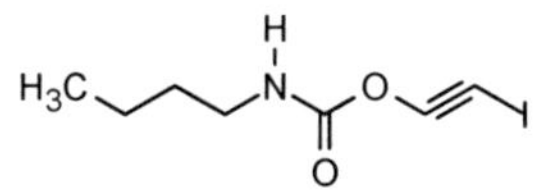
iodopropynyl butylcarbamate

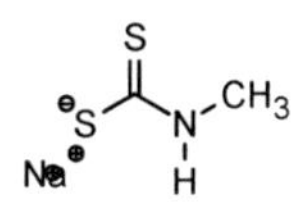
Metam-sodium

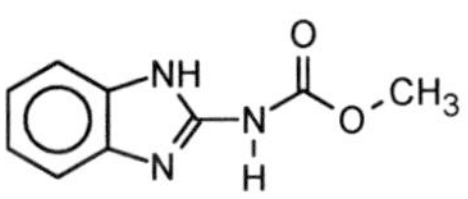
methyl 2-benzimidazolecarbamate

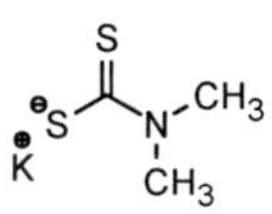
potassium dimethyldithiocarbamate

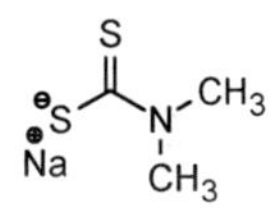
sodium dimethyldithiocarbamate

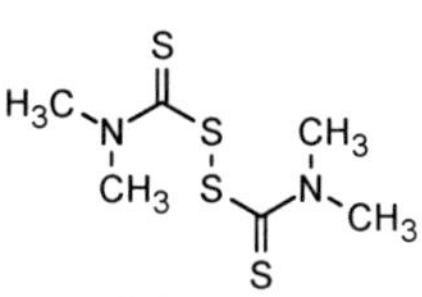
Thiram

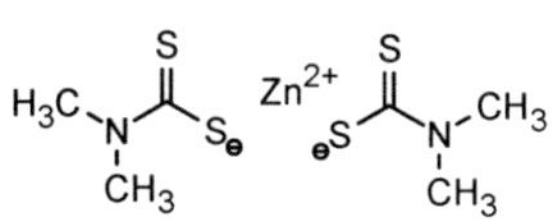
zinc dimethyldithiocarbamate

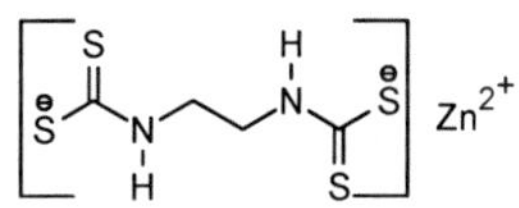
zinc ethylenebis(dithiocarbamate)

3.2.10 FORMALDEHYDE-RELEASING COMPOUNDS

Table 3.10. Properties of formaldehyde-releasing compounds.

Chemical name (CAS and EC #s): benzylhemiformal (14548-60-8, 238-588-8), 1,3-dimethylol-5,5-dimethylhydantoin (6440-58-0, 229-222-8), 3,5-dimethylpyrazole-1-methanol (85264-33-1, 286-553-0), 4,4-dimethyloxazolidine (51200-87-4, 257-048-2), dimorpholinomethane (5625-90-1, 227-062-3), dioxabicyclooctane (59720-42-2, –), 4,4'-(2-ethyl-2-nitrotrimethylene)dimorpholine (1854-23-5, 217-450-0), hexahydroxytriazine (4719-04-4, 225-208-0), hexamethylenetetramine cis-chloroallyl chloride (51229-78-8, 223-805-0), 2-((hydroxymethyl)amino)ethanol (34375-28-5, 251-974-0), 4-(2-nitrobutyl)morpholine (2224-44-4, 218-748-3), oxazolidine (66204-44-2, 266-235-8), Oxazolidine-E (7747-35-5, 231-810-4), sodium N-(hydroxymethyl)glycinate (70161-44-3, 274-357-8), 1,3,5-tetrahydro-3,5-dimethyl-2H-1,3,5-thiadiazine-2-thione (TDTT) (533-74-4, 208-576-7), tetrakis(hydroxymethyl)phosphonium sulphate (55566-30-8, 259-709-0), triethylhexahydro-s-triazine (7779-27-3, -), tris(hydroxymethyl)nitromethane (126-11-4, 204-769-5)			
Commercial name (manufacturer): Bioban, Dazomet, DMDMH, DMHMP, Dowicil, EDHO, MBM, Nuosept, Tris nitro, Troysan, Vancide (BASF, Buckman Laboratories, Clariant, Cytek, DOW, Hess & Clark, International Specialty Products, Kemira, Lanxess, Lonza, Lubrizol, Nalco, Rhodia, Stepan, Troy, Vanderbilt)			
Main product content, %: 50-100		**Molecular weight**: 75.11-406.28	
CHEMICAL PROPERTIES			
Water solubility, %: 0.22-127	**Chemical reactivity**: forms formaldehyde		**pH**: 4.5-11.5
PHYSICAL PROPERTIES			
Melting point, °C: -60-175	**Density**, g/cm^3: 0.985-1.30	**Refractive index**: 1.404-1.483	
Viscosity, mPas: 7.5-40	**Vapor pressure**, hPa (20°C): 0.0034-133		
Odor: formaldehyde	**Henry law constant**, atm m^3/mol: 7.1E-12 to 2.66E-10		
OPTICAL & ELECTRICAL PROPERTIES			
Color: colorless to white to yellow to brown			
HEALTH & SAFETY			
Autoignition temperature, °C: 250-400		**Flash point**, °C: 49-87	
LD50, oral, rat, mg/kg: 333-3,720		**LD50, dermal, rabbit**, mg/kg: 420-2,877	
ENVIRONMENTAL IMPACT			
Bluegill sunfish, 96-h LC50, mg/l: 1.3-320		**Daphnia magna**, 48-h LC50, mg/l: 1.9-45	
Rainbow trout, 96-h LC50, mg/l: 0.16-514			
log Kow: -2.37-0.63			
ANTIMICROBIAL ACTIVITY			
Product types: group 2, # 6, 7, 9, and 13		**Typical concentration**, %: 0.05-3.5	
Minimum inhibition concentration, mg/l: 0.01-2,500		**pH of use**: 4-12	
APPLICATIONS			
in-can preservatives; film preservatives; fiber, leather, rubber, and polymer biocides, metal-working fluids			

Chemical formulas:

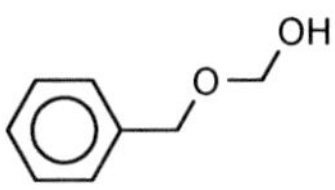

benzylhemiformal

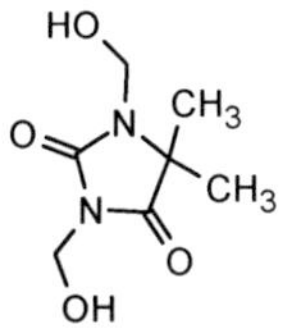

1,3-dimethylol-5,5-dimethylhydantoin

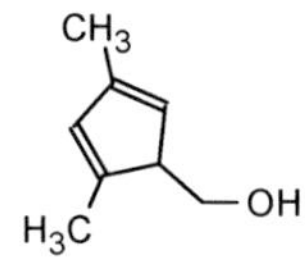

3,5-dimethylpyrazole-1-methanol

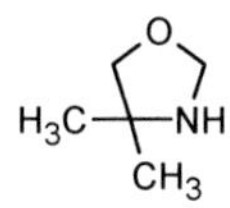

dimethyloxazolidine

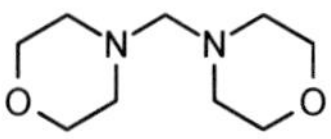

dimorpholinomethane

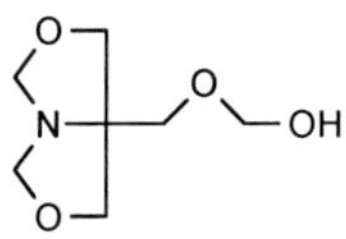

dioxabicyclooctane

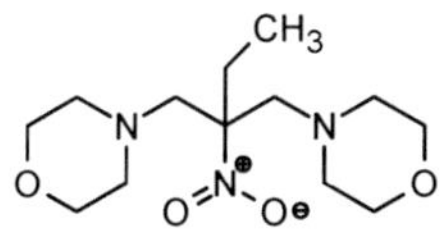

4,4'-(2-ethyl-2-nitrotrimethylene)dimorpholine

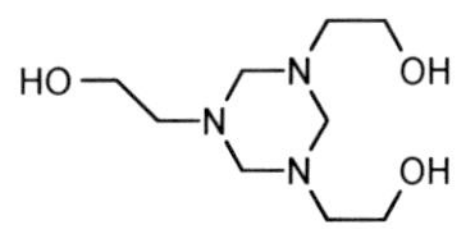

hexahydroxytriazine

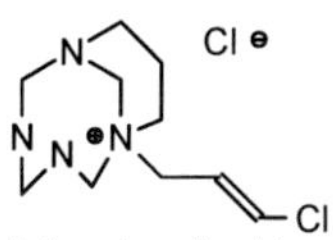

hexamethylenetetramine cis-chloroallyl chloride

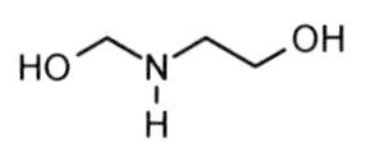

2-((hydroxymethyl)amino)ethanol

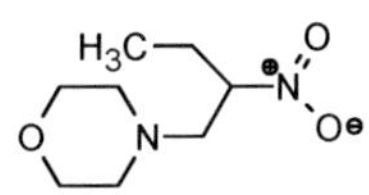

4-(2-nitrobutyl)morpholine

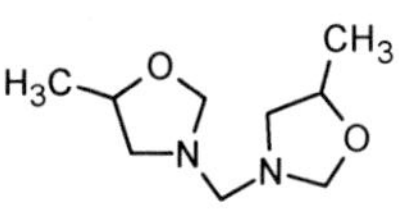

oxazolidine

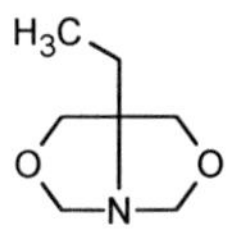

Oxazolidine-E

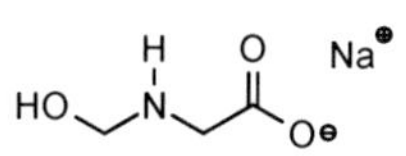

sodium N-(hydroxymethyl)glycinate

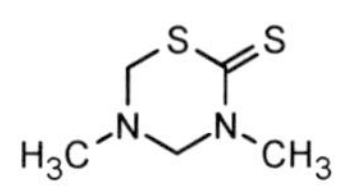

TDTT

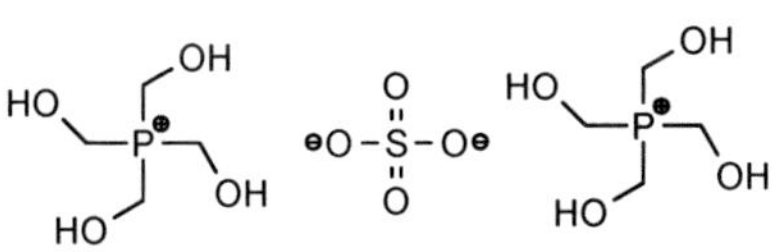

tetrakis(hydroxymethyl)phosphonium sulphate

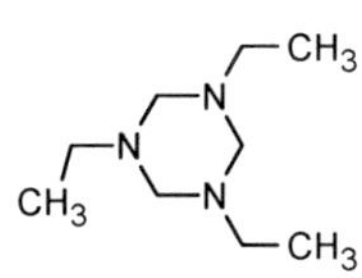

triethylhexahydro-s-triazine

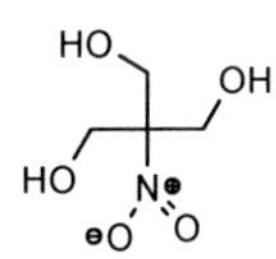

tris(hydroxymethyl)nitromethane

3.2.11 HALOALKYLTHIO COMPOUNDS

Table 3.11. Properties of haloalkylthio compounds.

Chemical name (CAS and EC #s): dichlofluanid (1085-98-9, 214-118-7), N-((trichloromethyl)thio)phthalimide (133-07-3, 205-088-6), N-trichloromethylthio tetrahydrophthalimide (133-06-2, 205-087-0), tolylfluanid (731-27-1, 211-986-9)	
Commercial name (manufacturer): Americide, Captan, Captex, Folpet, Folpan, Kaptan, Orthocide, Phthalan Vangard (Bayer Cropscience, Biesterfeld, Bonide Products, Chemtura, Drexel, International Specialty Products, Lanxess, Loveland Products, Makhteshim Chemical Works, PPG, Sherwin-Williams, TSGE, Wilbur Ellis)	
Main product content, %: 88-97	**Molecular weight**: 296.56-472.6
CHEMICAL PROPERTIES	
Water solubility, %: traces to 0.1	

Table 3.11. Properties of haloalkylthio compounds.

PHYSICAL PROPERTIES		
Melting point, °C: 93-178	**Density**, g/cm^3: 1.52-1.74	**Odor**: odorless to faint
pH: 5-6	**Vapor pressure**, hPa (20°C): 0.000002-.0000173	
Henry law constant, atm m^3/mol: 7.6E-7		
OPTICAL & ELECTRICAL PROPERTIES		
Color: off-white to buff		
HEALTH & SAFETY		
Autoignition temperature, °C: 290 to >400	**Flash point**, °C: 192	
LD50, oral, rat, mg/kg: 2636-9,000	**LD50, dermal, rabbit**, mg/kg: >5,000-22,600	
Threshold limiting value, NIOSH, mg/m^3: 5		
ENVIRONMENTAL IMPACT		
Bluegill sunfish, 96-h LC50, mg/l: 0.072-0.141	**Daphnia magna**, 48-h LC50, mg/l: 0.42-9.96	
Fathead minnow, 96-h LC50, mg/l: 0.2-0.298	**Rainbow trout**, 96-h LC50, mg/l: 0.036-0.073	
log Kow: 2.35-3.9		
ANTIMICROBIAL ACTIVITY		
Product types: group 2, #7; group 4, #21	**Typical concentration**, %: 0.25-10	
Minimum inhibition concentration, mg/l: 10-1000		
APPLICATIONS		
film preservatives and antifouling products		

Chemical formulas:

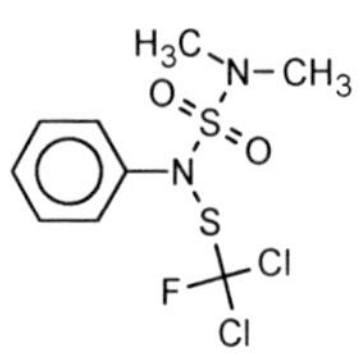

dichlofluanid

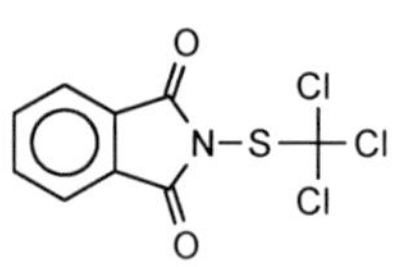

N-((trichloromethyl)thio)phthalimide

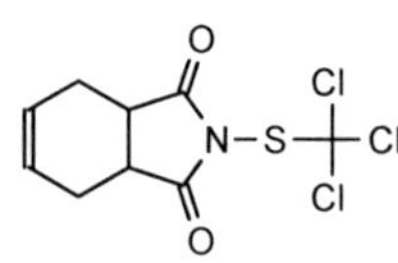

N-trichloromethylthio tetrahydrophthalimide

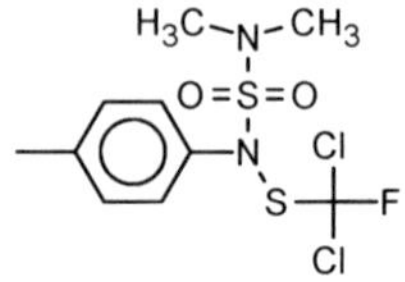

tolylfluanid

3.2.12 HETEROCYCLIC N,S COMPOUNDS

Table 3.12. Properties of heterocyclic compounds.

Chemical name (CAS and EC #s): 1,2-benzisothiazoline-3-one (2634-33-5, 220-120-9), 1,2-benzisothiazol-3(2H)-one, 2-butyl- (4299-07-4, 420-590-7), 5-chloro-2-methylisothiazol-3-one (26172-55-4, 247-500-7), dichlorooctylisothiazoline (64359-81-5, 264-843-8), methylisothiazolinone (2682-20-4, 220-239-6), octhilinone (26530-20-1, 247-761-7), terbutryn (886-50-0, 212-950-5), thiabendazole (148-79-8, 205-725-8), thiamethoxam (153719-23-4, 428-650-4), 2-(thiocyanmethylthio)benzothiazole (21564-17-0, 244-445-0)		
Commercial name (manufacturer): BIT, Kathon, MIT, OIT, Proxan, TCMBT (Abbott Laboratories, Arch Chemicals, Buckman Laboratories, Clariant, DOW, Hercules, International Specialty Products, Kemira, Lanxess, Lonza, Nalco, Rochester Midland, Rohm & Haas, Thor, Troy)		
Main product content, %: 20-100	**Molecular weight**: 115.16-282.24	
CHEMICAL PROPERTIES		
Water solubility, %: 0.0014-3.84		
PHYSICAL PROPERTIES		
Melting point, °C: -40-304	**Density**, g/cm^3: 1.038-1.44	**Odor**: odorless to pungent
Viscosity, mPas: 1.3-40	**Vapor pressure**, hPa (20°C): 0.000049-0.083	
pH: 2.4-13.5	**Henry's law constant**, atm m^3/mol: 1.15E-8	
OPTICAL & ELECTRICAL PROPERTIES		
Color: colorless to yellow to brown		
HEALTH & SAFETY		
Autoignition temperature, °C: 371-530	**Flash point**, °C: 28-148	
LD50, oral, rat, mg/kg: 285-4,400	**LD50, dermal, rabbit**, mg/kg: 690-10,000	
ENVIRONMENTAL IMPACT		
Bluegill sunfish, 96-h LC50, mg/l: 0.014-5.9	**Daphnia magna**, 48-h LC50, mg/l: 0.0052-4.71	
Fathead minnow, 96-h LC50, mg/l: 0.140	**Rainbow trout**, 96-h LC50, mg/l: 0.0027-3	
log Kow: 2.47-3.74		
ANTIMICROBIAL ACTIVITY		
Product types: group 1, #2; group 2, #6-13	**Typical concentration**, %: 0.005-0.25	
Minimum inhibition concentration, mg/l: 0.07-5,000	**pH in use**: 2-12	
APPLICATIONS		
private and public health area disinfectant, in-can preservatives, biocides in film, wood, fiber, leather, rubber, polymers, masonry, liquid-cooling systems, and metal-working fluids, and slimicides		

Chemical formulas:

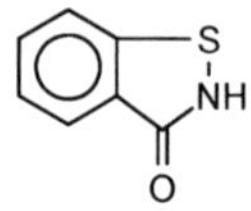

1,2-benzisothiazoline-3-one

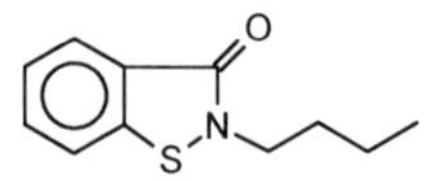

1,2-benzisothiazol-3(2H)-one, 2-butyl-

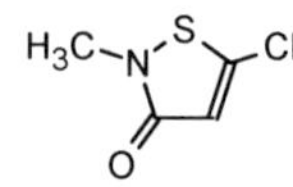

5-chloro-2-methylisothiazol-3-one

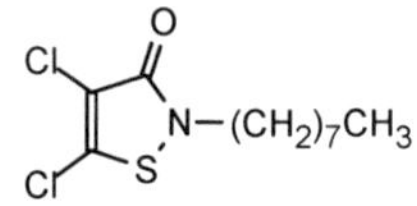

dichlorooctylisothiazolone

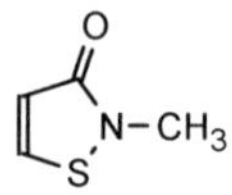

methylisothiazolinone

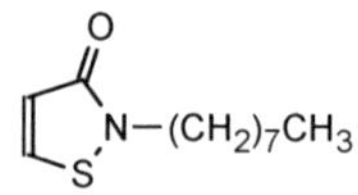

octhilinone

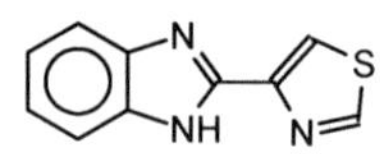

thiabendazole

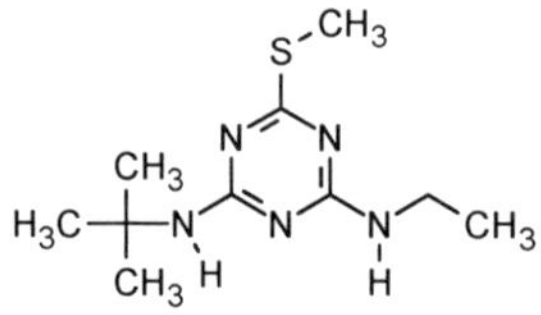

terbutryn

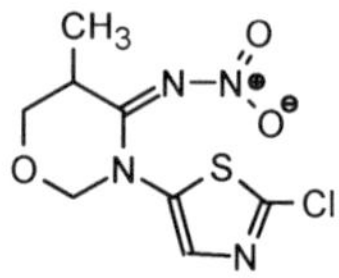

thiamethoxam

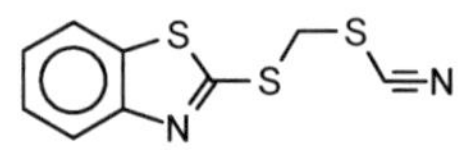

2-(thiocyanmethylthio)benzothiazole (TCMTB)

3.2.13 METAL-CONTAINING PRODUCTS

Table 3.13. Properties of metal containing products.

Chemical name (CAS and EC #s): bis-(N-cyclohexyldiazeniumdioxy)-copper (312600-89-8, –), copper (7440-50-8, 231-159-6), copper dihydroxide (20427-59-2, 243-815-9), copper oxide (1317-38-0, 215-269-1), copper pyrithione (14915-37-8, 238-984-0), copper sulfate (7758-98-7, 231-847-6), copper thiocyanate (1111-67-7, 214-183-1), dicopper oxide (1317-39-1, 215-270-7), disilver oxide (20667-12-3, 243-957-1), oxybisphenoxyarsine (58-36-6, 200-377-3), potassium 2-biphenylate (13707-65-8, 237-243-9), potassium sorbate (24634-61-5, 246-376-1), silver (-, 231-131-3), silver chloride (7783-90-6, 232-033-3), silver nitrate (7761-88-8, 231-853-9), silver zeolite A, (130328-18-6, –), sodium biphenylate (132-27-4, 205-055-6), sodium p-chloro-m-cresolate (15733-22-9, 239-825-8), sodium pyrithione (3811-73-2, 223-296-5), troclosene sodium (2893-78-9, 220-767-7), zinc pyrithione (13463-41-7, 236-671-3)		
Commercial name (manufacturer): Bactekiller, Cu-HDO, Dowcide, Omadine, Vancide (Arch Chemicals, Bayer, Bio-Lab, Ciba, Clariant, DOW, DuPont, Fuji Chemical, Harlan Laboratories, Ionx Technologies, King Technology, Lanxess, PPG, Rohm & Haas, Sherwin-Williams, Verichem)		
Main product content, %: 97-100	**Molecular weight**: 79.55-502.24	
CHEMICAL PROPERTIES		
Water solubility, %: traces to 216		
PHYSICAL PROPERTIES		
Melting point, °C: 184-1,326	**Boiling point**, °C: 2,212	**Density**, g/cm^3: 1.36-10.49
Refractive index: 1.729-2.705	**Vapor pressure**, hPa (20°C): 0.000001-0.001	
Odor: odorless to pleasant		**pH**: 6-8.3

Table 3.13. Properties of metal containing products.

OPTICAL & ELECTRICAL PROPERTIES	
Color: white to gray to black, also blue & green	**Dielectric constant**: 10.3-18.1
HEALTH & SAFETY	
Autoignition temperature, °C: 180-225	**Flash point**, °C: 150-225
LD50, oral, rat, mg/kg: 30-5,000	**LD50, dermal, rabbit**, mg/kg: >1,000 to >2,500
Threshold limiting value, NIOSH, mg/m^3: 0.01-1	**Max. concentration**, IDLH, mg/m^3: 10-100
ENVIRONMENTAL IMPACT	
Daphnia magna, 48-h LC50, mg/l: 0.0095-750	
Fathead minnow, 96-h LC50, mg/l: 0.014	**Rainbow trout**, 96-h LC50, mg/l: 0.001-0.29
ANTIMICROBIAL ACTIVITY	
Product types: group 1, #1-5, group 2, #6-13, group 4, #21	**Typical concentration**, %: 0.03-2
Minimum inhibition concentration, mg/l: 10-1,000	
APPLICATIONS	
full range of applications as disinfectants, preservatives, and acting components of antifouling products	

Chemical formulas:

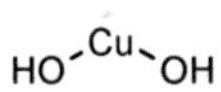

copper dihydroxide

Cu-O·Cu

dicopper oxide

NCSCu

copper thiocyanate

Cu=O

copper oxide

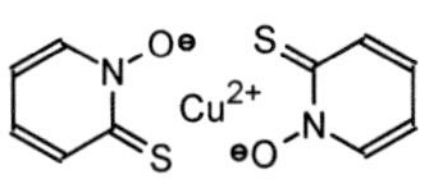

copper pyrithione

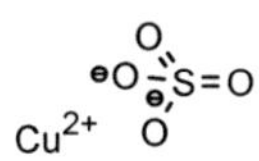

copper sulfate

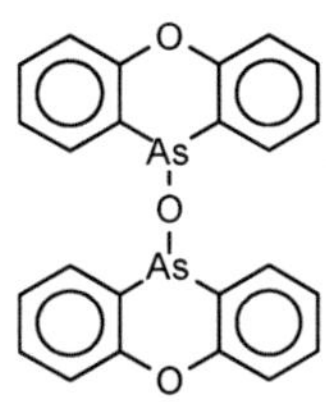

oxybisphenoxyarsine

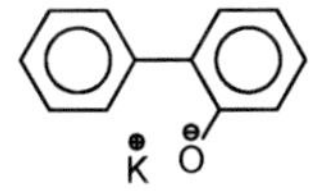

potassium 2-biphenylate

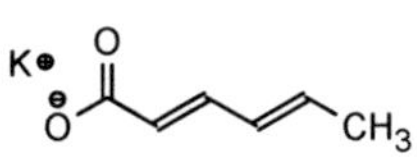

potassium sorbate

Ag-Cl

silver chloride

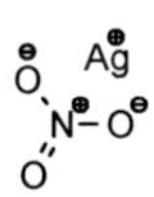

silver nitrate

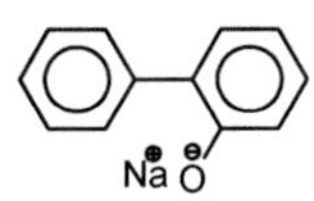

sodium biphenylate

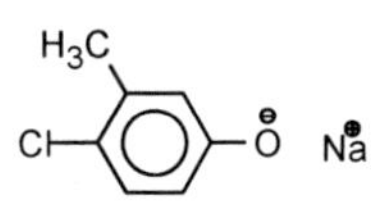

sodium p-chloro-m-cresolate

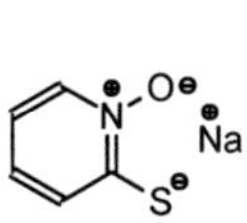

sodium pyrithione

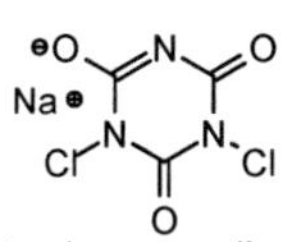

troclosene sodium

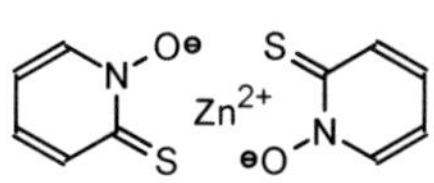

zinc pyrithione

Ag-O-Ag

disilver oxide

3.2.14 OXIDIZING AGENTS

Table 3.14. Properties of oxidizing agents.

Chemical name (CAS and EC #s): chlorine (7782-50-5, 231-959-5), 1,3-dichloro-5,5-dimethylhydantoin (118-52-5, 204-258-7), hydrogen peroxide (7722-84-1, 231-765-0), magnesium bis(2-carboxylate-monoperoxybenzoic acid) (84665-66-7, 279-013-0), phthaloyaminoperoxycaproic acid (128275-31-0, 410-850-8), trichloroisocyanuric acid (87-90-1, 201-782-8)		
Commercial name (manufacturer): DCDMH, MMPP, PAP, Symclosene (Bode, DOW, DuPont, Euro Chlor, Harcros Chemicals, Harlan Laboratories, Lonza, Occidental, PPG, Solvay, Wacker)		
Main product content, %: 30-98	**Molecular weight**: 34.01-494.65	
CHEMICAL PROPERTIES		
Water solubility, %: 0.5 to misc		
Chemical reactivity: oxidizers		
PHYSICAL PROPERTIES		
Melting point, °C: -101-249	**Boiling point**, °C: -34.04-152	**Density**, g/cm^3: 1.1-1.5
Refractive index: 1.383-1.406	**Vapor pressure**, hPa (20°C): 0.000024-6399	
Surface tension, mN/m: 18.4-80.4		**Odor**: odorless to pungent
Heat of combustion, MJ/kg: -2.83	**Viscosity**, mPas: 14	
pH: 2.7-5.5	**Henry law constant**, atm m^3/mol: 7.04E-9 to 0.478	
OPTICAL & ELECTRICAL PROPERTIES		
Color: colorless to white to green	**Dielectric constant**: 1.45	
HEALTH & SAFETY		
LD50, oral, rat, mg/kg: 542 to 15,802	**LD50, dermal, rabbit**, mg/kg: >2,000 to >20,000	
Threshold limiting value, NIOSH, mg/m^3: 0.2-1.45	**Max. concentration**, IDLH, mg/m^3: 5-105	
ENVIRONMENTAL IMPACT		
Bluegill sunfish, 96-h LC50, mg/l: 0.75-2.35	**Daphnia magna**, 48-h LC50, mg/l: 0.16-37	
Rainbow trout, 96-h LC50, mg/l: 0.91-56	**log Kow**: -0.94	
ANTIMICROBIAL ACTIVITY		
Product types: group 1, #1-3,5, group 2, #11,12	**Typical concentration**, %: 1-6	
APPLICATIONS		
disinfectants in human hygiene, private and public health areas, veterinary, and drinking water; preservatives for liquid-cooling systems and slimicides		

Chemical formulas:

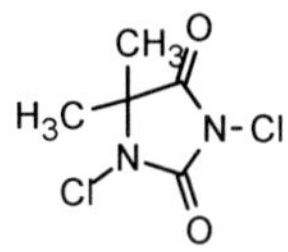

1,3-dichloro-5,5-dimethylhydantoin

HO - OH

hydrogen peroxide

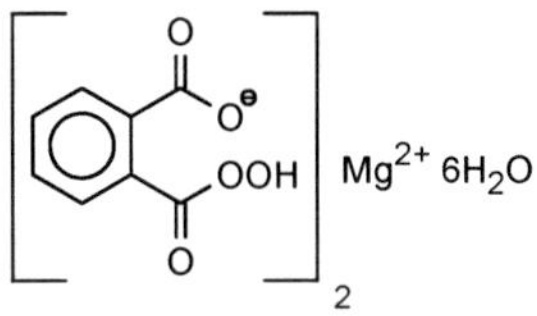

magnesium bis(2-carboxylate-monoperoxybenzoic acid)

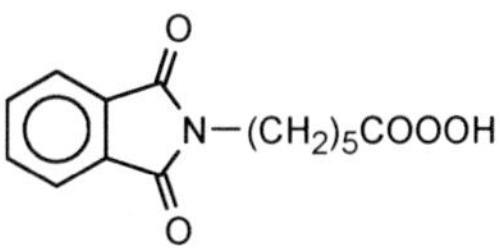

phthaloyaminoperoxycaproic acid

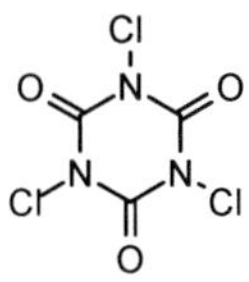

trichloroisocyanuric acid

Other products which may also be included in this group can be found in groups of acids, carbamates, and metal-containing materials.

3.2.15 PHENOLICS

Table 3.15. Properties of phenolics.

Chemical name (CAS and EC #s): benzylchlorophenol (120-32-1, 204-385-8), 5-chloro-2-(p-chlorophenoxy)phenol (3380-30-1, 429-290-0), 4-chloro-3-cresol (59-50-7, 200-431-6), parachlorometaxylenol (88-04-0, 201-793-8), phenol (108-95-2, 203-632-7), 4-tert-amylphenol (80-46-6, 201-280-9), thymol (89-83-8, 201-944-8)		
Commercial name (manufacturer): Chlorophene, DCPP, Dettol, Husept, Ketolin, Santophen, Snoclosan (Ciba, Clariant, Contec, Envirosystems, Lanxess, Lonza, OHSO, Swan Chemicals, World Pharmaceuticals)		
Main product content, %: 5-100	**Molecular weight**: 94.11-218.68	
CHEMICAL PROPERTIES		
Water solubility, %: 0.0149-6.7		
PHYSICAL PROPERTIES		
Melting point, °C: 41-246	**Boiling point**, °C: 181.8-262.5	**Density**, g/cm^3: 0.962-1.37
Refractive index: 1.523-1.541	**Vapor pressure**, hPa (20°C): 0.000000187-6.19	
Thermal expansion coef., 10^{-4} °C^{-1}: 10.9	**Heat of combustion**, MJ/kg: 33.1	
Surface tension, mN/m: 64.5	**Viscosity**, mPas: 3.44-250	**Odor**: phenolic
Henry law constant, atm m^3/mol: 2.7E-9-3.33E-7		**pH**: 4.4-9.7
OPTICAL & ELECTRICAL PROPERTIES		
Color: colorless to white to yellowish to brown		
HEALTH & SAFETY		
Autoignition temperature, °C: 490-595	**Flash point**, °C: 3-188	
LD50, oral, rat, mg/kg: 317-5,129	**LD50, dermal, rabbit**, mg/kg: 850 to >5,000	
Threshold limiting value, NIOSH, ppm: 5	**Max. concentration**, IDLH, ppm: 250	

Table 3.15. Properties of phenolics.

ENVIRONMENTAL IMPACT	
Bluegill sunfish, 96-h LC50, mg/l: 0.33	**Daphnia magna**, 48-h LC50, mg/l: 0.59-61
Fathead minnow, 96-h LC50, mg/l: 2.5-41	**Rainbow trout**, 96-h LC50, mg/l: 0.72-11.9
log Kow: 3.1-4.03	
ANTIMICROBIAL ACTIVITY	
Product types: group 1, #1-3, group 2, #6,7,10,13	**Typical concentration**, %: 0.2-5
Minimum inhibition concentration, mg/l: 10-5,000	**pH in use**: 1-14
APPLICATIONS	
disinfectants in human hygiene, private and public health areas, and veterinary products; in-can preservatives, film biocides, masonry preservatives, and metalworking fluid preservatives	

Chemical formulas:

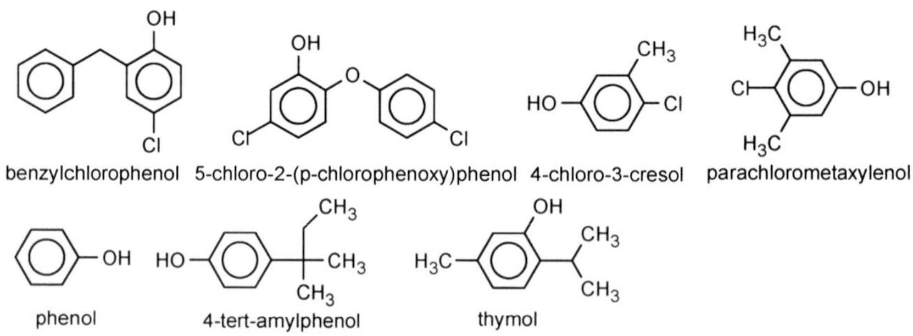

Phenolics reported in this section are more frequently used in North America. But some are also included on the European Commission list: Chlorophene, DCPP, sodium 2-biphenylate (see in Section 3.2.13), and Triclosan (see Section 3.2.7).

3.2.16 POLYMERIC MATERIALS

Table 3.16. Properties of polymeric materials.

Chemical name (CAS and EC #s): poly(hexamethylene-bicyanoguanide-hexamethylenediamine) hydrochloride, PHMB (27083-27-8 also 91403-50-8, –) (also known as Akacid forte CAS# 57028-96-3 and Akacid CAS# 374572-91-5), poly(2-hydroxypropyldimethylammonium chloride) (25988-97-0, –), 1-vinyl-2-pyrrolidinone polymer iodine complex (25655-41-8, –)	
Commercial name (manufacturer): Akacid forte, Baquacil, GreatAp, Povidone iodine, Reputex, Vantocil (Acros Organics, AKA Cetral Research Laboratories, Alcoholes Montplet, Arch Chemicals, Guangzhou WeiBo Chemical, Lanfang Greatap Chemicals, Lonza, Mareva)	
Main product content, %: 20-100	
ANTIMICROBIAL ACTIVITY	
Product types: group 1, #1-5; group 2, #6-7, 9-13	**Typical concentration**, %: 0.3-5

Table 3.16. Properties of polymeric materials.

APPLICATIONS
broad range of disinfectants and preservatives as per application groups above

Chemical formulas:

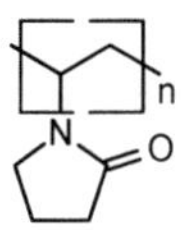

1-vinyl-2-pyrrolidinone polymer

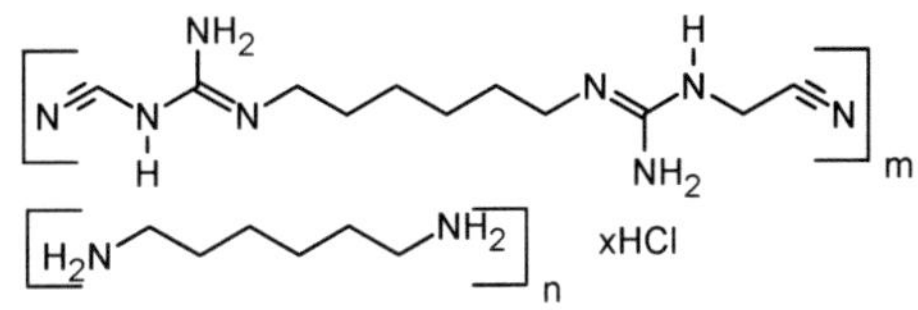

polyhexamethylenebicyanoguanide-hexamethylenediamine

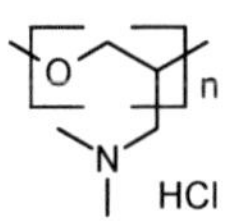

poly(2-hydroxypropyldimethylammonium chloride)

A recent invention[22] provides technologies to transform conventional and commercially important polymers into antimicrobial polymeric materials, which may find wide applications in medical devices, hospital equipment, water purification/delivery systems, food storage and packaging, hygienic products, consumer products, household items, and bio-protective applications. Commercially important piperidine derivatives (monomeric and polymeric) now are used as photo-stabilizers in polymeric materials but have also antimicrobial properties. Methods of compounding are discussed in the details of the invention.[22]

3.2.17 PYRIDINE DERIVATIVES

Table 3.17. Properties of pyridine derivative.

Chemical name (CAS and EC #s): copper 8-quinolinolate (10380-28-6, 233-841-9), copper, sodium, and zinc pyrithione are included in Section 3.2.13	
Commercial name (manufacturer): Bioquin, Dokivin (Genett Group, Kop-Coat, Rohm & Haas)	
Main product content, %: 100	**Molecular weight**: 351.85
CHEMICAL PROPERTIES	
Water solubility, %: traces	
PHYSICAL PROPERTIES	
Melting point, °C: 270 decomp.	**Density**, g/cm^3: 1.63
OPTICAL & ELECTRICAL PROPERTIES	
Color: greenish	
HEALTH & SAFETY	
LD50, oral, rat, mg/kg: 4,500	

Table 3.17. Properties of pyridine derivative.

ENVIRONMENTAL IMPACT	
log Pow: 2.46	
ANTIMICROBIAL ACTIVITY	
Product types: group 2, # 8, 9, 12	**Typical concentration**, %: 0.2-2
Minimum inhibition concentration, mg/l: 2-5	**pH in use**: 3-12
APPLICATIONS	
wood, fiber, leather, rubber, polymer preservatives and slimicides	

Chemical formula:

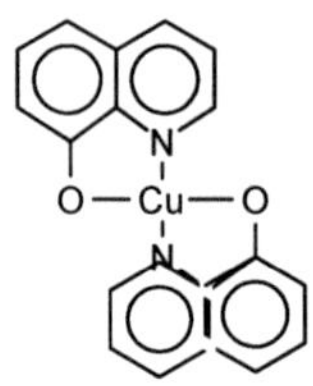

copper 8-quinolinolate

3.2.18 QUATERNARY AMMONIUM COMPOUNDS AND OTHER SURFACE ACTIVE AGENTS

Table 3.18. Properties of quaternary ammonium compounds and other surface active agents.

Chemical name (CAS and EC #s): alkyl(C12-16)dimethylbenzylammonium chloride (68424-85-1, 270-325-2), benzethonium chloride (121-54-0, 204-479-9), benzyl-C12-18-alkyldimethyl (68391-01-5, 269-919-4), benzyl (coconut oil alkyl) dimethylammonium chlorides (61789-71-7, 263-080-8), cetylpyridinium chloride (123-03-5, 204-593-9), decyl isononyl dimethyl ammonium chloride (138698-36-9, 270-331-5), dioctyl dimethyl ammonium chloride (5538-94-3, 226-901-0), didecyl dimethyl ammonium chloride (7173-51-5, 230-525-2), laurylamine dipropylenediamine (2372-82-9, 219-145-8), N-octadecyldimethyl{3-(trimethoxysilyl)propyl}ammonium chloride, OTPAC (27668-52-6, 248-595-8), tetradecyldimethyl(3-(trimethoxysilyl)propyl)ammonium chloride (41591-87-1, 255-451-8)		
Commercial name (manufacturer): ADBAC, Arquad, DDAC, Hyamine (Brulin & Co., Buckman Laboratories, Dow Corning, Lonza, Field Fisher, Proctor & Gamble, Stepan Company, Vertellus Specialty Chemicals)		
Main product content, %: 50-99	**Molecular weight**: 299-55-448.10	
CHEMICAL PROPERTIES		
Water solubility, %: 25		
PHYSICAL PROPERTIES		
Melting point, °C: 61-165	**Density**, g/cm^3: 0.865-0.978	**Refractive index**: 1.417-1.613
Surface tension, mN/m: 27-33	**Vapor pressure**, hPa (20°C): 0.0000075	
pH: 5-12	**Viscosity**, mPas: 25-150	**Odor**: mild

Table 3.18. Properties of quaternary ammonium compounds and other surface active agents.

OPTICAL & ELECTRICAL PROPERTIES	
Color: colorless to white to yellowish	
HEALTH & SAFETY	
Autoignition temperature, °C: 200-380	**Flash point**, °C: 11-100
LD50, oral, rat, mg/kg: 320-600	**LD50, dermal, rabbit**, mg/kg: 2,600
ENVIRONMENTAL IMPACT	
Bluegill sunfish, 96-h LC50, mg/l: 0.52-2	**Daphnia magna**, 48-h LC50, mg/l: 1
Fathead minnow, 96-h LC50, mg/l: 1-1.6	**Rainbow trout**, 96-h LC50, mg/l: 0.68-2
log Kow: 1.71-2.36	
ANTIMICROBIAL ACTIVITY	
Product types: group 1, #1-4, group 2, #8,10-13	**Typical concentration**, %: 0.002-7.95
Minimum inhibition concentration, mg/l: 1.5-170	**pH in use**: 1-12
APPLICATIONS	
disinfectants for human hygiene purposes, private and public health areas, veterinary applications, and food and feed areas; in-can preservatives, wood, masonry, and liquid-cooling system preservatives, slimicides, and metalworking fluid preservatives	

Chemical formulas:

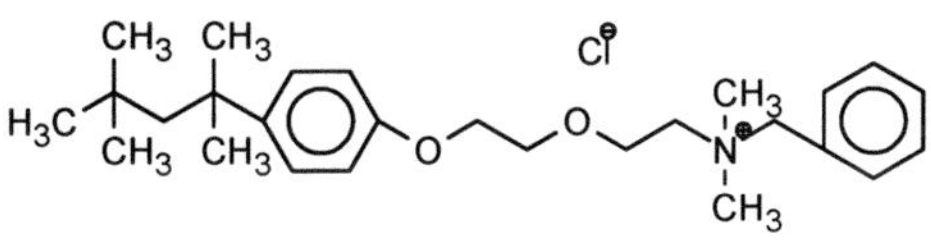

benzethonium chloride

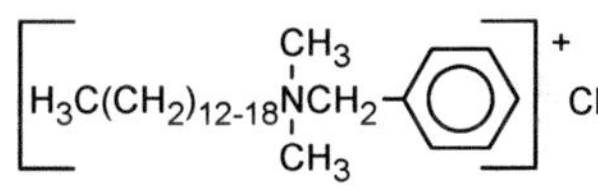

benzyl C_{12-18} alkyldimethyl

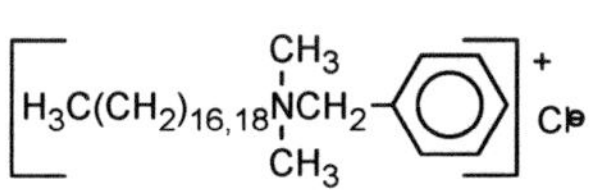

benzyl (coconut oil alkyl) dimethylammonium chlorides

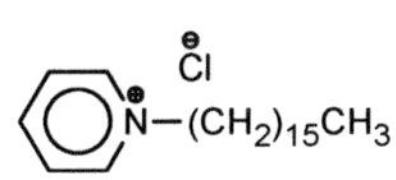

cetylpyridinium chloride

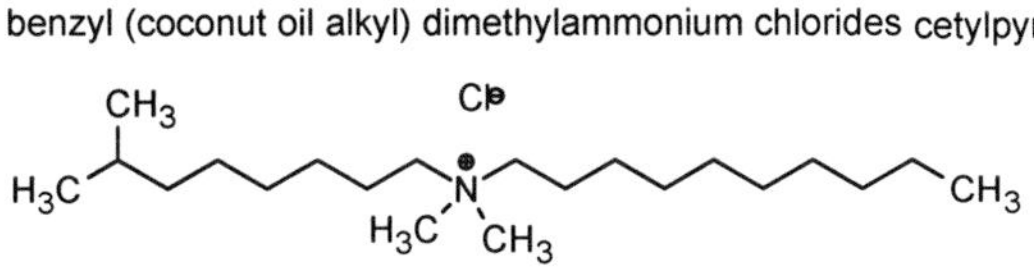

decyl isononyl dimethyl ammonium chloride

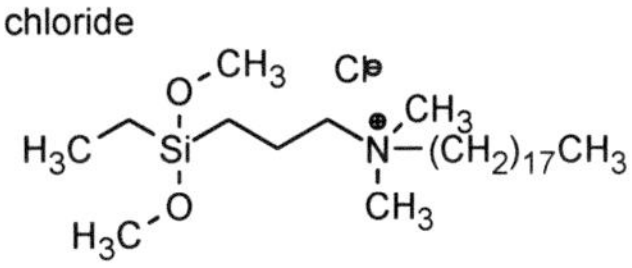

OTPAC

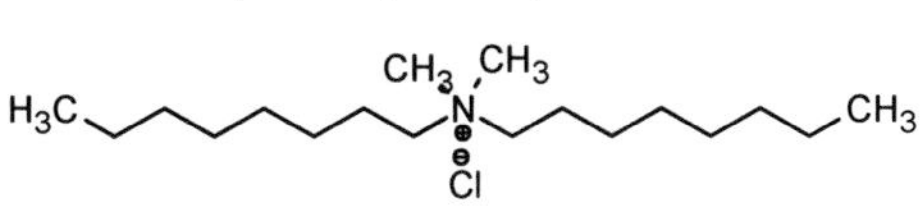

didecyl dimethyl ammonium chloride

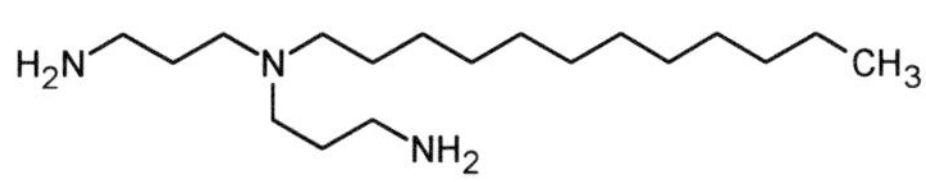

laurylamine dipropylenediamine

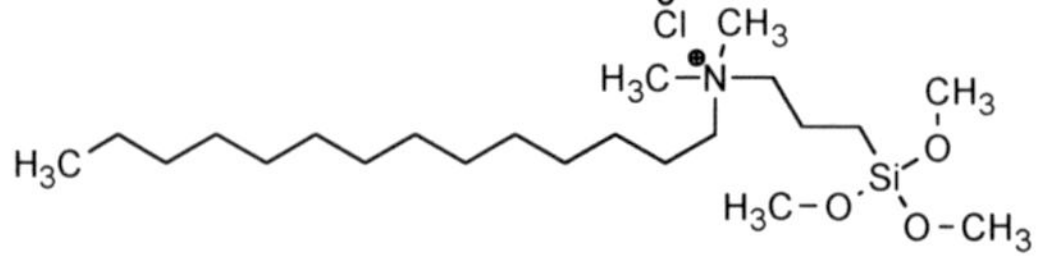

tetradecyldimethyl(3-(trimethoxysilyl)propyl)ammonium chloride

3.2.19 OTHER (NOT INCLUDED) PRODUCTS

Table 3.19. Properties of various products.

Chemical name (CAS and EC #s): creosote (8001-58-9, 232-287-5), 1,3-dichloro-5-ethyl-5-methylhydantoin (89415-87-2, 401-570-7), etofenprox (80844-07-1, 407-980-2), fenpropimorph (67564-91-4, 266-719-9),		
Commercial name (manufacturer): DCEMH, DTBMA (Lonza, PromChem, Rohm & Haas)		
CHEMICAL PROPERTIES		
Water solubility, %: traces		
PHYSICAL PROPERTIES		
Melting point, °C: 25-37	**Boiling point**, °C: 230	**Density**, g/cm^3: 1.03-1.1
Vapor pressure, hPa (20°C): 9E-9 to 6		
Henry law constant, atm m^3/mol: 2.26E-8 to 2.44E-6		
OPTICAL & ELECTRICAL PROPERTIES		
Color: brown to black		
HEALTH & SAFETY		
Autoignition temperature, °C: 336	**Flash point**, °C: 74-85	
LD50, oral, rat, mg/kg: 2,524-42,800		
ENVIRONMENTAL IMPACT		
Daphnia magna, 48-h LC50, mg/l: 3.07-4.3	**log Pow**: 4.93-7.05	
ANTIMICROBIAL ACTIVITY		
Product types: group 2, # 8 and 9	**Typical concentration**, %: 1-3	
APPLICATIONS		
wood preservatives and fiber, leather, rubber, and polymer biocides		

Chemical formulas:

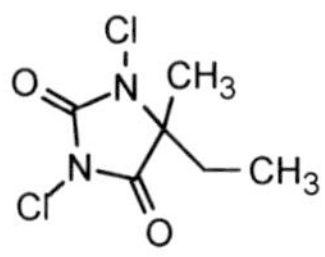

1,3-dichloro-5-ethyl-5-methylhydantoin

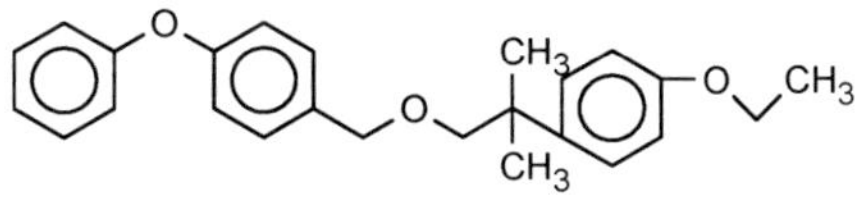

etofenprox

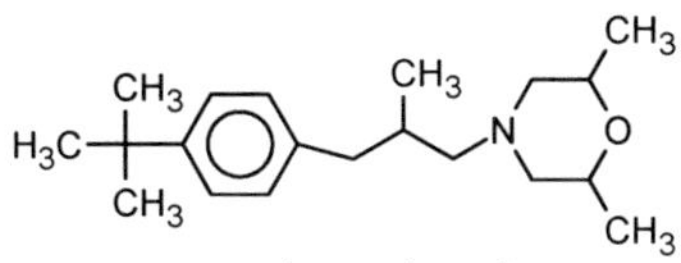

fenpropimorph

3.3 PRINCIPLES OF SELECTION OF BIOSTABILIZERS

Considering that it is possible to find out information on minimum inhibition concentration of biocides for different microorganisms and list of microorganisms which potentially affect most products, the task of biocide selection is more complex than it seems. The list of important check points or essential features seems to be a good indicator of complexity of biocide choice. The biocide selection depends on the following essential properties of biocide, material, and conditions of its use:

- biocidal efficiency towards microorganisms potentially affecting material to be protected
- wide microorganism killing ability and range
- prevention of protected material change beyond noticeable level of performance
- effect of pH on performance
- effect of temperature on performance
- bioavailability (effect of formulations components)
- contact time required
- concentration required
- potential for use of "booster" combinations
- permissible concentrations
- effect of regulations
- molecular weight
- longevity
- available delivery systems
- cost
- convenience of transportation, storage, and application
- good miscibility
- dosing frequency and rate (antifouling compositions)
- effect of processing conditions on biocide activity
- effect of sterilization on biocide activity

- effect of material shape and thickness on biocide efficiency
- compatibility with other materials in formulation
- effect on color of material to be protected (color change, staining, discoloration, degradation)
- effect on other components of formulation
- stability of biocide under conditions of exposure
- stability of biocide effectiveness under changing condition of material (e.g., pH, formation of reactive products of degradation, increased porosity, etc.)
- corrosiveness
- toxicity
- environmental impact.

With this number of important and practical questions, the biocide selection is certainly more complex than its simple addition based on suggestions from raw material supplier.

Biocidal efficiency perceived and real only meet if laboratory methods of determination can simulate conditions of the biocide performance under real conditions. This can be well argued based on experimental work on microbiologically influenced corrosion of mild steel.[23] The studies on the microbiologically influenced corrosion are most frequently based upon microbiological tests and gravimetric techniques, considering the use of isolated strain bacteria. In natural conditions, the microorganisms are found in consortia, and the corrosion processes are different, therefore the results with isolated strains do not predict real conditions very well. In this case, corrosion rate with and without biocides and their different concentrations can be conveniently studied by polarization resistance or electrochemical impedance spectroscopy. In addition to the monitoring, the corrosion rate of some biofilm characteristics can be studied.[23] This example underlines the importance of the methods used to monitor the effect of biocides, and needs to apply techniques which can resemble real life conditions.

Bioavailability of biocide is important for its performance (as well as its toxic effects). Kathon in paints has been studied for its effect on toxicity to humans.[24] Comparison was made between the biocide effect when applied from water dispersion, dispersion in emulsion used for paint, and dispersion in paint (contains pigments). It was found that only about 50% of bioavailable biocide from water dispersion is available from emulsion dispersion and about a third from full paint composition. It is because biocide in more complex formulations is engaged in interaction with resin, pigment, and potentially many other additives. This is an important example to consider that product testing has to be done on real formulation, as otherwise results will point to different required concentrations of biocide than those which are expected to work.

Various data on biocide bring information on minimum inhibition concentration which shows that biocides do not have equal performance in case of different microorganisms. This affects the required concentration under which biocide performs and frequently suggests potential studies which may lead to finding more effective ("booster") compositions of biocides which make up for deficiencies of a single component to some types of microorganisms. This is perceived to be the future of studies which may lead to more efficient products and permit adjustments to match changes made by microorganisms to conditions which affect them.

Biocide is considered effective until lack of product performance is detected. This can be a formation of mould stains on material surface, consumption of components of formulation which leads to cracking or change in mechanical performance, or alteration of fuel performance. Determination of these changes critical for performance should be used for determination of the effect of biocide on real life performance.

It is common knowledge that pH affects killing abilities of biocide. On one hand, pH depends on material composition but it also depends on many other processes, such as thermal and UV degradation, deposition of pollutants, and even action of some microorganisms (for example, sulfur reducing bacteria change pH of material on which the pray).

Temperature is a very essential factor. Low temperature was found to affect performance of water biocides. Too high temperature during processing or sterilization (or its environmental exposure) may cause degradation or its loss by evaporation of some or all biocide added to formulation. Temperature is also a factor in biocide diffusion, which may work in conjuction with a material's shape and thickness (thick samples may require less biocide than thin films for the same duration of protection).

Concentration of biocide depends on its inhibition (or killing) concentration, contact time required, its molecular weight, dosing frequency and rate (in antifouling products). It also depends on delivery systems available for a product which controls its diffusion and migration and thus longevity of protection. Concentration influences cost, and as such, it is usually the most important factor in product selection. Concentration is also affected by regulations which usually restrict the amounts of biocides which can be used in a particular product.

Convenience of transportation, storage, and application are also essential cost factors which usually determine product selection. It is always expected that biocide in its applied form has good miscibility with the material to be protected. Good compatibility also means that it will not react or otherwise alter properties of any component of formulation. It should also, in most cases, not affect color (some do affect, for example, wood color and still do not have a negative effect on their selection). Color change of materials in contact is very frequently the most essential requirement (e.g., historical documents, paintings, or stone monuments). In these instances, color change, discoloration, or staining are not permissible at all.

Finally, toxicity and environmental effects are probably the most important aspects of biocide selection. Many biocides, very popular in past, are not any longer in use, even though they were extremely effective, because they were toxic for humans or other living creatures. Recent stringent regulations have eliminated many biocides, which puts emphasis on efficient and intelligent use of those which are left to provide a continuous level of protection in an environment of ever-adjusting microorganisms according to their living conditions. Considering this, selection of biocide composition is not a one-time research but rather continuous monitoring of performance and early detection of needs of reinforcement or alterations.

3.4 LONGEVITY OF BIOSTABILIZED MATERIALS

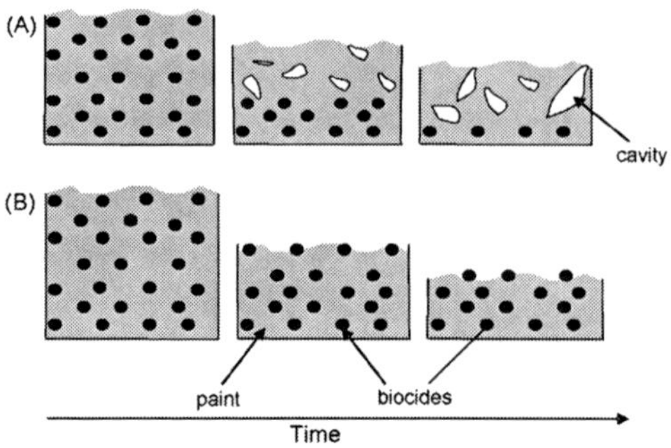

Figure 3.4. Bulk erosion (A) and surface erosion (B). [Adapted, by permission, from Fay, F; Linossier, I; Peron, J J; Langlois, V; Vallee-Rehel, K, *Prog. Organic Coat.*, **60**, 194-206, 2007.]

Longevity of biostabilized materials is one of the most important aspects of stabilizer incorporation but unfortunately it is the least understood and investigated. It is not caused by the lack of need, considering that many attempts are on the way to improve performance. Below we will discuss the available information on parameters which influence longevity.

In the case of leachable biocides and antibacterial agents, the zero rate order of release is usually desired but the first order release rate is typically found.[25] This means that most common biocides will tend to over-release in the beginning and under-release at a longer time scale. For this reason, many efforts are being made to modify a surface in such a manner that a biocide is bound to the surface. This can be done either by surface treatment or by reaction with functional groups of polymer, either readily available or specially introduced. These methods look attractive but studies are needed to assess their usefulness.

Figure 3.4 shows the mechanisms of surface erosion which is one of the roadblocks to overcome in order to make surface treatment work. The diagram also characterizes the effects involved in loss of conventional biocides.[26] In the case of bulk erosion, the degradation occurs in the entire film thickness and consequently the durability is reduced. In the case of surface erosion, the surface is renewed regularly and keeps being active during the lifetime of a product. Most products are degraded by bulk erosion mechanism because of the effect of moisture intrusion and oxygen and related phenomena which cause extraction or degradation of vital components of formulation and increase porosity of material. Surface erosion was a common mechanism in the case of self-polishing coatings (now phased out) and it is also common in titanium oxide stabilized materials which degrade on exposure to ultraviolet radiation. Any of these mechanisms leads to a loss of active biocide and increases potential for its extraction because material becomes more porous (A) or surface area is increased (A and B).

Studies of antifouling paints which contributed to the mechanism of erosion (Figure 3.4) also show that biocide loss depends on the nature of the extractant. Under conditions of this experiment, natural sea water caused the most severe surface loss, followed by laboratory simulated water and demineralized water.[26] In another study,[27] pH was found to determine stability of Kathon (5-chloro-2-methyl-4-isothiazolin-3-one), which, at higher pH, undergoes hydrolysis (Figure 3.5).

In antifouling paints, the release of biocide was increased with an increase in its water solubility.[28] Biocides which work in contact with water (but not in water solution) have low water solubility. Their rates of release from material can be well explained on the

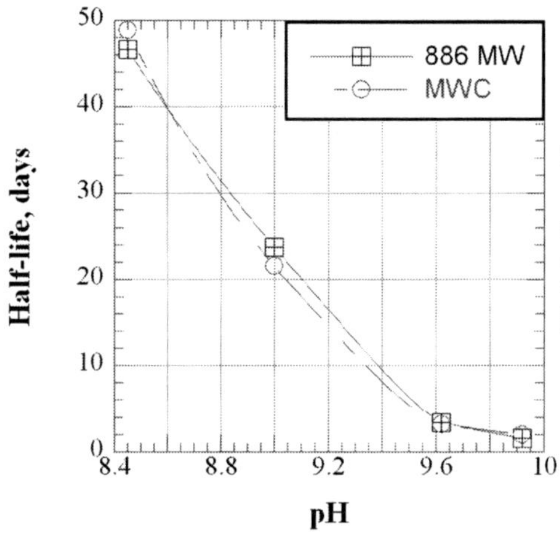

Figure 3.5. Effect of pH on half-life of two different grades of Kathon. [Data from Barman, B N; Preston, H G, *Tribology Intl.*, **25**, 4, 281-87, 1992.]

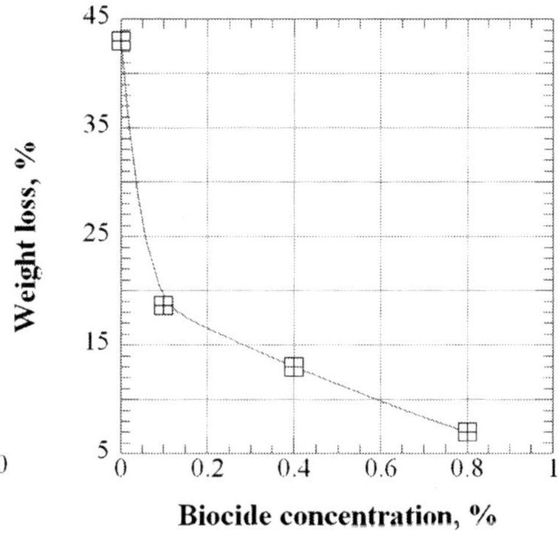

Figure 3.6. White rot resistance of birch treated with Tebuconazole-containing nanoparticles. [Data from Gozzelino, G: Dell'Aquila, G D; Romero Tobar, D, *J. Appl. Polym. Sci.*, **112**, 2334–2342, 2009.]

basis of Fickean diffusion equations.[29] These biocides have to migrate to the material's surface to work and their rate of migration has to be controlled to match short and long term requirements (sufficient concentration to perfom, combined with long-term presence to protect material).

Use of controlled release combinations is expected to provide sufficient balance. This is done either by incorporation of fillers which are known to delay migration of other additives (e.g., silica or zeolites) or by encapsulation of additive in the carrier polymer. Figure 3.6 shows the effect of concentration of biocide in nanoparticle protected wood on weight retention by the protected wood. Fungicide is dissolved in a minimal amount of solvent and incorporated in polyvinylpyridine. Nanosize-particles containing different amounts of biocide are used for pressure-treatment of wood. State of polymer also plays an essential role in the release rate. It was observed that Triclosan doped HPDE specimens which were uniaxially drawn had a lower release rate than undrawn samples. The higher the draw ratio, the higher the crystallinity and the lower the extraction of Triclosan.[30] This seems to be the obvious consequence of an increased number of obstacles on the pathway of diffusion, considering that crystalline areas do not swell.

Copper is a commonly occurring contaminant in harbors and marinas. One source of copper in these environs is copper-based antifouling coating used on vessel hulls.[31] Monthly average flux rates of dissolved copper from the hard vinyl and modified epoxy coatings were 3.7 and 4.3 $\mu g/cm^2/day$. Hull cleaning activities generated between 8.6 and 3.8 μg dissolved copper/cm^2/event, drastically increasing copper emission.[31]

Surface coating with concentrated biocide can be potentially considered as a method of protection of materials. Stainless steel was surface-treated with silver-zeolite matrix coating.[32] Durability of the coatings declined most markedly when the coating had been applied with a wet process and scrubbed between uses with a test tube brush. Powder-coated surfaces cleaned with a towel retained a high degree of activity after five cycles of use.[32] This shows that surface treatment can be easily removed from a material.

Antimicrobial properties of nylon-6 fabrics and washing durability were improved by grafting methacrylic acid to create carboxylic groups and react them with dimethylalkylbenzyl ammonium chloride.[33] Different innovative methods of improving retention of biocides in textiles and fibers are discussed in review paper,[34] indicating that retention is usually poor.

In additon to physical reasons for biocide retention in protected material there are also chemical reasons which may cause its degradation. Antioxidant butylated hydroxytoluene, BHT, increased the efficacy of chlorthalonil in protecting wood in laboratory and outdoor exposure tests.[35] It was suggested that the antioxidant helps in protection of organic biocides.[35] UV radiation must have a strong influence on many biocides since they are capable of absorption UV radiation and thus may potentially undergoe chemical changes.

The above review of existing literature shows that there is a need to increase understanding of durability of biocidal finishes as well as to increase the lifetime of protection.

REFERENCES

1 Denyer, S P, *Intl. Biodeterioration Biodeg.*, 227-45, 1995.
2 Russell, A D, *Intl. Biodeterioration Biodeg.*, 247-65, 1995.
3 McDonnell, G, in Sterilization and Disinfection, **Encyclopedia of Microbiology**, *Elsevier*, 2009, pages 529-48.
4 Simoes, M; Pereira, M O; Vieira, M J, *Water Res.*, **39**, 5142-52, 2005.
5 Rilling, M C, in Polymer and Microorganisms, **Encyclopedia of Soils in Environment**, *Elsevier*, 2004, pages, 287-94.
6 Ganesh Kumar, C; Anand, S K, *Intl. J. Food Microbiol.*, **42**, 9-27, 1998.
7 Cloete, T E, *Intl. Biodeterioration Biodeg.*, **51**, 277-82, 2003.
8 Russell, A D; Furr, R J; Maillard, J Y, *ASM News*, **63**, 481–487, 1997.
9 Chapman, J S, *Intl. Biodegradation Biostab.*, **51**, 133-38, 2003.
10 Power, E G M, in Aldehydes as Biocides, **Progress in Medicinal Chemistry**, *Elsevier*, 1997, Vol. 34, pages 150-201.
11 Faust,M; Altenburger, R; Backhaus, T; Blanck, H; Boedeker, W; Gramatica, P; Hamer, V; Scholze, M; Vighi, M; Grimme, L H, *Aquatic Toxicology*, **63**, 43-63, 2003.
12 Russell, A D, in Mechanisms of Bacterial Resistance to Antibiotic and Biocides, **Progress in Medicinal Chemistry**, *Elsevier*, 1998, Vol.35, pages 134-197.
13 Meyer, B, *Intl. J. Food Microbiol.*, **112**, 275-79, 2006.
14 Chapman, J S, *Intl. Biodeterioration Biodeg.*, **41**, 241-45, 1998.
15 Grabski, C; Gisi, U, *Crop Protection*, **6**, 1, 64-71, 1987.
16 Morato, J; Mir, J; Codony, F; Mas, J; Ribas, F, **Handbook of Water and Wastewater Microbiology**, *Elsevier*, 2003, pages 129-40.
17 Gatidou, G; Thomaidis, N S, *Aquatic Toxicology*, **85**, 184-91, 2007.
18 Nicoletti, G; Domalewska, E; Borland, R, Mycol. Res., 103, 9, 1073-84, 1999.
19 Gerke, T; Schaper, U-A; Faber, W, **US Patent** 7,098,178, Henkel AG, Aug. 29, 2006.
20 Pinori, M; Lattanzio, M; Modena, D; Mascagni, P, **US Patent** 7,427,620, Italfarmaco, Sep. 23, 2008.
21 Wenderoth, B; Dambach, S; Meszaros, L; Fidorra, U, **US Patent** 7,419,618, Bayer, Sep. 2, 2008.
22 Sun, Y; Chen, Z, **US Patent**, 7,541,398, University of Texas, June 2, 2009.
23 Hernandez Gayosso, M J; Zavala Olivares, G; Ruiz Ordaz, N; Garcia Esquivel, R, *Mater Corrosion*, **56**, 9, 624-29, 2005.
24 Joyjock, M A; Hazelton, G A; Lewis, P G; Wooder, M F, *Fd. Chem. Toxic.*, **34**, 3, 227-82, 1996.
25 Makala, U; Wood, L; Ohman, D E; Wynne, K J, *Biomater.*, **27**, 1316-26, 2006.

26 Fay, F; Linossier, I; Peron, J J; Langlois, V; Vallee-Rehel, K, *Prog. Org. Coat.*, **60**, 194-206, 2007.
27 Barman, B N; Preston, H G, *Tribology Intl.*, **25**, 4, 281-87, 1992.
28 Thouvenin, M; Peron, J J; Charreteur, C; Guerin, P; Langlois, J-V; Vallee-Rehel, K, *Prog. Org. Coat.*, **44**, 75-83, 2002.
29 Gozzelino, G: Dell'Aquila, G D; Romero Tobar, D, *J. Appl. Polym. Sci.*, **112**, 2334–2342, 2009.
30 Iconomopoulou, S M; Voyiatzis, G A, *J. Controlled Release*, **103**, 451-64, 2005.
31 Schif,K; Diehl, D; Valkirs, A, *Marine Pollution Bull.*, **48**, 371-77, 2004.
32 Cowan, M M; Abshire, K Z; Houk, S L; Evans, S M, *J. Ind. Microbiol. Biotechnol.*, **30**, 102-106, 2003.
33 Shalaby, S E; Al-Balakocy, N G; Abdel-Fatah, O M; Elshafei, A M, *J. Appl. Polym. Sci.*, **110**, 738–746, 2008.
34 Holme, I, *Color. Technol.*, **123**, 59-73, 2007.
35 Schultz, T P; Nicholas, D D; Kirker, G T; Prewitt, M L; Diehl, S V, *Intl. Biodet. Biodeg.*, **57**, 45-50, 2006.

4

Biodegradation, Biodeterioration, and Biostabilization of Industrial Products

This chapter is the core of the book. It contains information on biodeterioration of products and methods of its inhibition. Each section discusses a group of materials essential for our well-being. Materials in each section are discussed according to a consistent breakdown, which help one to find information and compare it for different materials. The following breakdown of information is used throughout the chapter (with some exceptions, e.g., leather and petroleum product sections):

1. Relevant microorganisms
2. Essential product components
3. Mechanisms of biodegradation and biodeterioration
4. Results of biodeterioration
5. Biostabilization
6. Formulations
7. References

The first section includes a list of microorganisms which were found to be involved in biodegradation of a particular material or contaminated material. The members of this list can be identified using information on taxonomy in Chapter 2. The second section contains information on components of formulations required to manufacture a particular group of products with special attention to their effect on the biodegradative process.

The third section discusses mechanisms of biodegradation and biodeterioration which have been found to play a role in biodeterioration of materials included in the group. Special attention is given to the role which these mechanisms play in shortening the useful life of the product. The fourth section enumerates, as demonstrated in the practice of use of the materials of a particular group of products, results of action of microorganisms on material properties which lead to its replacement and disposal.

The methods of biostabilization, known from open literature, patents, and information given by manufacturers of biocides, are given in the fifth section with some relevant formulations given in the sixth section. Each section has a separate list of references for further reading and information retrieval.

This chapter is intended for expedited retrieval of relevant information and therefore care is taken that it is concise and contains all available relevant information.

4.1 BUILDING MATERIALS

This is a very diverse group of materials. Some members of this group are discussed in a separate sections, including coatings and paints, electrical materials (cables), metals, roofing, sealants and adhesives, stones, and wood. The information on these products will not be discussed in the section. This section reviews the following products:

- brick
- cementitious matrix
- ceramic tiles
- gypsum board
- thermal insulation
- doors and windows (plastic profiles)
- piping
- siding
- wallpaper

This means discussion of two groups of materials: mineral-based (the first five groups) and polymeric materials (mostly PVC-based) are included in the last four groups (although some other than polymeric materials are also involved). Some information (especially mechanisms of degradation) regarding the last group will be broadly discussed in section 4.14 (polymers) and in the subsection devoted to PVC. This also regards some insulating materials.

4.1.1 RELEVANT MICROORGANISMS

Table 4.1 contains a list of microorganisms which cause deterioration of building materials.

Table 4.1. Microorganisms in building materials

Material	Microorganisms	Refs.
brick	*Alternaria alternata, Aspergillus, Aureobasidium pullulans, Candelariella aurella, Chlorella sp., Cladosporium spherospermum, Escherichia coli, Fusarium sp., Lecanora albescens, Lecanora dispersa, Penicillium sp., Protoparmeliopsis muralis, Marinococcus halophilus, Rhodotoruls sp., Serpula lacrymans, Stachybotrys chartarum, Ulocladium tuberculatum*	2-7
cementitious matrix	*Acidothiobacillus thiooxidans, Halothiobacillus neapolitanus, Thiobacillus thiooxidans*	9-11
ceramic tiles	*Navicula pelliculosa, Phaeodactylum tricornutum, Rhodotorula mucilaginosa, Sporobolomyces roseus*	13-14

Table 4.1. Microorganisms in building materials

Material	Microorganisms	Refs.
gypsum boards	*Acremonium, Actinobacteria, Alternaria, Aspergillus, Chrysonilia, Cladosporium, Eurotium, Memnoniella echinata, Monascus, Monocillium, Mucor, Oidiodendron, Paecilomyces, Penicillium, Scopulariopsis, Sphaeropsidales, Stachybotrys chartarum, Stachybotrys cylindrospora, Trichoderma viride, Trichoderma harzianum, Trichoderma longibrachiatum, Ulocladium, Yeasts*	15-17
thermal insulation	*Alternaria alternata, Aspergillus versicolor, Cladosporium sphaerospermum, Penicillium chrysogenum, Trichoderma viride*	20
doors and windows	*Aspergillus niger, Aureobasidium pullulans, Fusarium spp., Ochrobactrum, Pseudomonas chlororaphis, Pseudomonas fluorescens, Pseudomonas putida*	17, 21
piping	*Desulfovibrio gigas, Legionella pneumophila, Metallogenium spp., Microbacterium, Ochrobactrum anthropi, Pseudomonas aeruginosa, Pseudomonas fluorescens, Sulfolobus spp., Thiobacillus ferrooxidans, Thiobacillus neopolitabus, Thiobacillus sp., Thiobacillus thiooxidans, Variovorax sp.*	22-24, 28-30
siding	no reports in literature found	
wallpaper	*Aspergillus versicolor, Aspergillus ustus, Aureobasidium pullulans, Chaetomium sp., Cladosporium spherospermum, Penicillium chrysogenum, Penicillium verrucosum, Stachybotrys chartarum* (also known as *atra*), *Trichoderma spp., Trichoderma harzianum, Ulocladium tuberculatum*	34, 36-37

Fungi appear in a certain succession on building substrates according to the minimum moisture demands of the species involved.[19] The primary colonizers, species of *Penicillium* and *Aspergillus*, appear first, then secondary colonizers, *Cladosporium spp.*, and finally tertiary colonizers, *Ulocladium spp.* and *Stachybotrys spp.*, appear only on a very moist substrate.[19]

Studies of several building materials including wallpaper show that only very limited growth of microorganisms occurs on wallpaper when relative humidity is 80-82%.[19] When relative humidity is increased to 96-98%, microorganism growth rate dramatically increases.[19]

4.1.2 ESSENTIAL PRODUCT COMPONENTS

Mineral-based building materials, such as brick, mortar or concrete, are susceptible to microbial attack.[1] The degree of contamination depends on their pore size distribution and their alkalinity.[1]

Practically no microorganism should be able to grow below the equilibrium relative humidity of a material (approximately 65% for gypsum board). Growth of xerophylic fungi is possible above 70%, while hydrophilic microorganisms (various moulds, yeasts, decay fungi, and bacteria) require at least 90% (95% for germination and growth on nutrient-rich media).[17] On materials with limited nutrients, such as construction materials, or at temperatures below the optimum for growth, the minimum is higher than that mentioned so far.[17] At relative humidity near 95%, gypsum board supports fungal growth when covered with a biodegradable surface (paint or wallpaper).[18] A biodegradable material treated with a biocide (e.g., biocide-treated plasterboard) offers partial resistance to fungal growth on a biodegradable surface covering.[18]

Wallpapers differ in composition from other PVC materials (window/door profiles pipes, and siding) by extensive use of plasticizers and epoxidized soybean oil (plasticizer/stabilizer) both of which change PVC resistance to the growth of microorganisms. Wallpaper is vulnerable to microbiological attacks and therefore it should be protected by biocides (although in practice only some are protected).[32]

4.1.3 MECHANISMS OF BIODEGRADATION AND BIODETERIORATION

The mineralogical composition of the raw materials, the firing temperature, the location of the bricks in the buildings, the environments to which they had been exposed, the action of natural or polluted filtration water, the action of microorganisms and the environmental conditions, all contribute towards the deterioration rate of bricks.[2]

Bacterial action facilitates brick decay through the consumption of chemical compounds in the brick. Calcites are produced by the activity of *Escherichia coli*. On the other hand, lichens can protect brick surfaces from atmospheric weathering.[2] Lichens can also be destructive because they can form biofilms, and by exuding acids, cause etching and disaggregation of the brick. With normal variability of moisture level (and frost), lichens expand and contract, causing mechanical damage.[2]

4.1.4 RESULTS OF BIODETERIORATION

Corrosion of concrete sewer pipes was detected as early as 1900 and was attributed to the effect of hydrogen sulfide, an anaerobic decomposition product of sulfur-containing microorganisms in waste water.[27] The reaction involved was regarded as purely a chemical process, in which hydrogen sulfide produced under anaerobic conditions is oxidized to sulfuric acid in the presence of oxygen.[27] Later, it was established that the presence of acidophilic *Thiobacilli* is required for these degradative processes to occur.[27]

Bacterial-induced corrosion causes severe damage to pre-cast concrete sewer pipes. The rate of degradation may reach several millimeters per year.[10] It is estimated that 20% of the total damage to concrete structures in waste water conveyance and storage systems is caused by bacteria.[10] Sulfate reducing bacteria from the genera *Desulfovibrio* is the primary culprit for anaerobic corrosion of cementitious matrix.[10] Concrete initially exhibits a pH in the range of 11-13 (too alkaline for the growth of most bacteria).[31] The initial step in reducing the pH may involve the abiotic neutralization of the concrete by reaction of the cementitious material with carbon dioxide (carbonation) or reaction with other acidic gases such as hydrogen sulfite.[31] This abiotic neutralization can reduce the pH of the concrete to as low as 9.[31] At this pH, *Thiobacillus thiooxidans* can grow and further reduce pH to 4 and even 2.[31]

Sulfuric acid, the main oxidation product of sulfur-oxidizing bacteria (*Halothiobacillus neapolitanus*), induces corrosion and degradation of cementitious mixtures.[9] Sulfuric acid reacts with free lime, $Ca(OH)_2$, in the cement paste to form gypsum, $CaSO_4\,2H_2O$, which in turn produces a corroding layer, which may cause cracking and degradation due to the large difference in density.[9] In addition, calcium-aluminate-hydrate present in the cement paste may react with freshly formed gypsum producing ettringite ($3CaO\ Al_2O_3\ 3CaSO_4\ 32H_2O$), which increases the internal pressure, leading to the formation of cracks.[9]

Trichothecenes of the trichodermol (toxic substances) have been produced by *Stachybotrys chartarum*.[15]

Microbiologically induced corrosion of copper pipes caused the increase of the copper release to water and formation of pits on copper pipes.[30] A bacterial strain identified as *Variovorax sp.* was isolated from the biofilms present on the inner surfaces of the copper pipes.[30]

Literature on pipes does not give examples of similar processes with PVC pipes which are now the most common. Because PVC pipes are produced from unplasticized PVC they are sufficiently resistant to biodegradative attacks. The same comment applies to PVC siding. PVC products are rarely protected with biocides and if they are they either contain plasticizers and/or certain organic stabilizers or are produced for medical applications.[32]

Ergosterol and chlorophyll a are used as biological markers correlating with the presence of fungal and algal biomass.[33] Using ergosterol as a biomarker, it was determined that the fungal infections of wallpaper in damp rooms ranged from 1.7 g/m^2 (bathroom) to 3.5 g/m^2 (bedroom) near the values in the cellar rooms determined after the water damage.[33]

Contamination of different building materials was also compared using the ergosterol method and correlated with the number of mould count.[35] It was found to be very reliable criterion for mould determination. Wallpaper being one of building materials under study had medium range vulnerability to mould colonization (similar to glass wool, gypsum board, and emulsion paint).[35]

High quantities of the carcinogenic mycotoxin sterigmatocystin, especially on wallpaper, were produced by *Aspergillus versicolor*.[36] This demonstrates a potential health hazard in water damaged buildings.[36]

4.1.5 BIOSTABILIZATION

Table 4.2 contains a list of most suitable biocides used for protection or remediation condition of building materials.

Table 4.2 Biocides used in building materials.

Material	Biocide	Refs.
brick	alkyl-(3,4,-diclorobenzyl) dimethylammonium chloride, lauryl dimethyl benzylammonium bromide	2
cementitious matrix	silver coated zeolite	10
ceramic tiles	N-acryloyl-N-(phosphonomethyl) glycine	13
gypsum boards	2,2-dibromo-3-nitrilopropionamide	18
thermal insulation	usually not used on inorganic insulation	20
doors and windows	intrinsic biostability if not plasticized	

Table 4.2 Biocides used in building materials.

Material	Biocide	Refs.
piping	5-chloro-2-methyl-4-isothiazolin-3-one/2-methyl-4-isothianzoline-3-one (7 vol%) and glutaraldehyde (15 vol%) (stainless steel protection), 2,2 dibromo-3-nitrilopropionamide (heat exchangers), glutaraldehyde (oilfield pipe), silver-zinc zeolite (stainless steel pipe coating)	22-23, 25, 26
siding	none used	32
wallpaper	Na-hypochlorite (surface maintenance), pentachlorophenol	34, 38

Protective treatment of brick may involve cleaning to remove biological remains, treatment with a biocide to inhibit and/or eradicate future growth, and the application of a moisture barrier.[8] It should be considered that biocides had a limited effectiveness in this treatment.

Microbial mineral precipitation is one of the research areas considered to be capable to produce protective coating on concrete.[12] Different microorganisms have been used to increase the compressive strength of cement mortar and for the remediation of cracks in concrete such as ureolytic bacteria.[12] These bacteria influence the precipitation of calcium carbonate by its urease enzyme. This enzyme catalyzes the hydrolysis of urea to CO_2 and ammonia, resulting in an increase of the pH and carbonate concentration in the bacterial environment.[12] Precipitation of calcium carbonate crystals occurs by heterogeneous nucleation on the bacterial cell wall once supersaturation is achieved.[12] A surface treatment using microbial carbonate precipitation (biodeposition) with *Bacillus sphaericus* was even more effective in decreasing the permeation properties of mortar and concrete as compared to a treatment with ureolytic mixed cultures.[12]

Zeolite aluminium silicate ceramic coating containing 2.5 wt% silver and 14% zinc ions can be coated directly on stainless steel.[22] Silver ions inhibit bacterial enzymes, interfere with electron transport, and bond to DNA, increasing the stability of the double helix. Zinc inhibits nutrient uptake and interferes with proton transfer.[22] Stainless steel surfaces with Ag-Zn coatings significantly reduce *Legionella pneumophila* as compared to plain stainless steel surfaces.[22] Stainless steel with these antimicrobial ceramic coatings is available for use in air systems, including air conditioners and ventilation ducts and at air intake and exhaust sites.[22] The product is likely to be especially useful for any air-handling surface on which moisture or condensate occurs.[22]

REFERENCES

1 Warscheid, T; Braams, J, *Intl. Biodet. Biodeg.*, **46**, 343-68, 2000.
2 Lopez-Arcea, P; Garcia-Guinea, J, *Building Envir.*, **40**, 929-41, 2005.
3 Dobson, J; Power, J M; Singh, J; Watkinson, S C, *Intl. Biodet. Biodeg.*, **31**, 129-41, 1993.
4 Gazzano, C; Favero-Longoa, S E; Matteucci, E; Roccardi, A; Piervittori, R, *Intl. Biodet Biodeg.*, **63**, 836-43, 2009.
5 Gutarowska, B; Piotrowska, M, *Building Envir.*, **42**, 1843-50, 2007.
6 Meklin, T; Reponen, T; Toivola, M; Koponen, V; Husman, T; Hyvarinen, A; Nevalainen, A, *Atmospheric Environ.*, **36**, 6031-39, 2002.
7 Ortega-Calve, J J; Arino, X; Hernandez-Marine, M; Saiz-Jimenez, C, *Sci. Total Environ.*, **167**, 329-41, 1995.
8 Griffin, P S; Indictor, N; Koestler, *Intl. Biodet.*, **28**, 187-207, 1991.

9 Fisher, M; Bar-Nes, G; Zeiri, Y; Sivan, A, *Biodeg.*, **19**, 321-28, 2008.
10 Haile, T; Nakhla, G; Allouche, E, *Corrosion Sci.*, **50**, 713-20, 2008.
11 Rogers, R D; Knight, J J; Cheeseman, C R; Wolfram, J H;. Idachaba, M; Nyavor, K; Egiebor, N O, *Cement Concrete Res.*, **33**, 2069-76, 2003.
12 De Muynck, W; Debrouwer, D; De Belie, N; Verstraete, W, *Cement Concrete Res.*, **38**, 1005-14, 2008.
13 Piunova, V; Berger, D; Neckers, D C; Bullerjahn, G S; McKay, R M; Fedorov, A V, *Photochem. Photobiol. Sci.*, **7**, 1565-69, 2008.
14 Grant, C; Wright, I C; Springle, W R; Greenhalgh, M, *Intl. Biodet. Biodeg.*, **32**, 279-88, 1993.
15 Nielse, K F; Hansen, M O; Larsen, T O; Thrane, U, *Intl. Biodet. Biodeg.*, **42**, 1-7, 1998.
16 Hyvarinen, A; Meklin, T; Vepsalainen, A; Nevalainen, A, *Intl. Biodet. Biodeg.*, **49**, 27-37, 2002.
17 Pasanen, A-L; Kasanen, J-P; Rautiala, S; Ikaheimo, M; Rantamak, J; Kaariainen, H; Kalliokoski, P, *Intl. Biodet. Biodeg.*, **46**, 117-27, 2000.
18 Vacher, S; Hernandez, C; Bartschi, C; Poussereau, N, *Building and Envir.*, in press 1-6, 2009.
19 Pasanen, A-L; Juutinen, T; Jantunen, M J; Kalliokoski, P, *Intl. Biodet. Biodeg.*, **30**, 273-83, 1992.
20 Klamer, M; Morsing, E; Husemoen, T, *Intl. Biodet. Biodeg.*, **54**, 277-82, 2004.
21 Shah, A A; Hasan, F; Hameed, A; Ahmed, S, *Biotechnol. Adv.*, **26**, 246-65, 2008.
22 Rusin, P; Bright, K; Grba, C, *Lett. Appl. Microbiol.*, **36**, 69-72, 2003.
23 Grant, D M; Bott, T R, *Heat Transfer Eng.*, **26**, 1, 44-50, 2005.
24 Ranaldi, G; Zanardini, E; Sorlini, C, Eds., **Encyclopedia of Microbiology**, *Elsevier*, 2009.
25 Batista, J F; Pereira, R F C; Lopes, J M; Carvalho, M F M; Feio, M J; Reis, M A M, *Biodeg.*, **11**, 441-48, 2000.
26 Vilcaez, J; Miyazawa, S; Suto, K; Inoue, C, *J. Japan Petroleum Institute*, **50**, 4, 208-17, 2007.
27 Gu, J-D; Ford, T E; Berke, N S; Mitchell, R, *Intl. Biodet. Biodeg.*, **41**, 101-9, 1998.
28 Nica, D; Davis, J L; Kirby, L; Zuo, G; Roberts, D J, *Intl. Biodet. Biodeg.*, **46**, 61-8, 2000.
29 Steele, A; Goddard, D T; Beech, I B, *Intl. Biodet. Biodeg.*, 35-46, 1994.
30 Reyes, A; Leteliera, M V; De la Iglesia, R; Gonzalez, B; Lagos, G, *Intl. Biodet. Biodeg.*, **61**, 135-41, 2008.
31 Davis, J L; Nica, D; Shields, K; Roberts, D J, *Intl. Biodet. Biodeg.*, **42**, 75-84, 1998.
32 Wypych, G, **PVC Formulary**, *ChemTec Publishing*, Toronto, 2009.
33 Gors, S; Schumann, R; Haubner, N; Karsten, U, *Intl. Biodet. Biodeg.*, **60**, 50-59, 2007.
34 Pasanen, P; Kalliokoski, P; Pasanen, A-L, *Building Environ.*, **32**, 3, 281-87, 1997.
35 Gutarowska, B; Zakowska, Z, *Intl. Biodet. Biodeg.*, **49**, 299-305, 2002.
36 Nielsen, K F; Thrane, U; Larsen, T O; Nielsen, P A; Gravesen, S, *Intl. Biodet. Biodeg.*, **42**, 9-16, 1998.
37 Gutarowska, B; Piotrowska, M, *Building Environ.*, **42**, 1843-50, 2007.
38 Gebefuegi, I; Parlar, H; Korte, F, *Ecotoxicol. Environ. Safety*, **3**, 269-300, 1979.

4.2 COATINGS AND PAINTS

This discussion does not include antifouling paints, which are discussed in Section 4.8 (Marine transport and installations). There are four groups of problems related to protection of paints and coatings. These include:

- plant hygiene (measures to prevent contamination on the production level)
- in-can preservation (protection of paints and coatings against spoilage during storage)
- film preservation (protection of paints and coatings after paint was applied)
- sanitation (maintenance of painted surface to prevent growth of microorganisms)

All the above aspects are discussed below.

4.2.1 RELEVANT MICROORGANISMS

Table 4.3 contains a list of microorganisms found to cause deterioration of coatings and paints.

Table 4.3 Microorganisms in coatings and paints.

Material	Microorganisms	Refs.
plant hygiene	microorganisms coming with raw materials, contamination during storage of raw materials, and contamination from production line (generally species which are credited with spoilage during storage)	2
in-can	*Alcaligenes, Achrobacter, Aspergilus, Bacillus, Candida albicans, Escherichia coli, Fusarium, Geortrichum, Graphium putredinis, Micrococcus luteus, Penicillium, Proteus vulgaris, Pseudomonas, Rhodotorula rubra, Saccharomyces cerevisiae, Serratia marcescens*	20
film preservation	*Achromobacter sp., Acinetobacter calcoaceticus, Acremonium, Actinomyces israelii, Alternaria alternata, Arthrobacter globiformis, Aspergillus candidus, Aspergillus fumigatus, Aspergilus sp., Aspergillus versicolor, Aureobasidium pollulans, Bacillus sp., Bacteroides clostridiformis, Bacteroides fragilis, Bacteroides hypermegas, Bacteroides ovatus, Bacteroides sp., Bifidobacterium adloscentis, Candida sp., Cephalosporium sp., Chaetomium globosum, Chlorococcum, Chlorella, Citrobacter freundii, Cladosporium cladosporioides, Clostridium butyricum, Clostridium sphenoides, Clostridiurn subterminale, Clostridium sp., Cryptococcus laurentii, Curvularia lunata, Curvularia senegalensis, Desulfovibrio desulfuricans, Enterobacter cloacae, Enterobacter sp., Epicoccum nigrum, Eubacterium aerofaciens, Fusarium solani, Geotrichum sp., Graphium putredinis, Helminthosporium, Klebsiella pneumoniae, Lactobacillus fermentum, Lyngbya, Monascus, Mucor, Nigrospora, Nostoc, Paramecium, Penicillium sp., Peptococcus saccharolyticus, Plectonema, Proteus rettgeri, Providencia alcalifaciens, Providencia rettgeri, Pseudomonas aeruginosa, Pseudomonas cepacia, Pseudomonas fluorescens, Pseudomonas maltophilia, Pseudomonas sp., Rhizopus, Rhodotorula rubra, Rhodotorula sp., Sacchoromyces sp., Serratia marcescens, Stachybotris chartarum, Streptococcus intermedius, Trentepohlia odoratae, Trichoderma, Tripospermum, Ulocladium cucurbitae*	1, 4, 12, 14, 15, 18, 20, 25-26
sanitation	*Aspergillus niger, Aspergillus flavus, Aspergillus ustus, Chaetomium globosum, Cladosporium herbarum, Myrothecium verrucaria, Paecilomyces variotii, Penicillium funiculosum, Stachybotrys atra, Trichoderma viride*	29-30

In the analysis of the causes of spoilage, it is important to consider that paint is analyzed with a delay which may be sufficient for microbial composition to be changed. The first invaders (real causes of problems) may not be available anymore because activity of microorganisms have changed conditions such as oxygen concentration, pH, presence of some nutrients. The first (and even some secondary) invaders may not survive these changes. In a freshly painted building in Brazil, yeast and *Cladosporium* were the first microorganisms detected a week after painting.[4] The yeast population had fallen to undetectable levels after the third week, and this microbial group was not detected again until the 31st week, after which they increased to high levels on the 42nd week.[4] *Aureobasidium* showed a pattern similar to the yeasts.[4] The fungal population on biocide-containing surfaces was significantly lower than on non-biocide-containing paint, but there was no statistically significant difference in the level of fungal biodiversity.[4]

Porosity and roughness of the materials are the parameters of influence on algae and cyanobacteria colonization.[10] The presence of microorganisms on facades depends on humidity. The porosity of the materials increases water absorption and retention. Also, high roughness makes the retention and adhesion of microorganisms to the surface easier.[10]

4.2.2 ESSENTIAL PRODUCT COMPONENTS

Cellulose ether polymers are used by the paint industry as thickening agents for waterborne paints. Reduction in their degree of polymerization results in the loss of paint viscosity (paint thinning).[9] Endoglucanases hydrolyze at random the 1,4-linkage in the cellulose ether chains. These enzymes cause a rapid drop in substrate viscosity because they lower degree of polymerization in the cellulose ether polymers and cause loss of paint viscosity.[9]

Many emulsion raw materials are susceptible to attack by bacteria, yeast, and/or fungi.[2] These include surfactants, wetting agents, defoamers, and thickeners.[2] Production of microbial cellulase may result in viscosity changes.[2]

4.2.3 MECHANISMS OF BIODEGRADATION AND BIODETERIORATION

The mechanism of degradation of a complex product such as paint or coating cannot be adequately described because too many factors play role. Most important for understanding the degradation sequences is to analyze the effect of microorganisms on components of the formulation. The main component – polymer – responsible for properties has too large a molecular weight to pass through cellular membranes, therefore, in order to be biodegraded, it first has to be depolymerized (or changed to smaller fragments). This is usually too difficult a task for microorganisms, therefore they usually deal with polymer fragments which were formed by other degradative processes (thermal energy, UV, chemical reactions, etc.). Microorganisms may contribute to depolymerization of some polymers by excreting substances (e.g., acids) which have the ability to affect certain polymers (e. g., hydrolysis reaction). Microorganisms may also excrete extracellular enzymes, causing erosion of polymer, but these processes are polymer-specific and naturally slow.

For this reason, additives, which are in most cases small molecules, are the most frequent subject of influence. Additives perform certain functions vital to coating performance, and because of their gradual degradation, this function is no longer supported.

Physical forces, such as heating/cooling, freezing/thawing, or wetting/drying, can cause mechanical damage, demonstrated by cracking or flacking of a polymeric film. Microorganism also play role in these processes because they contain water which can change phase. Also, the growth of many fungi can cause small-scale swelling and bursting, as the fungi penetrate the polymer solids.

4.2.4 RESULTS OF BIODETERIORATION

Table 4.4 shows the most important results of microbiological attack.

Table 4.4. The results of microbiological attack.

Changes observed	Responsible microorganisms	References
odor (malodor)	*Escherichia, Enterobackter, Klebsiella, Proteus, Pseudomonas*	19
gas formation	fermentative microorganisms, sulfate-reducing bacteria	2, 19
viscosity change		19
pH change	fermentative microorganisms, sulfate-reducing bacteria	2, 19
discoloration	*Chlorophytes* (green algae) and *Cyanophytes* (cyanobacteria, also named bluegreen algae), *Chlorella cf. mirabilis, Chroococcidiopsis fissurarum, Flavobacteria, Klebsormidium flaccidum, Rhodotorula rubencens, Serratia, Stichococcus bacillaris*	2, 10, 17-18, 27
disfigurement	*Aspergillus, Aureobasidium, Bacillus Sp., Cladosporium, Penicillium*	13, 22
blistering	*Escherichia coli, Staphylococcus aureus*	11
delamination	*Escherichia coli, Staphylococcus aureus*	11
toxin formation	*Aspergillus, Cladosporium, Fusarium, Paecilomyces, Stachybotrys chartarum*	28,30
loss of mechanical properties	*Aspergillus terreus, Corynebacterium, Enterobacter agglomerans, Pseudomonas aeruginosa, Pseudomonas fluorescens, Serratia rubidaea*	21, 24

4.2.5 BIOSTABILIZATION

Table 4.5 contains a list of the most suitable biocides used for protection or remediation condition of coatings and paints.

Table 4.5 Biocides used in coatings and paints

Material	Biocide	Refs.
plant hygiene	cleaning of reactors, intermediate and bulk storage tanks as well as filters, filter holders, pipelines and hoses with a biocidal solution (biocides similar to those used as in-can preservatives, but in the case of biofilm formation use of strong oxidizing biocides in combination with non-oxidizing biocides may be necessary)	2-3
in-can	1,2-benzisothiazolin-3-one, 2-brom-nitropropane-1,3-diol, 5-chloro-methylisothiazolin-3-one, chloroacetamide, 2-methylisothiazolin-3-one, silver chloride	4, 6-7
film preservation	carbendazim (N-octyl-2H-isothiazolin-3-one) (0.25 wt%), dichlorooctylisothiazoline, iodopropynyl butyl carbamate, N-(3,4-dichlorophenyl)-N,N-dimethyl urea (0.25 wt%), octylisothiazolinone, 2-tert-butylamino-4-ethylamino-6-methylthio-1,3,5-triazine (terbutryn), zinc dimethyldithiocarbamate, zinc pyrithione	4, 6-7

Table 4.5 Biocides used in coatings and paints

Material	Biocide	Refs.
sanitation	thiabendazole ((thiazolyl-4)-2-benzimidazole) (thermal fogging in libraries and archives, 5 ml/^{3}, 10% solution), triclosan	29

Plasters and paints of a single building complex in Switzerland contained 7 active ingredients (BIT, OIT, CMIT/MIT, Irgarol 1051, zinc pyrithione, zinc oxide, and a formaldehyde releasing compound). In most buildings, contents ranged between 0.03 and 0.6 g/ m^2 for each biocide.[5] It was possible to monitor leaching of these biocides and trace the source of biocide leaching by analysis of water samples.[5] Some of the most persistent biocides (terbutryn, diuron, carbendazim, irgarol, and mecoprop) were found in a catchment of about 30 ha.[5]

Analysis of roofing paints shows that the rain intensity and cumulative raining time after application are the dominant factors influencing the concentration of the biocide leaching into rain water.[7]

Wood coatings were found to penetrate only 250 μm into the wood and showed an inhomogeneous distribution.[6] Natural weathering experiments after three months indicate that the amount of biocide (3-iodo-2-propynyl butylcarbamate) decreased to only 30% of the original content.[6]

N-chloro-2,2,6,6-tetramethyl-4-piperidinyl methacrylate was synthesized and used to prepare water-based polymeric N-halamines by emulsion polymerization.[8] The addition of a small amount of the polymeric N-halamine latex emulsions into commercial water-based latex paints served as an antimicrobial additive, having potent antimicrobial activity against *Staphylococcus aureus, Escherichia coli, Candida tropicalis*, and *Stachybotrys chartarum* spore.[8] The antimicrobial functions of the new paints were long-lasting for more than 1 year under normal in-use conditions.[8]

The costs of registration of new biocides are often prohibitive and micro-encapsulation technologies are usually expensive. On the contrary, the addition of an inorganic carrier is relatively simple, cheap, and has no detrimental effect on mechanical properties.[23] The controlled release of biocide from porous inorganic frameworks based on silica is feasible. A dynamic equilibrium is established, which maintains sufficient concentration of biocide at the coating interface over an extended time period.[23] There is evidence that biocide housed in such frameworks has a longer effective activity because it is protected from the environmental degradation processes.[23]

4.2.6 FORMULATIONS

Table 4.6. Improved stability of rheological properties[31]

Component	Concentration, weight parts
Water	243.2 - 2 x
Thickener, HM-EHEC	x
Defoamer, Byk 022	5

Table 4.6. Improved stability of rheological properties[31]

Component	Concentration, weight parts
Dispersing agent, Tamol 731	6.5
Preservative, Canguard	1
Filler, Hydrocarb	110
Pigment, Kronos 2190	180
Binder, Vinamul 3650	454.3

HM-EHEC - hydrophobically modified ethyl(hydroxyethyl) cellulose

Table 4.7. High solids, semi-transparent, alkyd wood stain[32]

Component	Concentration, parts
Long oil alkyd (Synolac 6005W)	52.82
Pigment red (Sicoflush L2817)	4.63
Pigment yellow (Sicoflush L1916)	2.30
Rheological additive (Bentone 34)	0.60
Calcium drier (Nuodex Ca)	2.77
Cobalt drier (Nuodex Co)	0.37
Zirconium drier (Nuodex Zr)	0.30
Biocide (Preventol A5)	0.72
HALS (Tinuvin 292)	0.45
Anti-skin (Exkin 2)	0.20
Solvent (Varsol 40)	34.84

Table 4.8. Flat interior paint (composition providing biocide with synergist activity)[33]

Component	Concentration, parts
Deionized water	150
Propylene glycol	25.9
Biostat (Proxel GXL)	0.5
Pigment dispersant (Colloid 226/35)	8.0
Non-ionic surfactant (Tergitol NP-9)	2.2
Defoamer (Colloid 640)	3.5
Thickener (Polyphobe 102)	15.0
Synergistic biocide of invention (butylaminoethanol)	5.0
Titanium dioxide (Ti-Pure R-902)	250.0
Ground marble (Snowflake PE)	125.0
Nepheline syenite (Minex 4)	125.0

Table 4.8. Flat interior paint (composition providing biocide with synergist activity)[33]

Component	Concentration, parts
Disperse at high speed then let down with:	
Latex (UCAR Latex 379G)	419.8
Coalescent solvent (UCAR Filmer IBT)	16.8
Colloid 640	3.5
Adjust pH to 9.0 and viscosity to 90 KU with:	
Thickener (Polyphobe 102)	14.5
Synergistic biocide of invention (butylaminoethanol)	2.0
Deionized water	51.3

N-alkylethanolamine and N,N-dialkylethanolamine, the substituted ethanolamines, are used in combination with a biocide for the anaerobic shelf stabilization of a latex paint, uncured coating, uncured sealant, uncured adhesive or related product. This invention provides a novel means of using biocides more effectively. By combining typical biocides with the alkanolamines, one can obtain much better microbial control per unit of biocide than is obtainable without the alkanolamine.[33] The main biocide may be any compound effective in stabilizing shelf stored paints, coatings, sealants and adhesives. Typical biocides include isothiazolinones, benzoisothiazolinones, formaldehyde, triazines, gluteraldehyde, bronopol and phenolics. Preferably, the biocide is something other than a quaternary ammonium biocide.[33] The amine is added to the formulation along with the other components and the biocide prior to sealing for shelf storage. The synergist alkanolamines should be used at a level between 1,000 ppm to 10,000 ppm. The synergist alkanolamine may be volatile (e.g., N-isopropylethanolamine) so that it will evaporate from the product after the package is opened, or the synergist alkanolamine may be non-volatile (e.g., didodecylaminoethanol) so that it will remain in the product after the package is opened.

Table 4.9. Microencapsulated biocide[34]

Component	Concentration, parts
Water	389.6
Polyacrylate (Coatex BR 3, Dimed)	1.5
Gum arabic	0.6
Silicone defoamer (Aspumit AP, Thor GmbH)	0.3
Zinc pyrithione powder or 4,5-dichloro-2-octylisothiazolin-3-one	60.0
Concentrated hydrochloric acid	4.0
Formaldehyde-melamine resin 144.0 (Quecodur DMQ, Thor GmbH)	600.0

The invention pertains in particular to plasters and paints that are to be protected with biocides against attack by microorganisms.[34] The encapsulated biocide gives long protection to coatings exposed to the effects of damp or water. Biocide bonds to solid particles of a carrier material and it is released in a delayed manner.[34]

Particulate carrier for the biocide formulations[35]

A particulate carrier material is impregnated with a biocidal formulation and serves as a vehicle for introduction of the biocide into a liquid-based media, such as a surface coating or surface cleaning compositions, in order to allow controlled release of the biocide to combat bacterial, fungal, or algal growth for an extend period of time.[35] Inorganic carrier particles have biocide adsorbed within the pore system having a retention factor of at least 0.8. In order to secure appropriate retention of biocide, a preferred inorganic carrier particle has a pore surface area of 25 to 200 m^2/g.[35] The carrier particles suitable may be chosen from amorphous silicas, amorphous aluminas, pseudoboehmites (a form of microcrystalline aluminium hydroxide), Y-zeolites, or dealuminated Y-zeolites.[35] The amount of biocide added to the porous inorganic carrier particles depends on the biocide type and its effectiveness at various activity levels (it is usually, in the range of 5 to 25% of the weight of carrier).

Table 4.10. Interior, flat house paint[36]

Component	Concentration, vol%
Grind	
Water	12.00
Dispersant (Tamol 681)	2.50
Surfactant (Carbowet DC01)	0.24
Defoamer (Foamaster SA-3)	0.14
Propylene glycol	0.50
Preservative (Kathon LX)	0.19
Titanium dioxide (Ti-Pure R-706)	5.99
Nepheline syenite (Minex 4)	4.59
Calcium carbonate (Omyacarb 5)	2.22
Calcium carbonate (Omyacarb 12)	2.22
Diatomaceous earth (Diafil 525)	0.69
Letdown	
Water	9.50
Hollow sphere, polymeric opaque pigment (Ropaque Ultra)	5.50
Binder (Rhoplex EZ Clean 200)	43.00
Dispersant and neutralizing amine (AMP-95)	0.32
Propylene glycol	0.50
Defoamer (Foamaster SA-3)	0.21
Non-ionic rheology modifier (Acrysol RM-7)	1.00

Table 4.10. Interior, flat house paint[36]

Component	Concentration, vol%
Hydrophobic anionic thickener (Acrysol TT-935)	0.6
Water	8.10

Kathon LX is used as preservative

Table 4.11. Fast-dry, white traffic paint[37]

Component	Concentration, vol%
Center	
Binder (Fastrack 2706) (50%)	51.99
Dispersant (Tamol 901) (30%)	0.68
Surfactant (Surfynol CT-136)	0.32
Defoamer (Drew L-493)	0.28
Titanium dioxide (Ti-Pure R-900)	2.92
Calcium carbonate (Omyacarb-5)	33.69
Mix the above at a sufficient speed to pull a slight vortex (about 1000 to 1500 rpm) for about 15 minutes, then add:	
Methanol	4.50
Coalescent (Texanol)	2.90
Defoamer (Drew L-493)	0.49
Thickener (Natrosol 250HR) (2%)	0.84
Water	1.39

Table 4.12. Universal primer[38]

Component	Concentration, vol%
Grind	
Water	11.50
Dispersant (Tamol 165A)	1.58
Surfactant (Tergitol 15-S-40)	0.11
Defoamer (Drewplus L-475)	0.13
Preservative (Kathon LX) 1.5%	0.12
Thickener (Natrosol 250 HR)	0.05
Titanium dioxide (Ti-Pure R-706)	4.49
Extender (Minex 4)	6.20
Extender (Attagel 50)	0.10

Table 4.12. Universal primer[38]

Component	Concentration, vol%
Letdown	
Binder (Rhoplex PR-295)	58.00
Coalescent (Texanol)	0.56
Propylene glycol	3.25
Defoamer (Drewplus L-475)	0.50
Rheology modifier (Acrysol RM-2020 NPR)	1.75
Rheology modifier (Acrysol RM-8W)	0.39
Ammonia (28 Be)	0.20
Water	11.08

Mildewcides such as Rozone 2000 (DCOIT) or Rocima 40 (IPBS) are recommended

REFERENCES

1 Tothill, I E; Seal, K J, *Intl. Biodet. Biodeg.*, **31**, 241-54, 1993.
2 Gillatt, J, *Intl. Biodet.*, **26**, 205-16, 1990.
3 Ludensky, M, *Intl. Biodet. Biodeg.*, **51**, 255-63, 2003.
4 Shirakawa, M A; Gaylarde; S C; Gaylarde, P M; John, V; Gambale, W, *FEMS Microbiol. Ecol.*, **39**, 165-73, 2002.
5 Burkhardt, M; Kupper, T; Hean, S; Haag, R; Schmid, P; Kohler, M; Boller, M, *Water Sci. Technol.*, **56**, 12, 63–67, 2007.
6 Volkmer, T; Schwarze, W M R, *Holz Roh. Werkst.*, **66**, 181-89, 2008.
7 Jungnickel, C; Stock, F; Brandsch, T; Ranke, J, *Env. Sci. Pollut. Res.*, **15**, 3, 258-65, 2008.
8 Cao, Z; Sun, Y, *Appl. Mat. Interfaces*, **1**, 2, 494-504, 2009.
9 Tothill, I E; Seal, K J, *Intl. Biodet. Biodeg.*, **31**, 241-54, 1993.
10 Barberoussea, H; Ruota, B;Yepremian, C; Boulon, G, *Building Environ.*, **42**, 2555-61, 2007.
11 Zafar, F; Ashraf, A M; ahmad, S, *Reactive Functional Polym.*, **67**, 928-35, 2007.
12 Andersen, B; Hollensted, M, *Intl. J. Food Microbiol.*, **126**, 172-79, 2008.
13 Morton, L H G; Surman, S B, *Intl. Biodet. Biodeg.*, 203-21, 1994.
14 El-Sayed, A H M M; Mahmoud, W M; Davis, E D; Coughlin, R W, *Intl. Biodet. Biodeg.*, 69-79, 1996.
15 Saad, D S; Kinsey, G C; Kim, S; Gaylarde, C C, *Intl. Biodet. Biodeg.*, **54**, 990-103, 2004.
16 Howard, G T, *Intl. Biodet. Biodeg.*, **49**, 245-52, 2002.
17 Gaylarde, C C; Gaylarde, P M, *Intl. Biodet. Biodeg.*, **55**, 131-39, 2005.
18 Gaylarde, PM; Gaylarde, C C, *Intl. Biodet. Biodeg.*, **46**, 93-97, 2000.
19 Tothill, I E; Best, D J; Seal, KJ, *Intl. Biodet.*, 359-65, 1988.
20 Hyvarinen; A; Meklina, T; Vepsalainen, A; Nevalainen, A, *Intl. Biodet. Biodeg.*, 49, 27-37, 2002.
21 Kay, M J; Morton, L H G; Prince, E L, *Intl. Biodet.*, **27**, 205-22, 1991.
22 Blake, R C; Norton, W N; Howard, G T, *Intl. Biodet. Biodeg.*, **42**, 63-73, 1998.
23 Edge, M; Allen, N S; Turner, D; Robinson, J; Seal, K, *Prog. Org. Coat.*, **43**, 10-17, 2001.
24 Howard, G T; Blake, R C, *Intl. Biodet. Biodeg.*, **42**, 213-20, 1998.
25 Lim, G; Tan, T K; Toh, A, *Intl. Biodet.*, **25**, 27-37, 1989.
26 Andersen, B; Thrane Nissen, A, *Intl. Biodet. Biodeg.*, **46**, 111-16, 2000.
27 Wee, Y C, *Intl. Biodet.*, **24**, 367-71, 1988.
28 Hossain, M A; Ahmed, M S; Ghannoum, M A, *J. Allergy Clin. Immunol.*, **113**, 200-8, 2004.
29 Rakotonirainy, M S; Fohrer, F; Flieder, F, *Intl. Biodet. Biodeg.*, **44**, 133-39, 1999.
30 Anearn, D G; Price, D L; Simmons, R; Noble-Wang, J; Crow, S A, *Adv. Appl. Microbiol.*, **55**, 113-38, 2004.
31 Karlson, L; Joabsson, F; Thuresson, K, *Carbohydrate Polym.*, **41**, 25-35, 2000.
32 Bulian, F; Graystone, J A, **Wood Coatings**, *Elsevier*, 2009, p. 137-54.
33 Gernon, M D, Taminco, Belgium, **US Patent**, 7,553,882, June 30, 2009.

34 Baum, R; Antoni-Zimmermann, D; Wunder, T; Schmidt, H-J, Thor GmbH, Germany, **US Patent**, 7,429,392, September 30, 2008.
35 Aldcroft, D; Jones, H; Turner, D; Edge, M; Robinson, J; Seal, K, Ineos Silicas Ltd., GB, **US Patent**, 6,905,698, June 14, 2005.
36 Rhoplex EZ Clean 200, Product brochure, Rohm & Haas/DOW, 2009.
37 Fastrack 2706, Product brochure, Rohm & Haas/DOW, 2007.
38 Rhoplex PR-295. Avanse technology emulsion polymer for aqueous stain blocking primers. Product brochure, Rohm & Haas/DOW, 2009.

4.3 CULTURAL HERITAGE

Stones and wood are discussed in the separate sections of this chapter.

4.3.1 RELEVANT MICROORGANISMS

Table 4.13 contains a list of microorganisms found to cause deterioration of cultural heritage objects.

Table 4.13. Microorganisms contaminating in cultural heritage objects.

Material	Microorganisms	Refs.
fresco (mural)	*Acremonium charticola, Acremonium strictum, Acremonium kiliense, Acremonium spp., Apatococcus lobatus, Aspergillus sydowii, Aureobasidium pullulans, Beauveria sp., Chlorella vulgaris, Chlorococcum sp., Chlorogloea microcystoides, Chroococcus lithophilus, Chrysosporium sp., Cladosporium sp., Cladosporium sphaerospermum, Engyodontium album, Gloeocapsa spp., Gloeothece rupestris, Muriella terrestris, Mycelia sterilia, Penicillium crysogenum, Pseudocapsa dubia, Rubrobacter radiotolerans, Scopulariopsis brevicaulus, Verticillium lecanii, Verticillium suchlasporium, Verticillium sp.*	5, 9, 15
glass	*Arthrobacter agilis, Arthrobacter crystallopoietes, Arthrobacter sp., Bacillus firmus, Bacillus licheniformis, Bacillus megaterium, Bacillus mojavensis, Bacillus pumilus, Bacillus simplex, Bacillus thuringiensis, Bacillus vallismortis, Brevundimonas subvibrioides, Brevundimonas variabilis, Cladosporium spp., Flexibacter sp., Frankia sp., Geodermatophilus sp., Leucobacter albus, Leucobacter komagatae, Micrococcus sp., Nitrosospira sp., Paenibacillus amylolyticus, Paenibacillus kribbensis, Paenibacillus pabuli, Paenibacillus polymyxa, Stenotrophomonas maltophilia, Streptomyces sp.*	13, 18, 19
leather	*Acremonium, Aspergillus flavus, Aspergillus luchuensis, Aspergillus niger, Aspergillus sulphureus, Chrysosporium evolceanui, Chrysosporium indicum, Chrysosporium keratinophilum, Chrysosporium sp., Chrysosporium tropicum, Fusarium, Penicillium citrinum, Penicillium chrysogenurn*	12
parchment	*Alternaria alternata, Aspergillus nidulans, Bacillus, Botryotrichum, Chaetomium globosum, Chromelosporium carneum, Cladosporium cladosporioides, Fusarium, Micrococcus, Mucor, Nocardia, Ophiostoma, Penicillium chrysogenum, Scopulariopsis, Serratia, Sporendonema, Streptomyces, Trichoderma*	3, 11

Table 4.13. Microorganisms contaminating in cultural heritage objects.

Material	Microorganisms	Refs.
painting	*Alternaria alternata, Aspergillus flavus, Aspergillus glaucus, Aspergillus niger, Aspergillus oryzae, Aspergillus penicilloides, Aspergillus versicolor, Cladosporium cladosporioides, Cladosporium sphaerospermum, Fusarium moniliforme, Fusarium sp., Pencillium chrysogenum, Penicillium frequentans, Penicillium funiculosum, Penicillium notatum, Penicillium purpurgenum, Penicillium sp., Pseudomonas spp., Ryizopus nigicans, Saccharomyces sp., Trichoderma harzianum, Trichoderma viride, Ulocladium atrum*	21-22
varnish	*Arthrobacter oxydans, Aureobasidium pulluluns, Bacillus amyloliquefaciens, Chrysonilia sitophila, Streptomyces celluloflavus*	7
paper	*Arthrinium urticae, Aspergillus carneus, Aspergillus flavus, Aspergillus nidulans, Aspergillus niger, Aspergillus versicolor, Aureobasidium pullulans, Botryotinia fuckeliana, Chaetomium globosum, Cladosporium cladosporioides, Cladosporium herbarum, Cunninghamella elegans, Debaryomyces hansenii, Doratomyces stemonitis, Epicoccum nigrum, Eurotium pseudoglaucum, Geomyces pannorum, Geosmithia putterilli, Mucor, Paecilomyces variotii, Penicillium brevicompactum, Penicillium canescens, Penicillium chrysogenum, Penicillium citrinum, Penicillium pinophilum, Penicillium restrictum, Penicillium spinulosum, Penicillium turbatum, Penicillium variabile, Peziza ostracoderma, Phlebiopsis gigantea, Phoma pomorum, Rhizopus oryzae, Thanatephorus cucumeris, Trichoderma pseudokoningii, Ulocladium botrytis*	6, 8, 11, 14, 16
papyrus	*Botryodiplodia theobromae, Chaetomium elatum, Chaetomium globosum, Chaetomium ochraceum, Emericellopsis minima*	3
photographic material	*Alternaria alternata, Aspergillus nidulans, Aspergillus ustus, Aspergillus versicolor, Bacillus amyloliquefaciens, Bacillus megaterium, Bacillus pichinotyi, Bacillus pumilus, Bacillus subtilis, Cladosporium cladosporioides, Cryptococcus albidus, Kocuria kristinae, Mucor racemosus, Pasteurella haemolytica, Penicillium chrysogenums, Phoma glomerata, Sphingomonas paucimobilis, Staphylococcus epidermidis, Staphylococcus haemolyticus, Staphylococcus hominis, Staphylococcus lentus, Staphylococcus lugdunensis, Trichoderma longibrachiatum*	2, 4
textiles	*Chaetomium sp., Cytophaga, Sorangium, Sporocytophaga, Stachybotris atra, Vibrio*	1, 3

Microbial colonization of the glass surface was heterogeneous at 0.8–7% areal coverage. A typical biofilm of the studied glasses displayed a total thickness of approximately 10–60 μm.[18] Information about the thickness and microbial composition of biofilms offer an essential background to optimize cleaning procedures or conservation strategies for stained glass windows.[18]

4.3.2 ESSENTIAL PRODUCT COMPONENTS

Brussels tapestries were subjected to conservation efforts, especially because they presented a high degree of deterioration (color changes and material loss). It was discovered that *Pyrrhalta hateola* beetles were responsible for excessive damage.[1] The beetles did not consume textile materials, and had no direct effects in tapestry destruction. Their presence indirectly affected the tapestry due to the higher mold-growth on samples with beetle remains.[1] This underlines complexity of problems and difficulties in finding solutions.

In cinematographic films, the organic components are potential carbon sources for growth of microorganisms, if environmental conditions permit.[2] Gelatin is the most biodegradable of all substrates used in past. Many prokaryotic and eukaryotic microorganisms exhibit proteolytic capacity.[2] Cellulose acetate biodegradation is mediated by the cooperative action of esterase(s), lipase(s) and cellulase(s).[2]

Colophony and Venetian turpentine are natural products that have been widely used since antiquity as protective varnishes and binding agents in paintings as well as polychromed sculptures.[7] The main components of resins (~90%) are abietane and pimarane tricyclic diterpenes.[7]

Hemicellulose-sized paper, as well as lignin-rich wood pulp, cannot be easily attacked by common, airborne fungi, whereas gelatin sizing facilitates the growth of fungi.[8]

Chemical composition and durability of the glass samples and availability of an additional organic layer are important factors influencing the extent of microbial growth.[18]

4.3.3 MECHANISMS OF BIODEGRADATION AND BIODETERIORATION

The oxidative processes in colophony caused by certain fungi and bacteria from collections occur mainly at positions C7 and C15 as well as dehydroabietic acid dehydrogenation.[7] GC-MS studies confirm this mechanisms of biodegradation.[7]

4.3.4 RESULTS OF BIODETERIORATION

In mural paintings in the Crypt of the Original Sin, biological colonization caused alteration of colors and rosy discoloration.[5]

Aging processes of varnishes constitute fundamental obstacles in preserving pictorial and sculptural works of art.[7] Resins change in composition and properties over time, these transformations take the form of yellowing, crackling, microfractures and changes in solubility.[7]

Fungi are serious degrading agents of bibliographic documents, particularly cellulolytic fungi. The change in the document aesthetics, either by discoloration by weak acids produced by fungi, or by the accumulation of pigments that may stain its support in a phenomenon referred to as foxing are major concerns related to fungal colonization.[11] Some foxing stains occur as reddish and/or brownish marks on the paper support, and are thought to have both biotic and abiotic origins, from the metabolic activity of microorganisms, to metal or ink oxidation.[11]

4.3.5 BIOSTABILIZATION

Table 4.14 contains a list of the most suitable biocides used for protection or remediation of condition of cultural heritage.

Table 4.14. Biocides used in treatment of cultural heritage objects.

Component	Biocide/Protection or remediation method	Refs.
fresco (mural)	alkyl-benzyl-dimethyl ammonium chloride in isopropyl alcohol (working solution – 2% in distilled water) (Preventol R80, Lanxess),[5] p-chloro-m-cresol (0.3%)[20]	5, 20
glass	not reported	
leather	radiation[10]	10
painting	2-(4 thiazolyl)-benezimidazal, pentachlorophenol, Vinyzene (220 ppm added to varnish solution)	22
paper	3-aminopropylmethyldimethoxysilane (6% uptake),[14] radiation (doses ranging from 14.5 to 25 kGy)[10,17]	10, 14, 17
photographic material	presence of metallic silver on the emulsion (black films) produces a biocide effect inhibiting the microbial growth,[23] formaldehyde and 4-chloro-3-cresol salt[24]	23-24
textiles	o-phenylphenol (silk) and p-chloro-m-cresol (cotton and canvas),[25] radiation[10]	10, 25

Restoration of mural paintings required use of biocide.[5] Several biocides were tested and alkyl-benzyl-dimethyl ammonium chloride was selected.[5] The biocide solution was applied three times at one-week intervals to permit a deeper penetration into the stone; after disinfection, biocide removal and cleaning by water were performed.[5]

Studies have shown that substrates of medieval paintings are very stable because they contain high concentrations of heavy metals such as lead, copper and mercury as well as organic substances of animal origin, such as egg yolk rich in lipids or plant origin such as fig latex containing fatty acids and fig leaves containing powerful triterpenoids which are persistent irritants.[9] High concentration of copper reduced the biodiversity of microorganisms.[9]

The γ-irradiation disinfestation treatment of the Ramses II mummy was performed by the NucleArt Laboratory, Grenoble, France.[10] Other cases of radiation treatment include documents damage by flood in the Alan Mason Chesney Medical Archives, Boston, USA, celluloid films infested by fungi in Romania, and prisoners shoes in the State Museum, Majdanek, Poland.[10] Five thousand wooden sculptures, parts of altars, pieces of furniture, musical instruments, and other wooden, paper, textile, and leather items were treated in special facility in Croatia.[10]

A new process using aminoalkylalkoxysilane was developed recently for paper and book mass deacidification.[14] Besides deacidifying, it also improves the mechanical properties of paper, and aminoalkyloxysilane acts as fungicide.[14]

REFERENCES

1 Sanchez-Pinero, F; Bolivar, F C, *Intl. Biodet. Biodeg.*, **54**, 297-302, 2004.
2 Abrusci, C; Martin-Gonzalez, A; Del Amo, A; Catalina, F; Collado, J; Platas, G, *Intl. Biodet. Biodeg.*, **56**,

58-68, 2005.
3 Ranalli, G; Zanardini, E; Sorlini, C, **Encyclopedia of Microbiology**, *Elsevier*, 2009, pgs. 191-205.
4 Abrusci, C; Marquina, D; Del Amo, A; Corrales, T; Catalina, F, *Intl. Biodet. Biodeg.*, **58**, 142-49, 2006.
5 Nugari, M P; Pietrini, A M; Caneva, G; Imperi, F; Visca, P, *Intl. Biodet. Biodeg.*, **63**, 705-11, 2009.
6 Michaelsen, A; Pinar, G; Montanari, M; Pinzari, F, *Intl. Biodet. Biodeg.*, **63**, 161-68, 2009.
7 Romero-Noguera, J; Bolivar-Galiano, F C; Ramos-Lopez, J M; Fernandez-Vivas, M A, Martin-Sanchez, I, *Intl. Biodet. Biodeg.*, **62**, 427-33, 2008.
8 Zotti, M; Ferroni, A; Calvini, P, *Intl. Biodet. Biodeg.*, **62**, 186-94, 2008.
9 Milanesi, C; Baldi, F; Vignani, R; Ciampolini, F; Faleri, C; Cresti, M, *Intl. Biodet. Biodeg.*, **57**, 7-13, 2006.
10 Katusin-Razem, B; Razem, D; Braun, M, *Radiat. Phys. Chem.*, **78**, 729-31, 2009.
11 Mesquita, N; Portugal, A; Videira, S; Rodriguez-Echeverria, S; Bandeira, A M L; Santos, M J A; Freitas, H, *Intl. Biodet. Biodeg.*, **63**, 626-29, 2009.
12 Nigam, N; Dhawan, S; Nair, M V, *Intl. Biodet. Biodeg.*, **33**, 145-52, 1994.
13 Roelleke, S; Gurtner, C; Drewello, U; Drewello, R; Lubitz, W; Weissmann, R, *J. Microbiol. Meth.*, **36**, 107-14, 1999.
14 Rakotonirainy, M S; Dupont, A-L; Lavedrine, B; Ipert, S; Cheradame, H, *J. Cultural Heritage*, **9**, 54-59, 2008.
15 Gorbushina, A A; Petersen, K, *Intl. Biodet. Biodeg.*, **46**, 277-84, 2000.
16 Montemartini Cortea, A; Ferroni, A; Salvo, V S, *Intl. Biodet. Biodeg.*, **51**, 167-73, 2003.
17 da Silva, M; Moraes, A M L; Nishikawa, M M; Gatti, M J A; Vallim de Alencar, M A; Brandao, L E; Nobrega, A, *Intl. Biodet. Biodeg.*, **57**, 163-67, 2006.
18 Müller, E; Drewello, U; Drewello, R; Weißmann, R; Wuertz, S, *J. Cultural Heritage*, **2**, 31-42, 2001.
19 Marvasi, M; Vedovato, E; Balsamo, C; Macherelli, A; Dei, L; Mastromei, G; Perito, B, *J. Cultural Heritage*, **10**, 124-33, 2009.
20 Garg, K L; Jain, K K; Mishra, A K, *Sci. Total Environ.*, **167**, 255-71, 1995.
21 Seves, A M; Sora, S; Ciferri, O, *Intl. Biodet. Biodeg.*, 215-24, 1996.
22 Inoue, M; Koyano, M, *Intl. Biodet.*, **28**, 23-35, 1991.
23 Abrusci, C; Marquina, D; Del Amo, A; Catalina, F, *Intl. Biodet. Biodeg.*, **60**, 137-43, 2007.
24 Lourenço, M J L; Sampaio, J P, *Intl. Biodet. Biodeg.*, **63**, 496-502, 2009.
25 Nugari, M P; Priori, G F; Mate, G; Scala, F, *Intl. Biodet.*, **23**, 295-306, 1987.

4.4 DENTAL MATERIALS

4.4.1 RELEVANT MICROORGANISMS

Table 4.15 contains a list of microorganisms found to cause deterioration of dental materials or contaminating their surfaces, by which the health of users of these materials can be endangered.

Table 4.15. Microorganisms causing deterioration of dental materials or contamination of their surfaces.

Material	Microorganisms	Refs.
dentures	*Candida albicans*	3
implants	*Bacillus stearothermophilus, Staphylococcus aureus, Staphylococcus epidermidis*	6, 12
impression material	*Staphylococcus aureus*	4
oral care	*Actinomyces viscosus, Azospirillum brasilense, Bacillus brevis, Streptococcus mutans, Streptococcus rattus, Streptococcus salivarius, Streptococcus sanguis, Streptococcus sobrinus*	10, 12

Table 4.15. Microorganisms causing deterioration of dental materials or contamination of their surfaces.

Material	Microorganisms	Refs.
restoration composites	*Actinomyces naeslundii, Fusobacterium nucleatum, Lactobacillus acidophilus, Lactobacillus rhamnosus, Neisseria subflava, Prevotella nigrescens, Streptococcus mutans, Streptococcus oralis, Streptococcus sanguis, Streptococcus sobrinus, Veillonella dispar*	5, 13
root canal core material	*Actinomyces strains, Candida albicans, Enterococcus faecalis, Porphyromonas endodontalis, Pseudomonas aeruginosa, Staphylococcus aureus*	7, 14
water systems	*Moraxella spp., Mycobacterium spp., Flavobacterium spp., Legionella pneumophila, Pseudomonas aeruginosa*	8

It has been estimated that dental biofilms contain >500 different bacterial taxa.[2] Plaque is colony of microorganisms protected by a biofilm. More plaque can adhere to dentures than to normal teeth because the surface area of the denture is substantially larger.[3] Denture stomatitis is a common problem with denture wearers.[3]

A single bacterium may have a diameter of 1 μm (typically between 0.1 to 50 μm) but oral biofilms can be more than millimeter thick.[12] The bacteria are organized in stacks, permeated by water cavities.[12]

The water used in dental unit water systems acts as a coolant for the high-speed drills and as an irrigant during dental procedures.[8] A wide variety of bacteria, protozoa, and fungi have been recovered from dental water systems. This includes the presence of opportunist pathogens which may endanger both patients and dentists.[8]

4.4.2 BIOSTABILIZATION

Table 4.16 contains a list of the most suitable biocides used for protection of dental materials.

Table 4.16. Biocides used in dental materials.

Material	Biocide	Refs.
adhesive patches	cetylpyridinium chloride	1
dentures	apatite-coated titanium dioxide (5% in acrylic resin)	3
implants	γ-irradiation,[6] organic ligands of silver,[6] silver-antibiotic compounds[6]	6
impression material	sodium N-chlorobenzenesulfonamide or sodium N-chloro-p-toluenesulfonamide[4]	4
oral care	cetylpyridinium chloride,[1] chlorhexidine,[12] parabens (cariogenic bacterium),[10] silver-containing zeolite (chewing gum)[18]	1, 10, 12, 18
restoration composites	silver-zinc glass powder (3,0%; Irgaguard B7000),[17] silver-zinc zeolite (0.4%; Irgaguard B5000)[17]	17

Table 4.16. Biocides used in dental materials.

Material	Biocide	Refs.
root canal core material	chlorhexidine (5.25%, disinfection of filling material),[15] glutaraldehyde (root canal decontamination),[9] gutta-percha,[7] Resilon (polyester containing bioactive glass, bismuth oxychloride, and Ba sulfate),[7] sodium hypochlorite (2%, disinfection of filling material)[15]	7, 9, 15
water systems	filtration,[11] hydrogen peroxide,[11] hydroxybenzoin acid ester,[11] iodine,[11] ozone,[11] peracetic acid,[11] silver ions,[11,16] sodium hypochlorite,[11] UV[11]	8, 11, 16

Titanium dioxide in the presence of UV light catalyzes degradation of organic chemical materials, especially its anatase grades.[3] This is frequently used in antibacterial activities such as water treatment or purification of air.[3] It can also be used for purification of dentures. The technological problem exists because titanium dioxide is also able to degrade acrylic resin, which is a binder in dentures. But if the surface of titanium dioxide is coated with apatite, it does not affect properties of acrylic resin but is still able to perform its antifungal functions.[3]

4.4.3 FORMULATIONS

Erodible adhesive patches

The controlled release of cetylpyridinium chloride has been applied in dental products in which low levels (less than 10 mg) of cetylpyridinium chloride were loaded into a polymer matrix. Erodible adhesive patches, resins, and varnishes containing cetylpyridinium chloride have been used for the treatment of plaque and gingivitis.[1] Adhesive patches were made from polyvinylalcohol, hydroxyethylcellulose, and chitosan.[1] Over 95% of cetylpyridinium chloride was released from polyvinylalcohol-based patches after 7 h, only 50% of cetylpyridinium chloride was released from hydroxyethylcellulose-based patches, and only a small amount from chitosan-based patches.[1] Hydrogels based on 2-hydroxyethyl methacrylate constitute another vehicle which can be used in this application.[1]

Table 4.17. Composition for water treatment in dental equipment.[16]

Component	Concentration, wt%
sodium percarbonate	8.19
sorbitol	10.00
sodium bicarbonate	24.98
tetrasodium EDTA dihydrate	9.42
citric acid	40.02
quaternary ammonium compound mixture	4.25
silver nitrate	0.14
polyethylene glycol	1.00
sodium benzoate	2.00

REFERENCES

1 Densmore, C G; Robison, T W; Smith, Lewis, R E, *J. Appl. Polym. Sci.*, **99**, 3153-62, 2006.
2 Mah, T F C; O'Toole, G A, *Trends Microbiol.*, **9**, 1, 34-39, 2001.
3 Shibata, T; Hamada, N; Kimito, K; Sawada, T; Sawada, T; Kumada, H; Umemoto, T; Toyoda, M, *Dental Mater. J.*, **26**, 3, 437-44, 2007.
4 Pusineri, C; Del Torto, M, **US Patent**, 6,559,199, Rhodia Chimie, May 6, 2003.
5 Finer,Y; Santerre, J P, *J. Biomed. Mater. Res.*, **69A**, 233-46, 2004.
6 Brunetto, P S; Fromm, K M, *Chimia*, **62**, 249-52, 2008.
7 Bodrumlu, E; Alaçam, T, *J. Am. Dent. Assoc.*, **138**,1228-32, 2007.
8 Walker, J T; Marsh, P D, *Intl. Biodet. Biodeg.*, **54**, 87-98, 2004.
9 Power, E G M, *Prog. Med. Chem.*, **34**, 150-201, 1997.
10 Soni, M G; Taylor, S L; Greenberg, N A; Burdock, G A, *Food Chem. Toxic.*, **40**, 1335-73, 2002.
11 Walker, J T; Marsh, P D, *J. Dentistry*, 721-30.
12 Jones, F H, *Surface Sci. Rep.*, **42**, 75-205, 2001.
13 Bradshaw, D J; Walker, J T; Burger, B; Gangnus, B; Marsh, P D, *Meth. Enzymology*, **337**, 416-25, 2001.
14 Tay, F R; Pashley, D H; Williams, M C; Raina, R; Loushine, R J; Weller, R N; Kimbrough, W F; King, N M, *JOE*, **31**, 8, 593-98, 2005.
15 Royal, M J; Williamson, A E; Drake, D R, *J. Endodontics*, **33**, 1, 42-44, 2007.
16 Zawada, J A; Moore, C L D, **US Patent**, 7,033,511, A-DEC, Inc., April 25, 2006.
17 Qian, X, **US Patent**, 6,924,325, Kerr Corp., August 2, 2005.
18 Barry, J E; Trogolo, J A, **US Patent**, 6,365,130, AgION Technologies, April 2, 2002.

4.5 ELECTRICAL AND ELECTRONIC PRODUCTS

4.5.1 RELEVANT MICROORGANISMS

Table 4.18 contains a list of microorganisms found to cause deterioration of electronic and/or electrical equipment.

Table 4.18. Microorganisms causing deterioration of electronic and/or electrical equipment.

Material	Microorganisms	Refs.
cable	*Alcaligenes denitrificans, Aspergillus, Bacillus cereus, Bacillus circulans, Bacillus subtilis, Ochrobactrum anthropi, Penicillium, Pseudomonas aeruginosa, Pseudomonas veronii, Trichoderma, Vibrio harveyi, Xanthomonas maltophila*	2, 7, 11
metal (corrosion)	*Desulfovibrio desulfuricans*	1
paper insulation	*Cladosporium, Penicillium spp.*	13
printed circuit	*Aspergillus flavus, Aspergilfus niger, Aspergillus versicolor, Chaetomium globosum, Gliocadium virens, Penicillium pinophilum*	10
transformer tank	*Mycobacterium, Pseudomonas*	13

4.5.2 MECHANISMS OF BIODEGRADATION AND BIODETERIORATION

Microorganisms may initiate or accelerate corrosion reactions by creating differences in electric potential by stimulating either the anodic or cathodic reactions at the metal surface, due to the products of their metabolic activities.[1] *D. desulfuricans* forms a biofilm that may cause some mass transfer resistance to a number of substances and may establish an electrochemical cell.[1] The activity of bacteria results in a deposition of ferric sulfates. Sulfate-reducing bacteria may alter the local environment in several ways: diminution of pH and reducing thiosulfate to sulfide.[1]

Surface electric resistance of insulation depends on properties of the material used as insulator and the thickness and composition of thin water layer on its surface. The last can be affected by microorganisms is several ways:[13]

- enzymatic oxidation (formation of hydrophylic groups on the surface)
- degradation of surface which is capable to retain more moisture
- conductivity of mycelium
- deposition of conductive substances (e.g., acid or salts)

Internal electric resistance deteriorates by similar actions, with the exception that the internal structure of the insulator is compromised by microorganisms.[13]

4.5.3 RESULTS OF BIODETERIORATION

The dielectric properties of polyimides can be altered following the growth of a microbial biofilm.[9] Initial deterioration may be slow under ambient conditions; however, the deterioration processes can be accelerated in humid conditions or in enclosed environments, such as submarines, space vehicles, airplains, and other closed facilities, which are places where polyimides are used as electrical insulation materials.[9]

Biofilms consist mainly of water (up to 95%) and, thus, provide an electrolyte that increases surface conductivity.[12] This has led to unexpected failures of electrical equipment.[12] Fungal growth on printed circuits affects the surface conductivity and causes short circuits; mold growth on computer motherboards has been reported, resulting in a failure of computer, radio and radar equipment in the tropics.[12]

4.5.4 BIOSTABILIZATION

Table 4.19 contains a list of the most suitable biocides used for protection or remediation of the condition of electronic or electrical equipment.

Table 4.19. Biocide used in electronic or electrical equipment and parts.

Material	Biocide	Refs.
cable	10, 10-oxybisphenoxyarsine[2]	2

Some bacteria may produce fluorescence, the intensity of which correlates with their concentration. *Pseudomonas fluorescence* (presently known as *Pseudomonas veronii*) is one such microorganism and it is also present on PVC articles which contain plasticizers, such as cable and wire.[2] The measurement of bioluminescence can be conveniently used to determine the effect of biocide action in a non-distractive, continuous method of monitoring.[2]

Less known is that the electric current can be used for prevention of biodegradation. One is an electrical disinfection method.[3] The surfaces of bacterial cells are usually negatively charged, therefore, a positive electrode draws the cells on its surface and is able to kill them by electric current at a low electric potential of around 1-1.5 V.[3] Electric current can also be used for the prevention of bacterial cell accumulation on the surface by applying alternating potentials.[3] This alternating potential of 1.2 and -0.6 V was applied to a fishing net made from nylon yarn, and the surface of the nylon yarn was coated with an electroconductive paint approximately 50 μm thick, which was used as a working electrode.[3] No increase in weight caused by attachment of biofouling organisms to the net was observed for 158 days, which was the duration of observation.[3]

In studies of various bacteria, it was determined that they differ in their electrokinetic potential which can be then used as described in the previous paragraph to eliminate them, but the value of their electrokinetic potential also determines adhesion forces of microorganisms to any given surface, depending on its electric properties.[4]

Electric charge can also be applied to deposit biocide on the surface. A stainless steel surface was protected by silver zeolite which was deposited by the powder coating method, attached to the surface by electric forces in its dry form, and then heated to melt, forming an active coating.[5]

The most important effect of electric charges is in the explanation of reasons for biocorrosion of metals. The so-called cathodic depolarization theory of biocorrosion was proposed in 1934 by Kuer and van der Vlugt, which was developed into the current understanding of biocorrosive processes.[6]

REFERENCES

1 Raman, V; Tamilselvi, S; Rajendran, N, *Mater. Corrosion*, **59**, 4, 329-34, 2008.
2 Jones, C; Robson, G D; Greenhaulgh, M; Eastwood, I; Handley, P S, *Biofouling*, **18**, 1, 21-27, 2002.
3 Omae, I, *Chem. Rev.*, **103**, 3431-48, 2003.
4 Kopteva, Z P; Zanina, V V; Kozlova, I A, *Surf. Eng.*, **20**, 4, 275-80, 2004.
5 Cowan, M M; Abshire, K Z; Houk, S L; Evans, S M, *J. Ind. Microbiol. Biotechnol.*, **30**, 102-6, 2003.
6 Iverson, W P; *Adv. Appl. Microbiol.*, **32**, 1-36, 1987.
7 Pintado, J L; Montero, F, *Intl. Biodet. Biodeg.*, **29**, 357-65, 1992.
8 Wainwright, M; Ali, T A; Barakah, F, *Intl. Biodet. Biodeg.*, **31**, 1-13, 1993.
9 Gu, J-D, *Intl. Biodet. Biodeg.*, **52**, 69-91, 2003.
10 Alava, J I; Oritz de Urbina, G; Solozabal, R; Valero, J M, Mater. Design, **17**, 1, 19-21, 1996.
11 Gu, J-D; Roman, M; Esselman, T; Mitchell, R, *Intl. Biodet. Biodeg.*, **41**, 25-33, 1998.
12 Flemming, H-C, *Polym. Deg. Stab.*, **59**, 309-15, 1998.
13 Wasserbauer, R, *Intl. Biodet. Biodeg.*, **53**, 171-76, 2004.

4.6 FIBERS AND TEXTILES

Some relevant information will be included in Section 4.14 of this chapter in order to avoid repeating the same information. This regards information on product components, mechanisms of biodegradation and results of biodegradation of synthetic polymers.

4.6.1 RELEVANT MICROORGANISMS

Table 4.20 contains a list of microorganisms found to cause deterioration of fibers and textiles.

Table 4.20. Microorganisms in deterioration of fibers and textiles.

Material	Microorganisms	Refs.
aromatic PA (Nomex)	*Escherichia coli, Staphylococcus aureus*	3
cotton	*Aspergillus fumigatus, Chaetomium funicola, Chaetomium globosum, Chaetomium indicum, Curvularia lunate, Escherichia coli, Fusarium lateritium, Fusarium sp., Memnoniella echinata, Myrothecium verrucaria, Sordaria fimicola, Staphylococcus aureus, Trichoderma viride*	5, 9

Table 4.20. Microorganisms in deterioration of fibers and textiles.

Material	Microorganisms	Refs.
flax (linen)	*Aspergillus terreus, Cladosporium herbarum, Fusarium culmorum, Penicillium funiculosum, Phoma sp., Trichoderma sp., Trichoderma viride*	20
polyamide-6	*Aspergillus puniceus, Aureobasidium pullulans, Bacillus mycoides, Bjerkandera adusta, Candida albicans, Escherichia coli, Mucor hiemalis, Penicillium sp., Phanerochaete chrysosporium, Trichotecium roseumUlocladium sp.*	10, 17
polypropylene	*Alternaria solani, Fusarium oxysporum lycopersici, Verticillium albo-atrum*	6
silk	*Streptomyces, Variovorax paradoxus*	18
wool	*Aspergillus* and *Penicillium* (29 species each*), Chrysosporium* (21 species), *Fusarium* (13 species), *Trichophyton, Microsporum,* and *Acremonium* (11 species each), *Mucor* (10 species), *Arthroderma* (8 species), *Malbranchea* (7 species), *Chaetomium, Ulocladium, and Verticillium* (6 species each), and *Paecilomyces* and *Scopulariopsis* (5 species each) – full list of isolated species in reference 14	14

4.6.2 ESSENTIAL PRODUCT COMPONENTS

Cellulose has a complex structure in which its components play different roles, as follows:[9]

- pectins act as a food source and promote microbiological attack
- lignins (an amorphous, cross-linked polymer which bonds to carbohydrate materials) slow down biodegradation
- naturally occurring waxes are also deterrents.

A silk fiber consists of two types of protein: fibroin and sericin. The fibroin is a major component of silk fiber (75%), serving as the core.[15] Fiber contains at least two fibroin proteins, light-chain (25 kDa) and heavy-chain (325 kDa) and the sericin is a minor component (25% of the silk fiber), serving as a glue-like protein coating on the two fibroin cores to conceal the unique luster of fibroin.[15] Both fibroin and sericin contain the same 18 amino acids, although in different concentrations.[15] A crystalline structure is found only in the fibroin and it contains repetitive amino acids (-Gly-Ala-Gly-Ala-Gly-Ser-) along its sequence, forming a large number of β-sheet microcrystallites.[15] This reinforcement contributes the strength and stiffness to the silk fiber.[15]

4.6.3 MECHANISMS OF BIODEGRADATION AND BIODETERIORATION

Cellulose biodegradation mechanism is quite well understood. Natural hydrolysis of cellulose, i.e., cleavage of the 1,4-glycosidic bonds in the presence of water, is a normally slow process, but if it is catalyzed by enzymes, a textile can be degraded in only a few days under the right conditions.[9] The enzymes continue degrading the fibers, even after the organism, which produced them, has been destroyed.[9] Cellulose passes through intermediate forms (oligomers. cellobiose, glucose) as it is broken down by a system of several enzymes referred to as the cellulase complex.[9] There is a general consensus in the literature as to the existence of three main enzyme component types: exoglutanases, endoglucanases, and enzymes with β-glucosidase capabilities.[9] Cellulases first attack exposed amorphous regions.[19] It appears that hydrolysis of amorphous regions results in fragmentation, exposing a larger surface of cellulose to the enzymes.[19]

Fungi and bacteria degrade cotton fabric in two different modes.[16] Fungi attack fibres from the inside towards the outer layer of fibres, whereas the degradation of cotton fibres by bacteria proceeds from the fibre surface towards the inner parts of the fibres.[16] Bacteria are of lesser importance in comparison to fungi, because they require a higher percentage of moisture and therefore they are limited in their activity.[16]

Biodegradation of wool is very much controlled by a sequence of events.[14] Microorganisms initially use non-keratinous substrate components such as membrane material, nucleic remnants, cytoplasm residues, endocuticle, or a cell membrane complex, as well as lipids present in the material.[14] Although the non-keratinous elements of wool can be utilized by a wide variety of fungi, the total decomposition of keratinous material can only be achieved by specialized microorganisms capable of generating a specific sequence of events.[14] The next stage – the keratinolysis process – not only involves keratinase activity but also sulfitolysis, which is considered to be the key reaction of keratinolysis, because proteases alone cannot break the disulfide bonds and generate the complete hydrolysis of the keratin.[14] In summary, biodegradation of keratin is the result of three reactions, namely: deamination, sulfitolysis, and proteolysis.[14]

Silks can be gradually biodegraded by proteolytic enzymes over a long period of time.[15]

4.6.4 RESULTS OF BIODETERIORATION

Cellulose biodegradation results in depolymerization of cellulose macromolecules, which is reflected in decreased molecular weight and strength, increased solubility and changed crystallinity.[16]

Microbial activity in wool causes several types of deterioration. They include:[14]

- surface changes (color, appearance)
- mould odor
- stains of various kinds, usually very difficult to remove
- structural decomposition
- chemical changes
- reduction in strength of the material as the final result

4.6.5 BIOSTABILIZATION

Table 4.21 contains a list of the most suitable biocides used for protection of fibers and textiles.

Table 4.21. Biocides used in fibers and textiles

Material	Biocide	Refs.
aromatic PA (Nomex)	N-chlorination by bleach (sodium hypochlorite)[3]	3

Table 4.21. Biocides used in fibers and textiles

Material	Biocide	Refs.
cotton	cetyltrimethylammoniumbromide,[8] chitosan,[6] dichloro-m-aminophenylhydantoinyl-s-triazine,[11] 1,3-dimethyl-4,5-dihydroxyethylene urea,[16] 3-methylol-2,2,5,5-tetramethylimidazolidin-4-one,[5] octenidine,[8] silver,[8] silver salts[8] water glass (sol-gel method)[4]	4, 5, 6, 8, 11, 16
fibers	5-chloro-2-(2,4-dichlorophenoxy)phenol (triclosan)	2
polyamide-6	dimethylalklbenzyl ammonium chloride,[10] immobilized -cyclodextrin derivative,[13] silver-loaded zirconium phosphate nanofibers[12]	10, 12, 13
polyester	cetyltrimethylammoniumbromide,[8] octenidine,[8] silver,[8] silver salts[8]	8
polypropylene	chitosan[6]	6

The temperature and humidity of the beds create appropriate conditions allowing for effective bacterial proliferation.[1] Personnel in contact with contaminated textiles were the source of transmission of the microorganisms to susceptible patients.[1] Bed-making in hospitals releases a large quantities of microorganisms into the air, which contaminate the immediate and non-immediate surroundings.[1] These risks can be reduced by use of biocidal textiles. The biocidal materials introduced into the hospital textiles should have the following key characteristics:[1]

- wide spectrum of antimicrobial, antifungal and antiviral properties
- be effective against the already existing antibiotic resistant microorganisms involved in nosocomial infections
- not permit the development of microorganisms which are resistant to the active component
- do not cause skin irritation or sensitization
- be safe to humans

The chlorination reactions of aromatic polyamides with sodium hypochlorite aqueous solutions forms stable polymeric N-halamines by a simple chlorine bleaching treatment of Nomex without the sacrifice of its other excellent properties.[3] The chlorinated Nomex fabrics provide potent, durable, and refreshable antibacterial efficiency against both Gram-positive and Gram-negative bacteria.[3]

3-methylol-2,2,5,5-tetramethylimidazolidin-4-one (MTMIO) can be synthesized by methylolation of 2,2,5,5-tetramethylimidazolidin-4-one (TMIO).[5] Because of chemical reaction of MTMIO with cellulose, TMIO rings can be grafted onto cellulose-containing fabrics.[5] Then, after a subsequent chlorination, the treated fabrics are converted to halamine structures, which then demonstrate effective antibacterial efficacy.[5] Such obtained halamine structure is much more stable.[5] The halamine structure can survive repeated home laundering and would require less frequent chlorine recharging to maintain the biocidal properties.[5]

An antifungal efficacy test of the silver laundry machine, which electrically generates silver ions, was carried out against four fungi – *Trichophyton rubrum, Candida albi-*

cans, Microsporum canis, and *Aspergillus flavus* – which cause major fungal infection in humans and animals.[7] Compared with the conventional laundry machine, washing with the silver laundry machine was effective against most of the fungi.[7]

Cetyltrimethylammoniumbromide, octenidine, silver, and silver salts were embedded in silica sol (10 wt%) and coated on polyester textiles.[8] All coatings gave good biocidal performance, but silver and octenidine coatings give exceptional performance and longevity.[8] Three main methods can be used in protecting cellulose fibers: deposition of biocide on their surface, altering cellulose chain in such a manner that it is not digestible by microorganisms, and creation of barrier coatings on the surface of textiles.[9] In addition, in some cases (e.g., museum), it is possible to stop progress of biological processes by climate control.[9]

Silver has long been recognized as a broad-spectrum and highly effective antimicrobial agent for treating wounds and burns.[12] Silver ion works by denaturating the proteins and nucleic acids of the bacteria by binding to their negatively charged components.[12] Silver also generates oxygen which destroys the cell wall membranes of bacteria.[12]

REFERENCES

1 Borkow, G; Gabbay, J, *Medical Hypotheses*, **70**, 990-94, 2008.
2 Tixier, C; Singer, H P; Canonica, S; Mueller, S R, *Environ. Sci. Technol.*, **36**, 3482-89, 2002.
3 Sun, Y; Sun, G, *Ind. Eng. Chem. Res.*, **43**, 5015-20, 2004.
4 Xing, Y; Yang, X; Dai, J, *J. Sol-Gel. Sci. Technol.*, **43**, 187-92, 2007.
5 Qian, L; Sun, G, *J. Appl. Polym. Sci.*, **89**, 2418-25, 2003.
6 Abdou,E S; Elkholy, S S; Elsabee, M Z; Mohamed, E, *J. Appl. Polym. Sci.*, **108**, 2290-96, 2008.
7 Jung, W K; Kim, S H; Koo, H C; Shin, S; Kim, J M; Park, Y K; Hwang, S Y; Yang, H; Park, Y H, *Mycoses*, **50**, 265-69, 2007.
8 Mahltig, B; Fiedler, D; Bottcher, H, *J. Sol-Gel Sci. Technol.*, **32**, 219-22, 2004.
9 Montegut, D; Indictor, N; Koestler, R J, *Intl. Biodet.*, **28**, 209-26, 1991.
10 Shalaby, S E; Al-Balakocy, N G; Abdel-Fatah, O M, Elshafei, A M, *J. Appl. Polym. Sci.*, **110**, 738-46, 2008.
11 Lee, J; Broughton, R M; Akdag, A; Worley, S D; Huang, T S, *Fibers Polym.*, **8**, 2, 148-54, 2007.
12 Duan, Y-y; Jia, J; Wang, S-h; Yan, W; Jin, W; Wang, Z-y, *J. Appl. Polym. Sci.*, **106**, 1208-14, 2007.
13 Gawish, S M; Ramadan, A M; Mosleh, S; Morcellet, M; Martel, B, *J. Appl. Polym. Sci.*, **99**, 2586-93, 2006.
14 Blyskal, B, *Intl. Biodet. Biodeg.*, **63**, 631-53, 2009.
15 Kojthung, A; Meesilpa, P; Sudatis, B; Treeratanapiboon, L; Udomsangpetch, R; Oonkhanond, B, *Intl. Biodet. Biodeg.*, **62**, 487-90, 2008.
16 Tomsic, B; Simoncic, B; Orel, B; Vilcnik, A; Spreizer, H, *Carbohydrate Polym.*, **69**, 478-88, 2007.
17 Friedrich, J; Zalar, P; Mohorcic, M; Klun, U; Krzan, A, *Chemosphere*, **67**, 2089-95, 2007.
18 Forlani, G; Seves, A M; Ciferri, O, *Intl. Biodet. Biodeg.*, 46, 271-75, 2000.
19 Saqib, A A N; Whitney, P J, *Intl. Biodet. Biodeg.*, **58**, 180-85, 2006.
20 Sharma, H S S, *Trans. Br. Mycol. Soc.*, **88**, 1, 122-25, 1987.

4.7 LEATHER AND LEATHER PRODUCTS

Michalina Falkiewicz-Dulík

4.7.1 GENERAL INFORMATION

Biodeterioration in the leather industry means aggressive activity of macro- and microorganisms on rawhides and skins during processing, storage of leather and leather products, and their use. Skin, as a biological material, creates an optimal environment for the development of various microorganisms. Because of a significant concentrations of proteins, lipids, and carbohydrates, it provides a suitable environment for a growth of a number of biological species. Bacteria, fungi, flies, beetles and their larvae may cause biodegradative processes and contaminate material which has not been previously affected.

Microbiology of the skin is the study of specific biochemical processes and skin microflora beginning with raw material through the processing and storage to the use of manufactured goods. It aims at detection and analysis of the reasons of skin lesions in order to adequately protect against such damage and find remedies which can effectively eliminate further action of microorganisms.

In the tanning and shoe industries, the financial losses caused by microbiological damage can reach 6-8%.[1] Information on the extent of these loses, based on the data from Food and Agriculture Organization of the United Nationals (FAO) and the International Council of Tanners (ICT), is included in Tables 4.22-4.25 and in Figures 4.1 and 4.2.

Table 4.22. Production of rawhides and skins, leather and leather footwear (averages 1999-2001). [Data from FAO.]

	Rawhide & skins production in 1000 tons	Leather production in billions sqf.		Footwear with leather uppers in million pairs
		heavy	light	
Bovine hides and skins				
World	1,510.4	495.2	10,999.6	4,343.9
Developing countries	1,180.5	313.0	6,534.4	3,270.7
Developed countries	329.9	182.2	4,465.2	1,073.2
Sheep and goat skins				
World	1,765.4		4,501.5	
Developing countries	1,351.6		3,245.4	
Developed countries	413.8		1,256.1	

Table 4.23. Top leather producing countries accounting for over 70% of finished leather production in million sqf. [Data from ICT]

	1997	1998	1999	2000	2001
China	2,964.4	3,285.3	3,900.0	4,141.8	4,493.2
Italy	2,149.4	1,981.6	1,820.0	2,060.4	2,065.7
India	1,369.3	1,379.6	1,392.0	1,428.3	1,423.7
Korea	1,268.8	869.2	1,000.0	1,086.4	973.0
USA	735.5	847.0	930.7	861.0	831.8
Former USSR	932.0	926.0	896.0	731.1	782.5
Brazil	672.3	705.2	725.0	784.6	756.0
Mexico	632.9	673.6	668.2	619.9	568.3
Spain	558.6	531.8	500.0	542.2	541.2
Argentina	403.1	430.6	461.6	485.2	487.2
Turkey	550.3	509.6	556.8	410.0	419.2

Table 4.23. Top leather producing countries accounting for over 70% of finished leather production in million sqf. [Data from ICT]

	1997	1998	1999	2000	2001
Pakistan	343.2	343.3	370.0	311.8	314.4

Table 4.24. Production of hides, skins, leather, and leather footwear (averages 1999-2001). [Data from FAO]

	Developing countries	Developed countries	World
	Million US$		
Rawhides and skins from bovine, sheep, and goat	557.8	3,859.4	4,417.1
Leather (rough-tanned and finished, all types)	7,524.3	7,221.8	14,746.2
Footwear with leather uppers	11,257.7	13,401.2	24,658.9
Total	**19,339.8**	**24,482.4**	**43,822.2**

Table 4.25. Word imports of leather and leather products, 2000-2004 (value in million US$). [Data from Council for Leather Exports (CLE) Statistics, Geneva]

	2000	2001	2002	2003	2004
Leather	14,979.29	16,208.67	15,909.70	17,532.66	18,653.14
Leather footwear	27,594.22	28,867.41	30,568.73	33,663.95	36,420.19
Leather garments	4,453.76	4,761.08	4,206.77	4,188.46	3,974.20
Leather goods	5,863.87	6,166.62	6,606.16	7,514.61	9,007.34
Leather gloves	1,390.65	1,352.89	1,361.24	1,480.83	1,638.76

4.7.2 RAWHIDE AND SKIN

4.7.2.1 Structure

The skin consists of three main layers: the epidermis (epidermis with hair), the dermis and subcutaneous tissue (Figure 4.3). The dermis contains dermal papillae and reticular. Hair follicles and sweat glands reside between these layers. Papillae is composed of thick collagen fibers whereas the reticular layer, situated near the surface of the skin, is formed from a thin and delicate fibers. Papillae, which adheres to the epidermis, has a very tight arrangement of fibers forming the face of the skin. Collagen (60-80% of dry weight)[2] is the primary protein which forms skin. All other skin components are not useful and they are removed in the process of tanning.

Any macroscopically visible collagen fiber (cross-section of about 0.2 mm and a length of several millimeters) is composed of 30-300 monofilaments (diameter of about 5 μm), which are built from 200-1,000 fibrils (cross-section of about 1,000 Å). Each fibril contains approximately 700-800 molecules of collagen. A collagen molecule consists of 3 peptide chains, each with 1,052 amino acids, which are twisted with each other to form triple coil, which has a length of about 3,000 Å, a diameter of about 14 Å, and a molecular

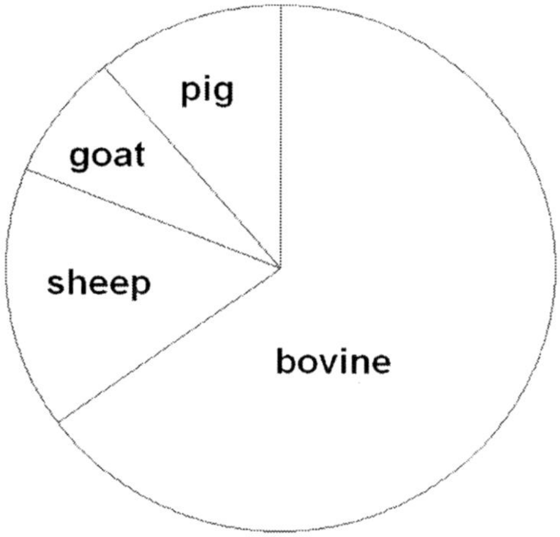

Figure 4.1. Market share of leather by animal type in %. [Data from ICT].

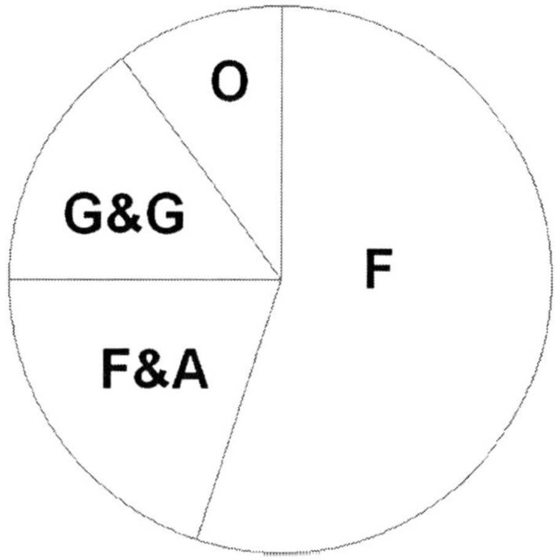

Figure 4.2. Market share of leather by the end use. F – footwear, F&A – furniture and automotive, G&G – garment and gloves, O – other. [Data from ICT].

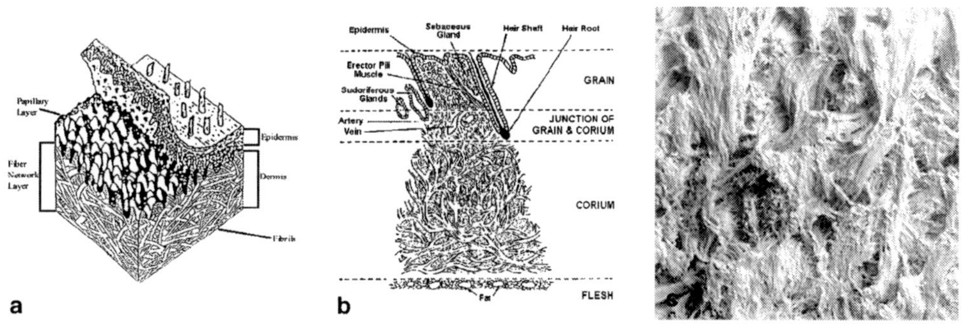

Figure 4.3. Skin structure. (a) Scheme of skin showing epidermis, basement membrane and the two layers of dermis. A portion of epidermis is peeled off to show papillae in the papillary layer. [Adapted, by permission, from Reed, R, Ancient Skins, Parchments and Leathers, Seminar Press, London, 1972.]; (b) Scheme of skin structure [Adapted, by permission, from Technical Encyclopedia, WNT, Warszawa,1986]; (c) SEM micrograph of leather cross-section showing fiber structure.

weight of about 300,000. Fibrous structure strongly affects the internal surface area. The total surface area of the fibrous structure weighing 1 kg is 1,000 ÷ 2,500 m^2.[3]

4.7.2.2 Composition of raw leather

Components of skin are inorganic substances, such as: water (60-70%), minerals (0.35-0.5%) and organic matter. Lipids are 2 ÷ 6% of bovine skins, 2 ÷ 10% of goat skins, 5 ÷ 30% of sheepskin. Carbohydrates (monosaccharides – simple sugars – glucose, homopolysaccharides – glycogen, mucopolysaccharides – hyaluronic acid, chondroitin sulfate B, chondroitin sulfate C) and nitrogen-containing substances that do not belong to a group of proteins, and proteins forming the backbone skin are other essential components which form its histological structure.[3,4]

For its biochemical functions, skin contains enzymes and vitamins. The composition of skin extract varies depending on the type, gender, age, and living conditions of the animal.

The pH of skin is close to neutral, and it is in the range of 6.2 ÷ 7.8.[4]

Protein substances are the major component of skin, they constitute about 33%. The most important skin proteins are fibrous proteins such as collagen, keratin, elastin, and reticulin. In addition, the skin contains spherical proteins, such as albumin, globulin, and complex proteins (approximately 3.5%).[3,4] These proteins are not a part of skin structure. Table 4.26 contains the elemental composition of proteins.[3]

Table 4.26. Elemental composition of skin proteins in %.[3]

C	H	O	N	S, P, Fe, I, Br, Cl
45-55	6-8	19-25	16-19	0.5-2.5

The amino acid composition of collagen depends on the structure of its molecular chains and groups actively involved in the creation of intermolecular bonds largely conditioned by the properties of collagen and its changes under the influence of different reagents. The collagen contains 19 amino acids that can be divided into aliphatic, aromatic, and heterocyclic. Irrespective of their origin, collagen contains two amino acids, namely, hydroxyproline and hydroxylysine, which do not occur in other proteins. Collagen has more glycine than any other protein. It does not contain cysteine, cystine and tryptophan.[2,5]

4.7.2.3 Damage caused to the animal

The skin can be colonized by microorganisms and ectoparasites during an animal's life. Damage caused to the raw material as a result of these settlements may constitute a significant portion of all the causes of decline in the quality of leather.

Skins damaged by a gadfly have visible holes on their surfaces. Ten to fifteen holes caused by a gadfly damage of 60-120 cm^2 of skin. Gadfly larvae living under the skin also reduce the quality of leather. Even healed holes caused by gadfly larvae cause damage to the finished leather in the form of numerous scars visible on the surface of a face.

Bovine skin damage by biting by fly *Stomoxys calcitrans* is similar to the damage caused by lice, although there are individual differences such as the damage pattern. Openings on the skin surface are grouped together and they are more uniform in their shape and size. Damages to the surface extend into the skin bulk.[6]

Skin damage, attributed to lice, is caused by sucking and biting by lice. Mixed infections are most common. Sucking lice are specific insects that live on the skin and suck blood. The most common sucking types are short-nosed cattle louse (*Haematopinos eurystermus*), long-nosed cattle louse (*Linognathus vituli*), small cattle louse (*Solenopotes capillatus*). Biting lice (*Bovicola bovis*), preferring thick, hairy covering, are found in different parts of the animal body. They introduce secretions into skin using their jaws. In the case of animals infected by lice, the following damage is observed: spots similar to pits and fissures (holes often run across the face, sometimes across the area of several pore follicles, so-called bite channels) or small to 3 mm wide spots without damage to the face (light spots). Leather finished with material infected by louse contains the damage in the form of round spots from the loss of face (nubuck spots), depressions, pits having the size of 1-3 mm (craters), heterogeneously rough grain, dilated pores, circular face deformation similar to pimples (pustules), finely cut face, and open scratches of mechanical origin.[7] (see Figure 4.4).

The decisive factors in the intensity of damage by lice are biochemical effects of parasites on the fresh skin (secretion of the glands), the potential of introduction pathogens directly into the skin by sucking activity of lice, environmental factors which are factors affecting the activity of pests and animal defense mechanism, and the genetic susceptibility of the individual characteristics of animals (immune resistance, the intensity of infection or allergy).[8-10]

Mangue Psoroptic larva attacks mainly sheep, causing a characteristic sheep scabs. These insects can also attack cattle. Bovine skin damage caused by scabies is most common in the U.S. The parasite attacks the epidermis, and the intense irritation and skin reaction ultimately leads to an increase in its thickness and the formation of a rough face.

Demodex folliculorum is a common insect of the family of mites found on the skin of cattle. It penetrates into the skin through a hair bag. It reproduces in bags of sebaceous glands and hair follicles. It creates cysts rarely seen on live animals, but often leads to the disintegration and the creation of small hollows on the surface of a face. Cysts lying below the surface of a face weaken the strength of the finished leather during finishing operations in the shoe industry.[11]

The most common parasites that cause damage to the skin of sheep and cattle are fungal infections of the skin (*Mycotic dermatitis*). *Dermatophilus congolensis* fungus grows in epidermal skin layer. Reproductive layer of the epidermis creates scabs. This leads to felting the wool. These organisms do not penetrate the skin, but the pile bags are enlarged and destroyed and this part of the face surface becomes unattractive, wrinkled, and rough. These changes are limited to the outer surface of the face, resulting in skin which may only be used as suede. In sheep, *Dermatophilus congolensis* grows mainly in the rear body parts, whereas in cattle, on the legs and abdomen, causing extensive changes.[12]

Ringworm is caused by fungi *Trichopyton* that grows on hair and skin of cattle. This is a highly infectious disease, and through contacts with the animals it can be transferred to humans. It produces a characteristic circular damage to the raw skin of infected animal.[6]

A defect on the skin of sheep known as the "cockle" is related to the occupation of ground by a sheep tick, *Melagophagus ovinus*. It appears on the surface of the skin in the form of a small dense pattern, which tends to create intense lines running from the spine to the abdomen. Nodules forming along these lines cause significant changes in the skin. Fibrous structure becomes particularly dense, which reduces the penetration of chemical compounds during the tanning, and results in characteristic defects of the skin. Skin is hard, suede density is low, color irregular, and some areas are excessively tanned. Ticks tend to gather in the natural folds of fleece, which is likely to lead to the creation of grooves running across the skin. The changes visible on the skin surface are caused by allergy. If parasites are early removed, skin gets back its natural look and structure.[13]

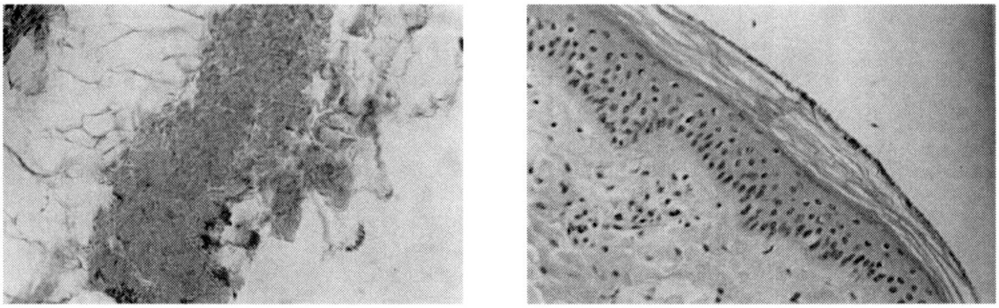

Figure 4.4. Histological picture of leather sample. (Left) Leather sample on agar culture from subcutaneous tissue side (12 h of exposure results in extensive colonization by *B. antracis*. Less intensive bacterial growth is seen on the periphery of subcutaneous tissue and in deeper layers of the loose collagen. (Right) The surface of horny epidermis layer occupied by bacteria after 24 h exposure to *B. antracis*.

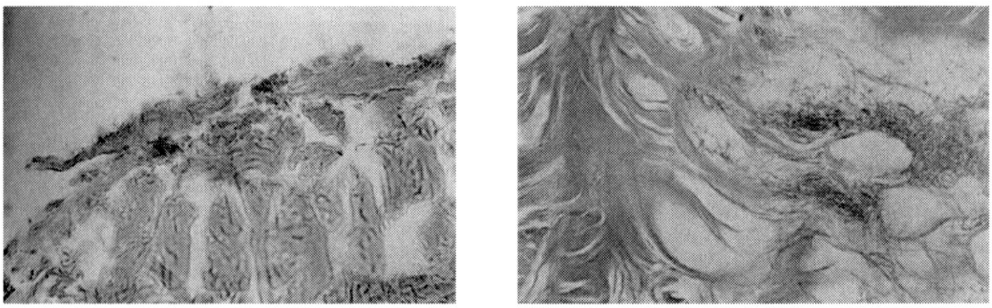

Figure 4.5. Histological picture of leather. (Left) After 24 h, there is a limited bacterial colonization. Colonization is focused on the periphery of leather sample. (Right) After 72 h there is extensive bacterial colonization in the area of the loose subcutaneous layer of collagen fibres.

The raw skins originating from animals which are sick or dead due to anthrax may contain *Bacillus anthracis,* causing a disease called anthrax, which is dangerous for people and animals. Laforce[14] found *Bacillus anthracis* in goat hides and in aerosols of milled goat skins. *Bacillus anthracis* is present in 33% of African goat skins, which were checked for the presence of anthrax.

The source of the anthrax threat in leather tanneries may be cattle, sheep, horses, and sometimes pig skins, and to a lesser extent, skins of infected wild animals such as elk and deer. In 1995 in Radom (Poland), the entire stock was infected by mixing good stock with the infected skin.

Freshly removed skin is a favorable ground for the rapid growth of bacteria. The large water content (about 65%), blood and serum, the nature of the structure of the subcutaneous tissue and epidermis, the size and density of hair follicles bags – all affect the rate of contamination by attacking bacteria, depending on their ability to penetrate skin depths.

Mierzejewski and Mendrycka,[15] examining the process of penetration and settling of pig skin by anthrax, both from the side of epidermis and subcutaneous tissue, observed a much faster penetration of bacteria from the subcutaneous layer. Penetration rate of *Bacillus anthracis* in the tissue of the skin depends on the environment in which they are located, and the liquid content facilitates the penetration of bacteria into the tissue and the colonization of the deeper layers of the dermis. Under natural conditions, the long storage of hides (e.g., longer transport) may cause germination of bacilli, their infiltration and vegetation in the skins of healthy animals (see Figures 4.4 and 4.5).

4.7.2.4 Damage to a hide by the autolysis

The horny layer of skin beyond the epidermis is a living tissue. Many physiological processes taking place, with an influence of hormones and enzymes, continue to function after skinning for a period of sometimes many days. In the living tissue, there may be about 1,000 different enzymes, catalyzing a series of complex chemical reactions. In addition to synthesis, a living system is also subjected to a controlled autolysis. This process is aimed at eliminating dead cells and the degradation of matter into simpler substances, such as breakdown products of proteins, carbohydrates, fats, nucleic acids, etc., which are necessary for the synthesis of new compounds or life processes. After removing skin from the animal, it loses its natural biological function, life processes are stopped and they are replaced by post-mortal processes, dependent largely on external factors. At the time of slaughter, the skin shows a neutral or slightly alkaline pH of 7.0-7.2. This continues for about 3 hours, after which the degradation of glycogen, contained in the tissue, occurs with formation of lactic acid. The skin becomes acidic, and the pH is reduced to approximately 6.

In the absence of oxygen (air) at high temperature and high humidity, when insufficiently cooled skins are together for a long time, there is a rapid restoration of protein under the influence of the skin's own enzymes without the participation of putrefactive bacteria, although the symptoms are similar to putrefraction. Such changes are called autolysis or self-digestion.

Deterioration of hides occurs when the natural enzymes present in the skins begin degradation of proteins and fats by autolysis. Digestion of proteins of the skin starts from globular proteins: albumin, globulin, and mucopolysaccharide. In this process, intracellular enzymes, such as cathepsin, play an essential role, resulting in hydrolytic breakdown of proteins into peptides and amino acids. Lysosomes, with exception of cathepsin, also contain enzymes that catalyze the breakdown of sugars, lipids, and other compounds. Amidase produces ammonia from amides of monocarboxylic acids. Deterioration rate can be increased by the presence of fatty acids, namely acetic and butyric acids. Fatty acids are produced by the enzyme lipase, which hydrolyzes fats and cysteine derived from decomposing epidermis. With the increase in fatty acid content, pH is reduced to 5.0. The period known the post-mortem stiffening, in which the tissue becomes slightly swollen and more rigid, begins. At this point, skins have the lowest pH and are then most resistant to bacterial digestion. As a result of the further process of autolysis the acidity gradually disappears and release of ammonia is observed. After 3 days, the pH of skin is raised by the alkaline breakdown products of protein and at pH above 8.0, protective activity of acids ceases. In the first stage of protein reconstruction, amorphous proteins are being hydrolyzed. Prolonged autolysis causes significant damage to the skin without the involvement of bacterial processes. Observed histological changes in the skin are caused by the reconstruction processes of proteins, leaching of cellular substance, cleavage of intercellular elastin fiber bundles, separation of collagen fibers by the decomposition of fat cells, and the migration of fat liberated from the cells and glands. Leather produced from the skin damaged by such autolytic processes of raw material is spongy, loose, malleable, sensitive to bending.

4.7.2.5 Bacterial deterioration of fresh skins

Skin is covered with an extremely large number of microorganisms, approximately 10^5-10^7 microorganisms per 1 g of skin, mainly from soil, natural fertilizer, and water. These microorganisms, if they are not pathogenic bacteria attacking the skin of a living organism, have a negligible impact on the skin of a living animal.[17]

The skin that was removed from the animal rapidly loses the natural protective properties, and, because of its chemical composition, it becomes a good medium for the development of various microorganisms. With favorable external conditions, i.e., high temperature and high humidity, bacterial growth is very fast. The latent period of development necessary for the regeneration of spores depends on the temperature as follows: at 5°C – 48 hrs, at 20°C – 10 hrs, at 30°C – 2 hrs.[18]

For skins, the main source of infection with microorganisms is the environment (air, floor, transportation, etc.) with which skin is contacted after removal from the animal. According to Stanley,[19] skin collected at the slaughterhouse may contain $6x10^5$ bacteria per 1 g of skin. Especially dangerous is skin stained with blood at the time of slaughter, which causes local rotting during storage. The second source of skin infection may be a physiological effect caused by pre-slaughter blood and lymph. Such cases occur in the pathological state in the course of long-term illnesses.

Epidermis is a barrier only in the initial period of bacterial growth. The flesh side surface of the skin becomes rapidly contaminated. Growing bacteria begin to penetrate into the skin. The subcutaneous layer, having a porous structure, is more permeable for the bacteria, which can readily reproduce and colonize skin. Bacteria growing in the subcutaneous layer gradually penetrate into the dermis layer of connective tissue between the bundles of fibers. It is easy to colonize around the walls and around the lymph nodes.

Bacteria from the epidermis can penetrate the dermis layer through the channels of glands and hair follicles bags.

The products of autolysis are a perfect breeding ground for rotting bacteria. Over 80 species of bacteria on the skin of the animal were identified, and there are probably many more of them. The most dangerous are those that break down protein substances.

In order to provide nutrients, the growing bacteria secrete enzymes (egzoenzymes), which digest the organic macromolecular constituents of the skin to simple absorbable substances. Type of enzymes depends on the type of bacteria and substrate. Bacteria present on the skin are Gram-positive and Gram-negative. On the raw skin the following genera of bacteria were identified: *Proteus, Micrococcus, Bacillus, Enterbacteriacea, Alkaligenes*.[20-23] These are typical bacteria secreting proteolytic enzymes that hydrolyze proteins into peptides or amino acids and thereby cause destruction of the skin.

Aerobic bacteria begin septic processes and create a friendly environment in the tissue, which facilitates further decay and promotes degradational changes caused by anaerobic bacteria. The process of rotting caused by bacterial activity occurs if the skin protection is inadequate. Under the influence of enzymes secreted by proteolytic bacteria, the most vulnerable are albumins, globulins, and to a lesser extent, reticulins, elastin and tropocollagen. Elastin, fibrous protein of skin, characterized by a high content of valine and a small content of acidic and basic amino acid, is relatively easy to be broken down by proteolytic bacteria. Hydrolyzed under the influence of elastase, as well as, to a large extent, such enzymes as papain, ficain, and bromelain. In the first days of decomposition of the skin under the influence of proteolytic bacteria, in addition to amorphous proteins, soluble forms of collagen (procollagen, tropocollagen) are degraded. Mature collagen protein and keratin are only marginally hydrolyzed and then only at very advanced stages of bacterial growth after several days of decay. Native collagen is resistant to proteolytic enzymes.

Symptoms of bacterial attack are sliminess, odor of ammonia and hydrogen sulfide, and loosening of hair in bags. Aerobic bacteria do not produce proteolytic collagenase, but are still able to induce changes around the collagen fibers. Under the influence of bacterial proteolytic enzymes, and to some extent of amylolytic enzymes, the following components of the skin undergo the partial or complete hydrolysis: hair bags, sweat, and sebaceous glands, reproductive cells of the Malpighi's layer of the epidermis, the cells of blood vessels, nerve endings, fibrins, some layers of fat cells of thermostatic layer, and other cells and elements of between-fibrilar substances of skin. The main symptom of these changes is the release of the hair roots, leaving the bags, and release of the hair bundles of collagen fibers from between-fibrilar substance, which results in loose fibers. Although the septic bacteria are mainly able to break down cellular and fibrous, non-collagen proteins, a group of anaerobic organisms of the genus *Clostridium*, with such species as *Clostridium histolyticum, Clostridium perfringens,* and *Clostridium capitovale*, producing the enzyme collagenase, are able to break down collagen. The growth of these bacteria and the activity of enzyme is particularly high in anaerobic conditions. Optimum pH for collagenase extracted from the bacteria decomposing skin (*Clostridium capitovale*) is 6.7-7.4. Enzyme activity increases with increasing temperature. Bacteria secreting collagenase is particularly dangerous because in a very short time (24 hrs.) it can cause formation of deep pits, and if deterioration continues, perforations or collagen tissue weakness is observed. Collagenase acts only with collagen and gelatin, but does not hydrolyze casein, hemoglobin, albumin and fibrin. Collagenase destroys the raw hides, relatively poorly dried or non-salted, extending its activity to the process of tanning. The effect of decay of fresh skin can be spotted in the finished leather in the form of a loose structure, the face peeling, blisters, pin punctures, and empty spaces.

Optimal conditions for bacterial growth and multiplication cause damage to occur only after several hours. Penetration time of bacteria from the skin flesh surface depends on temperature and humidity, e.g., at 20°C it is about 20 hours.[19,24] Soaking without preservatives accelerates penetration.[19]

Penetration of bacteria from the flesh side is the result of enzyme digestion of the structural components of skin, causing increased permeability of the face layer.[24] The time in which this process occurs depends not only on the temperature and humidity, but also on the type of contamination and the bacteria count on skin.[24] Oyen[24] found that face damage (matt surface) occurs after 24 hours at 20°C, 8 hours at 30°C, and 6 hours at 30°C for skin contaminated with blood.

Improperly cooled skin, stored in pile, can increase temperature to 40°C. The process of spontaneous temperature increase in the stack[19] is the result of exothermic reactions occurring during the biochemical decomposition of organic matter. The more bacteria involved in this process, the more energy is released into the environment and the higher the temperature observed in the stack. Increase of skin temperature to 30°C or 40°C causes increase in bacteria count from 10^6 to an uncountable number of bacteria per gram of skin.[25]

A large number of bacteria and high temperature in the stack causes that, in addition to the destructive biochemical changes by the bacteria, thermal changes also occur. Changes due to microbial activity are often difficult to see in the raw skins, but become evident in the final product.[25]

4.7.2.6 Bacterial deterioration of salted hides

To protect against microbiological degradation, hides and skins are commonly preserved by ordinary salt. However, salt has limited bactericidal properties and thus cannot destroy all bacteria, but it protects the skin against bacterial growth for some time. All bacteria of a skin are present on a freshly salted skin. As a result of the gradual penetration of salt into skin tissue and the displacement of water, microorganisms are getting worse living conditions. Many bacteria, including the proteolytic and pathogenic, which cannot adjust to the environment with higher salt concentration, die after several or more days. The action of sodium chloride, which inhibits the growth of bacteria in raw hides, is more or less by intense dehydration of skin tissue, which creates unfavorable conditions for the development of microorganisms, but encourages the growth of halophilic bacteria. Inadequate salting leads to the re-development of spores. The reason for the growth of a number of bacteria is their progressive adjustment to a higher concentration of salt, for example: *Micrococcus roesus* has best growth at 8% salt content in the substrate, and *Micrococcus pyogenus* and *Bacillus mesentericus* survive in substrates having a concentration of NaCl up to 25%.[20] Salt-tolerant microorganisms gradually adapt on the surface of salted hides, and, under these conditions, they begin to multiply.

Halophiles can easily grow on salted skins. Typically they need a minimum of 15% NaCl in the environment to grow. Orlita *et al.*[26] isolated from salted hides 100 strains of microorganisms, of which 36 have been identified as halophiles and 64 as Gram-negative and Gram-positive bacteria. The bacterial flora of fresh and salted hides differ, but *Bacillus, Staphylococcus, Micrococcus* are isolated in both cases.

Halophiles produce colored pigments, which cause formation of patches on raw skin on the rough side. Most halophiles stain skin to pink, red, red-orange, but some to blue or purple, although some forms do not cause color changes at all. Red spots usually appear on the calf and pig skins, rarely on cattle and sheep. Spots are visible as small dots, or may occupy larger areas of the skin. Sometimes the presence of red spots is connected with sliminess of the rough side and a clear scent of ammonia, which testifies to the processes of rotting. Some microorganisms involved in the production of colored stains, such as *Micrococcus roseus* (red color), *Sarcina auriantica* (red-orange color), *Sarcina lutea* (yellow color) have been identified. Red dye stain is very durable and, according to Babakina and Kutukowa,[17] not soluble in water, pure alcohol, ether, gasoline, sulfuric acid, glacial acetic acid, fuming hydrochloric acid, and bases.

Extensive colored spots produced by bacteria on the flesh side usually cause damage to the skin called "red heat". Optimum conditions for the growth of bacteria require moisture content of at least 65%, temperature 33-40°C, pH of 6,5-8,0. They can also develop in a highly alkaline environment, but their growth continues in an environment of pH below 6.0. Heat and humidity have a direct impact on the development of microflora in skins protected by salt.

The vast majority of microorganisms causing color spots develop on pig skins which contain 7-14% NaCl and water content ranging from 48-55%.[20]

Halophiles are usually present in the brine used for the protection of skin.[27-29] They have great abilities for settling and digestion of raw hides and skins. Some of them demonstrate the capacity of enzymatic hydrolysis of gelatin, some also cause degradation of casein. Halophilic bacteria growth is very slow at room temperature. The ideal conditions for their rapid growth is at 40°C.[27,30-31] Growth of these bacteria on a skin can be directly related to damage of the face, when the skin is kept for longer periods at higher temperature. This was shown by Birbir and Bailey[32] by examining their impact on the quality of the face during storage of raw hides and skins in contaminated brine. Damage was observed in samples of raw hides and skins stored for 7 weeks at 40°C. The damage occurred to the finished hides face associated with "red heat" and caused by halophiles. Lesions were easily seen with the naked eye and they looked similar to a suede face; it was called the "suede effect". Studies by electron microscopy of samples taken from the damaged area showed that the upper layer of the face is partially removed, and the collagen fibers are relaxed and visible on the surface. The cross-sectional area of damage was relative to destroyed fiber structure with clearly visible torn fibers (see Figures 4.6 and 4.7).[32]

The anatomy of color spot formation by halophiles during the development of infections was studied by Vreeland, Angelini and Bailey.[33]

Skin lesions begin before they are visible to the naked eye as a colored stains. A large number of halophile organisms inhabits skin from the hair and they are present in the layer of grain before red spots are visible on the flesh side. Infected hides and skins show detectable damage within 2 weeks of salting. The most visible symptoms of "red heat" (red colonies on rough side) do not occur over the next 2 weeks. This type of damage may not be visible on the finished skin. Red spots are only superficial and they can be easily wiped out. In this case, the hair is firmly attached and there is no smell of rotting. At this stage, the redness does not degrade the quality of leather hides. Long time storage of the contaminated, salted hides may affect their quality. With advanced redness, hair starts to fall out, the epidermis is damaged or completely destroyed, flesh site becomes slippery, and there is a strong ammoniacal odor. Halophilous bacteria can cause a clear separation between the face and the dermis layer of the skin raw. Microscopically observed characteristic changes include bundles of collagen fibers, which have lost their original structure, and elastic fiber fragments or their complete breakdown. Hair follicles and skin glands are almost completely destroyed.[17,33]

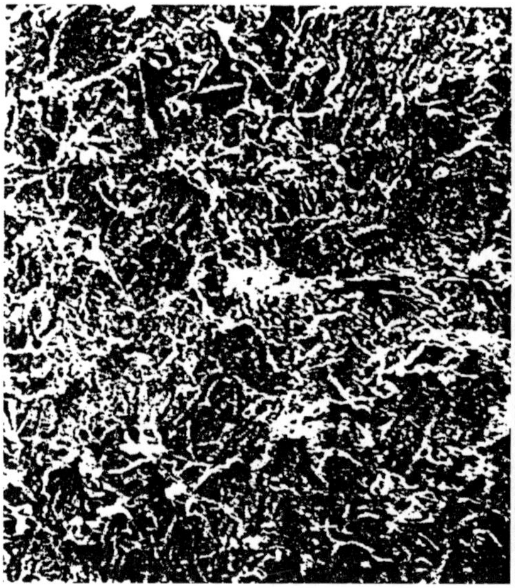

Figure 4.6. Scanning electron micrograph of grain surface of crust leather prepared from halophilic bacteria treated cattle hide samples held for seven weeks at 40°C. [Adapted, by permission, from Bailey, D; Birbir, B, *J. Amer. Leath. Chem. Assc.*, **91**, 47-51, 1996.]

Figure 4.7. Scanning electron micrograph of the cross-section of crust leather prepared from halophilic bacteria treated cattle hide samples held for seven weeks at 40°C. [Adapted, by permission, from Bailey, D; Birbir, B, *J. Amer. Leath. Chem. Assc.*, **91**, 47-51, 1996.]

In raw skins infected with halophilic bacteria, there are many modes of deterioration of fiber structure. The fibers are broken into smaller pieces. Destruction of fiber structure within the finished skin is associated with the type of damage caused by the bacteria in the skins which are protected by salt. This type of hide damage may eventually cause the skin defect called "break", which has not been previously attributed to halophiles. Red halophiles show varying ability to infect and digest hides. White bacteria (that do not stain), appear to have little impact on the destruction of raw hides and skins, even with intense growth.[33]

Quantity and type of damage by "red heat" of raw hides and finished leather is a direct result of both the collection of bacterial species which infect the skin and the place where the individual organisms grow on the skin. Leather is weak, with large dull spots on the face (see Figures 4.8 Figure 4.9).

According to Bitlishi,[34] halophilous bacteria leave traces of their presence in the form of skin damage. According to Stanley,[19] these lesions are caused by halophile-like bacteria which are tolerant to sodium chloride.

Salt Stains

The raw salted skins (mainly calf, horse, and sheep, less pig) have salt stains. They are insoluble in water, with mineral crusts occurring on the outer or inner layer of skin tissue, having diameter of one to several millimeters and color from creamy white to yellow-red-brown. Formation of salt stains on the skins is independent of the presence of microorganisms. Residue of flesh which was not removed from the rough side is very susceptible to

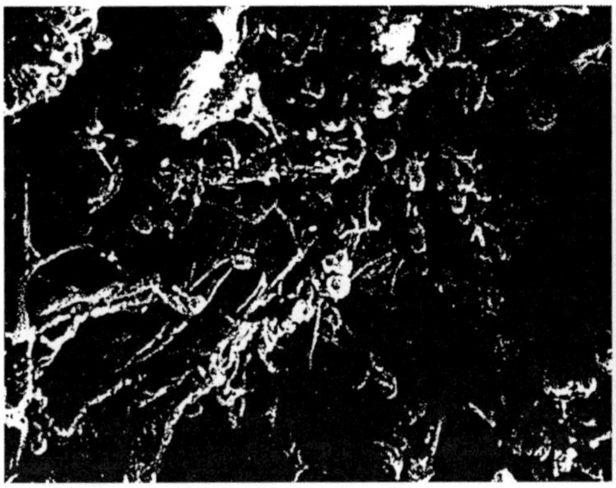

Figure 4.8. Interior of brine-cured hide sample held for 14 days at 40°C after treatment with *Haloferax gibbonsii* (2000x). [Adapted, by permission, from Vreeland, R H; Bailey, A, *J. Amer. Leath. Chem. Assc.*, **93**, 121-31, 1998.]

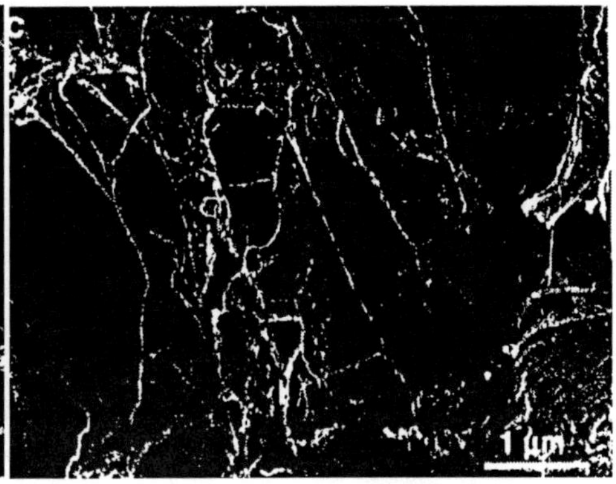

Figure 4.9. Interior of a brine-cured hide sample held for 43 days at 40°C after treatment with *Haloarcula hispanica*. [Adapted, by permission, from Vreeland, R H; Bailey, A, *J. Amer. Leath. Chem. Assc.*, **93**, 121-31, 1998.]

the formation of spots of salt, as opposed to fatty tissue, where there were no spots. Blood or lymph stains, acquired during slaughter and/or skinning, directly affect the formation of patches of salt, but blood remaining in vessels has a much greater impact. At temperatures below 10-12°C salt stains do not form at all. They appear only at temperatures above 13-15°C, and, at the higher temperatures (studied up to 40°C), this process occurs faster. Formation of salt stains is caused by diffusion into the skin and accumulation of hardly soluble salts, mainly calcium. This is possible only at high moisture content of skin tissue, which is in turn linked to an appropriate humidity (above 75%). In skins dry-salted and dried, the salt stains do not occur.

It was repeatedly confirmed that the type and amount of salt impurities such as calcium, magnesium, and iron salts, is of decisive importance for the formation of patches of salt. Salt stains are formed even in the case of salts containing relatively small amounts of impurities: 0.15% $CaSO_4$ and 0.03% $MgSO_4$. The spots do not occur even with the strong growth of bacteria when salt is free of Ca and Mg. Enzyme phosphatase plays a dominant role in the formation of salt stains. Although one may assume that phosphatases, together with other enzymes, may act in the formation of a salt stain, but results indicate that only phosphatase plays a role here.[17] Its exclusion by a suitable inhibitor fully protects the skin from forming salt stains. Phosphorous compounds released by enzymes escape with brine from the skin tissue but they are also partially bound by calcium salts. The presence of phosphorus in the spots with salts of calcium and magnesium has been found by many researchers. Small amount of alkaline phosphatase in the bovine skins (9.5 µg of p-nitrophenol compared to 80.0 µg in calf skins) is the main reason that the bovine skins rarely have patches of salt. One can prevent the formation of patches of salt, if salt used is free of alkali earth elements, or if it contains addition of phosphatase inhibitor. The salts of copper, zinc, phosphate, sodium fluorosilicate, and boric acid are the strongest inhibitors of these enzymes. Addition of 1-2% of these compounds protects skins from forming salt stains.[20]

4.7.2.7 Microorganisms in the skin processing

Microbes have some impact on production processes of leather. In some processes, microorganisms can be highly beneficial (dehairing). In other processes, the presence of microorganisms is quite harmful to the skin. This includes the preservation of raw hides and skins, soaking, pickling, and storage of finished leather. Species and abundance of microorganisms at different stages of the processing of skin are provided in Tables 4.27 and 4.28.

Table 4.27. Microorganisms occurring in tanning processes.[23,26]

Bacteria	Fungi
Bacillus anthracis, Bacillus brevis, Bacillus cereus, Bacillus firmus, Bacillus laterosporus, Bacillus licheniformis, Bacillus megaterium, Bacillus pumilus, Bacillus sphaericus, Bacillus subtilis, Micrococcus candidus, Micrococcus luteus, Micrococcus roseus, Micrococcus rubens, Pseudomonas aeruginosa, Staphylococcus aureus, Staphylococcus epidermidis	*Aspergillus flavus, Aspergillus fumigatus, Aspergillus ochraceus, Aspergillus niger, Aspergillus wentii, Mucor mucedo, Paecilomyces erlichii, Paecilomyces variotii, Penicillium aculeatum, Penicillium brevicompactum, Penicillium chrysogenum, Penicillium decumbens, Penicillium funiculosum, Penicillium luteum, Penicillium purpurogenum, Penicillum roseopurpureum, Penicillium rugulosum, Rhizopus nigricans, Scopulariopsis brevicaulis, Trichoderma viride, Verticillium glaucum*

Table 4.28. Number of species occurring at different stages of leather tanning. [Data from Birbir, M; Ilgaz, A, *J. Amer. Leath. Chem. Assc.*, **80**, 147-53, 1995.]

Stage of tanning process	Number of species	
	Bacteria	Fungi
Fresh hides	37	–
Salted hides	40	–
Soaking	12	–
Liming	12	–
Bating	8	–
Pickling	–	18
Chrome tanning	–	18
Drying	–	11
Finished leather	–	18

Bacteria, filamentous fungi, and yeast can be found in tanning industry processes and products. Defects caused by microorganisms in leather are created in the process of preservation and tanning at different stages. The wet processes, such as soaking, liming, decalcification, and etching, are bacteria-dominated, whereas the processes of pickling and tanning are dominated by fungi and yeast because of an acidic environment. The names of processes and their microflora are given in Figure 4.10.

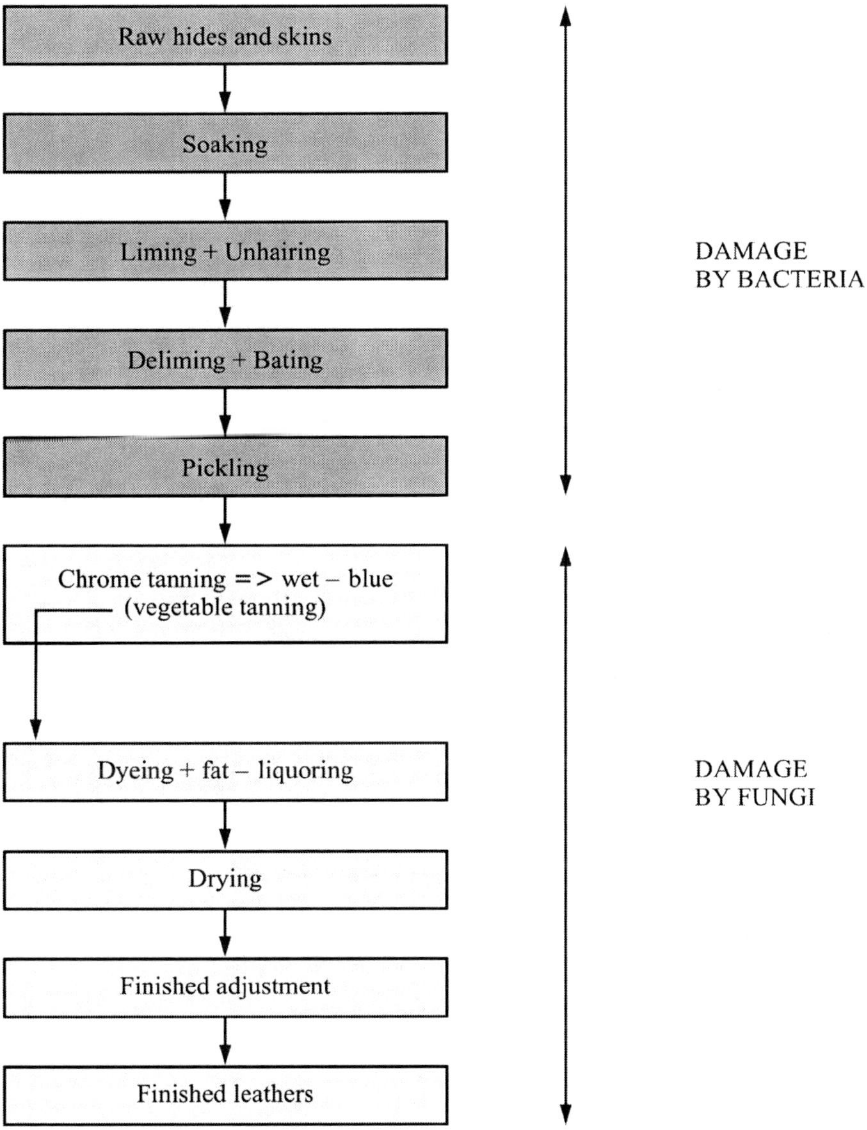

Figure 4.10. Technology of leather production. [Adapted, by permission, from Orlita, A, Int. Symp. Microbial Biodegradation and Biodeterioration of Technical Materials, Lódz, May 30-31, 2001, Technical University of Lódz, Lódz, 2001, pp. 41-54.]

Quantitative and qualitative microflora in each of the different processes is controlled by pH and temperature, as well as the presence of biocides. The graph (Figure 4.11) shows the pH at various stages of production.

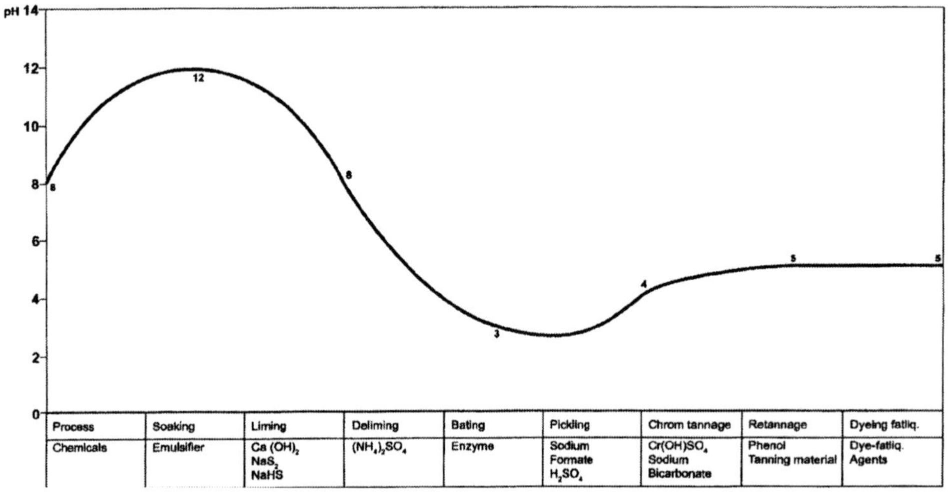

Figure 4.11. Chemicals and processes involved in leather production; pH values of the various stages. [Adapted, by permission, from J. Chemtech June 1992.]

Soaking

Skin soaking in water is the first tannery operation, performed to restore water lost by raw skin water during preservation. In addition, blood, dirt, salt and other soluble substances are removed. Proper conduct of soaking should assure hydration of skin tissue on the surface and through the entire thickness of skin with a minimum loss of cutaneous substance.

In the process of soaking, the salt content in skins is reduced, which makes skin vulnerable to bacterial attack and degradation. One reason, which may adversely affect the process of soaking and consequently the quality of the skin is the number and type of strains of microorganisms present in the water while soaking. Cleaning and flesh removal from hides significantly reduces the content of bacterial flora in water. Soaking temperature and time greatly influence the development of bacteria in soaking solution. Microbial activity is already visible at 5°C and if water temperature rises by 1-2°C, the number of bacteria in solution doubles. It is suggested that the temperature of the soaking solution should not be higher than 10°C.[23]

Most bacterial growth is caused by an increase in water temperature above 22°C or by extending the period of soaking during weekends and holidays. Typically, the process of soaking is conducted at 15-20°C with the addition of detergent, bacteriostatic agent, and sodium hydroxide.[35]

The adverse effect of soaking on the quality of skin is caused by the growth of bacteria capable of degradation of proteins using proteases, such as some anaerobes. Optimum conditions of spontaneous bacterial growth is observed after 40-48 hrs.[26] About 30-80% of bacteria isolated from the skins during the soaking has proteolitic activity. The bacteria invade the dermal substance and digest skin, causing perforations. Thin goat skin emits a bad smell and after 3 days on the surface of the flesh side rotten pits appear, which grow in the direction of grain until they cause perforation. In many cases, such damage occurs in a single day. Bovine leather perforation occurs in the same way but takes 5-7 days. The characteristic perforations of bovine skins start from the face side.[26]

The harmful effects of bacterial activity during the soaking are a loose and pimpled face of heavy cattle hides, small recesses on the face, loss of substance, and limp and loose fiber structure in the dermis, in both heavy and light finished skins.[14]

In order to limit the growth of bacteria during soaking of hides, it is necessary to frequently change the soaking solution, use clean water, and minimize the soaking time, as well as to control the bacterial population. It is possible to avoid digestion in the skins during soaking through addition of bactericidal substances, at least in some technological processes, especially in summer, when skins are soaked for a long time.

Liming

A process known as liming is used in the depilation of hides. It is usually carried out by using the chemical effect (sodium sulfide – calcium hydroxide – amines) and microorganisms or their enzymes, with or without the addition of alkali or buffer salt. Depilation by liming does not dependent on any bacterial action. This process is carried out at pH of 12-13, for 18 hours to 7 days. Limed hides are defleshed of excess tissue in order to impart appropriate thickness and the removal of muscle and fat. In some cases, defleshing is done after soaking (e.g., green skins).[36]

In the process of liming, biodegradation does not occur if raw materials are in a good condition.[26] Bacteria and their enzymes are irrelevant in the process of liming, because proteolytic and fat-degrading enzymes do not act in a solution of lime. Non-sporulating bacteria are killed in fresh lime solution, whereas the spore bacteria are capable of surviving in a passive form, but cannot reproduce. On the limed hides, proteolytic organisms, belonging to the genus *Bacillus* and *Microccocus*, can survive.[23]

Strains of aerobic bacteria isolated from skin, such as *Bacillus* and *Microccocus* species, produce enzymes having depilating action. Other examples of microorganisms whose products exhibit similar metabolic activity are *Aspergillus oryzae* and *Streptomyces fradiae*. In the choice of organisms for depilation, it is necessary to select species which give high depilation rate but have little effect on collagen, because the depilating enzymes encourage bacterial growth.

Bating

Liming is followed by the bating process, which is an enzymatic treatment in aqueous solution, resulting in a progressive and controlled degradation of proteins and removing unnecessary components that do not represent collagen fibers. Enzymatic process is carried out in order to make the skin soft, elastic, and flexible. This stage may last between 30 minutes to 12 hours. Environment during the etching operation is very conducive to bacterial growth due to the high temperatures (35 ÷ 40°C), the ideal pH of 8.0 (for microbial growth), and the presence of large amounts of protein, mostly in a solid form and partly in solution. Microorganisms present during the digestion process may come from commercial bating solutions, which are being reused.

The change in the microflora of the digestion solution is possible, depending on the different types of materials to be used, digestion time, and infections from the water, vat, etc., as well as the flora of delimed skin.

Currently, bating and deliming are usually carried out together by placing the hides in aqueous amine salt with proteolytic enzymes at a temperature of 27°C to 32°C.[36]

Pickling

This process is designed to stop the enzymatic digestion and to reduce the alkalinity of the chrome solution. Pickling is an acidification of skin conducted in such a way that it does not cause swelling under the influence of acid. In practice, this is achieved by adding salt. Solution containing acids (HCl, H_2SO_4, organic acids, NaCl) and salts is used to control the pH according to the requirements of the tanning process (leather stabilization, the process of digestion, and improvement of tannin penetration).[36]

It is generally accepted that pickled skins may be stored and protected for a significant period of time, because extremely acidic pickling solution (pH 2.0 ÷ 3.0) stops the growth of bacteria. Imported raw hides, infected with *Bacillus anthracis* were sterilized by pickling in a solution containing 3% HCl and 10% NaCl. Based on studies of impacts of tanning bath on survival of *Bacillus anthracis* spores in the skin and in the effluent, it was found that the liming process, despite the high pH (12 - 13) is unable to inactivate *B. anthracis* spores. Only the pickling process kills spores in the skin and effluents.[37]

The pickling process creates a suitable environment for the growth of molds. Their action is generally harmful. *Penicillium, Paecilomyces, Aspergillus,* and *Alternaria* are present on pickled skins. Fungicides such as thiobenzothiazol may be added during pickling.[36]

Tanning

Tanning is a process of transformation of raw animal hides to stable leather. During the process of tanning, semiproduct (wet-blue) is produced because of introduction of the tanning substances into skin tissue. This causes formation of additional bonds and collagen crosslinking and binding tannin functional groups with proteins. Vegetable, mineral, or synthetic tanning agents are used.

In most cases, the salts of chromium (III) and sometimes also of aluminum and zirconium salts are used as tannins. Leather for clothing, footwear, and industrial purposes is chrome tanned in drums for 4 to 24 hours. Vegetable tanning is the oldest process used in the leather tanning industry. It continues to be used in the processing of hard leather: soles, harness, and a special purpose leather. It lasts one day (in drums) or 6 weeks (in the tanning pits) for tannin (from bark) to penetrate into skin.[36]

Syntanes are synthetic tanning agents used only for certain, special leather or jointly with other tanning agents. Aldehyde and glutaraldehyde can be used alone as a tanning material, or together with chromium salts, improving the stability of the skin. After the process of tanning, the skin is called "wet blue" (chrome tanning), or "wet white" (if other tannins are used).

Molds pose a growing danger in the course of tanning, as well as in the case of leather storage after chrome or vegetable tanning. Acidic nature of the vegetable tanning wort does not permit development of putrefactive bacteria, but there are many microorganisms, for which the acids are an excellent nutritional material. Due to the oxygen requirements these microorganisms grow abundantly on the surface of wort, creating a thick coat. These include yeasts such as *Mycoderma* and *Torula*, and molds such as *Peni-*

cillium and *Aspergillus*. In vegetable tanning brines there are fungi which cause tannin degradation by enzyme tannase.

Enzymes secreted by fungi hydrolyze tannins, and the resulting decomposition products are deposited in the form of sludge and pyrogallic acid on the face of skin and on the surface of tanning vats.

The chrome tanned, vegetable tanned, and finished leathers are the perfect media for the growth of many types of microorganisms. Especially the presence of ammonium salts, phosphates, surfactants, and organic lubricating substances promote mold growth. Molds attacking wet-blue skin (chrome tanned leather) give rise to red spots. Orlita[26] isolated *Paecilomyces erlichii, Penicillium aculeatum, Penicillium purpurogenum, Penicillium roseopurpurum* from these spots. Spots on the skin appearing on the face and rough sides are caused by certain molds and yeast. These stains come from the decomposition of tannins, as well as the colors produced by these organisms. *Penicillium frequentans* produces pigment, which stains the skin to brown color. Pink spots on chrome tanned leather are caused by yeast *Torula muciloginosa* and *Torula glutinis*.

Drying

Drying process may also cause growth of fungi. Active substances in addition to microorganisms require adequate moisture and skin temperature, which may occur during the course of drying when the temperature and humidity rise in the oven or when air circulation is slow, or when the drying time is too long. Under these conditions there is a rapid growth of fungi. Microbiological decomposition of the skin is visible as colored patches, such as green, yellow-brown, dark-brown, gray, brown-green, and other colors. The colored patches formed on leather during drying are caused by fungi of the genera *Aspergillus, Penicillium, Peacilomyces,* and *Verticillium*.[26]

Growth and reproduction of bacteria may be caused by faulty procedures (e.g., skin contamination, the process without reducing the temperature, or contaminated salt). In the tanning processes, there are obvious differences in the composition of bacterial flora in skin salted, soaked, and during liming, or tanning. *Bacillus* species are characteristic of the first stage of hide deterioration. The processes of liming, tanning, and drying show less contamination than in the case of fresh, salted, and soaked skin. The presence of microorganisms at different stages of tanning process was studied by Birbir and Ilgazand.[23] Some of their results are included in Table 4.29.[23]

Table 4.29. Frequency of occurrence of various species of microorganisms at different stages of tanning. [Adapted, by permission, from Birbir, M; Ilgaz, A, *J. Amer. Leath. Chem. Assc.*, **80**, 147-53, 1995.]

Microorganism	Stage of tanning process & number of hide specimens (see explanation of stage number under the table)										
	1	2	3	4	5	6	7	8	9	10	11
Number of hide specimens	10	15	25	25	25	25	25	25	25	12	13
Bacillus brevis		2					1				
Bacillus cereus	4	8	3	1	2	14	3	3	8		
Bacillus firmus			1								

Table 4.29. Frequency of occurrence of various species of microorganisms at different stages of tanning. [Adapted, by permission, from Birbir, M; Ilgaz, A, *J. Amer. Leath. Chem. Assc.*, **80**, 147-53, 1995.]

Microorganism	Stage of tanning process & number of hide specimens (see explanation of stage number under the table)										
	1	**2**	**3**	**4**	**5**	**6**	**7**	**8**	**9**	**10**	**11**
Bacillus laterosporus				1							
Bacillus licheniformis		3	4	1	4	2		3	4		2
Bacillus megaterium	2	2	7	6	1						
Bacillus pumilus		3	4		6	2	13		6	9	1
Bacillus sphaericus	3	5	8								
Bacillus subtilis	7	10	24	9	7	1	12	6	4	9	5
Kurthia variabilis	1	3	1	1							
Micrococcus candidus		1	1								
Micrococcus luteus		10	5								
Micrococcus roseus	2	3	6						1		
Micrococcus rubens		6	3	2							
Pseudomonas aeruginosa				6							
Staphylococcus aureus	2	7	3	2			1				
Staphylococcus epidermidis		3	1	1							

1 – fresh hides, 2 – one week old salted hides, 3 – two month old salted hides, 4 – soaked hides, 5 – limed hides, 6 – chrome tanned hides, 7 – re-tanned hides, 8 – dried hides, 9 – finished hides, 10 – one day old leather, 11 – two months old leather.

It can be concluded from the above studies that aerobic bacteria (most likely *Bacillus*) are associated with most of the raw skins deterioration in both fresh and salted stock, and also during the wet processes.

All bacteria isolated by Birbir and Ilgaz[23] were proteolitic. After the drying stage, four different organisms were isolated from dried skins of which 3 belonged to the genus *Bacillus*. After completion of processing *B. cereus, B. lichenifofrmis, B. pumilus, B. subtilis* have been isolated from finished leather. Microbial flora of leather after one day were similar to two-month-old leather, and the species of *Bacillus* have been isolated from the skin.

Summary of deterioration of skin caused by microorganisms during processing in the leather industry is presented in Table 4.30.

Table 4.30. Microbial deterioration in leather industry. [Adapted, by permission, from Orlita, A, Int. Symp. Microbial Biodegradation and Biodeterioration of Technical Materials, Lódz, May 30-31, 2001, Technical University of Lódz, Lódz, 2001, pp. 41-54.]

Technology	Type of damage	Source of damage
	Biodegradation of hides and skin before tanning	
Raw hides: green hides salt hides Beamhouse: Soaking Liming Deliming Bating	Discoloration (colored stains) Putrefaction Hairslips Perforation	Halophilic and other bacteria Genus: *Micrococcus* *Bacillus* *Pseudomonas* *Proteus* *Escherichia* and other
	Biodegradation of leather after tanning	
Chrome-tanning (wet-blue) Vegetable tanning Fatliquoring Drying Finished leather	Colored stains Mold stains Lower quality Downgrade the economics Value of finished leather	Molds Genus: *Aspergillus* *Penicillium* *Paecilomyces* *Scopulariopsis* *Trichoderma* *Rhizopus* and others

Application of molecular techniques (16S ribosomal RNA sequencing techniques) to identify the bacteria isolated during the process permitted identification of 16 different types of bacteria contaminating the raw skin. Many species found were not previously identified in this environment. Several species identified for the first time, and some species commonly present in tanneries, are included in Table 4.31. Analysis of fresh skin revealed a large number of proteolytic and non-proteolytic bacteria colonizing the skin during processing. Proteolytic bacteria identified, including *Areomonas* species, *Proteus vulgaris, Shewanella putrefaciens, Pseudomonas* sp. were the most common, but only some of them showed proteolitic activity.[38]

Table 4.31. Bacteria present in the tanning processes of fresh hides as identified by RNA-based method. [Based on Oppong, D; Bryant, S; Rangarajan, R; Steele, S; Radwell, D; Hyllengren, L, *J. Amer. Leath. Chem. Assc.*, **101**, 4, 140-49, 2006.]

Bacteria	
Identified for the first time in the tanning environment	Commonly present in the tanning environment
Acidovorax sp., Acinetobacter sp., Aeromonas sp., Arthrobacte protophormiae, Brevibacterium lutescens, Comamonas sp., Delftia acidovorans, Dietzia maris, Janibacter terrae, Jeotgalicoccus psychrophilus, Nocardiopsis sp., Pseudomonas cannabina, Pseudomonas fulgida	*Aeromonas sp., Proteus vulgaris, Pseudomonas sp., Shewanella putrefaciens*

4.7.3 LEATHER

4.7.3.1 General data

Leather – the end product of the tanning industry – has properties required by its application. General characteristics include increased resistance to rotting and resistance to water.

Skin as a substance contains some amount of bound water, removing it causes structural changes in collagen, and thus changes the properties of the skin. Optimal performance has leather which contains 14-20% water. It can absorb much more water to about 50% without giving the feeling of being a wet material. Biological resistance of leather is better than that of hide, mainly because it contains collagen proteins, which are resistant to proteolytic enzymes except collagenase. Biological resistance depends greatly on the humidity and temperature. At a temperature of 30-40°C and relative humidity near 100% (tropical climate), the attack of microorganisms, especially fungi, occurs in a few days. Fungi cause collagen damage, although only slight, but even if removed, they leave permanent stains. The amount of non-collagen components in leather is drastically decreased after tanning, while tannins and a number of minerals become permanent components. In addition, during the tanning process, the removal or killing of vast numbers of microorganisms colonizing the raw skin takes place. But this process is not complete, because some microorganisms produce spores remaining in the hair pockets and skin bends. An important factor in the multiplication of fungi is the change in pH by 3-4 units toward the acidic. The result is that the conditions for the growth of bacteria become much worse, but better for the growth of filamentous fungi and yeast. The typical components of the microflora of buildings and human environments are aerobes of acidophilic character. A general rule of occurrence of individual species of microbes does not exist, because their growth depends on climate conditions and environment. Breeding ground for them is not collagen, but the accompanying substances in the form of protein residues, resulting from the destruction of keratin, globulin, albumin, sugars, fats, and mineral salts. The chemical composition and physical factors determine the type of microflora, which settles on surfaces. Complete removal of non-collagen substances is not technologically feasible, as it is impossible to decrease microflora in tannery to such an extent that no skin infections occur during and after tanning. Partial removal of contaminants does not lead to the reduction of infections, but changes the quality of microflora. Instead of some fungal species, others appear. Treatment of skin with fungicidal substances does not make it completely resistant to the action of microflora, and thus it becomes possible that even the best-protected skin during tanning process is attacked by molds. Preservatives for "wet-blue" skins (i.e., chrome tanned skins and hides of small animals sold in a wet state, lightly preserved and packed in polyethylene bags for further processing to obtain the finished leather, which requires additional tanning, dyeing, finishing, and oiling) do not give full resistance to the skin but cause the period of adaptation of microorganisms to be considerably extended to the limit of safe storage. There is also selection of microorganisms, since some are not able to adapt to the conditions, and there are others whose development in the absence of a preservative is probably suppressed by competing organisms. However, by subjecting the skin to the effect of fungicides, sufficient resistance is obtained from the technological standpoint.[39]

4.7.3.2 The microbiological decomposition of tanned leather

Leather, the result of technological process of tanning, is an excellent substrate for the growth of many types of microorganisms. The main damage to the finished leather is caused by fungi. The most common fungi on the leather, as identified by Orlita during studies of fungicidal activity,[26] can be found in Table 4.32.

Table 4.32. Fungal species isolated from finished leather.[26]

Aspergillus fumigatus, Aspergillus ochraceus, Aspergillus niger, Aspergillus wentii, Aspergillus flavus, Penicillium aculeatum, Penicillium brevicompactum, Penicillium chrysogenum, Penicillium decumbens, Penicillium funiculosum, Penicillium luteum, Penicillium rugulosum, Mucor mucedo, Rhizopus nigricans, Paecilomyces variotii, Scopulariopsis brevicaulis, Trichoderma viride, Verticillium glaucum

Apart from leather discoloration on the surface inhabited by most filamentous fungi (molds), they do not cause major visible damage. However, they may cause formation of depressions on the surface of leather.[40] Fungal growth on pickled skin or "wet-blue" begins with an infected surface, but molds can penetrate the entire thickness of the skin. Then the mycelium penetrates the weave of collagen fibers. And because the fungus is made primarily of polysaccharides, aminosaccharides and globular proteins, it has a very different response to the substances used in finishing operations than collagen. Second, a no less important reason for the deteriorating quality of the product, fat is metabolized by the fungus contained in the skin or added in the production process. Fungi break down fats, forming fatty acids, which may migrate to the skin surface, forming white, salty deposits. Fatty acids can react with the chrome tanning salts, giving chromium soaps, which cause uneven skin color. Presence of fat may lead to changes in the absorptive surface of the skin so that the coating is unevenly absorbed and the final gloss of the leather is not uniform. The quality of the product obtained from moldy skin "wet-blue" is reduced because patches of dull skin appear, which grow as mold growth progresses.

Zyska[41] presents a summary of species of fungi isolated from leather and leather products based on various publications until 1974. According to Kowalik[42] the most common and most active fungi, using by-products in the skin, include *Penicillium commune* and *P. glaucum, P. wortmannii, P. frequentans*, and *Aspergillus* species, such as *A. niger, A. flavus, A. candidus, A. oryzae A. fumigatus*. Also, *Alternaria, Cladosporium, Trichoderma, Fusarium, Neurospora, Aureobasidium pullulans, Scopulariopsis* and *Streptomyces coelicolor*. Some of these microorganisms have enzymes protease, lipase, and tannase.

Microflora of the mold with a clear dominance of the species of the genus *Penicillium* and *Mucor* was isolated by van der Werff from dust in warehouse of finished leather, and shoe shops, and from the surface of wet leather shoes (see Table 4.33).

Table 4.33. Fungi occurring on leather products.[43]

Warehouse dust and shoes shops	Moist footwear
Aspergillus versicolor, Cladosporium herbarum, Mucor hiemalis, Mucor racemosus, Mucor ramannianus, Penicillium brevi-compactum, Penicillium chrysogenum, Penicillium expansum, Penicillium phoeniceum, Penicillium roquefortii, Penicillium solitum, Pullularia pullulans, Rhizopus nigricans	*Penicillium corylophilum* *Penicillium solitum* *Penicillium spinulosum*

More than 70 species of fungi, including 17 *Aspergillus* species and 20 species of the genus *Penicillium* were isolated from leather and leather goods. Other species of fungi that were found, include *Chaetomium cupreum, Circinella simplex, Pichia rhodanensis*, and *Tritirachium cinnamoneum*.[44]

According to Zyska,[41] fungi and bacteria may cause the following stains:

- white-gray, under the influence of *Bacillus megatherium* on vegetable tanned leather
- gray-brown or violet, under the influence of fungi *Aureobasidium* and *Paecilomyces sp.* with vegetable tannins
- violet, under the influence of the fungus *Aspergillus nidulans* on leathers tanned by vegetable tannins
- red, under the influence of fungi of the genus *Aureobasidium* and *Paecilomyces* on skins tanned with Quebracho extract
- gray or black, under the influence of *Aspergillus niger*
- pink or red, under the influence of fungi of the genus *Chaetosphaeria, Rhodotorula rubra, Penicillium islandicum, Penicillium purpurogenum* or other unspecified types
- green, under the influence of the fungus *Verticillium glaucum* on leathers tanned with chrome and vegetable tanning agents;
- yellow-brown, dark brown, gray or green-brown, under the influence of *Aspergillus* and *Penicillium* on beige leather.

Grela-Jaskowska and Duda,[45] in the studies of the action of native microflora and fungi on the basic properties of leather, found that the dominant fungi invading the skin are *Penicillium sp.* and *Aspergillus sp.* They noted that the strength of the skin decreases with duration of action of fungi. The biggest drop in strength (from 5 to 8%) was noted after 3 months for undersole leather having vegetable tanning. Sharma[46] reported that 75% relative humidity caused only a slight loss of strength in leather tanned by chrome or vegetable agents. At 90% relative humidity, after 180 days, vegetable tanned leather has lost 13.6% of its strength, and chrome tanned leather only 1.27% of its strength. Under conditions of 70% relative humidity, vegetable tanned leather has lost only 9.6%, and chromium tanned skin only 0.72% of its strength. Fungi do not directly damage the collagen of the leather. Even long mold exposure does not change the properties of collagen. However, it is known that hydrolysis of tannins by fungi at high humidity leads to a weakening of fibers of the skin, formation of permanent stains, covering leather surface with mold, and reduction of aesthetic values and durability of leather goods.

4.7.4 CONTAMINATION OF SHOE MATERIALS

4.7.4.1 General data

According to Zyska,[44] 20 species of bacteria were isolated from leather shoes. In the analysis of footwear, attention should be given to the presence of pathogenic fungi and bacteria. Numerous studies demonstrated that microorganisms causing fungal infection were found in the footwear or garment parts directly in contact with human skin. Footwear, which does not properly fit increases the incidence of microinjuries which facilitate fungal infections. Also, daily use of shoes made from incorrectly chosen materials does not provide feet with the proper microclimate during use, and this is aggravated by wearing synthetic socks. Footwear products are exposed continuously to various microorganisms, which more readily grow because of presence of the sweat, fats, exfoliated epidermis. These secretions are deposited on the inner parts of footwear, where they form an ideal medium for microorganisms. Their presence may affect foot health because fungi found in shoes cause infection.

Dermatophytes and yeast fungi of the genus *Candida* are the etiological factors for fungal infection rates. The most common species of dermatophytes that cause athlete's foot are *Trichophyton rubrum, Trichophyton mentagrophytes,* and *Epidermophyton floccosum*. These are typical of pathogenic fungi, which attack the skin of the feet and nails. Nails yellow under their influence, then become darker, and cause nail stratification. Among the yeast, fungus *Candida albicans*, as an opportunistic pathogen, can cause fungal infection of nails with symptoms similar to those caused by dermatophytes. Among the filamentous fungi, *Aspergillus*, especially *Aspergillus fumigatus*, and the species of the genus *Scopulariopsis*, such as *Scopulariopsis brevicaulis* are also opportunistic pathogens.[47,48] In bacterial infections, the most important role plays *Staphylococcus aureus*, causing purulent skin changes. From lesions on the skin *Streptococcus β-hemoliticus, Streptococcus faecalis, Pseudomonas aeruginosa, Proteus vulgaris*, and *Clostridium* anaerobes were isolated.

4.7.4.2 Microbiological aspects of the use of footwear

Particularly favorable conditions for the development of microorganisms are found in fully enclosed sport shoes of wear all-day type. An unpleasant odor and discoloration appear first, followed by spots. These are only external signs of metabolic activity of microorganisms. Degradation changes all properties of footwear, including structural materials. Also physical-mechanical properties and hygienic properties degenerate. Deterioration of individual parts of footwear depends on the type of material, environmental conditions (temperature, humidity), conditions and manner of use of footwear, as well as foot microflora. Footwear is exposed to undergoing action of microorganisms, both during use and storage, conditions of which are often adverse in respect of high humidity and high temperature.

Studies[49-50] show that bacterial count inside the shoe increases from the first day of its use. Amount of bacteria accumulated in insoles within 3 weeks reaches $10^4/cm^2$, achieving a state of equilibrium, stable over the next 3 months, especially in closed shoes (see Figure 4.12).[50] The open shoes have a continuous slow growth of bacteria. Bacterial count after 5 months exceeded $10^4/cm^2$ (see Figure 4.13).[50] While the number of bacteria in closed footwear is the result of their development, depending on the microclimate of an enclosed space, the number of bacteria determined in open shoes mainly depends on contamination from the surrounding environment, as evidenced by the steady increase in the number of microorganisms depending on the period of use.

Analyzing the dynamics of multiplication of bacteria on insoles during use and storage of footwear, there was apparent decrease in the quantity of bacteria during rest, when the shoes were not used, and the growth of microorganisms during their use.

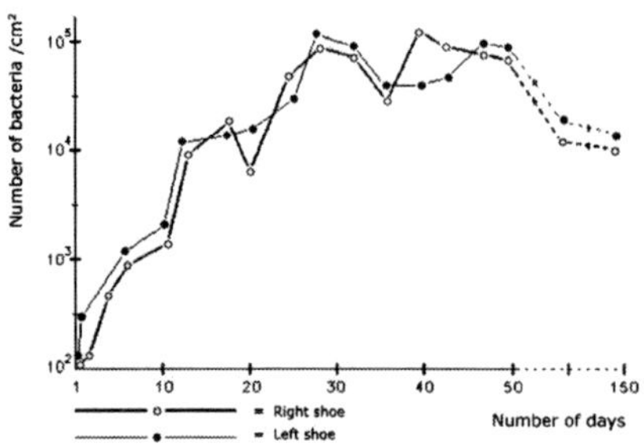

Figure 4.12. Bacterial count in insoles in fully closed leather footwear. [Adapted, by permission, from Orlita, A, Int. Symp. New materials and methods of shoe production, Kraków, November 16-17, 1998, CLPO, Kraków, 1998, pp.107-129.]

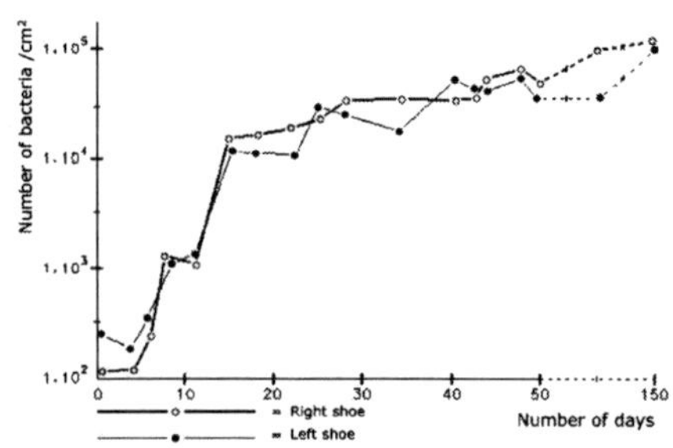

Figure 4.13. Bacterial occupation of open leather footwear. [Adapted, by permission, from Orlita, A, Int. Symp. New materials and methods of shoe production, Kraków, November 16-17, 1998, CLPO, Kraków, 1998, pp.107-129.]

Bacteria are a dominant group of microorganisms colonizing subsoles. Fungi and yeast are represented in a minimum quantity. On the basis of 30 studies, it can be concluded that spores of filamentous fungi are present by average amount of 50/cm^2 of subsoles, while the yeast fungi are present by average amount of 46/cm^2. In shoes stored for a period of time, there is an increase in filamentous fungi spores as a result of contamination from the surrounding environment. Number of yeast fungi and bacteria gradually decreased (see Figure 4.14).[50] Saprophytic fungi are a small fraction of microorganisms.

Comparison of colonization of subsoles in closed shoes made out of genuine leather and imitation leather shows faster bacterial growth in footwear from plastic substitutes (see Figure 4.15).[50] In a humidity saturated environment inside the shoe (90-100% relative humidity) organisms reproduce very well because there are sufficient sources of food from condensing sweat. Effect of relative humidity on the growth and development of the bacteria *Staphylococcus aureus* and *Bacillus megatherium* at 37°C, in the presence of artificial sweat, on cotton fabric was studied by Orlita.[49] Maximum bacterial growth is observed at a relative humidity of 100%. Relative humidity lower than 90% had little influence on the growth of bacteria, even at optimum temperature of incubation. The 96-100% humidity level is an essential condition for the propagation of bacteria and decomposition of sweat, and thus it is one of the factors worsening the feeling of comfort in footwear use.

Effects of sweat on the development of microorganisms depends on the type of materials used in the manufacture of footwear. Grela-Jaskowska *et al.*[51-52] found that sweat affects the growth of bacteria on nonwoven and cellulose-based materials, which was much less evident in the case of natural leather and artificial leather. The degree of infection of subsole materials during use of footwear showed an increase in the total number of bacteria and fungi, and the proliferation of fungi and yeast-like pathogenic bacteria, which may indicate an unfavorable microclimate for the user in the case of the shoes tested. At the same time, studies have demonstrated that the clinical status of feet worsened with

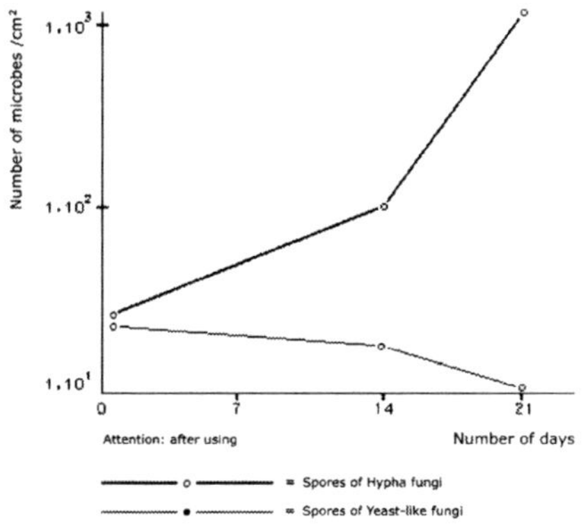

Figure 4.14. Hypha fungi and yeast-like fungi which occupy insoles in fully closed leather footwear, which were stored after use. [Adapted, by permission, from Orlita, A, Int. Symp. New materials and methods of shoe production, Kraków, November 16-17, 1998, CLPO, Kraków,1998, pp.107-129.]

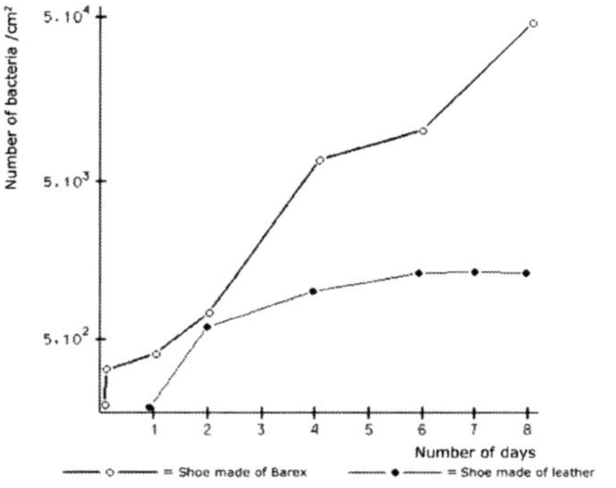

Figure 4.15. Bacterial count in insoles of fully closed leather footwear and synthetic leather Barex footwear at the beginning of use. [Adapted, by permission, from Orlita, A, Int. Symp. New materials and methods of shoe production, Kraków, November 16-17, 1998, CLPO, Kraków,1998, pp.107-129.]

time of use of non-hygienic footwear. Parasitic group of filamentous fungi that normally grow in wet areas of toes was similar to the bacteria – it was driven by increased sweating, primarily as a result of wearing shoes which did not permit permeation of air and water vapor. Fungi isolated from new and used footwear made out of leather and substitute materials are summarized in Table 4.34.[53-55]

Table 4.34. Fungi occurring on leather and substitute materials footwear.[53-55]

New footwear	Used footwear
Molds: *Alternaria sp., Aspergillus flavus, Aspergillus sp., Cladosporium sp., Mucor sp., Mycelia sterilia, Rhizopus sp., Penicillium sp.* **Yeast-like fungi**: *Candida sp., Rodotorula sp.*	**Molds**: *Alternaria sp., Aspergillus flavus, Aspergillus fumigatus, Aspergillus niger, Aspergillus sp., Cladosporium sp., Chrysosporium sp., Epicocum sp., Fusarium sp., Mucor sp., Mycelia sterilia, Penicillium sp., Scopulariopsis brevicaulis, Stemphulium sp., Trichothecium roseum* **Yeast-like fungi**: *Candida glabrata, Candida krusei, Candida kefyr, Candida sp., Rodotorula sp., Torulopsis sp., Sacharomyces cerevisiae* **Dermatophytes**: *Trichophyton mentagrophytes*

In the interior of used footwear, 23 species of fungi, including 9 species of pathogens that cause fungal infections of the skin and toe nails, were found. Fungi *Trichophyton mentagrophytes* and yeast of the genus *Candida,* such as, *C. krusei, C. glabrata, C. kefyr* are the most pathogenic species. Other pathogenic species belong to *Aspergillus,* especially *A. fumigatus*, and also *Scopulariopsis brevicaulis.*

From new shoes, yeast and mold fungi were isolated (see Table 4.35).[53-55] The development of these fungi may create favorable conditions for the development of other microorganisms inside the shoe during use.[54]

Microbiological assessment of the feet of footwear users for the period of 3 months showed the dynamics of colonization of the interior of footwear by fungi. Fungi were isolated from feet swabs in 27.3% of cases and shoe lining swabs gave 75% positive isolations. In two cases, isolated fungi were pathogenic, such as *Trichophyton mentagrophytes*, which was isolated from both feet and shoes. By contrast, *Candida albicans* and *Candida sp.* were isolated in a large number of colonies grown only on feet. Other species of fungi isolated from both swabs of linings and swabs of between-toe spaces formed only a single colony. Their presence cannot be associated with pathogenicity.

Use of proper footwear design and good hygienic qualities can cause a marked reduction in both quantity and quality of microflora inhabiting the foot. Fungi isolated from the footwear used by 63 children aged 8-14 years and their occurrence on the feet and shoes are given in Table 4.35.[54]

Table 4.35. Frequency of fungi occurrence on children feet and their footwear. [54]

Fungal species	Presence on feet		Presence in footwear
	before using	after 5 weeks use	
Alternaria sp.	7	0	4
Aspergillus flavus	1	0	1
Aspergillus fumigatus	1	4	2
Candida albicans	6	0	0
Candida guilliermondii	3	0	0
Candida kefyr (pseudotropicalis)	2	2	2
Candida krusei	0	1	1
Candida sp.	5	11	15
Cladosporium sp.	7	1	1
Geotrichum candidum	1	0	0
Mucor sp.	1	0	1
Penicillium sp.	19	9	9
Rodotorula sp.	6	4	5
Saccharomyces cerevisiae	0	2	3
Scopulariopsis brevicaulis	3	2	3
Syncyphalastrum sp.	1	0	0
Torulopsis glabrata	0	0	2
Torulopsis insconspicua	4	0	0
Torulopsis sp.	6	3	4
Trichophyton mentagrophytes	2	0	0

Table 4.35. Frequency of fungi occurrence on children feet and their footwear. [54]

Fungal species	Presence on feet		Presence in footwear
	before using	after 5 weeks use	
Trichotecium roseum	2	0	0
Verticilium sp.	1	0	0

Analysis of incidence of pathogenic fungi before the studies of the effect of shoes led to detection of *Trichophyton mentagrophytes* on both feet of one child. In five children, fungus *Candida albicans* was found, which, under favorable conditions, causes tinea. Tests after trials of shoes showed reduction of fungi on the feet of children. In the case of a child having *T. mentagrophytes,* foot therapy was used with full success. *Candida albicans, Candida guilliermondii, Alternaria sp., Torulopsis insconspicua* were not found in these studies. Population of *Penicillium sp., Cladosporium sp.,* and *Torulopsis sp.* declined.

Sometimes filamentous fungi are the cause of accelerated deterioration of footwear during use. High boots for women, used in a relatively short period of time in the autumn and winter, obtained from customer complaints, were examined by Falkiewicz.[56,57] Complaints included damage to the leather upper in the form of black and colored stains and cracking and delamination of polyurethane coating. Studies of materials from damaged boots showed that they were contaminated with filamentous fungi (see Table 4.36).

Table 4.36. Fungi isolated from new footwear materials and footwear materials taken from female heat-insulated footwear damaged by fungi (four boots were investigated). [54]

Fungal species isolated from footwear materials	Frequency of occurrence	
	footwear material from female heat-insulated footwear, damaged by fungi (12 specimens were investigated)	new footwear materials used in footwear production (8 specimens were investigated)
Alternaria sp.	7	4
Aspergillus flavus	3	2
Aspergillus niger	7	3
Aspergillus sp.	1	2
Cladosporium sp.	7	3
Chrysosporium sp.	1	1
Epicocum sp.	1	1
Fusarium sp.	1	1
Mucor sp.	4	0
Mycelia sterilia	5	2
Penicillium sp.	9	7
Rhizopus sp.	9	4
Stemphylium sp.	6	1
Trichothecium roseum	1	1
Ulocladium sp.	2	1

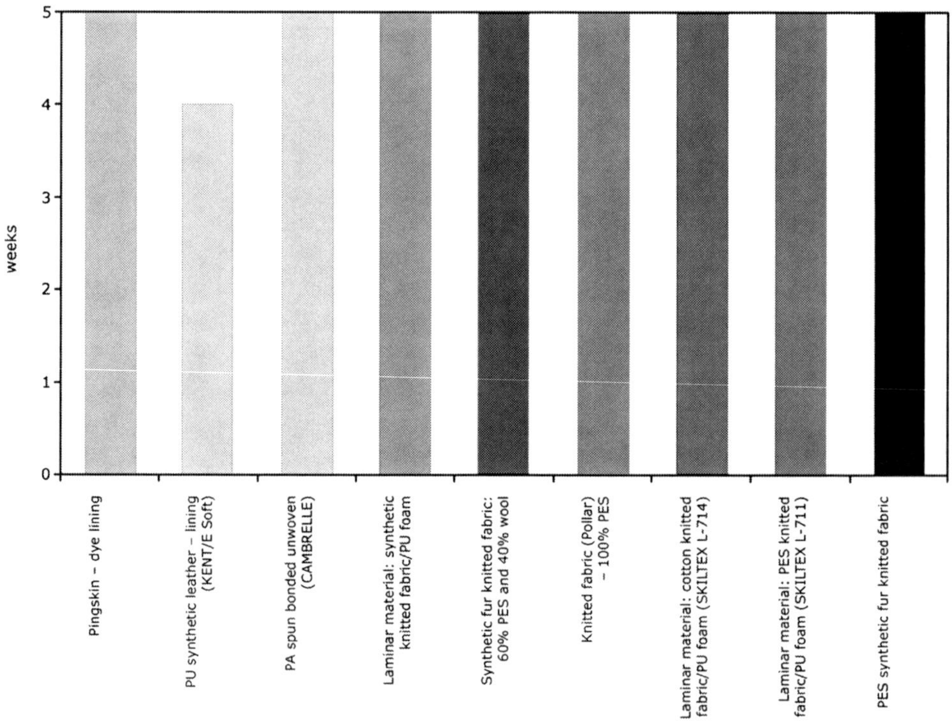

Figure 4.16. Bacterial (*Staphylococcus aureus*) survival rate on lining materials. [Adapted, by permission, from Falkiewicz, M; Pawlik, B; Macura, A B, *J. Mikrobiologia Lekarska*, **11**, 4, 291-96, 2004.]

Microbiological testing of new materials shows that the test materials used for production of footwear show varying degrees of mycological contamination. Filamentous fungi were the most frequently isolated microorganisms (see Table 4.36).

The identification of spores in production materials used in footwear manufacture showed a similarity between fungi species found in production materials and returned shoes. *Aspergillus niger, A. flavus, Rhizopus sp.* were dominant species. In both cases, black mold has also been isolated from the genera: *Alternaria, Stemphylium*, and fungi of the genus *Penicillium*. Dominant fungi *Aspergillus* and *Rhizopus* are important from a medical point of view, because of their pathogenicity for humans. *Aspergillus niger* and *A. flavus* cause skin lesions and attack nails, causing onychomycosis.

Microbiological materials used in the manufacture of footwear have a significant impact on the hygiene and service life of footwear. Molds are among the most dominant groups of microorganisms responsible for biodeterioration of natural polymers. Different strains of fungi attack materials with a different intensity, which depends on the composition of the raw material base, the specific physiological needs, and the metabolic mechanism of the strain. In considering the life processes of microorganisms isolated from the damaged shoe uppers, it should be noted that in the rivalry between different species,

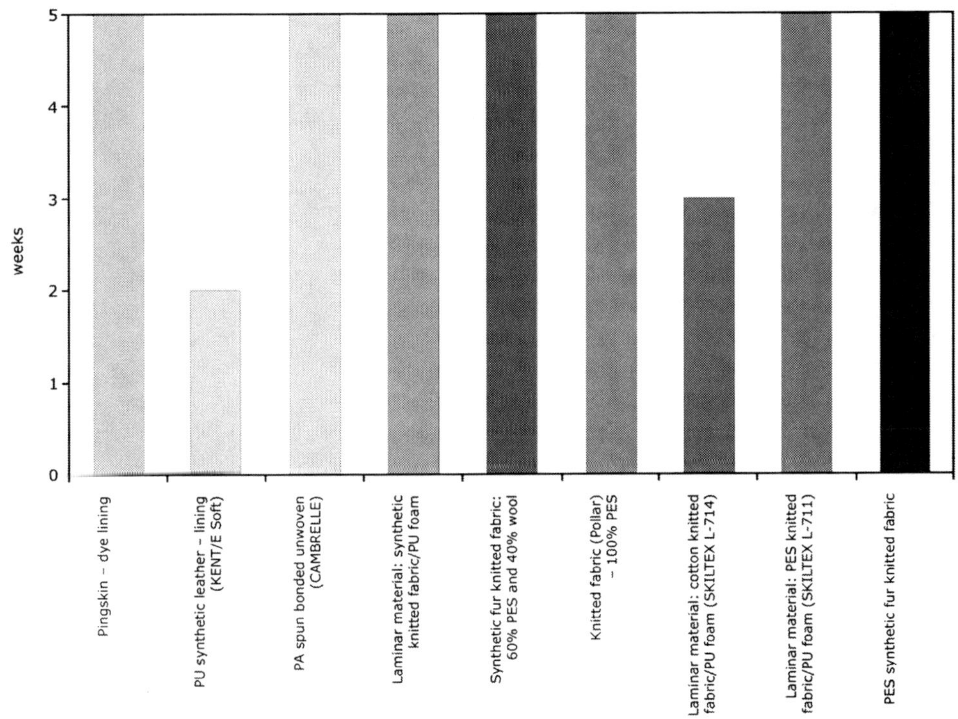

Figure 4.17. Fungal (*Candida albicans*) survival rate on lining materials. [Adapted, by permission, from Falkiewicz, M; Pawlik, B; Macura, A B, *J. Mikrobiologia Lekarska*, **11**, 4, 291-96, 2004.]

fungi of the genera *Aspergillus* and *Rhizopus* dominate, because of their great vitality and rapid growth rate compared to other isolated fungi. Microbiological characteristics of materials can be significantly improved by the use of antimicrobial additives.[56]

Microbiological testing of materials for shoes is conducted to verify the need for antimicrobial protection and to find ways to reduce the detrimental impact on foot health, which depends the microorganisms growth. Survival of the pathogens causing the infections was observed on different materials used in the manufacture of footwear, diversified in terms of chemical composition and structural construction. Every week for 1-5 weeks intensity of microbial growth was evaluated. Infected test samples (with sweat present) were incubated at 27°C and relative humidity of 70-80%. Samples showed an increase in concentration of many species of pathogenic fungi and bacteria. Natural materials, polyurethane foams, and leather-like materials were the most susceptible to the development of microorganisms.[58,59] Figures 4.16 and 4.17 show survival of *Staphylococcus aureus* and *Candida albicans* on lining.

4.7.5 LEATHER AND PARCHMENT IN LIBRARY MATERIALS

4.7.5.1 General data

Leather is a material of high durability, as evidenced by, for example, shoes from Roman times found in the twentieth century. On the wreck of the ship *Mary Rose* leather objects were found to be in good condition, and these were for years under water, and from leather found in *Metta Catharine*, which was sunk over 230 years ago, high quality leather accessories were made. However, skin aging is influenced by external factors that cause changes in physical, chemical, and mechanical properties. The external environmental factors are air, temperature, humidity, microorganisms, insects and rodents, pollutants, as well as disasters and accidents in the libraries. The method and place of its storage has a significant impact on the aging of the leather. It was observed that bound books have a much greater lifespan if they are stored, for example, in Aberystwyth, than in London. This implies a more harmful environment in the capital of Great Britain than in the less polluted countryside in Wales. It has been proven that various kinds of leather have different resistance to air pollution. Leather produced with a vegetable tanning process (catechu) is less durable than leather produced with tannins such as tare and sumak.[60]

Library materials made of leather are often massively attacked by microorganisms, which increase with time of moisture exposure. Food, moisture, and ambient temperature are complementary; shortage of one of these conditions can be offset by the remaining two.[61] In historical objects from leather, microorganisms find protein substances. All leather products, such as binding elements, costumes, and travel equipment contain collagen. The book bindings used in the past also contain animal glue produced from skin and bone obtained from cattle, as well as glues from rabbits and fish. In general, molds and yeast grow profusely on the library objects made out of skin. Parchments favor growth of proteolytic fungi, such as *Mucor*, *Chaetonium*, and *Gymnoascus* as well as imperfect fungi of the genera *Trichoderma, Aureobasidium, Epicoccum*, and *Verticillium*. Microflora on leather products is represented by similar types of fungi, however, enriched by the members of genus *Paecilomyces* famous for its lipolytic taste. Actinomycetes and other bacteria pose the biggest threat to parchment (see Figure 4.18).

Figure 4.18. Parchment bookbinding deteriorated by *Actinomycetes*. [Adapted, by permission, from Strzelczyk, A B, Int. Symp. Microbial Biodegradation and Biodeterioration of technical Materials, Lódz, May 30-31, 2001, Technical University of Lódz, Lódz, 2001, pp. 28-40.]

The skin of young animals, which was vegetable tanned, falls into shreds under the action of microorganisms. By contrast, objects made out of thick and tough skin, used for harness and heavy shoes, are more resistant to microbiological degradation. Similarly, resistant are skins, which for a long time were buried in the ground, such as archaeological leather. This is due to the reaction of vegetable tannins with iron salts present in soil, which produces complexes of iron with tannins, and these are very difficult to degrade. Because of microbiological processes, the museum collections abound in fragments of shoes, harnesses, luggage, etc.[61]

4.7.5.2 Archaeological leather

Wet excavated leather from archaeological sites (lakes, seas, and other environments), degraded by water, is usually black, wet, soft and spongy. These objects survived because of the anaerobic conditions that inhibit growth of aerobic bacteria. Also, such an environment stimulates formation of toxic substances such as methanol, hydrogen, indole, skatole, which are all toxic to microorganisms. Plant residues facilitate further penetration of the skin with tanning agents, which prolong the life of buried materials. Skin items found in archaeological excavations are subject to numerous physical processes such as extraction and reverse migration. Ground water penetrating into the skin extracts soluble substances such as short chain proteins, water-soluble sugars, salts of tanning components, and pigments. At the same time, the skin may become saturated with the substances present in the surrounding environment. Circulation of groundwater causes leaching of organic substances from decaying plants and other organic residues and transports them into leather. In this environment, the leather is saturated with minerals forming insoluble salts or complexes. Iron ions are the most reactive and cause formation of ferrous complexes with tannins. Black iron sulfides are formed from H_2S – the main product of microbiological degradation of sulfur contained in proteins (e.g., wool, hair, etc.). Often skin is saturated with calcium salts, which, when exposed to the atmosphere, crystallize on the surface. Archaeological objects made out of leather found above the water surface are saturated with minerals from the groundwater solution. Leather goods excavated from dry archeological sites are not always properly protected. Clammy leather becomes slightly discolored on the surface, covered with light-colored bloom, dark inside, hard and stiff when it dries. Often it is contaminated with mineral salts that crystallize on the surface. Analysis shows the presence of crystalline impurities containing Ca^{2+}, K^{+}, Fe^{2+}, Mg^{2+}, Mn^{2+}, Zn^{2+}, PO_4^{3-}, OH^{-}, CO_3^{2-}, SO_4^{2-}. Leather items lacking maintenance are easily attacked by microorganisms. Leather goods extracted from wet environments, which are black, swollen, and soft, are stored refrigerated until the process of conservation begins.

Archaeological leather is prone to microbiological growth manifested by white-gray and colored coating, which often covers the entire surface. Microorganisms growing on the moist leather can cause its biodeterioration, although the process is slow. Archaeological leathers are more deficient in easily degradable organic matter than other leathers and therefore are slower to biodeteriorate. Excavated leathers have the same types of fungi (*Penicillium, Paecilomyces, Aspergillus, Trichoderma*) and *Ascomycetes* that inhabit the skin of historical, leather-bound books in libraries.[62]

The most common germs on stuffed animals (fish, reptiles, and birds) are fungi belonging to *Aspergillus* and *Penicillium*, giving the black and green surface stains.[63] Nigam, Dhawan, and Nair found that museum objects containing keratin and dust are rich in keratinophilic and saprophytic fungi, which in the museum environment are responsible for deterioration of these objects (see Table 4.37).[63]

Table 4.37. Fungi isolated from deteriorating museum objects composed of feather and leather and their surrounding dust. [Adapted, by permission, from Nigam, N; Dhawan, S; Nair, N V, *J. Int. Biodet. Biodeg*, **33**, 145-52, 1994.]

Fungi	No. of isolates	Museum objects	Dust
Acremonium sp.	2	–	2
Aspergillus flavus	3	2	1
Aspergillus niger	3	2	1
Aspergillus sulphureus	1	–	1
Chrysosporium evolceanui	2	–	2
Chrysosporium indicum	2	–	2
Chrysosporium keratinophilum	2	1	1
Chrysosporium tropicum	3	2	1
Chrysosporium sp.	1	–	1
Fusarium sp.	2	1	1
Penicillium citrinum	1	–	1
Penicillium chrysogenum	1	1	–

4.7.5.3 Hard leather bindings

Bookbinding with leather was popular because of its flexural resistance, which permitted multiple opening and closing of the books without fear of breaking the spine. In addition, on the spine, the author's name and title and any other required information can easily be embossed in gold, silver, or other color, which remains visible even when embossing has been mechanically removed or damaged. Vegetable tanned leather is the most frequently used, because of its high elasticity. At the same time, the best bookbinding material can be obtained by combined chrome and vegetable tanning methods. Leather from such a process has higher water resistance and good stability to heat and light. It is also more resistant to biodeterioration than vegetable tanned leather. Because of long-term storage conditions, which are often inadequate (moisture, temperature, light), a significant number of leather bookbindings suffer from microbiological corrosion and aging under the influence of the environment. The following processes are responsible for biodeterioration:

- hydrolytic degradation of collagen proteins to free acids, namely sulfuric acid
- catalytic decomposition of collagen protein by iron and copper ions (Fe^{2+}, Fe^{3+}, Cu^{2+}).

Atmospheric sulfur dioxide absorbed by leather is oxidized to sulfur trioxide (SO_3) by reaction with atmospheric oxygen. With moisture present in leather, a free sulfuric acid (H_2SO_4) is produced.

Acids are present in leather bookbinding as a consequence of some decoration techniques, matting, and wax staining. Most cations of iron and copper are introduced into leather in the form of sulfites. Aging is accelerated by direct exposure to sunlight, dust, and storage in a humid, non-ventilated environment. It was found that in humid storage conditions, the concentration of free fatty acids in the book bindings can reach 8%.

Chemical destruction, the so-called "red decomposition", causes deterioration of vegetable tanned leather by affecting collagen bundles and depolymerization of protein molecules. In this case, the leather color changes to red and its surface flakes off, exposing the inner part of the dermis (see Figure 4.19.). On advanced stages, the entire skin begins

Figure 4.19. Chemical damage of leather bookbinding – red rot. [*Courtesy of KB Nat. Library of the Netherlands*]

Figure 4.20. Biological damage of leather bookbinding.[*Courtessy of KB Nat. Library of the Netherlands*]

to crumble and turns to dark powder. Leather sensitivity to the red deterioration depends largely on the type of tannins used.

Biological factors cause approximately 20% of the damage of the book covers by many representatives of both insects (*Dermestidae*, *Anobiidae*) and microorganisms. They damage the valuable material, not only by the formation of patches of varying color intensity and size, but they hydrolyze leather. Binding gradually breaks down because its physical and chemical properties have been degraded (see Figure 4.20). Bacterial growth is usually supported by the secretion of various metabolic products, of which organic acids have a particularly adverse impact on the material, partly because of their digestive action and partly because acids change pH permitting further colonization.

Microflora on the historical vegetable tanned leather used for book bindings consists mainly of microscopic fungi and Actinomycetes (Table 4.38).[64-66] The skin is a very good substrate for its nutritional value, because it contains a protein, enriched in the production with lipids and tannins.

Table 4.38. Microorganisms occurring on historical vegetable tanned leather bindings.[64-66]

Fungi:
Aspergillus candidus, Aspergillus fumigatus, Aspergillus niger, Aspergillus ochraceus, Aspergillus ustus, Aspergillus versicolor, Chaetomium globosum, Cephalosporium sp., Cladosporium epiphyllum, Fusarium sp., Geotrichum sp., Humicola sp., Mucor racemosus, Paecilomyces variotii, Penicillium chrysogenum, Penicillium decumbens, Penicillium frequentans, Penicillium funiculosum, Penicillium nigricans, Penicillium roquefortii, Pseudogymnoascus sp., Rhizopus nigricans, Scopulariopsis brevicaulis, Scopulariopsis sp., Spicaria divaricata, Spicaria silvatica, Spicaria violacea, Syncephalastrum racemosum, Thamnidium elegans, Trichoderma glaucum, Trichoderma viride, Verticillium sp.
Actinomycetes:
Streptomyces sp.

These components of leather under specific conditions may enhance mold growth. Also, adhesives, binders, and natural pigments can be used as a food source. In addition, earlier hydrolytic degradation of leather by, for example, "red digestion" makes a ground for further colonization. The population of microorganisms on the historical vegetable tanned leather varies depending on leather type. In many cases, fungi dominate the leather bindings of books, because acidity of the material creates proper environment for their growth. For this reason, they are able to grow more vigorously than Actinomycetes, whose growth is initially reduced by fungi and, most likely, the acidity of the skin. In the process of leather biodeterioration, competition for food and space and antibiosis play essential roles. Fungi and Actinomycetes can interact with each other. Fungi usually attack non-collagen materials (tannins, fats), while Actinomycetes destroy mycelium and collagen fiber.[64] In the first phase of deterioration *Penicillium* and *Zygomycetes* dominate. Tannins

Figure 4.21. Bookbinding made of white leather, deteriorated by microorganisms. Photographed by Vibolla Khlott. [Adapted, by permission, from Strzelczyk, A B; Fedrizzi-Szostok, A; Karbowska-Berent, J, Int. Symp. Microbial Biodegradation and Biodeterioration of Technical Materials, Lódz, September 8-10, 2003, Technical University of Lódz, Lódz, 2003, pp. 194-201.]

are their main source of food. These are followed by *Chaetomium globosum, Scopulariopcis brevicaulis*, and *Streptomyces sp.* Both fungi and Actinomycetes are capable of dissolving the collagen fibers, the latter can decompose collagen fibers from inside.[65] Table 4.39 lists microbes colonizing historical bookbindings made out of vegetable tanned leather.[64-66]

In the 16th and 17th centuries, parchment was used as a white binding material. Books were decorated by ornamental embossing.[67,68] To obtain a plain white leather, alum ($KAl(SO_4)_2 12H_2O$) was used. Pseudo-tanning with alum was used together with the procedure of oil oxidation, and sometimes other compounds were added to alum. This technology was recreated based on historic records.[66,69-70] In the Middle Ages, thoroughly cleaned skins were placed into a mixture of alum and oil, possibly with addition of sea salt, egg yolks, butter or flour. Skin was trampled by foot to incorporate the slurry into its structure, and then left in piles overnight. The next day, skin was tumbled again: trampled upon, rolled and struck against the surface of a countercoup, then smoothed with a flat tool and hung to dry in a warm oven. After drying, the appropriate softness and flexibility was achieved by strong rubbing with a hard blunt knife. Leather made in this manner has a white-cream shade, and it is smooth, soft, and pleasant to the touch. It is elastic, ductile, mechanically strong and resistant to chemical pollution and light. It was further protected against mechanical damage and the effect of the environment by priming the surface with natural resin or animal glue. All these operations resulted with very durable white leather used for bookbinding, but it was no less vulnerable to long-term degradation. Degradation processes are driven both by the chemical processes taking place under the influence of abiotic factors as well as the activity of certain types of microbes. During these processes, degradation of collagen fibrils occurs with formation of lower peptides and polypeptides. Insects (*Anobiidea* and *Dermestidae*) play a role in degradation, but also microorganisms including bacteria, actinomycetes and fungi (see Figure 4.21.). Despite of the fact that white leather was tanned with inorganic compound, which is not a source of food for microorganisms, the presence of lipids made it susceptible to mold attack. At high humidity, collagen was more vulnerable to biodegradation stimulated by bacteria. Leather sensitivity to water is caused by the weak bonds between aluminum salts and collagen fibers and this increases susceptibility to colonization by microflora.[71] White leather contains a small amount of bound aluminum (2-4%) in comparison with vegetable tanned leather (18-24% of tannins).[72,73]

Regardless of the tanning method, there are some similarities in respect of their susceptibility to microbial damage. Thick and stiff pig skin is more resilient than delicate deer leather. Also more susceptible to degradation are cattle and goat leathers because of their thickness. Microorganisms readily consume fat and water-soluble substances. The fat content of the old leathers is small, and therefore water-soluble substances are a more important source of food for microorganisms. These are products of the natural aging of collagen. The fat content, although initially around 6%, after years of hydrolysis and oxidation with formation of fatty acids, decreases to about 1%. These leathers, similar to vegetable tanned, become rigid, hard and brittle, and when damp or wet they are easily microdegradable.

Both the skins tanned by aluminum and vegetable processes have similar microbial succession during incubation. Common fungi of the genera *Penicillium, Aspergillus, Cladosporium* appear after a few days of incubation. Then, *Mucor sp.* and *Rhizopus sp.* assume dominance, often for several weeks (they use the remaining fat). At a later stage, bacteria begin to grow, especially actinomycetes of the genus *Streptomyces*, which contribute to the overall degradation of leather. Most samples of deer skin and calf, both control and treated with the various treatments, fall apart after 3-5 months. Significant differences in microbial susceptibility were observed for leather from different animals (various species and ages) with the same method of tanning. In studies of microbial destruction of historic parchment skin, *Streptomyces* plays special role because it is capable to degrade collagen. Also, some species of *Aspergillus* have collagenolytic enzymes.[74,75] The presence of hydroxyproline in the solutions during the incubation of samples shows that the micro structure of collagen is breached.[73]

Orlita *et al.*,[66] analyzing the fragments of historical white leather from the sixteenth century, found 16 species of fungal strains belonging to 9 genera. *Penicillium* strains of 7 species and 2 species of *Aspergillus* were the most common, while other fungi were represented by a single species.

4.7.5.4 Parchment

Writing materials made from animal hides in any period of history were not in exclusive use. One of them – parchment – was used for these purposes for over 2000 years. On parchment, there were created some of the greatest works of art in Middle Ages in the form of beautiful, illuminated books and manuscripts, which are today among the most valuable collections of libraries, archives, and museums.[67]

Parchment is a lightweight leather made with no tanning. After soaking, skin undergoes a process of liming, dehairing, and deliming. Then it is stretched up on the frames in order to obtain the flattened arrangement of collagen fibers characteristic of parchment. Chalk and other finishing additives are rubbed in. After drying, an opaque film is obtained, having a white or slightly yellow color. Drying in a strained state is the most important process of parchment manufacture. Uniform tension prevents the transparency of parchment. Methods of treatment in parchment used in the Middle East are known from the descriptions in the rabbinic Jewish literature. Primitive methods of depilation, after skin was removed from the animal, were carried out by trampling the wet skin or hitting it with sticks. Sometimes the skin was preserved with salt. Dehairing often was conducted after skin was soaked in extracts from parts of plants (branches, stems, leaves, fruits) or in aqueous solutions of flour or animal manure, using the enzymatic activity of growing bacteria. This way the cleansing and relaxing of collagen fibers was achieved.

Among the materials used by the ancient Jewish writers, animal skin parchments were not tanned or lightly surface tanned, with gallnuts or other vegetable tanning, making parchment fairly well or completely saturated with vegetable tannin. Tanning made parchment and writing more resistant, because the inks containing iron salts combined with vegetable tannins, which were, in turn, bound to the fibers of parchment.

Medieval methods of parchment manufacture differed from the ancient methods mainly because of use of slaked lime to dehair animal skins. The processes of liming and rubbing chalk gave parchment an alkaline reaction that made it less sensitive to the effects of acidic pollutants.[67] Finishing treatments included degreasing, softening, giving adequate shade, and smoothing the surface, so that it was easy to apply inks and paints. The original face was preserved, because it gave a nice background effect, especially for manuscripts which were partially or fully colored. In the Middle Ages, beginning in the twelfth century, parchment was mainly manufactured by monks in monasteries. Currently active in Europe are a few shops that produce the parchment for the elegant binding of books, documents, and repair of historical manuscripts.[77] Parchment is also used for orthopedic purposes. Parchment has a very high tensile strength. The main component of parchment is a protein fiber – collagen. Also, the small quantities of other proteins, sugar, fat, and impurities are present in parchment.[78] The parchment, which is both thin and strong, is considered to have the highest quality. These qualities are found in the parchment of calfskin, known as vellum (from latin vitulus – calf). In northern Europe, calf skin was the most frequently used skin, and sheep and goat skins in Southern Europe.[79] But horse, cow, and even deer, lion, and camel skins were also used for parchment production.[67] The pH of parchment is either neutral or slightly alkaline.

Parchment is a very durable material, if properly stored in dry conditions. In good condition, it can survive for up to 2000 years, as exemplified by the scrolls found in dry caves of Qumran near the Dead Sea.[67] However, over many years of storage, there are changes in the structure of parchment due to aging effects of heat, ultraviolet, oxygen (free radical oxidation), pollutants of air, and humidity.[80-81] As the consequence of these factors the structure of collagen is changed by its partial depolymerization and alterations in aminoacid composition. However, moisture poses the greatest threat to the parchment. The optimum relative humidity for the storage of parchment is 40-50% at temperature of 15-20°C.[82] Parchment is a material highly hygroscopic, and therefore increase in the relative humidity from 10% to 95% increases the water content of parchment by about 30%.[83] Under the influence of moisture, the strain induced during processing is relaxed. Penetration of moisture between the fibers causes hydration of collagen. This, in turn, results in swelling and flexibilization of the parchment sheets. Relaxation of collagen fibers is the cause of their disordered bonding during drying. Parchment becomes hard, stiff, transparent, and horny, and its surface is uneven, wrinkled and bent (see Figure 4.22.). Gelatinization effects are reversible, but their removal requires special treatment.[85] Changes in the structure resulting from parchment aging can increase its susceptibility to microbial

Figure 4.23. A fragment of a legal parchment document (15th century) The outer margins and top edges of the leaves bear signs of *Streptomycete* growth: the disappearing edge of the sheet, numerous stains, some with perforations in the centres. [Adapted, by permission, from Karbowska-Berent, J; Strzelczyk, A, **The Role of Streptomycetes in the Biodeterioration of historic Parchment**, *Wydawnictwo Uniwersytetu M. Kopernika*, Torun, 2000.]

Figure 4.22. Gelatinized parchment after contact with water. [Courtesy of KB Nat. Library of the Netherlands.]

attack. In the damp parchment, microorganisms initially settle at the edges (spine or sides), places where there is the best access to oxygen and moisture. The degree of damage depends on the rate of moisture absorption and drying by parchment at a given location. Signs of destruction occur on a series of adjacent pages. As deterioration progresses, book's edges become thinner or disappear, leading to increased access of oxygen to the deeper parts of the book and shifting sites of attack of microorganisms into the deeper layers of the book. Shrinking edges of sheets are usually ragged, darkened, show traces of stains, and are often covered with colorful stains (see Figure 4.23). Places attacked and located in the vicinity are saturated with the products of parchment decomposition.

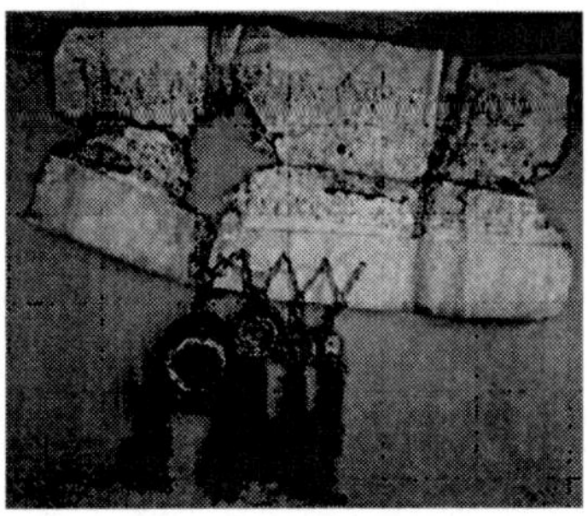

Figure 4.24. A parchment document from 1337, deteriorated by *Streptomycetes*. See the significant mass loss, numerous white stains and deformations of the sheet. [Adapted by permission, from Karbowska-Berent, J; Strzelczyk, A, **The Role of Streptomycetes in the Biodeterioration of historic Parchment**, *Wydawnictwo Uniwersytetu M. Kopernika*, Torun, 2000.]

Unlike parchment books, on the surface of parchment documents colonies of microorganisms are usually distributed fairly evenly, and the most common symptoms include intense color staining (see Figure 4.24). These are signs of a slow and long-term growth of microorganisms in conditions of relative humidity slightly exceeding the limits.

Strzelczyk and Karbowska[84] proposed a model showing the sequence of the destruction of parchment:

- gelatinization is a result of short-term increase in moisture content of the parchment, which consists of hydration of collagen molecules and relaxation of fiber networks. Drying pages without strain leads to disordered adherence of collagen fibers, resulting in deformations and stiffness of parchment
- prolonged dampness of parchment leads to a slow growth of microorganisms. Their presence is seen because of colorful, irregular spots. Often after drying, mycelium dies, but the spots remain
- initially, microorganisms grow at the expense of non-collagen components of parchment

- collagenic abilities of microbes cause enzymatic degradation of collagen, which results in shortening chains of proteins and production of peptides. Some of these nutrients are consumed by microorganisms to support their growth and development, while the remaining excess of degradation products is left in the surrounding space, and, after drying, it creates deformed and rigid areas
- at very advanced stages of destruction of parchment, infested places disappear or are transformed into a darkened, viscous mass, which shrinks on drying and hardens, causing pages to stick together

The phenomenon of microbial degradation of historic parchments was the subject of studies by several authors who have identified a number of strains of filamentous fungi and bacteria. Table 4.39 contains microorganisms found on parchments by Karbowski-Berent and Strzelczyk.[85] Microbes isolated in the period 1962 to 1969 from contemporary and nineteenth-century parchments were compiled by Zyska.[41]

Table 4.39. Microorganisms identified on parchment. [Adapted, by permission, from Karbowska-Berent, J; Strzelczyk, A, **The Role of Streptomycetes in the Biodeterioration of Historic Parchment**, *Wydawnictwo Uniwersytetu Mikolaja Kopernika*, Torun, 2000.]

Fungi:
Alternaria sp., Aspergillus flavus, Aspergillus versicolor, Aspergillus sp., Botryotrichum piluliferum, Chaetomium sp., Cladosporium cladosporoides, Epicoccum sp., Fusarium sp., Mucor scinrinus, Mucor sp., Ophiostoma sp., Penicillium chrysogenum, Penicillium commune, Penicillium meleagrinum, Penicillium notatum, Penicillium sp., Penicillium variable, Scopulariopsis sp., Sepedonium sp., Stemphylium sp., Stysanus sp., Trichoderma sp., Verticillium sp.
Sensu stricto bakteria:
Bacillus subtilis, Bacillus licheniformis, Bacillus megaterium, Bacillus pumilus, Bacillus sp., Bacterium sp., Micrococcus sp., Streptococcus sp.
Actinomycetes:
Nocardia sp., Streptomyces fimicarius, Streptomyces rochei, Streptomyces anulatus, Streptomyces griseoruber, Streptomyces diastaticus

Microorganisms grow with different rates on both sides of parchment. Destruction of a sheet of parchment usually begins from the flesh side and damage on the side of the symptoms is often more intense. In the advanced stages of destruction, the face side also is attacked, even though the collagen and elastin fibers have a very dense compact structure[86] (see Figure 4.25).[85]

Direct effect on the dominance of the microbial attack on the parchment has a relative humidity, the residence time in a humid climate, and the rate of drying. The most active species which colonize wet parchment are bacteria *sensu stricto*, such as *Streptococcus sp., Micrococcus sp., Bacillus sp.,* and *Bacterium sp.* As a result of their strong attack, wet parchment may change partly or wholly into a gelatinous, stinking mass, and the infested place after drying shrinks and sticks to its neighbors. The process of destruction of parchment, which was not wet but kept at too humid conditions is less dramatic. As the drying of parchment continues, the activity of bacteria disappears, and their place is occupied by fungi and actinomycetes. Activity of fungi from classes of *Ascomycetes* and *Fungi imperfecti* causes extensive degradation of parchment. If a small amount of moisture is present, fungus spotting occurs, caused by production of colored spores, mycelium, and color pigments. The resulting spots are colored, ranging from light yellow to various

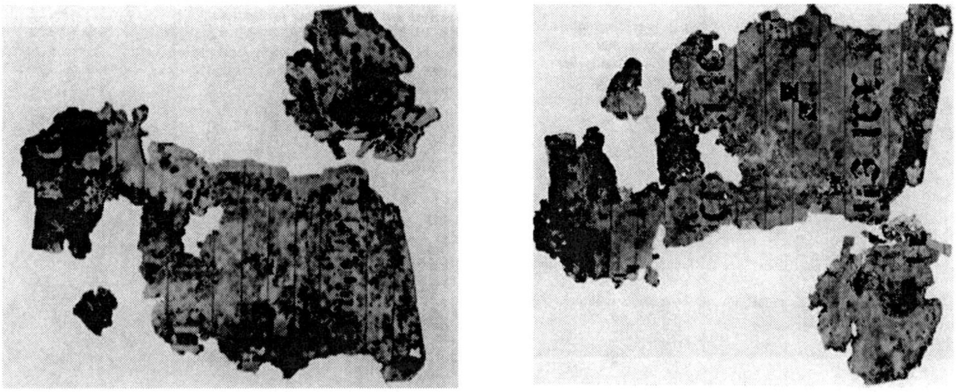

Figure 4.25. Fragments of extensive deterioration of Italian antiphonal parchment. Apart from the weakening of the parchment and numerous gaps, violet staining is visible. There is also a difference in the intensity of violet stains on both sides of the parchment. [Adapted by permission, from Karbowska-Berent, J; Strzelczyk, A, **The Role of Streptomycetes in the Biodeterioration of historic Parchment**, *Wydawnictwo Uniwersytetu M. Kopernika,* Torun, 2000.]

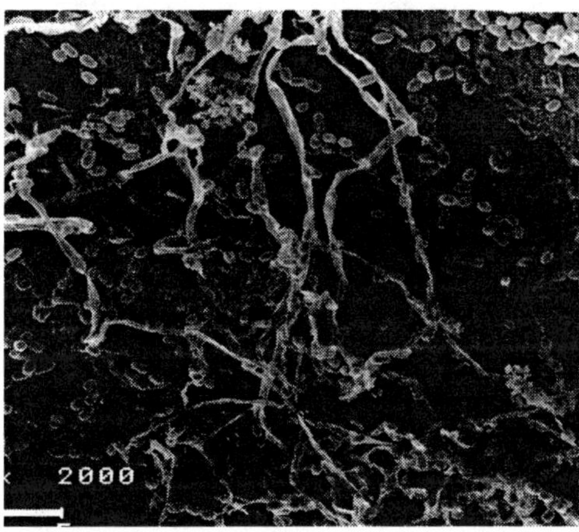

Figure 4.26. *Streptomycete pseudohyphae* and chains of pseudoconidia covering abundantly the parchment surface. [Adapted by permission, from Karbowska-Berent, J; Strzelczyk, A, **The Role of Streptomycetes in the Biodeterioration of historic Parchment**, *Wydawnictwo Uniwersytetu M. Kopernika,* Torun, 2000.]

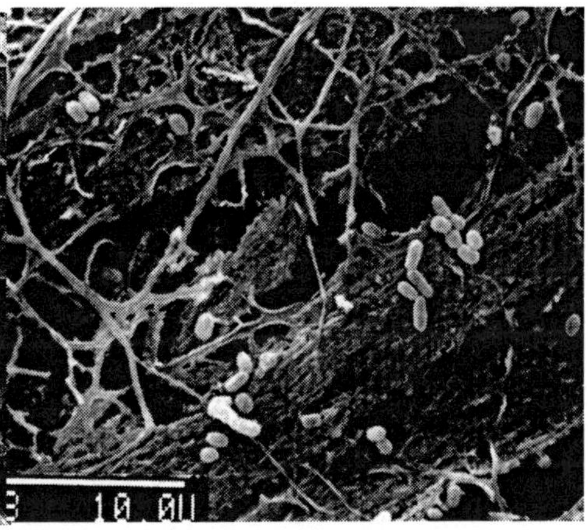

Figure 4.27. Splits in collagen fibres accompanied by *Streptomycete pseudohyphae* and pseudoconidia. [Adapted by permission, from Karbowska-Berent, J; Strzelczyk, A, **The Role of Streptomycetes in the Biodeterioration of historic Parchment**, *Wydawnictwo Uniwersytetu M. Kopernika,* Torun, 2000.]

shades of brown, gray, black, and they usually have irregular shapes. *Actinomycetes* of the genus *Streptomyces* are the major threat to the parchment. Their presence is evident in the form of small (diameters of around 5 mm) round patches of gray, white, cream, pink, or beetroot, and the appearance of transparent and fragile places. Depending on the stage of deterioration, in the center of some spots perforations can be observed, which can sometimes result in substantial loss of surface area of parchment (see Figures 4.24 and Figure 4.25).[85]

Figure 4.28. Fragments of degraded, cracked collagen fibers. [Adapted by permission, from Karbowska-Berent, J; Strzelczyk, A, **The Role of Streptomycetes in the Biodeterioration of historic Parchment**, *Wydawnictwo Uniwersytetu M. Kopernika,* Torun, 2000.]

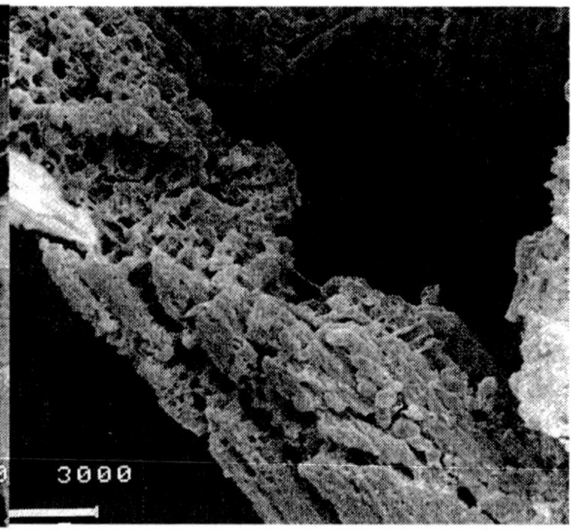

Figure 4.29. A fragment of a heavily damaged collagen fiber with the spongy structure and large pit. [Adapted by permission, from Karbowska-Berent, J; Strzelczyk, A, **The Role of Streptomycetes in the Biodeterioration of historic Parchment**, *Wydawnictwo Uniwersytetu M. Kopernika,* Torun, 2000.]

Actinomycetes of the genus *Streptomyces*, including *Streptomyces amulatus, S. rochei, S. griseoruber,* and *S. diastaticus* are the most active in deterioration of parchment.[85] They have a number of characteristics that enable them to grow on historic parchments. They are mesophilic aerobes, have spores resistant to drying, grow in neutral or slightly alkaline pH, and demonstrate the ability to degrade proteins in the parchment, especially collagen. Due to these features, actinomycetes of the genus *Streptomyces* destroy the parchment by degrading the structure of collagen and form color spots. *Streptomycetes* produce a protease having collagenic activity. Some of them produce pigments on parchment, causing pink staining. Parchment darkening, ease of detachment of fragments, and formation of perforations are evidence of proteolitic activity. SEM observations of extensive damage by Karbowska-Berent and Strzelczyk[85] revealed destruction of collagen fibers (Figure 4.26). Mucopolysaccharides and progelatin are first digested, which exposes collagen fibers, which are then slowly digested by collagenase, and later a non-specific protease, resulting in the formation of pits and fissures splitting and cracking of parchment (Figures 4.27 and 4.28). The remaining spongy, amorphous mass is likely made of decomposition products of collagen fibers (Figure 4.29). Some strains, such as *S. anulatus* spread very rapidly (1-2 months), causing extensive damage of parchment structure and formation of clear macroscopic symptoms of degradation, other strains make equally extensive damage but more covertly and slowly (e.g., *S. diastaticus*) or degrade parchment only partially (e.g., *S. rochei* and *S. griseoruber*). Strains of *S. rochei* and *S. griseoruber* damage structure and cause pink staining, presumably by chinoid dyes.[85]

4.7.6 PROTECTION OF SKIN AND LEATHER FROM MICROFLORA

4.7.6.1 General data

The type and amount of microbial flora present on raw hides, skins, or finished leather depend on the condition of raw material. Modern tanning procedures, including the processing of green leather, the use of vegetable tanning enzymes, use of fat emulsions and natural materials for finishing leather, and long time storage and transportation of leather "wet-blue" all increase the possibility of microorganisms growth.

To prevent loss of material in the leather industry, caused by bacteria and fungi, salt and chemical preservatives are used. Preservation has a decisive influence on obtaining the required skin pH. During neutralization (pH=7), conditions conducive to microbial contamination are created, and this may result with fermentation. In the later stages of the skin processing, there is an increase in acidity and biological processes slow down. The pH=2 inhibits microbial growth. Also, pH=11 completely stops the development of microorganisms and related fermentation processes.

4.7.6.2 Preservation of raw hides

Preservation provides protection against skin deterioration during transport and storage. In most cases, hides are preserved by dehydration in the process of salting or drying. In all cases, the goal is to slow down or completely prevent growth of microorganisms on the skin. However, some bacteria may exist in an environment that contain very little water. Therefore, the protective effect is improved by addition of biocides. Inhibition of the growth of microorganisms, including prevention of the activity of natural skin enzymes can be achieved by the use of appropriately low temperature. Use of these factors and their combinations for particular application, leads to creation of a number of methods of preservation. There are following preservation techniques in common use:

short-term preservation:

- cooling/freezing
- dynamic short-term preservation with NaCl
- short-term preservation without NaCl

long-term preservation:

- wet salting preservation (salting, salt solution soaking, dynamic method)
- preservation by drying (without NaCl)
- dry salt preservation (salting, drying)
- irradiation
- dehydration with organic solvents
- lyophilization
- preservation by pickling

Immediately after slaughter, especially during the warm season, the skin should be cooled in order to protect against autolytic processes. Skins should be processed immediately on the day of collection, if not stored at low temperatures, because after 20 hours. skins will have a rotten smell.

Skins having temperature of 30°C stacked and stored for more than 10 hours have permanent damage,[87] irreparable by tanning processes. Already after 4 hours, deterioration of skin at room temperature begins.[88] Cooling at this stage by placing containers with hides in the cooling chamber does not help because bacterial growth continues and skin lesions are initiated. Achieving good quality leather requires cooling of each skin separately to a temperature of 5-6°C, immediately after removing it from the carcass. Waiting time for cooling should not exceed 1 hour after skinning.[89]

Activity of bacteria on raw skin is highly dependent on temperature. Examining the effects of microorganisms on raw skin at different temperatures, it was found that at 10°C it is possible to store skin for 4 days without deterioration. At -1°C, storage time can be extended to 3-4 weeks.[26] Good results were obtained after 3 weeks of storage when raw hides and skins were rapidly cooled to 2-4°C. It is common in Australia that skins are chilled to 2°C and kept for 3 weeks before they arrive in the tannery for processing or preservation by brine. In the United States and Britain, skins are cooled to a temperature of 13°C. Also, in Czechoslovakia since 1981, a system is implemented to deliver to tanneries raw pig and cattle hides. Fresh pig skin shall be delivered to the tannery directly from the slaughterhouse. For bovine leather a short-term cooling and storage at 5-8°C for 2-3 days is a norm.[90] In Denmark and Ireland, almost all skins are cooled in slaughterhouses, and in Germany, only 50%.[87,91] Skin cooling decreases the intensity of bacterial growth and thus the possibility of bacterial skin damage, and it is the simplest form of short-term preservation.[92] However, until recently (2001) in Europe, only 2% of raw skins were protected with cold air or ice (2-6°C) for a maximum storage of up to 3 weeks.[36] In connection with the growing demands of environmental protection, it is increasingly important that fresh skin processing is used without chemical preservatives. But this requires to counter the growth of bacteria by cooling. There are various cooling techniques used:[25]

- **spraying the skin with cold water**[89] **or cooling with ice.**[87,91] Cooling takes place directly on the slaughter line. It is done by placing skin flat in a container, one after the other, with a special machine throwing pieces of ice. To cool one calfskin about 8 kg of ice is needed,[87] and to cool one raw hide requires 10 to 15 kg of ice.[91] Consumption depends on weight, the method of skin preparation for for cooling, and the processing line used. This method is used with good results in the Netherlands[87] and Ireland.[91] In Denmark, the company Hude-Centralen does short-term preservation by layering skins with ice shaves in quantity of 5 kg per skin for storage from Friday to Monday.[93] Cooling with crashed ice (1.1 kg/ m^2 of skin) requires that the procedure is repeated each day, which produces large amounts of waste water.
- **leather chilling with cold air, water with ice, or glycol**. After removing the skin from the carcass it hangs on the conveyor, and it passes, one after the other, either through cold water with ice or a cold glycol, or through a tunnel with cold air.[87] Ice water cooling is more efficient than air, because within 15 minutes the temperature of the skin is reduced from 36 to 5°C in water, while in the cooling air it is reduced to only 27°C.
- **freezing of fresh skins**. Freezing is an effective method, which, in many countries, has already been used frequently, for example, in Germany, 25% of the time.[94] If skin is frozen, it would effectively stop bacterial growth, but it is possi-

ble that some sensitive structures of skin, such as the face, can be damaged[95] and tear or tensile resistance will be reduced.[96] This is because irreversible physico-chemical changes of skin occur under the influence of sub-zero temperatures.[96] Storage time of skins preserved by these techniques is as follows:[97-99]
3 days at 10 to 15°C
3 weeks at 2 to -2°C
a few weeks at -4°C
several months at -10°C
unlimited time in the case of skin frozen at temperatures from -10 to -30°C

- **skin chilling by carbon dioxide gas or dry ice**. Carbon dioxide chilling is an adaptation of the techniques used in the food industry.[100] In contrast to the ice, skin moisture is not increased and waste water is not produced because phase change of carbon dioxide from solid takes place without a liquid phase. The method was developed for small Austrian slaughter houses.[95] Cooling skins is conducted with a controlled amount of CO_2 needed to achieve -3°C at the bottom of the container with the skins to -1°C in its upper part. This requires about 30 kg of dry ice for 250 kg of leather. Cooling to lower temperature resulted in a rough leather face or the effect of "goose bumps". Skins stored in CO_2 atmosphere had resistance to rot during storage, depending on temperature.[95] Increase in room temperature reduced time of safe storage. For example, the skin can be safely stored for 8 weeks at 2°C, 2 weeks at 15°C, 10 days at 18°C, and up to 5 days at 22°C. In New Zealand,[100] for CO_2 cooling, a special table was designed to move skins over tubes equipped with a dispenser of dry ice. Chilled skins fall by gravity into a container with cooled CO_2. If the skin is to be stored longer than 24 hours, the chilling is done at a slower pace, and CO_2 is blown to the storage container. Another technical solution for application of dry ice, called "super-chilling", was developed in the UK by Ralph Pownall.[100] Hides placed on a conveyor press are fed to the device which sprays them with CO_2 derived from dry ice. Spraying takes place at a temperature of -35°C, which causes almost immediate lowering of the temperature of hides and enables safe storage for at least 48 hours.[25]

Preservation of skins may also be performed using γ-radiation or electron beam, as follows:

- electron beam – a method developed,[101-103] requires rapid skin irradiation, no more than a few hours after obtaining skin. Placed on the conveyor, defleshed and cleaned skins were immersed in the solution of precursor, and then, to enhance the effectiveness of radiation, moisture was reduced by wringing. Removing excess moisture resulted in a balanced distribution of radiation dose to the entire surface of skin, reduction of weight, and thus decreased transport costs. Prepared skins were spread flat and packed in thermoplastic film. The wrapped skins were placed on moving flat conveyor belt, which transported them into the chamber with an electron accelerator. Irradiated skins were packaged in containers lined with polymeric material. Packaging facilitates transport, prevents mechanical damage and infection of skins by bacteria. Such prepared and preserved skins can be stored for 6 months. The method was patented by a Canadian Ionizing Energy Company. Skin preserved by electron beam after tanning and use for shoe uppers was well received.[104]

- γ-rays – the method has been studied since late 1960/70, but not applied in practice,[25] most likely because of technical difficulties and the destruction of collagen structure, manifested by a decrease in tensile strength and puncture resistance.[102] Only in 1999, Baily[104] announced that good quality leather finish for car seats can be obtained with preservation with γ-rays. However, no further detail about preservation by γ-rays and storage conditions were given therefore the method cannot be verified.

Inhibitory effect on bacterial growth can be achieved using physical or/and chemical methods. Chemical methods consist of skin saturation with substances that influence osmotic pressure and the structure and metabolism of bacteria. The most popular is salt enriched with broad-spectrum bacteriostatic substances. Preservative effect of salt is almost exclusively based on changing a quantity of free water, which inhibits the growth of microorganisms and causes their autolysis. Salt penetrates skin by diffusion and the water diffuses from its interior to the surface. In the preservation of hides and skins, sodium chloride is mostly used. It lowers the water content in the fresh skin by 25-30%. Sodium chloride may be replaced by potassium chloride.

Physico-chemical, biochemical, and histopathological changes which occur in fresh goat skin were compared with the behavior of salt-preserved goat skins at ambient conditions at 25-30°C and humidity 50-80%. The results clearly demonstrate the vulnerability of fresh skin to microbes and enzymatic degradation.[105]

However, the simple removal of water is not always effective, because halophilous bacteria thrive in an environment that contains up to 25% salt of leather weight. Salt does not eliminate the action of bacteria at temperatures above 10°C but only limits their growth, and therefore different methods of preservation using sodium chloride did not always ensure good quality raw material. Red and salt patches, accompanied by changes in chemical composition and structure of the dermis, were observed in stored skins. Double side finished leather had defects corresponding to 35% of its total surface on both sides of the skin, caused by microorganisms.[106] Infection by halophile bacteria can come from salt used for preservation. For these reason, other chemicals are used in addition to salt, enhancing its action through a significant reduction or complete inhibition of bacterial growth. Among the chemicals used for preservation of hides are antiseptics, which are compounds with oxidant and redox activity. Bacteriostatic or inhibitory properties imply that preservatives are applied immediately after skin was removed from animal carcasses, especially if skin is not chilled.

Adding adjuvants to the salt and the use of a dynamic preservation method by mixing skin in a drum with a mixture of salt and chemicals: In this method, skins are salted and dehydrated in 20-30 minutes. Also, salt consumption is significantly reduced from 50-75% to 30% to preserve skins for 6 months, and to 5-10% in the case of short-term preservation (10-15 days). Sodium metabisulfite, boric acid, naphthalene (at 1-2% of the salt weight) are most commonly added to salt, for protection against red spots. In addition, biocides are used as an alternative to salt. The skin may be submerged in biocide or biocide may be sprayed on the skin's flesh-side. Preservation often requires insecticide treatment (e.g., pyrethrum, pormethrin, p-dichloro-benzene, etc.) to repel insects during storage and transportation.[36]

Due to the protection of the environment, more and more attention is paid to the possibility of processing hides without preservation. However, the processing of hides and green skins is not always feasible because of the need for rapid transport, therefore skins must be protected at least for short-term storage and transportation.

In short-term preservation, the following methods are used:

- one- or two-sided spraying
- short immersion in appropriately selected solutions
- immersion in a rotating drum, leaving skins in a bath containing preservatives
- sprinkling skins with bacteriostatic salts (salt in this mixture is used for uniform distribution of preservatives; up to 2.5 wt% of skin is added).

High efficiency in short-term preservation is achieved with dynamic mixing methods in which skins are mixed with preservatives. Saturation of skin tissue with preservatives is achieved in 20-30 min. This is especially important in the warm season, when practically there are no conditions for cooling the warm skin after slaughter. In short-term preservation, a better effect is achieved by using different combinations of biocidal products, exploiting their synergistic effects.

Drying is the oldest method of skin preservation. Today, it is almost exclusively used in tropical countries, especially for those skins in which hair provides a major value. Fur skins are suitable for preservation by drying, because drying ensures the best quality of the hair cover. Drying is used in the air, as well as in drying equipment. It must take place under conditions giving full air access to the entire skin on both sides, ensuring adequate aeration by air circulation with temperature and humidity controlled. Drying is suitable for small, thin, fast-drying skins. Thick skins are not suitable for drying, because slow drying does not protect all layers of the skin from the initial decomposition processes. Adverse changes result during high temperature drying in the sun. Moisture contained in the middle layer of skin does not diffuse to the surface, resulting in skin damage. In skins preserved by drying, some damage is not easily seen, hence the difficulty in sorting and evaluating their quality. Dry skin can be kept indefinitely long.

Salting-drying method is used primarily in tropical climates, especially in the case hides for export. Initially, skins are salted using less salt (15-20% of the skin weight) and subsequently dried. Drying of salted hides is faster, because sticky substances, consisting of amorphous proteins, which hinder drying were somewhat removed in the process of salting. The skin preserved in this manner is better for processing, because it contains more moisture than skin preserved by drying, in which the water is lost, causing formation of additional bonds in collagen molecules and hindering its hydration and swelling.

Freeze-drying, i.e., evaporation of water by sublimation from skins in a frozen state under high vacuum is used only for valuable fur pelts, due to the high cost.

Skin dehydration with acetone is another method having, at the moment, rather theoretical importance.

Effectiveness of conservation depends on the type of preservatives and their method of application. There are various techniques for chemical preservation:

- **static conditions method** is based on manual sprinkling of defleshed skin with compounded preservatives. This technique is one of the least effective, because it uses a large amount of the preservative mixture even for sprinkling the entire surface of skin.[107-108] To limit use of preservatives, they must be mixed with some filler. The amount of filler depends on weight and volume of the preservative; for example, a mixture called "Liricure" contains 20-25% ethylene-diamino-tetra-sodium acetate (EDTA), 40% NaCl, and 35-40% filler, which is a saw dust.[108] In dynamic application, 2% EDTA and 3% NaCl without filler was sufficient.[108] In the static preservation of hides with NaCl 30-50% salt by weight of hides is required. In practice, more salt is usually used. The use of an excess of salt has no effect on the quality. Excess salt partially goes to waste water and part is recovered as waste from mechanical "shaking" of skins. Recovered salt is infected with bacteria and cannot be reused. It has to be sterilized at 340°C and then separated from organic matter by using appropriate membranes, which in practice is not feasible. The use of a commercial mixture (Liricure) requires 3-4 days,[108] and in the case of salt, even 10-14 days are needed because of slow processes of saturation and dehydration.[107] Slow preservation in static method can lead to bacterial skin lesions, especially if immediate cooling of skin after its removal from animal carcasses was not used. This method does not create the possibility of rational management of preservatives. It also does not allow use of small quantities of preservative for short-term storage.[107]
- **NaCl brine saturation** with possible addition of other preservatives is a method used by large exporters in the USA.[109-110] Skins after cleaning and defleshing are placed in the tumbler containing brine for 48 hours or more. When the salt is absorbed into the skin, it removes the excess water and skin is sprinkled with a little salt before transportation. One can speed up absorption of salt by use of the dynamic method.
- **dry salting** in which skin, preferably cleaned and defleshed, is mixed with rock salt and other preservatives in a low speed mixer. Steel mixing drums are better than wooden, because wood absorbs some brine, increasing its consumption. Duration of the process and the amount of preservative depends on requirements for skin dehydration and storage time. This method permits rapid dehydration of skin and saturation with preservatives. This results in an almost immediate termination of bacterial growth. The skin after removal from the drum is placed in piles for further draining of liquids. For short-term preservation, it is possible to place skin directly in transport containers[107]
- **spraying with or dipping in solutions of biocides** is recommended for short-term preservation of skin for a period of 3-8 days because action of biocides is limited to the surface.[98] The method was first reported in the 1960s. However, its

usefulness appears to be questionable, because there is no information in the literature which indicates commercial applications.[107]

4.7.6.3 Chemical preservatives

Preservatives used in short-term preservation of hides

There is a number of useful formulas for short-term preservation of hides, only a few of them comply with environmental requirements. According to the modern concept of environmental protection, preserving preparations must be degraded by chemical or biochemical means, without formation of harmful metabolites.

The selection measures should be considered not only from the point of preserving properties of skin, but also considering potential hazards resulting from their use, production, and disposal.[107] These criteria have not always been applied. Examples of such negligence include use of pentachlorophenol and its derivatives, which have effective biocidal action,[111-112] but are toxic to living organisms, including causing allergies and leukemia.[113] Toxicity of pentachlorophenol resulted in the gradual introduction of bans on its production, distribution, and use. In Germany, it was banned since 1989, not only because of the toxicity of PCP, but also because its incineration led to formation of polychlorinated dibenzodioxins and furans.[113] The U.S. took action in 1980 to ban the use PCP in skin preservation. The Committee of the American Association of Producers of Raw Hides and Leather (USHSLA), as a result of intervention of the Environmental Protection Agency (EPA) asked the manufacturers and suppliers of chemicals to develop effective biocides for use in the preservation of skin, but without harmful side effects. As a result, a formulation has been developed in which the active substance is 1,2-benzisothiazoline-3-one (BIT).[113] This compound was approved for use by the EPA and the U.S. Department of Agriculture. It has a broad spectrum of antimicrobial activity and a neutral pH and it is sensitive to redox agents, which permits its degradation in depilation solutions after their useful life. BIT formulation applied in solution protects skin for 48 hours. Addition of boric acid to BIT extended the preservation period to 12 days without the need to use brine. Boric acid applied alone is not so effective.[107]

Tolcide from Albright & Wilson UK Ltd[113-114] fulfills requirements for environmental protection and work safety. Methylene bis(thiocyanate) (MBT) and 2-(thiocyanmethylthio)benzothiazole (TCMTB) are the active substances in these preparations. MBT is a perfect biocide having a broad spectrum of action. It also has fungicidal properties. It works very quickly, and because of the low molecular weight, penetrates deeper into the skin as well. It was developed in 1967 as a replacement for mercury-based fungicide.[115] TCMB has fungicidal effects and it is suitable for use in long-term preservation. It inactivates growth of fungi commonly encountered in tanneries such as *Aspergillus flavus, A. niger, Peacilocyces variotii, Trichoderma viridi*, and *Penicillium* strains. It is active mainly on the surface of skin. The most important preparation of a Tolcide series is Tolcide FSM, which contains 10% MBT and 10% TCMB, which ensures its antibacterial and antifungal activity. Tolcide C30 is a strong fungicide, containing 30% TCMB; Tolcide MC contains 10% MBT and it is used as antibacterial in pre-tanned skins.[107]

Heterocyclic TCMB is also an active ingredient in the preparation of Mortanol 30, offered by the German company Zschimmer & Schwartz. Additional ingredients are surfactants, which increase the penetration of antibacterial and antifungal preparation TCMB into the skin. Mortanol 30 is environmentally-friendly. Biodegradation tests carried out according to the modified Zahn-Wellens test (OECD 33028) showed that after 28 days, 67% of the preparation was degraded. In addition, it has reactivity towards sulfide, which resulted in loss of activity in the presence of lime sulfide and intensified its biodegradation.[113]

Busan preparations, manufactured by Buckman Laboratories International, [113] may be used for the preservation of hides and intermediates. TCMB is an antifungal substance in some of them.

Non-phenol preparation, Mergal QB-50, offered by the German company Riedel-de Haen, was successfully used in short-term storage for many years. It rapidly inactivates skin microorganisms. The other formulation, called Mergal KM 265, has a broad spectrum of antibacterial and antifungal activity. It is used especially for pickled and tanned skin. Mergal KM 265 contains two active substances, one of which quickly penetrates into the skin, while the other remains on the surface of skin to protect it from fungal growth.[107]

Aracit 1A,[116] developed by TFL, does not contain TCMB and chlorinated organic components. It is biodegradable and highly effective.

Liricure,[108,117] developed by the American company LIRI, contains ethylene-diamino-tetra-sodium acetate (EDTA) and neutral salt such as NaCl. To increase the volume, sawdust is used as filler. Blend showed good antibacterial activity, and preserved skins after 3-4 days of static diffusion. After treatment, skins could be palletized and transported. Skin samples kept at 35°C did not show any signs of decay for 4 weeks. However, this mixture cannot be used anymore because of the harmful effects on biocenosis of sludge and on natural water ecosystems.[107]

Another way to preserve skin is by use of simple chemical compounds in the form of gas released from reaction. The gaseous products include sulfur dioxide, which was the subject of years of research by Hopkins and Bailey.[118-124] Sulfur dioxide was obtained from sodium sulfite and acetic acid. Hydrogen chloride gas,[125] a product of sulfuric acid sodium chloride, was also used.

Acid preservation of hides using acetic acid was studied by Sweet.[126] Skins were conditioned in a mixer in a 4% solution of acetic acid. After removal of water excess, skins were sprinkled with 55% NaCl containing 2.5% of preservatives called White and Yellow Absorption Flakes. Antibacterial protection lasted 11 days.[107]

Boric acid and sodium carbonate has long been recommended. Skin, after soaking in a saturated boric acid solution, can be stored for 5 days at 30°C.[127] Boric acid solution can also be applied in a drum (15 minutes, using 1% boric acid and 5% NaCl).[107,128] Skins preserved in this manner were stored without signs of decay for 7 days. Sodium carbonate applied by soaking in a 20% solution stabilizes the bacterial flora and molds for 8 days[98] or for 7 days after the dynamic processing in a low speed drum with 2% sodium carbonate and 5% NaCl.[107] Table 4.40 contains summary of results with different chemical treatments.[26,107,129-131]

Table 4.40. Chemical treatment of raw hides.[26,129-131]

Chemical name	Preservation method	Concentration, % by weight of raw hide	Protection duration
sodium sulfite	dynamic in drum, 60 min	10-20	up to 7 days
sodium sulfite acetic acid	dynamic in drum, 60 min	1 1-2	17-20 days
sodium sulfite sodium fluorosilicate	dynamic in drum, 60 min	10 0.1	17-20 days
sodium pyrosulfite	dynamic in drum 30 min 10% NaCl mixture 30% NaCl mixture	 0.5 1	 10-15 days 6 months
boric acid	soaking in saturated solution for 60 min dynamic in drum, 15 min 5% NaCl mixture dynamic in drum, 30 min 10% NaCl mixture 30% NaCl mixture	1.2 1 1 1	5 days at 30°C 7 days 10-15 days 6 months
potassium nitrate	dynamic in drum, 30 min 10% NaCl mixture	2	10-15 days
sodium carbonate	dynamic in drum, 30 min 10% NaCl mixture	2	10-15 days
zinc chloride trichlorophenol Na	dynamic in drum, 2-4 hrs	0.3-0.5 0.05-1	5-6 days at 25°C
sodium hypochlorite sodium tetraborate	dynamic	2 4	up to 4 days
zinc dimethyldithio-carbamate	mixture with salt	2% salt	6 months
naphthalene sodium carbonate	mixture with salt	2% salt 2% salt	6 months
naphthalene boric acid	mixture with salt	0.5% salt 1% salt	6 months

Preservatives used in long-term preservation of hides

The primary measure used in the long-term preservation of hides is the application of sodium chloride. Bacteriostatic action shows that it disrupts the osmotic pressure in the bacterial cell and skin cells. It causes dehydration of the skin to a level below which most bacteria cannot grow. The action is temporary, because if salt is removed from the skin conditions for bacterial growth are recreated.[87-88]

Bacteriostatic effect of NaCl can be seen from this study: fresh skin contained $2.9x10^5$ bacteria/g of skin, and skin, salted, stored 4 weeks at 20°C contained $4.1x10^5$ bacteria/g of skin.[88] Excessive humidity may cause the development of invasive bacterial spores. Thus, conditions of storage of skins preserved with sodium chloride are essential, particularly humidity and temperature.[88] "Red heat", caused by halophiles and the like, which secrete a red dye, indicates excessive humidity in the storage room.[88,101] These bacteria come from salt. In recycled and sea salt, there are 108 bacteria/g and only 1 bacteria/kg of rock salt, and therefore this form of salt is recommended for preservation of skins.[101,109,132]

Potassium chloride, used as a partial or total replacement of sodium chloride,[101,109] has a similar efficacy to sodium chloride preservative, but also inhibits halophile bacteria (10% of potassium chloride in place of sodium is sufficient).[109] Potassium chloride preserved hides showed no signs of infection by halophile bacteria, and after the finish did not differ significantly from the skins preserved with sodium chloride. Potassium chloride is a preferred preservative due to ecological requirements, since the excess can be used for fertilizing purposes.

Another way to circumvent the problem of halophile growth is to use bactericide having a broad spectrum of antibacterial action, such as dimetyldithiocarbamate of potassium.[133] This measure has been used with good results for several years in Australia as a supplement to brine.[109-110] It was also approved by the EPA.[133]

A chemical compound called BIT has been accepted by the EPA and the U.S. Department of Agriculture, not only for short-term preservation but also for long-term storage in addition to brine. The preparation has a broad-spectrum antimicrobial activity in an inert environment, including halophiles. It is stable in saturated brine, and increases the effectiveness of sodium chloride. Skin preserved with the addition of BIT retained good quality during one-year-long storage. Damage to the surface of the skin has been estimated at 10%, while skins preserved with brine alone had damage in 75% of cases.[113]

In the group of adjuvants for sodium chloride, Busan 1193, 84, and 1009 are suggested by Buckman Laboratories International for long-term storage. The preparations also prevent the formation of red spots on the skins, meaning that halophiles were inactivated.[113]

The addition of 3% boric acid eliminates purple spots on calf skins, and also inhibits the development of salt stains.[107] The skins preserved by brine had visible defects in the form of stains but after conservation with a mixture having the following composition: 50% NaCl and 3% of H_3BO_3 stored at 8-10°C and relative humidity 75-80%, purple spots were eliminated after two weeks, and other damage was inhibited.[17]

Table 4.41 summarizes examples of commercial preparations containing active substances in the preservation of leather.

Table 4.41. Some microbiocides used for preservation of skin and leather [data from Avecia, Bayer, Rokita, Sanitized]

Biocide	Application	Concentration, wt% skin
Preventol WB-L, Bayer) mixture of phenol-based biocides	raw hides mixture with salt addition to brine soaking	 0.5-1.5 0.5-1 0.03-0.05
Preventol WB plus-L, Bayer) mixture of phenol-based biocides	protection of pickled skin chrome tanning (wet blue) retanning protection "wet blue" vegetable tanning static drum	0.2-0.5 0.2-0.5 0.2-0.4 0.-0.5 50-100 g/m^3 wort 0.3-0.5
Preventol CR Plus-L, Bayer mixture of benzothiazol and surfactant	pickled skin chrome tanning additional tanning preservation of "wet blue"	0.05-0.15 0.1-0.15 0.05-0.1 0.05-0.15
Preventol Z-L carbamate	raw hide mixture with salt addition to salt solution soaking	 0.5-2 (salt) 0.5-1.5 0.05-1 (solution)
Preventol CMK, Bayer mixture of phenol derivative and chlorocresol	soaking chrome tanning ("wet-blue") retanning vegetable tanning drum static leather dyeing (casein-base)	0.03-0.06 (water) 0.1-0.2 0.2-0.3 0.20-0.40 0.03-0.06 (solution) 0.20-0.40
Preventol O Extra, Bayer phenol derivative	chrome tanning finished leather	0.2 0.2-0.5
Roksol KF1, Rokita phenol derivative	chrome tanning preservation of "wet-blue" finished, dry leather	 0.2-0.3 0.2-0.5
Roksol, Rokita mixture of aryl compounds and phenol derivatives	preservation of "wet-blue" finished, dry leather	0.2-0.3 0.2
Vantocil IB, Avecia guanidine derivative	preservation of raw hides spraying of flesh side soaking of flesh side drum processing for 30 min drum process with brine	 2-8 (solution, 24 h, boric acid 1:1) 0.5-2 (sol., 24 h, boric acid 3:1) 0.2-0.3 (boric acid 1:1, 24 h) 0.3 (boric acid, 6-12 months)
Proxel GXL, Avecia benzotriazole derivative (BIT)	preservation of raw hides drum for 30 min drum with brine	 0.2-0.3 (boric acid 1:1) 0.2 (boric acid 1:1, 6-12 months)

Preservation of hides by the use of a combination of silicon gel (environmentally friendly and easy to use) with 5% salt and with or without the addition of 0.1% p-chloro-meta-cresol (PCMC) per weight of leather can be used as an alternative method. The aim was to develop a method of skin dehydration to 30% water content, and use of bacteriostatic properties of salt. The effectiveness of the method was estimated by the moisture content in the skin, total nitrogen in aqueous extract, the number of bacteria, and the pollution load generated in the process of skin tanning. The method was as effective as the salt methods. The quality of the skin obtained was at par with the skin obtained from the raw material preserved with salt.[134]

4.7.6.4 Skin protection against microorganisms in the tanning processes

Hides and "wet-blue" are subject to attack by various kinds of microbes. Filamentous fungi grow in tanneries in the weakly acidic natural substrates containing carbohydrates, fats, and proteins. Medium containing 12-15% water provides ideal conditions for fungal growth. The majority of spores of filamentous fungi are able to survive in dry conditions, then spread in the air and multiply in favorable environmental conditions. The skin should be protected in the process of pickling by fungicides, as well as in the tanning phase, long-term storage, and transportation of tanned material. Complexes of phenols, and combinations of thiazine compounds (MBT are TCMTB) are useful biocides. Adding biocides to the tanning bath often has a dual role of preservative added to protect skins for a short time (less than 24 hours) during tanning to stabilize the process if the wort or pits are used in a continuous manner, without treatment between operations. In the leather industry, the whole palette of aids used in the form of suspensions and emulsions must be protected from spoilage by microorganisms. Given the great diversity of these additives (colorants, fats, finishing compositions, and varnishes), a variety of preservatives must be used. Isothiazolins, TCMTB, thions and triazines operate most effectively.[135] Fungicides used today are the same as those used many years ago (the same active compounds), but there are also some new preparations. New biocides are more sensitive to the conditions of use, which requires strict adherence to production technology during their dosage in the production process.

Biocides used to protect the skin against microbial attack, in addition to the typical characteristics such as broad spectrum and low toxicity, should be characterized by chemical affinity to tannins, good stability, and the lack of influence on dyeing process.[136-137]

Preservative dosage depends on many factors, including:[138]

- climatic conditions, namely temperature and humidity
- storage time
- cleanliness during production and storage of "wet-blue"
- presence of substances that promote the development of filamentous fungi on "wet-blue" (ammonium salts, phosphates, surfactants, and oil and organic finishes)
- ratio of skin surface area and its weight (which depends on the thickness of the skin)

Until recently, chlorophenol-type biocides were used for protection of "wet-blue", but they are now slowly being replaced by other active substances.

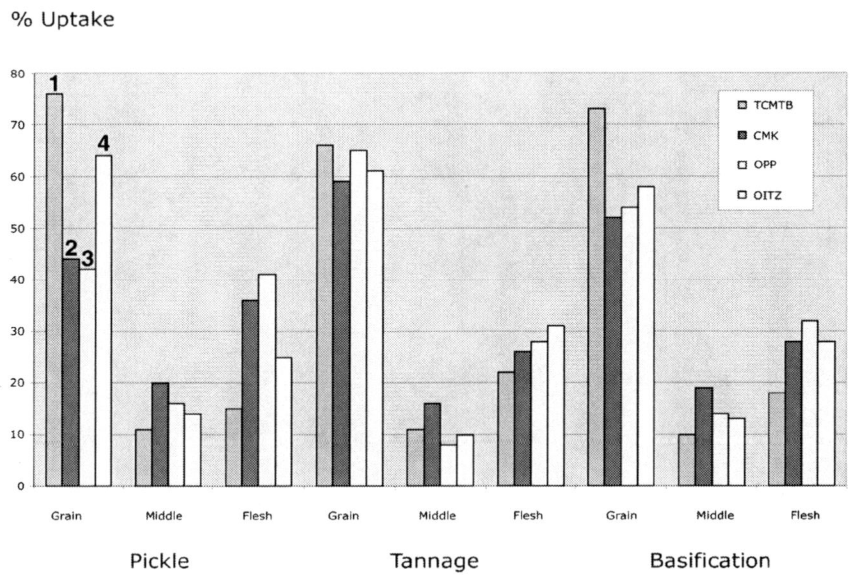

Figure 4.30. Butt: distribution of fungicides in grain, middle, and flesh layers as% of uptake. [Adapted, by permission, from Huber, Ch; German, H, P; *J. World Leather*, May, 75-82, 1997.] (1) TCMTB – 2-(thiocyanomethylthio)benzothiazole; (2) CMK – para-chloro-meta-cresol; (3) OPP – ortho-phenylphenol; (4) OITZ – 2-n-octylisothiazoline-3-one.

Deren and Veiss[139] showed that the effective control of fungal growth on the skin is possible by applying TCMTB. The most effective method of protection of "wet-blue" was addition of fungicide during pickling or at the initial stages of chrome tanning, which permits penetration of the preparation deep into skin. As a result, the "wet-blue" containing 100 ppm TCMTB showed no increase in filamentous fungi under tropical conditions (37°C, relative humidity 90%) for a period exceeding 25 days. The fungicide demonstrated compatibility with the skin, as well as with pickling solutions, and the ability to adapt to the chrome tanning and vegetable tanning conditions.

In the eighties, Orlita[26] tested TCMTB on the laboratory and production scales. Commercial product containing 30% TCMTB was used to protect the leather for lining. Fungicide was added in the pickling process or 30 minutes prior to the addition of chrome tanning ingredients. Tests showed that it is necessary to add 0.2% TCMTB per weight of skin to preserve it for a period of 3-4 months of storage as "wet-blue". Hauber and German[142] announced that for the prevention of fungal growth on "wet-blue" the following minimum concentrations of fungicides should be used: TCMTB – 250 ppm, p-chloro-m-cresol (CMK)/o-phenylphenol (OPP) – 58/280 ppm, and 2-n-octylisothiazolin-3-one (OITZ) – 80 ppm. However, fungicides at the recommended minimum concentration can be used only to protect skin for a short-term storage. In practice, for the three month period of storage concentrations, ranging between the minimum and the maximum in the

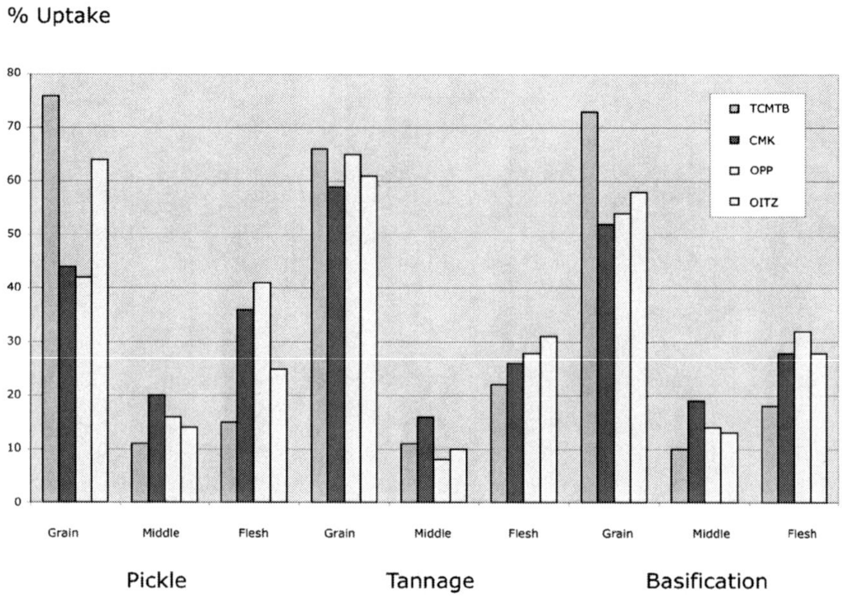

Figure 4.31. Flank: distribution of fungicides in grain, middle, and flesh layers as the % of uptake. For labels see Figure 4.30. [Adapted, by permission, from Huber, Ch; German, H, P; *J. World Leather*, May, 75-82, 1997.]

manufacturer's recommended doses should be used. The aforementioned authors have published the results on the addition of fungicides to chrome tanning, and their penetration, absorption and distribution in the "wet-blue". The results showed that the absorption of all the fungicides were high in the grain layer, and the lowest in the middle layer of bovine hides, butt, and both sides. Results are given in Figures 4.30 and 4.31.[142]

The following observations can be reported:

- in the face layer, TCMTB shows the highest level of absorption. Values for other biocides are comparable.
- in the middle layer, CMK and OPP penetrate the most and show the highest degree of absorption. Levels of absorption for OITZ in the face and middle layers are similar to TCMTB.
- in split layer (splits are obtained by mechanical separation into two layers of skin), OITZ shows a high percentage of absorption. TCMTB active compound is least absorbed by the flesh layer

Comparison of effectiveness of fungicides in "wet-blue" in prevention of growth of filamentous fungi, and evaluation of the most effective solutions for the various stages of technology was conducted in Australian tanneries.[143] Results are shown in Figure 4.32.

Research shows that the majority of systems preserved skin for 4-6 weeks. However, the results show only the relative effectiveness of protection according to tests on a small scale. TCMTB alone is not sufficient in both test concentrations in any phase of the addi-

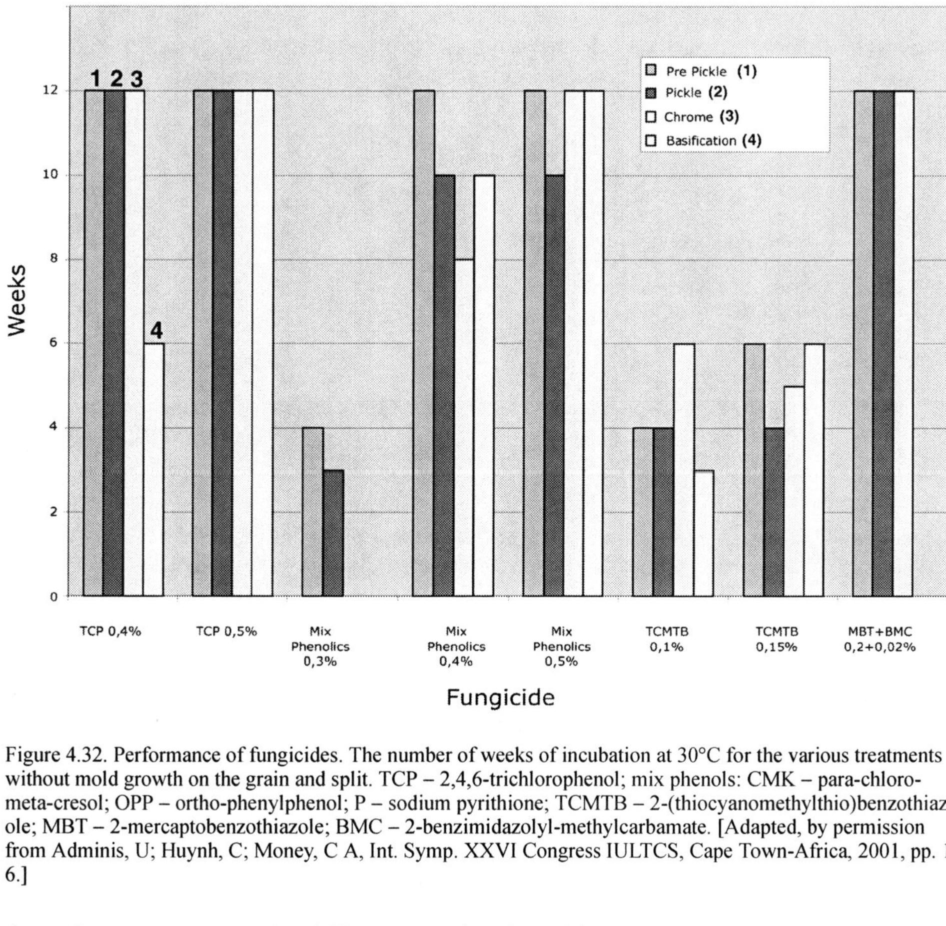

Figure 4.32. Performance of fungicides. The number of weeks of incubation at 30°C for the various treatments without mold growth on the grain and split. TCP – 2,4,6-trichlorophenol; mix phenols: CMK – para-chloro-meta-cresol; OPP – ortho-phenylphenol; P – sodium pyrithione; TCMTB – 2-(thiocyanomethylthio)benzothiazole; MBT – 2-mercaptobenzothiazole; BMC – 2-benzimidazolyl-methylcarbamate. [Adapted, by permission from Adminis, U; Huynh, C; Money, C A, Int. Symp. XXVI Congress IULTCS, Cape Town-Africa, 2001, pp. 1-6.]

tion. There was no growth of filamentous fungi on skins protected by 2-mercapto-benzothiazole (MBT) and 2-benzimidazolilo-methylcarbamate (BMC). A good result was obtained with TCP and a mixture of phenols used in the concentration of 0.4-0.5%.

Didato and Yanek[139] showed that TCMTB is interchangeable with p-nitrophenol as a fungicide for military footwear uppers. For several decades, fungicide has been recommended for use in the leather factories producing for the U.S. military; however, it has not been recently registered by the U.S. Environmental Protection Agency for this application. Toxic properties of PNP and poor biodegradability caused its demise. TCMTB showed growth inhibition of filamentous fungi for 2 months in tropical conditions in the same military application. Active substances used for skin protection during tanning are summarized in Table 4.42.

Table 4.42. Active substances used for protection in the leather tanning process.[26,143]

Antifungal agents used in the leather industry in the last 20 years:	
Phenolics:	
CMC	para-chloro-meta-cresol
OPP	ortho-phenylphenol
TCP	2,4,6-trichlorophenol
Heterocycles:	
TCMTB	2-(thiocyanomethylthio)benzothiazole
OITZ	2-n-octylisothiazol-3-one
BMC	2-bezimidazolyl-methylcabamate
MBT	2-mercaptobenzothiazole
P	sodium pyrithione
Other:	
DIMTS	diiodomethyltolylsulfone
The use these fungicides is uncertain due to increasing health and environmental regulations and restrictions. Consequently there is a need for cost effective fungicides with low human toxicity and minimum environmental impact.	

Table 4.43. Sequence of process for different hides. [Adapted, by permission, from Tissier, Ch; Chesnais, M, **Supplement to the methodology for risk evaluation of biocides**, *INERIS*, France, 2001.]

Cowhides	Sheepskin	Pigskin
soaking	soaking	soaking
unhairing	fleshing	unhairing
liming	dewooling*	liming
deliming		deliming
bating		bating
pickling	pickling	pickling
	washing	
	degreasing*	degreasing*
tanning	tanning	tanning
splitting		splitting
shaving		shaving
retanning	retanning	retanning
dyeing	dyeing	dyeing
fatliquoring	fatliquoring	fatliquoring
drying	drying	drying
buffing		buffing
finishing	finishing	finishing
plating	ironing	plating

*specific to the skin process

Search for biocides which are ecologically sound and safe for humans is directed towards natural preparations. Usefulness of oregano oil, showing antimicrobial action, was studied in fungal applications for skin during tanning and pickling processes. These studies demonstrated that antifungal activity increases with concentration. Skin treated with oregano oil was even more resistant to fungi than leather treated with chemical fungicides, such as 2-thiocyanomethylthiobenzotiazole (TCMTB) and N-octyl-isothiazolinone (NOITZ) at their recommended doses.[144] Chlorophyll extracted from spinach was another

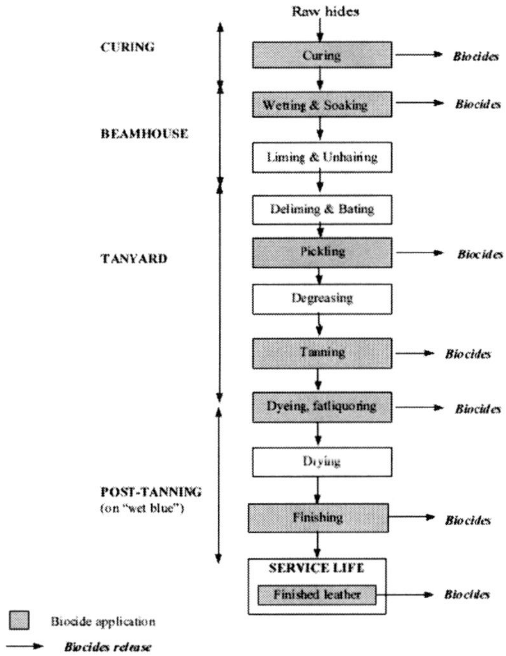

Figure 4.33. Leather tanning and biocide use. [Adapted, by permission, from Tissier, Ch; Chesnais, M, **Supplement to the methodology for risk evaluation of biocides**, *INERIS*, France, 2001.]

example of natural biocide.[145] Examples of currently available modern bacterial preparations and fungicides are summarized in Table 4.41, using their chemical and trade names, as well as showing their optimum concentrations. Biocides are applied in all stages of skin tanning (see Figure 4.33.).[36] Not all process steps are performed in every tannery and their sequence may change depending on the type of skin, as shown in Table 4.43.

The application of biocides depends on skin type and/or stages involved (see Table 4.44).[36] Storage prior to tanning and finishing is the main stage of biocide application. In the small tanneries, where skins are stored outside, there is no need to use biocides or they are used in very limited quantities. Thin skins are more susceptible to biological degradation, so they need more protection than thicker material. Preferred is the use of various biocides, in order to avoid that microorganisms adjust to their presence.[146] Usually two biocidal preparations can be used alternatively in the same year. The best practice is to change the active ingredients in biocides on a regular basis, preferably every six months.[36]

Table 4.44. Performance chemicals, in kg/100 kg hides. [Adapted, by permission, from Tissier, Ch; Chesnais, M, **Supplement to the methodology for risk evaluation of biocides**, *INERIS*, France, 2001.]

Performance chemicals	Heavy leather*	Light leather*
bates	0.8	0.8
biocides	0.1-0.4**	0.02-0.2**
syntans		3
fat liquors		4
dyeing auxiliaries		3.8
dyes		0.6
finishes		4

* The distinction between heavy and light leather is due to animal size. Representative hide weights are as follows (UNEP/IEO, 1994):

bovine (heavy)	15-40 kg/animal
sheep (light)	1-3 kg/animal
pig (light)	2-4 kg/animal

** based on a 30% active substance content (information from Rhodia)

Biocides may be added in any of the following steps; preservation (salting), soaking, pickling, tanning, dyeing, and finishing. Stages directly preceding storage or transportation are the most critical. Also, preservation, tanning, and finishing are critical. In closed systems of tanning, biocides are directly introduced into water in soaking drums, pickling, tanning, dyeing, and wet finishing. In general, biocides are likely added to dye solutions, emulsions containing oils, and casein finishes.[36] Doses of biocides vary to a great extent depending on the tannery. Examples of doses are given in Table 4.45.[36] Currently, there are tendencies to reduce biocides.

Table 4.45. Quantities of common biocides. [Adapted, by permission, from Tissier, Ch; Chesnais, M, **Supplement to the methodology for risk evaluation of biocides**, *INERIS*, France, 2001.]

Step	Amount of biocide per mass of raw hides, wt%
curing	0.01-0.5
soaking	0.01-0.5
pickling	0.01-0.5
tanning	0.01-0.3 and 0.3-0.5 (small hides)
finishing	0.3

Biocides used in the tanning industry are generally non-oxidizing biocides. They can be classified as quaternary ammonium compounds, isothiazoles, halogenated organic compounds, such as Bronopol (2-bromo-2-nitro-propane-1,3-diol), isothiazoles and halogenated organic compounds containing heterocycles, such as derivatives of benzothiazole.[36]

4.7.6.5 The use of biocides in the leather industry, and legislation, health and environment

Biocides differ as to the effects on humans, animals and the environment, but none is without negative impact. The future use of chemical components in biocides is uncertain because of possible regulations regarding health and the environment. In assessment of the impact of biocides on the environment, their biodegradability, stability (persistence), toxicity to fish, and effect on the seaweeds are considered. Products of their decomposition are also evaluated in the same respect. Availability of biocides in commercial applications depends on regulations, such as European Directives or regulations issued by the EPA. In Europe, the Directive 98/8/EC is a single document listing active substances and biocidal products permitted for use in the countries of the European Union, as well as Switzerland, Norway and Iceland. Under this directive, biocide manufacturers have to obtain authorization from each EU Member State prior to the sale of its product. The basis is to harmonize the market and strengthen health and environmental protection. The main requirements for biocidal substances are as follows:

- all active ingredients of biocidal products must be recorded, including areas of application and the manufacturer
- biocidal products, including their composition of active substances, must be recorded in each EU Member State
- biocides should be developed based on the estimated risk for the intended applications
- European Directives targeting biocidal products must have been adopted by the national laws of each member country beginning in 1999. In accordance with Article 16 of Directives 98/8/EC, there was introduced a ten-year period to 14 May 2014, in which the current system can be applied
- products of biocides have been divided into 23 groups and four categories. Category II contains preservatives and group 9 contains products used to protect skin, rubber, polymers, and fibers
- identified and notified active substances are given in the Annexes to Commission Regulation EC No 2032/2003 of 4 November 2003 and Commission Regulation EC No 1048/2005 of 13 June 2005, Amending Regulation No. 2032/2003
- the competent authority of the Rapporteur Member State must obtain complete documentation no later than 31 October 2008 on the existing active substances for which notification has been accepted within product type 9
- from 1 September 2006, Member States were required to repeal the authorization for biocidal products containing active substances listed in Annexes III and VII of these Regulations. In addition, preparations containing biocidal composition of its active substances contained in Annex II, but intended for use in other groups than those in which they will be evaluated, will not be permitted for use
- in European policy, in relation to biocidal products, the following rules apply:
  - the marketing and use will be permitted only for approved active substances and biocidal products authorized
  - active substances are regulated at EU level
  - biocidal products are regulated at the national level using the principle of mutual recognition of licenses by the Member States.

Consequently there is a need for new, inexpensive, effective fungicides with low toxicity effects on humans and minimal environmental impact. Biodegradability is an important demand. A number of widely used fungicides may not be used for skins in the future, due to regulatory or other restrictions. Widely used organochlorine compounds are among the most toxic. Heidemann[147] reported, in 1988, that in the preservation of leather, wood, and textiles there was about 30-40 substances used in various combinations, among them toxic and environment unfriendly substances which include chlorinated phenols, especially pentachlorophenol (PCP) and trichlorophenol (TCP), mercury compounds, tributyltin, metylhydroxyquinoline, diiodomethyl-para-toluenesulfone, benzothiazole derivatives, tetrametylthiuram disulfide, zinc dimethyldithiocarbamate, quaternary amine, for example, guanidine compounds, acid amides, derivatives of phenols, such as o-phenylphenol, and others. Gelo[148] pointed out that the focus on toxic compounds harmful to human health and environment, gained a special significance in Europe since 1989, i.e., after Germany banned the use, purchase, and sales of skin and other products containing more than 5 mg/kg (5 ppm) of pentachlorophenol (PCP), taking effect in March 1990. A large influx of imported raw hides, wet-blue, crust, footwear, and clothing containing PCP was the reason for issuing this decree. PCP is a cheap preservative with a broad spectrum of activity, even in warm climates, but also highly toxic, which may contain dioxins as contamination, and it is not biodegradable. According to the opinion of the legislature, the use of PCP as a preservative for leather is the main source of environment pollution by particularly dangerous dibenzo-p-dioxin (PCDDs) from dibenzofuran (PCDF), which are formed during combustion of materials containing PCP.[149] Gelo and Urbaniak[150] indicate that harmful action of PCP was not yet unambiguously proven, but the observations of 20 years has shown that even slight concentrations may affect human health. PCP can be absorbed not only by the respiratory system, but also through the gastrointestinal tract and skin. According to Kurbat and Kucharski,[151] phenol and its derivatives (e.g., chlorophenols) are compounds with documented toxicity.

Similar to Germany, limits on use of PCP is enforced in many countries around the world. In the U.S. the use of PCP is prohibited by law. In Canada and Switzerland, regulations are similar to those in force in Germany. Many countries only have regulations regarding safety shoes. Until recently, the use of trichlorophenol (TCP) was permitted, although it was known to contain impurities of PCP. Currently, the eco-labels have begun to influence the choice of fungicides and some German companies restrict the content of TCP in leather to 1 ppm. Ortho-phenylphenol (OPP) as a relatively safe for humans and the environment. It is also widely used in agriculture, but now is considered as a carcinogen in California, although it is not classified as carcinogenic in other parts of the United States. Leather sold in California containing more than agreed levels of OPP requires warnings. Tanneries need to choose and realize that in future, far-reaching restrictions on the use of fungicides might be expected.

Money[143] says that in Australia before German legislation, phenol-formaldehyde resin components were dominant in "wet-blue" preservation, but after regulation, TCMTB become a widely used fungicide. Methylene-bis-thiocyanate (MTB) is both a product of decomposition and an alternative product for TCMTB. It is less active, so it must be used at higher concentrations. The 2-benzimidazole-methylcarbamate (BMC), also known as carbendazim, has been widely used in agriculture. However, it can be used for "wet-blue" with no other active ingredients. According to Directive 2004/73/EC it is classified as a carcinogen category 2, which causes an obligation to mark the finished products and biocides, containing over 1000 ppm of this substance, with the symbol T (toxic). Diiodomethyltolyl-sulfone (DIMTS) is widely accepted but it is toxic to fish. It is quite expensive in Australia, but the mixtures were applied successfully in the past in many tanneries.

The tanning industry is a known polluter. The large amount of salt, chrome, and variety of other substances used in processing, and biocides used for production are the main effluents. The presence of biocides in effluents sometimes causes difficulties in the use of biological sewage treatment.[152-153] Consumption of water in the leather industry is given in Table 4.46.

Table 4.46. Average water consumption and characteristics of wastewater from leather production and finishing (EU, 1996). [Adapted, by permission, from Tissier, Ch; Chesnais, M, **Supplement to the methodology for risk evaluation of biocides,** *INERIS*, France, 2001.]

Process	Operation	Water use, m^3/ton raw hides
Beamhouse	soaking	6
	liming	9
	unliming and pickling	5
Tanning	pickling and chrome tannery	1
	washing after chrome tannery	1.5
Squeezing water out		0.5
Wet finishing	washing	0.5
	neutralization	0.3
	washing after neutralization	0.5
	second tanning, dyeing, greasing	0.3
	washing	0.7
Wet finishing		1.5
Squeezing water out		0.2
Cleaning water		5.0
Total amount of waste water		**32.0**

Potential sources of air pollution are solvents, spray application of biocides,[36] and evaporation during use of leather products.

Permitted active substances according to the Commission of the European Communities and EPA, which are used for the preservation of skin, fibers and polymeric materials are given in Chapter 3.

4.7.6.6 Antimicrobial shoe materials and footwear

The current state of knowledge on shoe materials indicates that protection against biodeterioration of raw materials for production of footwear is necessary. Some materials must be protected against activity of microorganisms already in production process, in other to prolong product life during use. The biodeterioration of shoe materials is caused by filamentous fungi. They are found in different natural environments. They belong to chemoorganotrophic materials for which leather or cotton are the sources of carbon. They can grow in a wide range of temperatures from -10°C to 60°C and pH from 1.0 to 12, although optimum temperature range is 20-30°C and pH 5-6. They are also characterized by their high tolerance to substrate moisture. The most important feature of filamentous fungi, important for development opportunities on different surfaces, is the secretion of hydrolytic enzymes (amylase, cellulase, protease, and lipase), and their oligotrophic character (low nutritional requirement). Natural materials, such as leather, fabrics made of cellulose fibers, and proteins are useful for them as a source of carbon. Destruction processes begin from the stage of transportation and storage.

Detected biodeterioration of shoe materials is the basis for further analysis of the conditions of their use and selection of the appropriate biocide to enhance the biological resistance on which sustainability of footwear depends. An important role in the prophylaxis of fungal foot infections is protection of shoe interiors against attack by microorganisms, especially by pathogenic fungi.

Among the various ways to prevent fungal growth rates in the forefront are sanitization materials used in the manufacture of footwear and footwear disinfection by the user during use. Sanitization involves the impregnation of materials with solutions that contain chemicals with biocidal or biostatic properties. The internal shoe components should be protected.[55,57,154] Antimicrobial finishes have positive impact on the quality of footwear, which has been proven through many years of practice. The following finishing systems are frequently used: Ultrafresch, Durafresch, Purista, finishing by Adidas (e.g., Ag ion), the Swiss company Sanitized AG, and others. In production of cellulose-based undersoles fungicide used is also a vulcanization accelerator. In rubber footwear for work use, preparations[7] Mergal and Sanitized S-130 are applied on textile material used in undersoles.

In Poland, production technology was developed with the participation of nonwoven fabric protected with biostatic preparations, such as Sanitized, Reputex, and Nano Silver.[58,155-156]

In some grades of hides, intended for use in tropical and subtropical zones, special treatment is made for tropical protection of leather from the accelerated biodegradation caused by microorganisms in these harsh conditions. The goal of tropicalizing is to slow down the development of saprophytic and pathogenic microorganisms during use of footwear. To achieve this goal, substances which are harmless to human health should be used. Silver ions are very effective at small concentrations (100-200 ppm),[55,57,157] but for yeast fungi, higher concentration of active ingredient and a longer duration of action are needed. Preparations containing silver ions are applied on textiles and padding inside the lining of gloves and footwear. A modern solution includes production of polymers, which are active against microorganisms.

Considering problems with growing resistance of microorganisms (mainly filamentous fungi and Gram-negative bacteria) to biocides, it is very important to use the appropriate doses of biocides. Mechanisms of tolerance of microorganisms are associated with adaptation to life in the presence of the so-called sublethal doses of biocides. Sometimes it is necessary to use several biocides in combination or use the new technological solutions, such as immobilization of biocides by functional groups, which are introduced into the structure of the material, or encapsulation of biocides, which controls slow release and long-term use of the product. Table 4.47 provides examples of active chemical substances used as the finishing materials for production of hygienic shoes.[57]

Table 4.47. Active substances in preparations used in hygienic finishes of footwear materials.[57]

Biocide/producer	Active substance	EC number	CAS number	Product type	Annex IV*
Sanitized T 99-19/ Sanitized AG	dimetyltetradecyl-3-(trimethoxylylpro-pyl) ammonium chloride	255-451-8	41591-87-1	9	x
Sanitized T 96-20/ Sanitized AG	5-chloro-2-(2,4-dichlorophenoxy)-phenol	222-182-2	3380-34-5	9	x
Sanitized T 90-04/ Sanitized AG	5-chloro-2-(2,4-dichlorophenoxy)-phenol 2-octyl-2H-izothiazole-3-one	222-182-2 247-761-7	3380-34-5 26530-20-1	9 6,9	x
Reputex 20/ Arch Chemicals	poly(hexamethylene biguanide hydro-chloride)	polymer	27083-27-8	6,9	x
Nano Silver/ NANOCO or NANOPAC	Ag	231-131-3	7440-22-4	9	x
AlphaSan RC 5000/ Milicen Chemical,	silver sodium hydrogen zirconium phos-phorate	-	265647-11-8	9	x
AgION/ AgIon Technol.	silver sodium hydrogen zirconium phos-phorate+silver zeolite	-	265647-11-8	9	x
Irgaguard B 5000/ Ciba	silver-zinc-zeolite	-	130328-20-0	9	x

Effective disinfection of footwear, particularly in the treatment of patients with fungal infections, is a different problem. Physical methods (e.g., cleaning, ventilation, laundry inserts, replacement of inserts) only eliminate the favorable conditions for development of certain microorganisms, but do not remove the more resistant bacteria and fungi found in shoes, particularly their spores. Solution to the problem could be disinfecting-freshening shoes. Systematic spraying of the inside of shoes causes inhibition of microbial growth, and thus prevents the formation of unpleasant odor, caused by decomposing bacteria.[59,158-160] Deodorants for feet and shoes often contain Triclosan [5-chloro-(2.4-dichlorophenoxy) phenol)]. In preparations for personal protection, essential oils are increasingly used, which contain biologically active substances. Components of essential oils inhibit the growth of microorganisms, but their activity is quite diverse. Mold spores and Gram-negative bacteria are hardly affected. The best antimicrobial properties have essential oils from thyme, savory, tea tree, pine, cinnamon, eucalyptus, and lavender. Synergistic effect can be obtained by using mixtures of chemical compounds and essential oils. An example might be a combination of Triclosan with tea tree oil. Deodorant for shoes containing the composition exhibits effective activity against bacteria, yeasts, and dermatophytes, but does not adversely affect the properties of material and durability of the shoe.[160-161] Shoes are also well protected by preparations containing silver, which has biocidal activity because of its ions.

4.7.6.7 Protection of heritage leather products against germs

Infection of sites of exposure and storage of museum objects of cultural value and their deterioration are a real problem in the protection of cultural heritage. Massive infestation of library collections can be caused by uncontrolled, excessive supply of moisture by a ventilation system. Just several days at 18-20°C and high humidity is sufficient for substantial increase in moisture of leather articles. Another factor conducive to the growth of fungi and bacteria is dust.

Protection of cultural heritage requires continuous monitoring of climate and storage conditions, i.e., the number of recommendations and rules for proper storage of collections, as well as more effective conservation techniques. Generally, in the scope of the conservation of monuments, one can identify three areas where microbiology is very important. These are:

- prevention of storage contamination and growth
- disinfection procedures
- restoration of the individual objects

Prevention of collections is determined by:

- maintaining cleanliness in the storage, regular vacuuming of the premises and collections by properly trained personnel
- monitoring and maintaining stable parameters, such as temperature and relative humidity in storage areas; the optimum conditions are: temperature 16-18°C, relative humidity 40-50%
- carrying out the reviews of collections by conservation services for biological purity and conducting systematic microbiological tests of air and material samples
- carrying out disinfection of materials acquired from antique shops, or other sources, before taking them to the warehouse
- meals by the library staff outside the warehouse.

Infection by microorganisms often leads to irreversible damage of the biological nature. A collection, once infected, must be subjected to systematic reviews. Taking into account the possible losses as a result of years of degradation, biological research is carried out and methods of safe and effective disinfection are explored.

Archaeological leather

Restoration of wet archaeological leather covers the mechanical and chemical cleaning (removal of fine particles of sand, organic debris, decomplexation of iron-tannin compounds), strengthening the skin, re-tanning and oiling, and its protection against microbes. Restoration of the skin as a result of these operations results in the regeneration of the color pattern of a face, restoring flexibility, and durability.

Comparison of microbial susceptibility of unprotected and protected by a progressive conservation of archaeological leather has been the subject of research by Strzelczyk, Bannach and Kurowska.[62] Procedures carried out in successive stages of restoration increased the microbiological susceptibility of protected leather. Saturating leather with a solution of silicone in the final stage of preservation is the most effective in delaying microbial growth on archaeological leather. The best resistance to microbial attack has been recorded for leather protected by a comprehensive conservation method, including

- mechanical brushing
- wet cleaning in distilled water

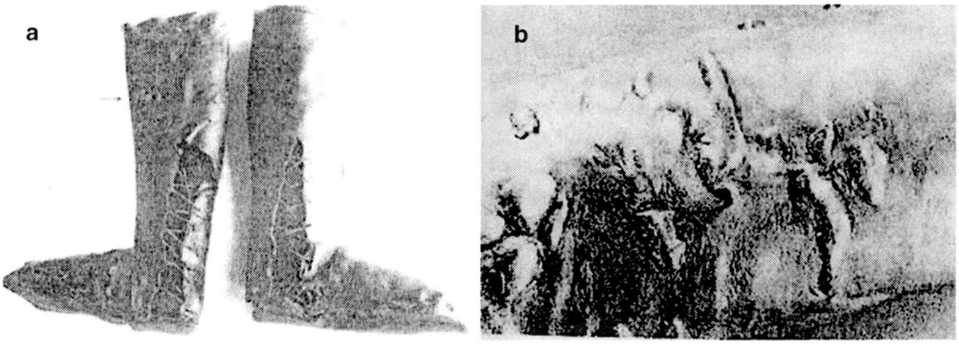

Figure 4.34. (a) Parasitical defects of boots, (b) Hypodermatosis on boot detail. [Adapted, by permission, from Orlita, A, Int. Symp. Microbial Biodegradation and Biodeterioration of Technical Materials, Lódz, September 8-10, 2003, Technical University of Lódz, Lódz, 2003, pp. 179-185.]

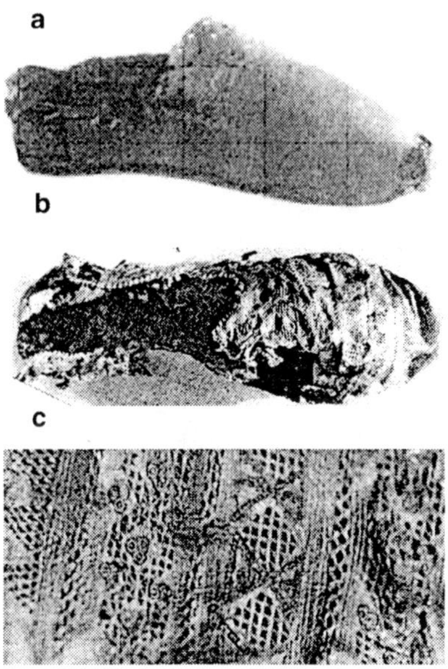

Figure 4.35. The footwear excavated from the well of All Saints Church in the Prague Castle. (a) Footwear after shaping, (b) Slippers as found, (c) Punched geometrical pattern on the surface of slippers. [Adapted, by permission, from Orlita, A, Int. Symp. Microbial Biodegradation and Biodeterioration of Technical Materials, Lódz, September 8-10, 2003, Technical University of Lódz, Lódz, 2003, pp. 179-185.]

- wet cleaning in a solution of distilled water with Canpac alkaline soap (pH=7.9) with ultrasound
- cleaning with acid (ortho) phosphate in order to dissociate iron-tannin complexes
- re-tanning with sumac
- tanning with alum ($KAl(SO_4)_2$ $12H_2O$)
- silicone impregnation to hydrophobize and soften the leather

Comprehensive procedure followed during restoration and conservation of archaeological leather shoes was presented by Orlita, Bravermanova, Brezinová, Hiavacek.[162-165] The work was done on footwear from the sixteenth century, buried in a well in All Saints Church in Prague Castle. Footwear was in the form of flat, molded items, mostly covered with mud and mold, soaked in fat. Multistage preservation was carried out, including

- disinfection with butanol vapors
- cleaning (greasy dirt and dry petroleum jelly were removed in a solution of solvent and surfactants with water to simultaneously soften shoe uppers. Softening of the soles was not sufficient, because they were very deformed.)
- mechanical cleaning by wiping with a cheese cloth
- neutralization was not required because the skin had a pH of 5

- softening with a set of softening ingredients (aqueous solution with a special grease and solvent) was carried out in two ways, by soaking and wrapping in cloth soaked in a special softening substance and placed in a plastic bag. Softening decreased brittleness and fragility of the skin
- shaping the softened leather footwear
- greasing with composition containing liquid oiling components and biocide to protect shoes against molds
- polishing shoes

One restored collection of footwear, newly formed, is a unique collection of 73 pieces of Renaissance luxury footwear found in the Czech Republic (see Figures 4.34 and 4.35).[162]

Vegetable tanned leather

To protect vegetable tanned leather for covers of books, it is desirable to develop procedures for conservation, which inhibits corrosion caused by environmental factors and microorganisms. The treatment is multifunctional and it should:

- neutralize free acids
- block the heavy metal ions
- stabilize and replace deficient substances
- inhibit decay and microbiological corrosion

There are two separate procedures for the conservation of leather covers, designed to satisfy the above required actions:

1. the use of an aqueous solution of potassium fluoride (KF) and potassium bicarbonate ($KHCO_3$) to neutralize acids and block heavy ions
2. the use of preparations of fatty compounds containing antimicrobial preparations to prevent multiplication of the filamentous fungi on protected leather bindings of books (e.g., a preparations containing o-phenylphenol).

The use of oiling compounds containing biocidal component has been generally accepted by conservators and in the practice of restoration of heritage leather.

The search for methods of effective protection of historic leather materials is in progress for a long time. At various times, a number of techniques and measures has been used for these purposes. Measures which attempt to act on the microorganisms occurring in the bindings of books include chemical and physical methods, such as gamma rays, microwaves, and electric current.

Historically, ammonia vapors were widely used in the preservation of ancient leather bindings, which, after detailed studies, proved to be dangerous for leather. Application of ammonia vapors increases susceptibility of leather to damage by microorganisms, because some bacteria and actinomycetes demonstrate ability to acquire this form of nitrogen as well as atmospheric nitrogen. Actinomycetes can grow on leather's collagen, even when the acidity of the leather is reduced or neutralized by ammonia.[64]

Orlita[66] tested o-phenylphenol (Preventol O) and 5-methyl-2-isopropylphenol (Thymol) in the form of vapor. Both preparations showed a satisfactory effect at higher temperatures (37°C) when the evaporation is increased. Even at ambient temperature (22°C), 48-hour exposure is sufficient to kill the mildew in the infected samples.

However, the preservation of historical leather in bookbinding by biocidal vapors is not comprehensive. The use of biocidal formulations in the form of vapors is enough to kill fungi in infected samples, but does not protect the leather objects against future inections. Tests confirmed that the protection by thymol is short-lived, which is a consequence of its high volatility.[66] Similarly, one cannot get protection for leather fittings, using radioactive irradiation or electric current. Long protection against microbial activity can be secured only by introducing effective substances directly to the material being hedged. Comprehensive protection and conservation of historic, vegetable tanned leather bindings includes inhibition of the effects of chemical corrosion and inhibition of the harmful biological processes. Elimination of mold spores is crucial since the growth of mold can damage not only the surface of leather bindings but also may cause breakdown of the oiling substances.

Leather binding is best protected by storing books in a dry environment, unfavorable for microbial growth. The effectiveness of conservation also depends on climatic conditions of storage.

Alum tanned leather

Procedure for restoration of the historic white leather covers of books (alum tanned leather) should include:

- cleaning
- actual conservation (re-tanning and possible neutralization)
- oiling (introduction of an active preparation of a biocidal agent with oiling substance, which, together with oil, can penetrate into the internal structure of the leather and thus create a protective film on the surface of the cover)

Antimicrobial compounds considered as components of oiling preparations must meet a series of requirements:

- be effective in biocidal activity
- be photostable in order to avoid discoloration of white leather
- have no adverse effect on the different components and properties of the skin
- be non-toxic

Many well-known biocides cannot be used to protect white leather bindings, because of their instability to light. Also, a large number of biocides are ineffective in action against fungi and bacteria. Orlita[66] recommends the following chemical compounds for the preservation of white leather bindings:

- isopropanol for cleaning
- formaldehyde for re-tanning
- N-(carbethoxy-pentadecyl)-trimethyl ammonium bromide, 5% solution (trade name Septonex) is used as a component of the oiling preparation

In applying these compounds, the skin is protected in all three phases of the procedure, i.e., by disinfection with isopropanol and formaldehyde, and the introduction of Septonex during coating, which protects for a long time.

Parchment

Parchment is protected mechanically and by mineral filling and degreasing. Parchment was also, in some cases, protected against moisture and scratches by covering with transparent varnish. For decontamination of historical documents from parchment, vaporized biocides are used. Volatile biocides must meet the following requirements to obtain the desired effect:

- be effective biocides, i.e., kill filamentous fungi, spores, and bacteria
- have a high partial pressure for vapors to have sufficient concentration for rapid destruction of microorganisms
- should not deteriorate the quality of protected objects (e.g., cause color changes, decrease in tensile strength, or cause corrosion)
- not be toxic with regard to personnel
- biocidal effectiveness should be obtained within a short period of time (40-48 hours after application)

Biocides proposed by Orlita that inhibit growth of microorganisms on the leather bindings of books and parchments are summarized in Table 4.48.[66]

Table 4.48. Biocides inhibiting growth of microorganisms on book leather bindings and parchment[66]

Chemical name	Trade name
p-chloro-m-cresol	Preventol CMK
o-phenylphenol	Preventol O
5-methyl-2-(1-methylethyl)phenol	thymol
formaldehyde	formaldehyde
glutaraldehyde	glutaraldehyde
butanol	butanol
propanol	propanol
amyl alcohol	amyl alcohol

Research conducted by Orlita[66] showed that p-chloro-m-cresol is very good for protection of parchment. Also, alcohols (amyl, butyl, propyl) are effective in elimination of filamentous fungi from parchments, if the documents are wet and left for 1-4 days between two sheets of filter paper. Alcohols exhibit the greatest activity in the form of 80% aqueous solutions. Their biocidal activity is related to the cellular protein denaturation and it is determined by the presence of water, which facilitates the penetration of alcohol into the cells. Also, alcohols used as volatile biocides exhibit toxicity against molds, while not influencing the change in hue or physical characteristics of protected materials. However, they may be used only in cases where there is no risk of dissolving ink, varnish, or colored decorations.

Many biocides are biologically effective, but toxic as well as causing a variety of changes of shades of a light-colored surfaces, causing some blurring. For this reason, they are not considered.

REFERENCES

1 Zapletal, P; Wodziak, K; Wojnicki, A, *Przegl. Skórz.*, **10**, 238-240, 1991.
2 Bienkiewicz, K, J, **Fizykochemia wyprawy skór**, *Wydaw. Nauk. Tech.*, Warszawa, 1977.
3 Smirnow, W, I; Pawlowa, M; Smiechowski, K; Gajewski, M, **Vademecum garbarza**, *Inst. Technol. Ekspl., Radom*, 1996.
4 Persz, T, **Zwiazki wieloczasteczkowe w przemysle skórzanym**, *Wydaw. Nauk. Tech.*, Warszawa, 1970.
5 Lasek, W, **Wykonczalnictwo skór miekkich**, *Wydaw. Nauk. Tech.*, Warszawa, 1971.

6 Hadley, P, J, *J. World Leather*, **5**, 33-36, 2000.
7 Corda, C, Untersuchunger zur ätologischen Abklärung erregerbedingter Schäden auf Rindleder, Diss., Vienna 1993.
8 Schröer, Th, *J. Leather*, **202**, 4703, 36-42, 2000.
9 Schröer, Th, *Journal G+P*, 48-65, 1992.
10 Branscheid, W; Schröer, TH, Journal G+P, 172-182, 1997.
11 Baker, K, P, *J. Soc. Leath. Technol. Chem.*, **62**, 5, 103-107, 1978.
12 Zapletal, P, *Przegl. Skórz.*, **3**, 84-86, 1993.
13 Halligan, G, *J. Soc. Leath. Technol. Chem.*, **75**, 4, 94, 1991.
14 Venkatesan, R, A, *J. Soc. Leath. Technol. Chem.*, **63**, 55-60, 1979.
15 Mierzejewski, J; Mendrycka, M, Int. Symp. Microbal Biodegradation and Biodeterioration of Technical Materials, Lódz, May 30-31, 2001, Technical University of Lódz, Lódz, 2001, pp. 179-187.
16 Stather, F, **Gerbereichemie und Gerbereitechnologie**, Berlin, 1951, p. 113.
17 Myjak, W, *Przegl. Wlókien. WOS*, **3**, 22-26, 2005.
18 Hausam, W, **Bakteriologie in der Lederindustrie**, *Wisenschaftliche Verlagsgesellschaft M.H.D.*, Stuttgart, 1946.
19 Stanley, A, J, *Leather*, **195**, 4622, 27, 1993.
20 Pietrzykowski, W, *Przegl. Skórz.*, **153**, 5-7, 1988.
21 Sweet, L; Henrickson, R, L, *J. Amer. Leath. Chem. Assc.*, **77**, 192, 1982.
22 Wada, K; Uchara, K; Kawamura, A, *J. Das Leder*, **26**, 3, 45, 1975.
23 Birbir, M; Ilgaz, A, *J. Soc. Leath. Technol. Chem.*, **80**, 147-153, 1995.
24 Oyen, V, *J. Leather*, **194**, 4604, 43, 1992.
25 Felicjaniak, B; Kosinska, K, *Przegl. Wlókien. WOS*, **10**, 32-34, 2005.
26 Orlita, A, Int. Symp. Microbal Biodegradation and Biodeterioration of Technical Materials, Lódz, May 30-31, 2001, Technical University of Lódz, Lódz, 2001, pp. 41-54.
27 Bailey, D; Birbir, M, *J. Amer. Leath. Chem. Assc.*, **88**, 291-299, 1993.
28 Kallenberger, W, E, *J. Amer. Leath. Chem. Assc.*, **79**, 104-114, 1984.
29 Kallenberger, E, W; Lollar, M, R, *J. Amer. Leath. Chem. Assc.*, **81**, 248, 1986.
30 Taylor, M, M; Diefendorf, E, J; Hannigan, M, V; Artymyshyn, B; Phillips, J,G; Feairheller, S, H; Bailey, D, B, *J. Amer. Leath. Chem. Assc.*, **81**, 43-61,1986.
31 Vreeland, R, H; Hochsten, L, I, The Biology of halophylic Bacteria, *CRC Press*, Boca Raton, 1993.
32 Bailey, D; Birbir, M, *J. Amer. Leath. Chem. Assc.*, **91**, 47-51,1996.
33 Vreeland, R, H; Angelini, S; Bailey, D, G, *J. Amer. Leath. Chem. Assc.*, **93**, 121-131, 1998.
34 Bitlishi, B, O; Karavana, H, A; Basaran, B; Sari, O; Yasa, I; Birbir, M, *J. Amer. Leath. Chem. Assc.*, **99**, 12, 494-501, 2004.
35 Przybyl, M, *Przegl. Wlókien. WOS*, **6**, 38-40, 2005.
36 Tissier, Ch; Chesnais, M, **Supplement to the methodology for risk evaluation of biocides**, *INERIS*, France, 2001.
37 Mendrycka, M; Mierzejewski, J, Int. Symp. Microbal Biodegradation and Biodeterioration of Technical Materials, Lódz, May 30-31, 2001, Technical University of Lódz, Lódz, 2001, pp. 207-210.
38 Oppong, D; Bryant, S; Rangarajan, R; Steele, S; Radwell, D; Hyllengren, L, *J. Amer. Leath. Chem. Assc.*, **101**, 4, 140-149, 2006.
39 Bienkiewicz, K, J, *Przegl. Skórz.*, **1**, 10-11, 1983.
40 Haines B, M; Rose A, H, **Microbial Biodeterioration**, *Academic Press*, Londyn, 1981, pp. 131-147.
41 Zyska, B, **Mikrobiologiczna korozja materialów**, *Wydaw. Nauk. Tech.*, Warszawa, 1977.
42 Kowalik, R, *Postepy Mikrobiol.*, **1**, 8, 105-125, 1969.
43 Van der Werff, P, J, **Mould fungi and bronnchial asthma. A Mycological and Clinical Study**, Leiden, 1958.
44 Zyska, B, **Mikrobiologiczny rozklad i korozja materialów technicznych**, *Technical University of Lódz*, Lódz, 2000.
45 Grela-Jaskowska, J; Duda, I, *Przegl. Skórz.*, **1**, 7-9,1983.
46 Sharma, O, P; Sharma, K, D, *J. Inter. Biodeter. Bull.*, **73**, 1980.
47 Macura, A, B; Pawlik, B, Int. Symp. New Materials and Finishing and Conservation Metods of Footwear, Kraków, November 16-17, 1998, Central Laboratory of Footwear Industry, Kraków,1998, pp. 101-105.
48 Macura, A, B; Pawlik, B, *Przegl. Skórz.*, **10**, 418-419, 1999.
49 Orlita, A; Ruzicka, J, *Przegl. Skórz.*, **11**, 267-270, 1988.
50 Orlita, A, Int. Symp. New Materials and Finishing and Conservation Metods of Footwear, Kraków, November 16-17, 1998, Central Laboratory of Footwear Industry, Kraków, 1998, pp. 107-129.

51 Grela-Jaskowska, J; Duda, I; Marcinkowska, E, Int. Symp. Healthy Footwear, Kraków, 1985, Economic University, Kraków, 1985. pp.46-49.
52 Grela-Jaskowska, J; Duda, I; Marcinkowska, E, Int. Symp. Modern Directions of Scientific Research and Evaluation of Quality Footwear Materials and Shoes, Kraków, June 16-17, 1988, Economic University, Kraków,1988, p. 20.
53 Falkiewicz-Dulik, M, Int Symp. Microbal Biodegradation and Biodeterioration of Technical Materials, Lódz, September 8-10, 2003, Technical University of Lódz, Lódz, 2003, pp. 296-305.
54 Falkiewicz-Dulik, M; Macura, A, B, *Mikol. Lek.*, **13**, 4, 265-271, 2006.
55 Falkiewicz-Dulik, M, *Ochrona przed Korozja*, **9s**/A, 75-84, 2006.
56 Falkiewicz-Dulik, M, Int. Symp. Innovation and Environmental Materials and Technologies for Footwear Industry and Other Light Industry Branchs, Kraków, November 25-26, Central Laboratory of Footwear Industry, Kraków, 2004, pp. 86-96.
57 Falkiewicz-Dulik, M, Int. Symp. Mat-Eco-Shoes, Kraków, November 23-24, 2006, CLPO, Kraków, 2006, pp. 24-30.
58 Falkiewicz-Dulik, M, Int. Symp. Modern Directions of Footwear Industry and Evaluation, with Particular Consideration Given to Comfort and Ecology of Product, Kraków, November 20-21, 1997, Central Laboratory of Footwear Industry, Kraków,1997, pp. 95-102, 186-201.
59 Falkiewicz-Dulik, M; Pawlik, B; Macura, A, B, *Mikol. Lek.*, **11**, 4, 291-296, 2004.
60 *J. Leather World*, **9**, 1996.
61 Strzelczyk, A, B, Int. Symp. Microbal Biodegradation and Biodeterioration of Technical Materials, Lódz, May 30-31, 2001, Technical University of Lódz, Lódz, 2001, pp. 28-40.
62 Strzelczyk, A,. B; Bannach, L; Kurowska, A, *J. Int. Biodet. Biodeg.*, **39**, 4, 301-309, 1997.
63 Nigam, N; Dhawan, S; Nair, M, V, *J. Int. Biodet. Biodeg.*, **33**, 145-152, 1994.
64 Strzelczyk, A, B; Kuroczkin, J, *J. Int. Biodet.*, **23**, 3-27, 1987.
65 Strzelczyk, A, B; Kuroczkin, J, *J. Int. Biodet.*, **25**, 39-47, 1989.
66 Orlita A; Garg, K, L; Garg, N; Mukerji, K, G, **Recent Advances in Biodeterioration and Biodegradation**, Calcutta, 1993, pp. 259-299.
67 Reed, R, Ancient Skins, **Parchments and Leathers**, *Seminar Press*, London, 1972.
68 Waterer, J, W, **A guide to the conservation and restoration of pojects made wholly or in part of leather**, New York, 1972.
69 Rehak, P; Orlita A, Int. Sym. 4 Celostatniho seminare konzervatoru knihovich sbitek, State Library of the CSR, Praha, 1986.
70 Thomson R, S; Calnan, C; Haines, B, **Leather: Its Composition and Changes with Time**, 1991.
71 Sykes R,L; Calnan, C; Haines, B, **Leather: Its Composition and Changes with Time**, 1991, pp. 10-11.
72 Lasek, W; Persz, T, **Technologia wyprawy skór, cz. 2: Wykonczenie**, *Wydaw. Nauk. Tech.*, Warszawa, 1981.
73 Strzelczyk, A, B; Fedrizzi-Szostok, A; Karbowska-Berent, J, Int. Symp. Microbal Biodegradation and Biodeterioration of Technical Materials, Lódz, September 8-10, 2003, Technical University of Lódz, Lódz, 2003, pp. 194-201.
74 Nordwig, A; Jahn, W, F, *J. Biochem*, **3**, 519-529, 1968.
75 Barthomeuf, C; Pourrat, H; Pourrat, A, *J. Ferm. Bioeng.*, **73**, 233-236, 1992.
76 Ryder M, L; Rück, P, **Pergament. Geschichte, Struktur, Restaurierung, Herstellung**, *Jan Thorbecke Verlag*, Sigmaringen, 1991, pp. 25-33.
77 Wildbrett M, E; Rück P, **Pergament. Geschichte, Struktur, Restaurierung, Herstellung**, *Jan Thorbecke Verlag*, Sigmaringen, 1991, pp. 359-363.
78 Zyska, B, **The Protection of Librarian Collecions against Deterioration. The Characteristics of Librarian Materials**, *University of Silesia*, Katowice 1991.
79 Gullick M; Rück, P, **Pergament. Geschichte, Struktur, Restaurierung**, Herstellung, *Jan Thorbecke Verlag*, Sigmaringen, 1991, pp. 145-157.
80 Rebrikova, N, L; Muldiyarov, P, Ya, *J. Restaurator*, **5**, 183-190, 1983.
81 Sobel, H; Ajie, H, *J. Free Radical Biol. Med.*, **13**, 701-702, 1992.
82 Zyska, B, **The Protection of Librarian Collecions against Deterioration. Protective Activities in Library**, *University of Silesia*, Katowice 1994.
83 Tanasi, M, T; Impagliazzo, G; Ruggiero D; Rück, P., **Pergament. Geschichte, Struktur, Restaurierung, Herstellung**, *Jan Thorbecke Verlag*, Sigmaringen, 1991, pp.203-215.
84 Strzelczyk, A; Karbowska, J, *Post. Mikrobiol.*, **32**, 19-32, 1993.
85 Karbowska-Berent, J; Strzelczyk, A, **The Role of Streptomycetes in the Biodeterioration of Historic Parchment**, *Wydawnictwo Uniwersytetu Mikolaja Kopernika*, Torun, 2000.

86 Gallo, F; Strzelczyk, A, *J. Bolett. Ist. Patol. Libro. "Alfonso Gallo"*, **30**, 71-87, 1971.
87 Oyen, V, *J. Leather*, **194**, 4606, 43, 1992.
88 Stanley, A, *J. Leather*, **195**, 4622, 27, 1993.
89 Daniels, R, *J. World Leather*, **15**, 2, 2002.
90 Orlita, A; Ludvik, J; Brabenes, M, Int. Symp. Congres on the Leather Industry, Budapest, October 13-19, 1986, Omikk-Technoinform, Budapest,1986, pp. 325-333.
91 Hodges, S, *J. World Leather*, **15**, 5, 2002.
92 Anonymous, *J. World Leather*, **8**, 30, 40, 1997.
93 Brabenec, M; Kubicek, F, *J. Kozarstvi*, **8**, 228, 1988.
94 Smiechowski, K, *Przegl. Skórz.*, **11**, 32-38, 2001.
95 Andres, H, *J. Das Leder*, **12**, 309 ,1992.
96 Sikora, J, *Przegl. Skórz.*, **1**, 22-23 , 2002.
97 Adzet, J, M, *J. Leather*, **198**, 4655, 114, 1996.
98 Adzet, J, M, *J. Word Leather*, **10**, 3, 59, 1997.
99 Beamhouse Technology, *J. Leather*, **194**, 4607, 1992.
100 Anonymous, *J. World Leather*, **8**, 3, 38-41, 1995.
101 Bailey, D,G, *J. Leather*, **196**, 4626, 82, 1994.
102 Ross, G, D, *J. Leather*, **194**, 4607, 139, 1992.
103 Ross, G, D, *J. Leather*, **199**, 4662, 17, 1997.
104 Bailey, D,G, Int. Symp. Kongres XXV JULTCS, Chennai Indie, 1999.
105 Preethi, V; Jeyameenakshi, R; Senthil Kumar, M; Aishwarya S; Sehgal, P, K, *J. Amer. Leath. Chem. Assc.*, **101**, 2, 51-57, 2006.
106 Bitlisli B,U;. Karavana, H A; Basaran, B; Sari, O; Yasa, I; Birbir, M, *J. Amer. Leath. Chem. Assoc*, **99**, 12, 494-501, 2004.
107 Felicjaniak, B; Kosinska, K, *Przegl. Włókien. WOS*, **12**, 23-27, 2005.
108 Russell, A, E, *J. World Leather*, **11**, 3, 43, 1998.
109 Lampart, G, *J. Leather*, **197**, 4637, 44, 1995.
110 Money, C, *J. Amer. Leath. Chem. Assc.*, **86**, 5, 227, 1991.
111 Cooper, D, R; Galloway, A, G, *J. Soc. Leath. Technol. Chem.*, **58**, 120, 1974.
112 Margold, F; Heidemann, E, *J. Das Leder*, **28**, 65, 1977.
113 Anonymous, *J. Leather*, **194**, 4600, 25 1992.
114 Albriht, Wilson, *J. Leather*, **90**, 4669, 17, 1997.
115 **US Patent**, 3,463,785.
116 Anonymous, *J. Leather*, Supl. 1, 1998.
117 Anonymous, **S. A. Patent**, 9,510,559.
118 Hopkins, W, J; Bailey, D, G, *J. Amer. Leath. Chem. Assc.*, **70**, 248, 1975.
119 Hopkins, W, J, *J. Amer. Leath. Chem. Assc.*, **75**, 389, 1980.
120 Hopkins, W, J; Bailey, D, G, *J. Amer. Leath. Chem. Assc.*, **76**, 134, 1981.
121 Hopkins, W, J; Bailey, D, G, *J. Amer. Leath. Chem. Assc.*, **78**, 6, 1983.
122 Bailey, D, G; Hopkins, W, J, *J. Amer. Leath. Chem. Assc.*, **70**, 372, 1975.
123 Bailey, D, G, *J. Amer. Leath. Chem. Assc.*, **71**, 400, 1976.
124 Bailey, D, G; Hopkins, W, J, *J. Amer. Leath. Chem. Assc.*, **76**, 334, 1977.
125 Grasso, G; Manzo, G; Bufalo, G; Fedele, G, Int. Symp. Congres Proceedings JULTCS, London, 1997.
126 Sweet, L, A, C; Henrikson, R, L, *J. Amer. Leath. Chem. Assc.*, **77**, 192, 1982.
127 Hughes, I, R, *J. Soc. Leath. Technol. Chem.*, **58**, 100, 1974.
128 Cwiakala, M; Felicjaniak, B; Szewczyk, R, *Przemysl Lekki*, **33**, 42, 1986.
129 Staniewski, J, *Przegl. Skórz.*, **4**, 117-120, 1979.
130 Cislo, B; Felicjaniak, B; Pietrzykowski, W, *Przegl. Skór.*, **1**, 19-20, 1989.
131 Preševa-Caceva, M, *Przegl. Skórz.*, **1**, 20-22, 1989.
132 Watts, A, Int. Symp. SLTC, Wetherby, 1994.
133 Bailey, D, G; Miller, R, H, *J. Amer. Leath. Chem. Assc.*, **86**, 5, 185, 1991.
134 J. Kanagaraj, N. K. Chandra Babu, S. Sadulla, G. Suseeka Rajkumar, V. Visalshi, N. Chandrakumar, *J. Amer. Leath. Chem. Assc.*, **95**, 10, 368, 2000.
135 Lloyd, G, R, *Przegl. Skórz.*, **5**, 149-152, 1992.
136 Lindner, W; Neuber, H, U, *J. Int. Biodeter.*, **26**, 195-203, 1990.
137 Lindner, W, *J. World Leather*, **5**, 61-65, 1998.
138 Rother, H, J, *J. World Leather*, May, 48-50, 1995.
139 Didato, D, T; Yanek, S, S, *J. Amer. Leath. Chem. Assc.*, **94**,7, 245- 258, 1999.

140 Stosic, R, G, *J. Amer. Leath. Chem. Assc.*, **88**, 171, 1993.
141 Christner, J, *J. Das Leder,* **47**, 44, 1996.
142 Huber, Ch; German, H, P, *J. World Leather*, May, 75-82, 1997.
143 Adminis, U; Huynh, C; Money, C A, Int. Symp. XXVI Congres IULTCS, Cape Town-Africa, 2001, pp. 1-6.
144 Bayramoglu, E, E; Gülümser, G; Karaboz, I, *J. Amer. Leath. Chem. Assc.*, **101**, 3, 96-104, 2006.
145 Colak, S, M; Bayramoglu, E, E; Uluc, D, *J. Amer. Leath. Chem. Assc.*, **101**, 2,66-72, 2006.
146 Kings, V, M; Bryant, S, D; Haque, M, N, *J. Amer. Leath. Chem. Assc.*, **96**, 5, 162-168, 2001.
147 Heidemann, E, *J. Das Leder*, **39**, 9, 1988.
148 Gelo, H; Wionczyk, B; Korowajczyk, L, *Przegl. Skórz.*, **6**, 196-198, 1996.
149 Fischer, W; Nickolaus, G, *J. Das Leder*, **41**, 174, 1990.
150 Gelo, H; Urbaniak, M, *Przegl. Skórz.*, **12**, 498-503, 1998.
151 Kurbat, E; Kucharski, M, Int. Symp. PTCH SIT-P-Chem, Gdansk, 1997, p. 51.
152 Smiechowski, K, Int. Symp. Microbal Biodegradation and Biodeterioration of Technical Materials, Lódz, May 30-31, 2000, Technical University of Lódz, Lódz, 2000.
153 Smiechowski, K, Int. Symp. Microbal Biodegradation and Biodeterioration of Technical Materials, Lódz, May 30-31, 2001, Technical University of Lódz, Lódz, 2001, pp.202-206.
154 Falkiewicz-Dulik, M; Rudzka, E, *Przegl. Skórz.*, **3**, 79-80, 1996.
155 Falkiewicz-Dulik, M, *Rynek Chem.*, **9**, 60-63, 2001.
156 Falkiewicz-Dulik, M, Int. Symp. New Material and Technological Solves In Footwear Industy and Economy of Waste Materials, Kraków, November 21-22, 2002, Central Laboratory of Footwear Industry, Kraków, 2002, pp.136-144.
157 Falkiewicz-Dulik, M; Macura, A, B; Pawlik, B, Int. Symp. 16th Congres of the International Society for Human and Animal Mycology, Paris, June 25-29, 2006, Paris, 2006, pp.P-0099.
158 Falkiewicz-Dulik, M, *Przegl. Skórz.*, **10**, 12-15, 2001.
159 Pawlik, B; Macura, A, B; Sopota, B; Falkiewicz-Dulik, M, *Mikol. Lek.*, **10**, 4, 285-294, 2003.
160 Falkiewicz-Dulik, M; Pawlik, B; Macura, A, B, *Mikol. Lek.*, **12**, 4, 273-278, 2005.
161 Falkiewicz-Dulik, M, *Przegl.Wlókien. WOS*, **4**, 38-42, 2005.
162 Orlita, A, Int. Symp. Microbal Biodegradation and Biodeterioration of technical Materials, Lódz, September 8-10, 2003, Technical University of Lódz, Lódz, 2003, pp. 179-185.
163 Bravetrmanova, M, Brezinova, H, Hiavacek, P, Int. Conference The Shoes in History '97", Zlin, 1998, pp 90-92.
164 Bravetrmanova, M, Brezinova, H, Hiavacek, P, *Archeologia historica*, **23**, 471-492, 1998.
165 Orlita, A, Vaculik, J, Int. Conference The Shoes in History '97", Zlin, 1998, pp 125-130.

4.8 MARINE TRANSPORT AND INSTALLATIONS

Considering that this book is concerned with the use of products which are currently applicable based on existing technology and regulations, organotin compounds and their technology of use are not included here. Interested readers may find many useful data on this subject in a quite recently published monographic source.[1]

4.8.1 RELEVANT MICROORGANISMS

Table 4.49 contains a list of fouling organisms.

Table 4.49. Microorganisms of concern in development of antifouling compositions.

Material	Microorganisms	Refs.
cages and netting in fish farming	*Ascidiella aspersa, Balanus sp., Biddulphia sp., Botrylloides sp., Brongniartella australis, Bryopsis sp., Bugula neritina, Campylodiscus sp., Caprella sp., Ceramium tasmanicum, Ciona intestinalis, Cladophora sp., Dendronotus frondosus, Ectocarpus siliculosus, Ectocarpus sp., Enteromorpha sp., Fragilaria sp., Gracilaria sp., Hiatella arctica, Hiatella spp., Littorina spp., Modiolus sp., Molgula ficus, Mytilus edulis, Obelia australis, Oscillatoria sp., Perna viridis, Pinctada sp., Polysiphonia abscissa, Scrupocellaria bertholetti, Scytosiphon lomentaria, Thais spp., Tubularia larynx, Ulothrix sp., Ulva nematoidea, Ulva spp., Vorticella sp.*	18
pearl culture	*Balanus amphitrite variegates, Balanus amphitrite communis, Bugula sp., Crassostrea sp., Dicarpa sp., Didemnum sp., Diplosoma sp., Lithophaga sp., Martesia sp., Membranipora sp., Mytilus sp., Parasmittina sp., Pinctada sp., Pinna sp., Pteria sp., Saccostrea sp.*	21
PVC based material	*Antedon bifida, Botryllus schlosseri, Chondrus crispus, Delesseria sanguine, Fucus serratus, Fucus spiralis, Haliclona viscosa, Halichondria bowerbanki, Halichondria panacea, Laminaria sp., Myxilla incrustans, Palmaria palmata, Pachymatisma johnstonia, Ulva lactuca*	6
sea water reverse osmosis membrane for desalination	Rhizobiales is dominating genera including *Bosea, Ochrobactrum, Shinella*, and *Rhodopseudomonas*	19
ships	*Amphora coffeaeformis, Balanus amphitrite larvae, Enteromorpha intestinalis, Eucalyptus resinifera, Eucalyptus rubida, Fragilaria, Mallotus japonicus, Mallotus japonicus, Navicula, Phyllidia pustulosa, Pseudomonas atlantica, Pseudomonas nautica, Synedra*	2, 5, 9

In total, about 160 marine species have been analyzed in one source[14] based on literature containing some identified species which cause fouling. They belong to the following phyla (the population of organisms in a particular phylum is the percentage of total):[14]

- Algae 19.5%
- Bacteria 1.95
- Bryozoa 2.6
- Cnidaria 17.95
- Echinodermata 3.2
- Mollusca 7.05
- Multi-phyla 0.25
- Parazoa 38.5
- Sea grass 1.95
- Tunicate 7.05

It has been estimated[15] that over 4,000 species are involved in the fouling of ship hulls if no protection is provided. It is known that many of these species do not become

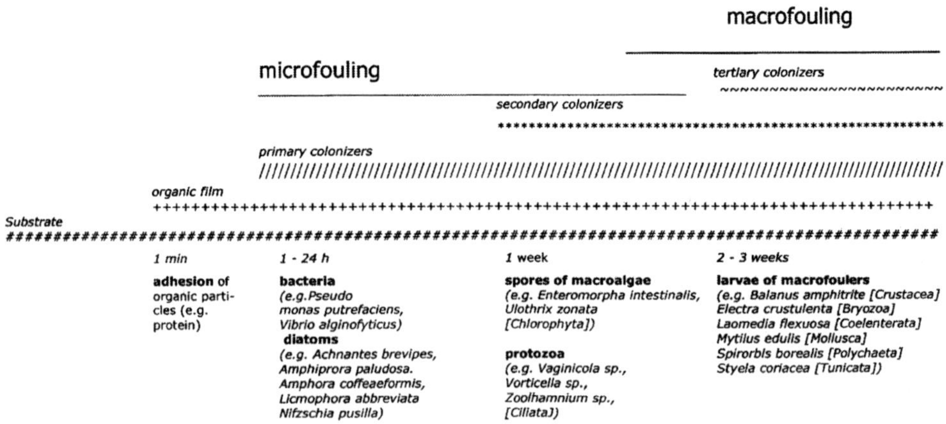

Figure 4.36. Colonization sequence. [Adapted, by permission, from Yebra, D M, Kill, S; Dam-Johansen, K, *Prog. Org. Coat.*, **50**, 75-104, 2004.]

fixed to ships that travel at speeds of above 4–5 knots.[15] It is therefore considerably easier to protect the hulls of high speed ships and those that do not remain for a long time in port.[15]

4.8.2 ESSENTIAL PRODUCT COMPONENTS

Antifouling products are used to control fouling organisms (microbes and higher forms of plant or animal species) on vessels, aquaculture equipment, or other structures used in water. Fouling occurs in both salt and fresh water, but it is more rapid in salt water and more fouling organisms are found in salt water.[2]

Antifouling products are used in the following applications: ship hulls, nets in fish farms, lobster pots, buoys, sluice doors, harbour constructions, inlet pipes, marine sensors, and offshore constructions.[2] The most important are ship hulls for pleasure craft and commercial ships (approx. 95% of the total demand).[2] Drilling platforms are, after ship hulls, the most important application (approx. 2.5% of the total demand).[2] The antifouling paint is applied below the waterline, and the active agents are continuously released to the water during use.[2] By this, antifouling compositions differ in utilization of biocide from any other application, which are designed to retain active components for as long as possible.

4.8.3 MECHANISMS OF BIODEGRADATION AND BIODETERIORATION

Biofouling is initiated by microfouling, which begins with formation of bacterial biofilms. Microfouling precedes macrofouling, which is the accumulation of invertebrates (i.e., barnacles) on submerged objects.[8] If a biofilm does not form on the hull of a ship, invertebrate aggregation does not occur.[8] Figure 4.36 shows the sequence of colonization. Bacterial biofilms are also responsible for causing biocorrosion of submerged surfaces.[8] The exopolymeric substances isolate the organisms from the biocide.[9] Many marine bacteria produce organic secretions which capture nutrients from the environment and protect the bacteria living within the biofilm from antibacterial compounds.[9] Diatoms secrete

Figure 4.37. Examples of heavily fouled hulls. [Adapted, by permission, from Yebra, D M, Kill, S; Dam-Johansen, K, *Prog. Org. Coat.*, **50**, 75-104, 2004.]

large amounts of exopolymeric substances from a slit (raphe) or apical pore in the frustule.[9]

Mathematical modelling data suggest that microbial biofilms do not exert a significant influence on the net biocide leaching and paint polishing processes unless they are strongly limited by mass transport resistances (e.g., isothiazolones).[10] Some published empirical studies, which reported a decrease in the release rates, have most likely encountered biocide trapping within exopolymeric substances.[10] It should be noted that absorption of biocides in exopolymeric substances might also have a positive effect on the performance of antifouling paints by acting as permanent biocide capacitors at the paint surface.[10]

4.8.4 RESULTS OF BIODETERIORATION

When fouling organisms attach to a ship's hull, the increased hydrodynamic drag results in decreased speed, higher fuel consumption, and more frequent removal from service for hull cleaning. A hull of a ship which is not protected by antifouling paint may accumulate up to 150 kg of biological material per m^2 after 6 month at sea.[6] A tanker has about 40,000 m^2 of underwater hull, thus weight increases are enormous. Fuel consumption may be increased by 40-50%[6] and cost of transportation by as much as 77%.[11] Biofouling is also responsible for metal corrosion, which decreases the useful life of ship. Figure 4.47 gives an example of biofouling.

4.8.5 BIOSTABILIZATION

Table 4.50 contains a list of most suitable biocides used for protection of antifouling compositions.

Table 4.50. Biocides used in antifouling compositions.

Material	Biocide	Refs.
cages and netting in fish farming	chlorpthalonil, cuprous oxide, dichlofluanid, Sea-Nine 211, zinc omadine	18
cooling water pipes	fluorosilicone elastomer Bioclean SPG	5

Table 4.50. Biocides used in antifouling compositions.

Material	Biocide	Refs.
EU list of applications	copper, copper pyrithione, copper thiocyanate, cybutryne, dichlofluanid, DCOIT (4,5-dichloro-2-n-octyl-3(2H)-isothiazolone), dicopper oxide, tolylfluanid, zinc pyrithione, zineb	7
heat exchanger cooled with seawater	sodium hypochlorite	16
pearl culture	epoxy coating, wax-based emulsion	21
PVC based materials	cuprous thiocyanate	6
sea water reverse osmosis membrane for desalination	chloramide (NH_2Cl), chlorine dioxide (ClO_2), hypochlorous acid (HOCl) or hypochlorites, ozone (O_3), also UV and sand filtration can be used	19
ship coatings	4,5-dichloro-2-n-octyl-4-isothiazolin-3-one (Sea-Nine 211), 5-chloro-2-(2, 4-dichlorophenoxyphenol attached to silicone	2, 3

The first protection of wooden ships against biofouling was done 4,000 years ago using copper sheet.

Studies seek to develop coatings containing non-leachable biocides fixed to a polymer in a permanent way. One such option is provided by silicone reacted with 5-chloro-2-(2, 4-dichlorophenoxyphenol) (Triclosan).[3] Silicone is also a low surface-energy polymer which does not support surface attachment of organisms. Barnacles start to detach from silicones at moderate speeds of 15 – 20 knots, but even the best commercially available fouling release silicones need hull speeds of around 30 knots to completely self-clean.[3] Therefore, research is under way to improve the surface properties of silicone.

Release rate of biocide and its effectiveness against fouling substance determine the end result of biocide action. At the same time, release rate determines the lifespan of antifouling action and potential polluting effect by biocide. Rates of release of several biocides were compared: cuprous oxide – 25-40, Irgarol 1051 – 5.0, dichlofluanid – 0.6, Seanine 211 – 2.9, TCMTB – 0.0, zinc pyrithione – 3.3 $\mu g\ cm^{-2}\ d^{-1}$.[4]

One efficient approach to prevent diffusion depleting the paint film's antifouling activity is to adsorb it strongly on a specie with low translational mobility (e.g, a high molecular weight polymer).[15] In one example, medetomidine was adsorbed onto a sulfonated polystyrene-block-poly(ethylene-ran-butylene)-block-polystyrene, generating a medetomidine–SDPS ion pair.[15] This permits control of release rate.

Copper forms Cu^+ and Cu^{2+} ions in an aqueous solution.[5] The Cu^{2+} ion is more stable than the Cu^+ ion.[5] The Cu^+ ion is converted to metallic copper and the Cu^{2+} ion by a disproportionation reaction.[5] Copper metal in water, is converted to the Cu^{2+} ion by an oxidation with oxygen in air.[5] Copper exhibits antifouling activities against organisms such as barnacles, tube worms and the majority of algal fouling species.[5] Several algal species (e.g., *Enteromorpha spp., Ectocarpus spp., Achnanthes spp.*) are resistant.[5] To achieve a protection against these tolerant species, various organic biocides are used in conjunction with copper.[5] Copper is used with a leaching rate of at least 10-15 $\mu g\ cm^{-2}\ day^{-1}$.[5]

Figure 4.38 shows the effect of biocide diffusion on deterioration of properties of antifouling paint. Sea water has to penetrate into the paint to dissolve biocides so that they

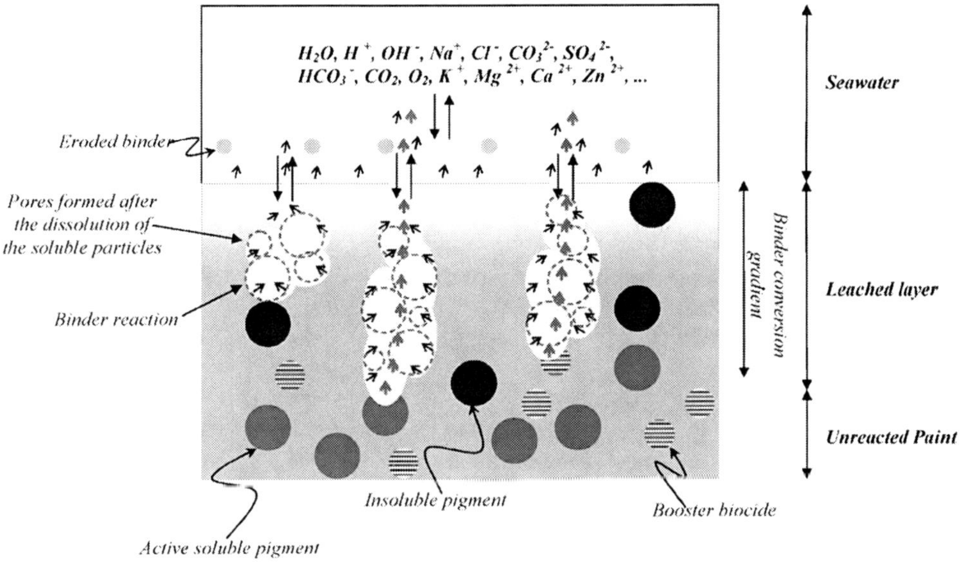

Figure 4.38. Schematic illustration of the behavior of a biocide-based antifouling system exposed to sea water. [Yebra, D M, Kill, S; Dam-Johansen, K, *Prog. Org. Coat.*, **50**, 75-104, 2004.]

can diffuse to the surface. Removal of biocides and the effect of sea water changes the characteristics of paint.[11]

Several types of biocide-free antifouling paints have been tested. These include non-eroding coatings, made out of silicones, fiber coats (short fibres applied in a dense pattern, 200–500 fibres/mm^2) which deters settlement of fouling organisms (an epoxy is applied and serves as glue, then electrostatically charged fiber are sprayed into the wet epoxy to be perpendicularly oriented), epoxy, polyurethane, and vinyl (all three are simple, anticorrosive coatings); and eroding coatings (specialty polymer coatings and ablative coatings, usually made out of epoxy or rosin).[12] Comparison of these coatings shows that silicone coatings which do not contain biocides have a potential to be tailored to requirements.

The presence of *Ceramium botryocarpum* (Rhodophyta, Ceramiales) in tank cultures significantly limits the establishment of opportunist green algae such as *Cladophora, Bryopsis*, or *Ulva* usually disturbing the cultures by producing substances which repel other organisms.[13] Addition of extracts of *Bacillus licheniformis, Bacillus pumilus, Bacillus subtilis*, and *Pseudomonas sp.* to paints had also some antifouling effect.[14] These microorganisms act through their enzymes which may be used as additives to the paint system to replace biocides (when efficient methods will be found). A large number of enzymes were studied and proposed as viable alternatives.[17] These include:[17] aminopepsidases, carboxypepsidases, esterases, glycolases, hydrolases, isomerases, lyases, oxidoreductases, peroxidases, and many other. It can be deduced from this list that the enzymes are directed to degradation of organic matter. This organic matter is related to formation of biofilm, as follows:[17]

- degradation of adhesives required by first settling organisms

- disrupting biofilm matrix
- generation of deterrents or biocides (e.g., non-polar metabolites residing on surface which repel larvae which wants to colonize the surface, or polar metabolites liberated to the surrounding water which trigger avoidance behavior by larval receptors; biocides are various enzymes capable of damaging settling cells)
- interfering with intercellular communication (settling organisms generate sensors informing other organism about the possibility of colonization. These can be enzymatically damaged as a means of prevention)

Crab (*C. pagurus*) carapace, mussel (*M. edulis* and *Mytilus galloprovincialis*) shell, whale skin (*Globicephala melas*), and lotus leaf were other natural products investigated, but they did not produce technical solution, although they had some repelling and growth inhibiting properties.[14]

Three general mechanisms are used to obtain antifouling coating:[20]

- Type 1: insoluble binder-soluble toxic compound (antifouling compound dissolves slowly, leaving behind a porous film, which further releases active compound)
- Type 2: soluble binder-soluble toxic compound (binder and toxic compound slowly dissolve; their combined action supplies fresh toxins to the seawater)
- Type 3: soluble toxic binder (binder is chemically combined with the toxic compound. Seawater slowly attacks the link between the toxic compound and the binder, which results in release of the toxin and dissolution of the coating)

The most recent advances in toxic antifouling coatings have been in coatings having the type 3 mechanism.[20] Type 3 coatings have been developed based on both copper oxide and tributyltin compounds.[20] These coatings, commonly referred to as ablative or self-polishing coatings, are designed to erode slowly in service.[20] As they erode they become smoother and release toxins that prevent fouling. A typical erosion rate is 50 to 75 micrometers per year.[20] In principle, the service life of these coatings is limited by the amount of coating applied.[20] If a five-year service life is needed, a coating thickness of 375 μm (5x75 μm/year) is applied to give the desired performance.[20]

4.8.6 FORMULATIONS

Coating on PVC-based material[6]

Revacryl 380	22.50
Water	40.12
Cuprous thiocyanate	8.91
Tophelex	5.36
BHA	2.97
NUDMB50-11	3.96
Titanium dioxide	16.18

Addition of NUMB50-11 (bacterial extract) helps to decrease the required concentration of copper-based biocide.[6]

Antifouling coating composition with excellent storage stability[22]

Zinc pyrithione	3
Sodium-2-pyridyloxy-N-oxide	0.15
Cuprous oxide	35
Red iron oxide	8
Talc	5
Varnish	10
Amide wax 20% in xylene	2
Xylene	36.85

Varnish was made by reacting 7.7 parts of methacrylic acid, 64.4 parts of methyl methacrylate, 28 parts of 2-ethylhexyl acrylate, and 3 parts of azobisisobutyronitrile in 100 parts of xylene and 20 parts of butanol.[22]

Encapsulated DCOIT compositions which provides in-can stability[23]

DCOIT (Kathon 287T)	150
Methyl methacrylate	10
1,4-butadenediol diacrylate	10
Trimethylpropane trimethacrylate	10
Tetrbutyl perpivalate	1
Water	254
Elvanol 50-442 (PVAl)	6
Aluminum sulfate (TG 8.3%)	60
Sulfuric acid (10 vol%)	34
Urea	6
Resorcinol	4.5
Water	65
Glutaraldehyde	5

DCOIT (4,5-dichloro-2-n-octyl-3(2H)-isothiazolone) has a dual coating (shell) with the first layer made out of acrylic polymer. After the first layer was formed, DCOIT had mean particle size of 19 μm. The second layer of coating is made out of urea-resorcinol-glutaraldehyde.

Antifouling paint[24]

Aldehyde resin	101
Butyl acetate	101
Aluminum diisopropoxide acetoacetic ethylester chelate	8
Tricresyl phosphate	15
Titanium 2-propylato, tris(dioctyl) phosphate (Ken-React 9S)	6
Zinc oxide	117
Iron oxide (red)	69
Dicopperoxide	350

Copper pyrithione	30
Turpentine	65
Montan wax	3
Asbestine	63
Butyl acetate	340

The paint, having a dry film thickness of 360-400 μm, was immersed in the port of Hamburg for 30 months and no weed or animal fouling was observed.

Antifouling paint with storage stability[25]

Cu acrylate resin	61.35
Cuprous oxide	35
Zinc pyrithione	3
Wax (thixotrope)	0.5
Zn 2-hydroxypyridine-N-oxide	0.15

Zn 2-hydroxypyridine-N-oxide is a viscosity stabilizing additive. Resin is obtained by polymerization of 600 parts of ethyl acrylate, 250 parts of 2-ethylbenzyl acrylate, 72 parts of acrylic acid and 25 parts of azobisisobutyronitrile. After polymerization is complete, 259 parts of naphthenic acid and 98 parts of copper hydroxide are added.

Silicone base for antifouling paints[26,27]

Synthesis of organosilicone compound having one end-blocked diorganopolysiloxane group and a condensation-curable silyl group was described.[26] This organosilicone compound can be used with or without biocide to obtain an antifouling silicone coating.

An organopolysiloxane having functional groups capable of condensation reaction at both molecular terminals and a hydrophobic silica form a base for a coating composition which has antifouling properties.[27]

Titanium-based catalyst for UV degradation of fouling microflora[28]

When buildings are to be newly coated with paint, the old paint may be scraped off the wall, and then stains such as mold growing in the undercoat are washed out with water before the new paint is applied onto the wall surface.[28] In washing the wall with water, a detergent which contains an antibacterial agent can be used. Such a detergent sterilizes the undercoat surface by which a certain level of aseptisation is achieved between the undercoat surface and the film of paint.[28]

The coating is susceptible to contamination by mold attack from exposed surfaces if there is no treatment made on the exposed surfaces of the paint coating. Once mold grows on the coating, then the coated surface can be stained.[28] Further, once mold grows on the coating, the mold may attack not only the coating but eventually the wall material itself, making the problem worse.[28]

In order to prevent or reduce stains and erosions by microorganisms such as mold, an optical catalyst can be added, such as titanium oxide.[28] Absorption of light which has a level of energy equivalent to a band gap between the valence band and the conduction band causes transition of electrons from the valence band to the conduction band.[28] Due to this transition of electrons, the valence band has electron holes.[28] The electrons in the conduction band move to matter adsorbed on the surface of the optically catalytic semiconductor, and this movement can chemically reduce the adsorbed matter.[28] The electron holes in the valence band get electrons from the matter which is adsorbed on the surface of the optically catalytic semiconductor, and this behavior can oxidize the adsorbed matter.[28]

If the material which is adsorbed by the optical catalytic titanium oxide is an organic matter for instance, action of the hydroxy radicals may eventually decompose the organic matter into water and carbon dioxide.[28] Among many semiconductor materials which have an optical catalyst capability, titanium oxide in particular works as a superior catalyst in such an oxidation-decomposition reaction as the above, and therefore it is widely used in antibacterial agents, deodorants, environmental purification agents, and so on.[28]

A problem, however, is that the titanium oxide optical catalyst itself can only work as a catalyst by absorbing light.[28] For this reason, even if a wall material is coated with a paint which contains titanium oxide that has the optical catalyst capability, antibacterial or antifouling effect based on the optically catalytic decomposition cannot be expected if the wall material is used in a dark place in the building or stored at a dark place where titanium oxide can absorb little or no light.[28] Further, even if the wall material is used at a sunny place, titanium oxide cannot absorb light or enough light during the night time, and so an antibacterial effect based on the optical catalyst capability cannot be expected, either.[28]

Further, titanium oxide itself does not have a strong capability to adsorb matter on its surfaces.[28] Therefore, in order to make titanium oxide exhibit its catalytic capability sufficiently, it is necessary to improve contact efficiency between titanium oxide and the target matter which is to be oxidized and decomposed.[28]

An adsorbent matter known in such an application is calcium hydroxyapatite, which can exchange ions with both cations and anions, and therefore it is highly adsorbent.[28] In particular, it is superb in adsorbing organic matter such as protein.[28] However, adding it with titanium oxide to a paint and dispersing of the matter sufficiently is not efficient in manufacture of the paint.[28] Further, making titanium oxide and an adsorbent matter simply coexist in a paint can only make small improvement in contact efficiency between titanium oxide and target matters to decompose.[28] In this invention, calcium hydroxyapatite has part of its Ca atoms substituted by Ti, which makes the resultant material to be both a good catalyst and have surface properties which help to absorb proteins.[28]

Antifouling paint[29]

Vinyl resin (Vinylite VYHH)	12
Rosin (ROM Rosin)	2
Xylene	3
Zinc oxide	13
Cuprous oxide	50
Polymethylguanidine salt	7
Gelling agent (polyamide wax)	3
Methylisobutyl ketone	10

REFERENCES

1 Konstantinou, I, Ed., **Antifouling Paint Biocides**, *Springer*, 2006.
2 Jacobson, A H; Willingham, G L, *Sci. Total Environ.*, **258**, 103-110, 2000.
3 Thomas, J; Choi, S-B; Fjeldheim, R; Boudjouk, P, *Biofouling*, **20**, 4/5, 227-36, 2004.
4 Thomas, K V, *Biofouling*, **17**, 1, 73-86, 2001.
5 Omae, I, *Chem. Rev.*, **103**, 3431-3448, 2003.
6 Eguia, E; Trueba, A, *J. Coat. Technol. Res.*, **4**, 2, 191–202, 2007.
7 List of participants/applicants to the Review Programme of existing active substances used in biocidal products in accordance with Article 6(5) of Commission Regulation (EC) No 2032/2003. Information as per 06/03/2009.
8 Melander, C; Moeller, P D R; Ballard, T E; Richards, J J; Huigens, R W; Cavanagh, J, *Intl. Biodet. Biodeg.*, **63**, 529-32, 2009.
9 Casse, F; Swain, G W, *Intl. Biodet. Biodeg.*, **57**, 179-85, 2006.
10 Yebra, D M; Kiil, S; Weinell, C E; Dam-Johansen, K, *Prog. Org. Coat.*, **57**, 56-66, 2006.
11 Yebra, D M, Kiil, S; Dam-Johansen, K, *Prog. Org. Coat.*, **50**, 75-104, 2004.
12 Watermann, B T; Daehne, B; Sievers, S; Dannenberg, R; Overbeke, J C; Klijnstra, J W; Heemkenm O, *Chemosphere*, **60**, 1530-41, 2005.
13 Bazes, A; Silkina, A; Defer, D; Bernède-Bauduin, C; Quéméner, E; Braud, J-P; Bourgougnon, N, *Aquaculture*, **258**, 664-74, 2006.
14 Chambers, L D; Stokes, K R; Walsh, F C; Wood, R K J, *Surface Coat. Technol.*, **201**, 3642-52, 2006.
15 Handa, P; Fant, C; Nyden, M, *Prog. Org. Coat.*, **57**, 376-82, 2006.
16 Eguia, E; Trueba, A; Rio-Calonge, B; Giron, A; Bielva, C, *Intl. Biodet. Biodeg.*, **62**, 79-87, 2008.
17 Kristensen, J B; Meyer, R L; Laursen, B S; Shipovskov, S; Besenbacher, F; Poulsen, C H, *Biotechnol. Adv.*, **26**, 471-81, 2008.
18 Braithwaite, R A; McEvoy, L A, *Adv., Marine Biol.*, **47**, 215-52, 2005.
19 Kim, D; Jung, S; Sohn, J; Kim, H; Lee, S, *Desalination*, **238**, 43-52, 2009.
20 Bleile, H; Rodgers, S D, in **Marine Coatings, Encyclopedia of Materials: Science and Technology**, *Elsevier*, 2008, pp 5174-85.
21 de Nys, R; Ison, O, in **Biofouling, The Perl Oyster**, *Elsevier*, 2008, pp 527-53.
22 Kosaza, Y; Koga, Y, **US Patent**, 7,553,481, API Corporation, June 30, 2009.
23 Hart, R L; Virgallito, D R; Work, D E, **US Patent**, 7,550,200, Microtek Laboratories, Inc., June 23, 2009.
24 Blum, H, **US Patent**, 7,482,394, Chemical Investment Ltd., January 27, 2009.
25 Lei, D; Waldron, C; Davis, M, **US Patent**, 7,435, 771, Arch Chemicals, Inc., October 14, 2008.
26 Yamaya, M; Sato, K, **US Patent**, 7,351,477, Shin-Etsu Chemical Co., Ltd., April 1, 2008.
27 Amidaiji, K; Yamamoto, T; Muramatsu, H, **US Patent**, 7,297,745, Chugoku Marine Patents, Ltd. and Shin-Etsu Chemical Co., Ltd., November 20, 2007.
28 Wakamura, M, **US Patent**, 7,157,503, Fujitsu Ltd., January 2, 2007.
29 Yang, K-W; Park, N-S; Choi, K-S; Hahu, S-J; Kim, J-M; Hong, W-P, **US Patent**, 7,108,741, SK Chemical Co., Ltd., September 19, 2006.

4.9 MEDICAL APPLICATIONS

4.9.1 RELEVANT MICROORGANISMS

Table 4.51 contains a list of microorganisms found to either cause deterioration of medical equipment or to cause dangerous infections.

Table 4.51. Microorganisms in medical applications.

Material	Microorganisms	Refs.
endoscope tubing	*Legionella, Mycobacteria spp.*	12
HDPE tubing	*Escherichia coli, Pseudomonas aeruginosa, Staphylococcus aureus, Staphylococcus epidermidis*	11

Table 4.51. Microorganisms in medical applications.

Material	Microorganisms	Refs.
indwelling devices	*Acinetobacter calcoaceticus, Escherichia coli, Escherichia faecalis, Klebsiella pneumoniae, Pseudomonas aeruginosa, Pseudomonas mirabilis, Serratia marcescens, Staphylococcus aureus, Staphylococcus epidermidis, Streptococcus viridans*	12
intravascular catheters	*Candida spp.* including *C. albicans, C. dubliniensis, C. glabrata, C. krusei, C. lusitaniae, C. parapsilosis, C. tropicalis*	12
textiles	*Absidia corymbifera, Aspergillus flavus, Aspergillus fumigatus, Candida albicans, Cladosporium cladosporioides, Epidermophyton, Escherichia coli, Microsporum canis, Penicillium verrucosum, Pseudomonas, Staphylococcus aureus, Trichophyton rubrum*	3, 4, 7
vascular catheter	*Candida albicans, Staphylococcus aureus*	1, 2
wounds and burns	*Acinetobacter spp., E. coli, Klebsiella spp., P. aeruginosa, S. aureus*	12

4.9.2 RESULTS OF BIODETERIORATION

Nosocomial infections are estimated to occur in at least 5% of all patients hospitalized.[3] It has been estimated that biofilms are associated with 65% of nosocomial infections and that treatment of these biofilm-based infections costs >$1 billion annually.[10] Textiles are an excellent substrate for bacterial growth under appropriate moisture and temperature conditions. The use of antimicrobial textiles, especially in those textiles that are in close contact with the patients, may significantly reduce the bioburden in clinical settings and consequently reduce the risk of nosocomial infections.[3]

4.9.3 BIOSTABILIZATION

Table 4.52 contains a list of the most suitable biocides used for protection of equipment and users of medical applications.

Table 4.52. Biocides used in medical applications.

Material	Biocide	Refs.
burns & wounds	silver nanoparticles[5]	5
HDPE tubing	silver	11
PVC-based devices	sodium azide,[1] silver[1]	1
polyurethane devices	chitin	2
textiles	benzalkonium chloride (0.25%),[4] cetyltrimethylammoniumbromide,[8] dimethylol-5,5-dimethylhydantoin,[6] glutaraldehyde (1.6%),[4] monomethylol-5,5-dimethylhydantoin,[6] octenidine,[8] polyvinylpyrrolidone iodine (1% available iodine),[4] polyester glycol iodine (0.18% available iodine),[4] silver[7]	4, 6, 7, 8

Comparison of Figures 4.39 and 4.40 shows extensive damage induced by the presence of diluted solution of fungicide micafungin on growing hyphal cells of *Aspergillus fumigatus*.[9]

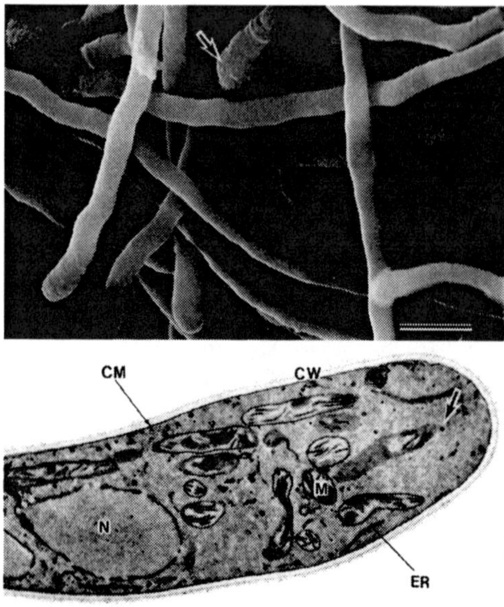

Figure 4.39. Untreated control hypae of *Aspergillus fumigatus*. Upper: SEM image of strong hypae with smooth surface; lower: TEM image of hypal cell. N - nucleus, M - mitochondria, ER - endoplastic reticulum, CM - cell membrane, CW - cell wall. [Adapted, by permission, from Nishiyama, Y; Hasumi, Y; Ueda, K; Uchida, K; Yamaguchi, H, *J. Electron Microscopy*, **54**, 1, 67-77, 2005.]

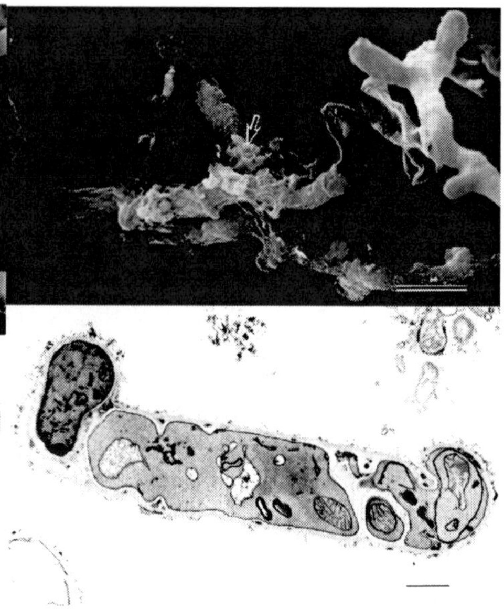

Figure 4.40. Hypae grown for 5 h with 0.1 μg ml^{-1} micafungin. Upper: SEM images shows crushing and distortion of hypae cells (bar 5 μm); lower: TEM image showing abnormal formation of septal walls. The hypha is segmented disproportionately by irregular induced septa. [Adapted, by permission, from Nishiyama, Y; Hasumi, Y; Ueda, K; Uchida, K; Yamaguchi, H, *J. Electron Microscopy*, **54**, 1, 67-77, 2005.]

4.9.4 FORMULATIONS

Composition for cleaning medical instruments[13]

Benzyl dimethyl ammonium chloride	0.1
Sodium tetraborate decahydrate	150
Glycerin	30
Ethoxylated nonylphenol (Terric GN9)	25
Dipropylene glycol methyl ether	200
Water	594.9

REFERENCES

1 Lakshmi, S; Pradeep Kumar, S S; Jayakrishnan, A, *J. Biomed. Mater. Res.*, **61**, 26–32, 2002.
2 Zilberman, M; Navon, A; Sandovsky-Losica, H; Segal, E, *J. Biomed. Mater. Res.*, **81A**, 392–398, 2007.
3 Borkow, G; Gabbay, J, *Med. Hypotheses*, **70**, 990-94, 2008.
4 Tortorano, A M; Viviani, M A; Biraghi, E; Rigoni, A L; Prigitano, A; Grillot, R, *J. Med. Biol.*, **54**, 955-57, 2005.
5 Jain, J; Arora, S; Rajwade, J; Khandelwal, S; Paknikar, K M, *Mol. Pharmaceutics*, just accepted, 2009.
6 Qian, L; Sun, G, *J. Appl. Polym. Sci.*, **89**, 2418-25, 2003.
7 Jung, W K; Kim, S H; Koo, H C; Shin, S; Kim, J M; Park, Y K; Whang, S Y; Yang, H; Park, Y H, *Mycoses*, **50**, 265-69, 2007.
8 Mahltig, B; Fiedler, D; Bottcher, H, *J. Sol-Gel Sci. Technol.*, **32**, 219-22, 2004.

9 Nishiyama, Y; Hasumi, Y; Ueda, K; Uchida, K; Yamaguchi, H, *J. Electron Microscopy*, **54**, 1, 67-77, 2005.
10 Mah, T-F G; O'Toole, G, *Trends Microbiol.*, **9**, 1, 34-39, 2001.
11 Lin, T-L; Lu, F-M; Conroy, S; Sheu, M-S; Su, S-H; Tang, L, *Medical Device Technol.*, 26-30, 2001.
12 Lindsay, D; von Holy, A, *J. Hospital Infection*, **64**, 313-25, 2006.
13 Sava, A; Kritzler, S, **US Patent**, 7,217,684, May 15, 2007.

4.10 METALS

4.10.1 RELEVANT MICROORGANISMS

Table 4.53 contains a list of microorganisms found to cause corrosion of metals.

Table 4.53. Microorganisms in metal deterioration.

Material	Microorganisms	Refs.
aerobic corrosion	*Aerobacter, Archangium, Acinetobacter, Bacillus, Clonothrix, Clostridium, Crenothrix, Desulfovibrio, Desulfotomaculum, Escherichia, Flavobacterium, Gallionella ferruginea, Herpetosyphon, Leptothrix sp., Lieskeella, Naumanniella, Metallogenium, Ochrobium, Pedomicrobium, Pseudomicrobium, Pseudomonas, Seliberia, Siderocapsa, Siderocapsa, Siderococcus, Sphaerotilus, Thiobacillus, Toxothrix*	3, 4
anaerobic corrosion	*Alteromonas, Chlorophyta, Cyanophyta, Desulfobacter, Desulfobulbus, Desulfococcus, Desulfomonas, Desulfonema, Desulfosarcina, Desulfotomaculum, Desulfovibrio, Shewanella*	3
iron-reducing microorganisms	*Acidiphilium acidophilum, Archaeoglobus fulgidus, Bacillus infernus, Clostridium beijerinckii, Deferribacter thermophilus, Desulfotomaculum reducens, Desulfovibrio profundus, Ferrimonas balearica, Geobacter metallireducens, Geothrix fermentans, Ferribacterium limneticum, Methanococcus thermolithotrophicus, Pyrococcus furiosus, Shewanella oneidensis, Sulfolobus acidocaldarius, Sulfurospirillum barnesii, Thermodesulfobacterium commune, Thermoterrabacterium ferrireducens, Thermotoga maritima, Thermus str.* SA-01, *Vibrio alginolyticus*	7
aluminum	*Desulfovibrio desulfuricans, Pseudomonas fluorescens, Pseudomonas sp.*	13, 14
carbon steel	*Desulfovibrio alaskensis, Desulfovibrio capillatus, Desulfovibrio vulgaris, Shewanella putrefaciens*	6, 15, 17
copper	*Acidovorax delafieldii, Acidovorax sp., Cytophaga johnsonae, Flavobacterium sp., Micrococcus kristinae, Pseudomonas fluorescens, Pseudomonas paucimobilis, Pseudomonas solanacearum, Rhodotorula sp., Sphingomonas sp., Variovorax sp.*	11, 12
stainless steel	*Gallionella sp., Leptothrix cholodnii, Leptothrix discophora, Leptothrix mobilis, Siderocapsa sp.*	8

4.10.2 ESSENTIAL PRODUCT COMPONENTS

For the bacteria–aluminum interaction, there is a sudden increase in the attraction forces as the bacterium approaches the substrate.[13] This "snap-in" phenomenon occurs when the bacterial cells are at a certain distance to the aluminum. The distances at which the bacteria start interacting with aluminum (due to the electrostatic forces) are larger than for stainless steel, mild steel, and copper.[13] The hydrophobicity and surface charges on the metal substrates and bacteria have a significant effect on cell–metal adhesion.[13] The surface charge greatly influences the adhesion force by controlling the electrostatic interaction.[13] The bacterial adhesion forces are enhanced by the increasing of the surface hydrophobicity.[13]

4.10.3 MECHANISMS OF BIODEGRADATION AND BIODETERIORATION

There are several mechanisms of metal corrosion, which are believed to operate, such as:[3]

- aerobic corrosion (the metal covered with microorganism colonies acts as an anode, and the area further away has higher oxygen concentrations and acts as a cathode site. Electrons flow between the anode and cathode, causing a corrosion process)
- anaerobic corrosion (when oxygen is not available to accept electrons produced, SO_4^{-2} or other compounds are used as electron acceptors. Each type of electron acceptor is unique in the pathway of metabolism)
- alternating corrosion (aerobic-anaerobic; this mechanism involves alternating action of two groups of microorganisms; resultant corrosion rates are often higher than those observed under either continuous oxic or anoxic conditions)
- exopolymer induced (proteins in bacterial exopolymeric materials use their disulfide-rich bonds to induce corrosion by reduction reactions)
- deposition of corrosive substances (formic, acetic, butyric acids are produced, for example, by fermentative microorganisms)
- by electron transfer (bacterial quinone, an effective electron transferring molecule, can be excreted by the bacterium *Shewanella putrefaciens*, suggesting that extracellular process of electron transport may be initiated and participate actively in the corrosion of metals).

Complex microbial ecosystems consisting of *Leptothrix*-related manganese-oxidizing bacteria, sulfur-oxidizing, sulfate-reducing, and slime-producing bacteria in the biofilm were responsible for corrosion of stainless steel.[8] Figure 4.41 shows filamentous bacteria and micronodules adhering to the surface of stainless steel.[8] EDX spot analysis indicated elevated values of Mn, Fe, Ca and O on the filaments, and the occurrence of manganese-rich micronodules.[8] These micronodules were about 10 μm in diameter and had an apparent outlook with a central hole of 1-2 μm.[8] EPMA analysis indicated that these micronodules were high in Mn, Ca and O.[8] The bacteria found were identified as *Leptothrix* manganese-oxidizing bacteria.[8]

The bacteria–metal adhesion force was reduced in the presence of the nutrients, because a trace organic film was formed, which decreased the metal surface wettability.[9] Stronger ionic strength in solution results in a larger bacteria–metal adhesion force, which is due to the stronger electrostatic attraction force between the positively charged metal surface and negatively charged bacterial surface.[9] Solution pH influences the interaction between the bacterial cells and the metal surface; the bacteria–metal adhesion force reached its highest value when the pH of the solution was near the isoelectric point of the bacteria, i.e., at the zero point charge.[9]

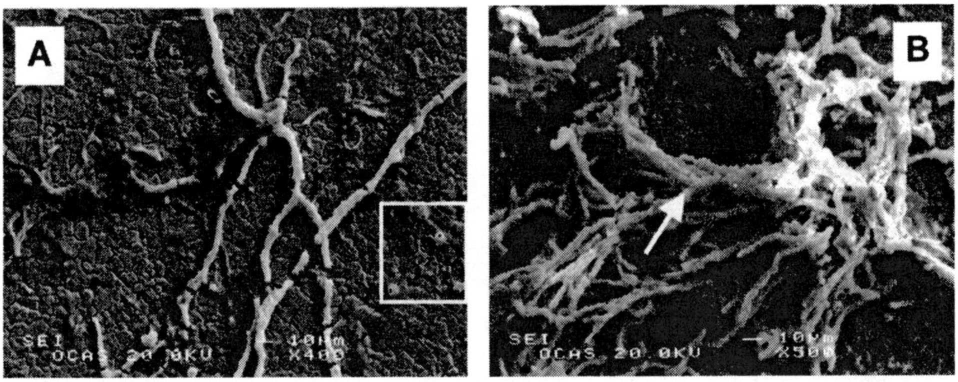

Figure 4.41. (A) adherence of filamentous bacteria and micronodules on a 316L stainless steel surface (400x magnification); (B) entanglement of filamentous bacteria on a 316L stainless steel surface (500x magnification). [Adapted, by permission, from Kielemoes, J; Bultinck, I; Storms, H; Boon, N; Verstraete, W, *FEMS Microbiol. Ecology*, **39**, 41-55, 2002.]

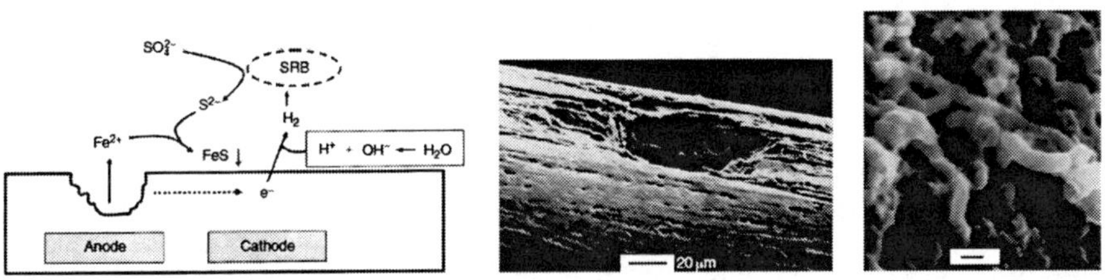

Figure 4.42. Model, result, microorganisms causing corrosion. Left – hydrogen consumed by microorganisms causes cathodic depolarization, and results in pitting corrosion. Middle – ferrous wire with pit caused by corrosion. Right – biofilm of *Desulfovibrio desulfuricans*. [Adapted, by permission, from Barton, L L; Fauque, G D, *Adv. Appl. Microbiol*, **68**, 41-98, 2009.]

4.10.4 RESULTS OF BIODETERIORATION

Microorganism participate in weakening of metal structure by:[3]

- formation and depositing on metal surface corrosive products
- cathodic polarization (consumption of hydrogen accumulated on the surface of metals; this causes formation of electrochemical potential)
- hydrogen embrittlement (permeation of molecular hydrogen into the metal matrix, resulting in cracking and structure deformation)

It is estimated that 70% of damage to gas transmission pipelines is caused by microorganisms.[3] Figure 4.42 shows the mechanism of cathodic polarization, result of damage caused by its action, and microorganism responsible for corrosion.[5]

Excessive metal corrosion with by-product release is the "blue water" phenomenon observed in copper tubing. The occurrences have been reported in Europe, USA, New Zealand, Australia, and Japan.[10]

4.10.5 BIOSTABILIZATION

Table 4.54 contains a list of the most suitable biocides used for protection of metals.

Table 4.54. Biocides used in metal protection

Material	Biocide	Refs.
cooling water	chlorination and glutaraldehyde	3
low carbon steel	imidazoline	16
oil drilling	chlorination and glutaraldehyde	3
mild steel	dithiocarbamate[2]	2

Addition of 5-chloro-2-methyl-4-isothiazolin-3-one/2-methyl-4-isothiazolin-3-one to control corrosion in mild steel pipes of cooling towers did not have effect.[1] The rate of corrosion of 0.324 mm/year was similar to the rate without biocide. The rate of corrosion was reduced to 0.024 mm/year after zinc chloride-based corrosion inhibitor was added.[1]

REFERENCES

1 Batista, J F; Pereira, R F C; Lopes, J M; Carvalho, M F M; Feio, M J; Reis, M A M, *Biodeg.*, **11**, 441-48, 2000.
2 Raman, V; Tamilselvi, S; Rajendran, N, *Mater. Corrosion*, **59**, 4, 329-34, 2008.
3 Gu, J-D, in **Corrosion, microbial, Encyclopedia of Microbiology**, *Elsevier*, 2009, pp 259-69.
4 Annuk, H; Moran, A P, **Microbial Glycobiology**, *Elsevier*, 2010, pp 781-801.
5 Barton, L L; Fauque, G D, *Adv. Appl. Microbiol*, **68**, 41-98, 2009.
6 Miranda, E; Bethencourt, M; Botana, F J; Cano, M J, Sanchez-Amaya, J M; Corzo, A; Garcia de Lomas, J; Fardeau, M L; Ollivier, B, *Corrosion Sci.*, **48**, 2417-31, 2006.
7 Herrera, L K; Videla, H A, *Intl. Biodet. Biodeg.*, **63**, 891-95, 2009.
8 Kielemoes, J; Bultinck, I; Storms, H; Boon, N; Verstraete, W, *FEMS Microbiol. Ecology*, **39**, 41-55, 2002.
9 Sheng, X; Ting, Y P; Pehkonen, S O, *J. Colloid. Interface Sci.*, **321**, 256-64, 2008.
10 Beech, I B; Sunner, J A, *Interface Sci. Drinking Water Treatment*, 245-55, 2006.
11 Angell, P; Chamberlain, A H L, *Intl. Biodet.*, **27**, 135-43, 1991.
12 Reyes, A; Leteliera, M V; De la Iglesia, R; Gonzalez, B; Lagos, G, *Intl. Biodet. Biodeg.*, **61**, 135-41, 2008.
13 Sheng, X; Ting, Y P; Pehkonen, S O, *J. Colloid. Interface Sci.*, **310**, 661-69, 2007.
14 Bott, T R in **Fouling of Heat Exchangers**, *Elsevier*, 1995, pp 223-67.
15 Herrera, L K; Videla, H A, *Intl. Biodet. Biodeg.*, 63, 891-95, 2009.
16 Gonzalez-Rodriguez., C A; Rodriguez-Gomez, F J; Genesca-Llongueras, J, *Electrochim. Acta,* **54**, 86-90, 2008.
17 Padilla-Viveros, A Garcia-Ochoa, E; Alazard, D, *Electrochim. Acta*, **51**, 3841-47, 2006.

4.11 MINERAL DISPERSIONS

4.11.1 RELEVANT MICROORGANISMS

Table 4.55 contains a list of microorganisms found to cause deterioration of mineral dispersions.

Table 4.55 Microorganisms in mineral dispersions.

Material	Microorganisms	Refs.
aerobic	*Acidovarad, Agrobacterium, Alcaligenes, Bdellovibrio, Flavobacterium, Methylbacterium, Pseudomonas, Rhizobium, Sphingobacterium*	1
anaerobic	*Aeromonas, Chromobacterium, Vibrio*	1

Table 4.55 Microorganisms in mineral dispersions.

Material	Microorganisms	Refs.
other	*Agromyces, Hyphomicrobium, Leptothrix, Micrococcus, Staphylococcus*	1
calcium carbonate	*Methylobacterium extorquens, Pseudomonas putida* (formaldehyde-degrading microorganisms)	2

Minerals such as pigments and fillers are frequently delivered to plants, which use them in the form of water dispersion. Some of these dispersions may be heated during grinding to temperatures as high as 110°C, which effectively sterilizes them, with the exception of some remaining spores which cannot be destroyed under these conditions.[1] This means that even dispersion which has undergone high temperature treatment may be potentially infected with microorganisms.

Dispersions usually contain some organic admixtures from their raw materials. In addition, they contain dispersants (most likely polyacrylic acid derivative)[1] which constitute a potential source of nutrients essential for microorganism to grow.

4.11.2 RESULTS OF BIODETERIORATION

Presence of microorganisms in mineral dispersion causes unpleasant odor, discoloration, increase in viscosity, and acidity.[1]

4.11.3 BIOSTABILIZATION

Table 4.56 contains a list of the most suitable biocides used for protection of mineral dispersions.

Table 4.56. Biocides used in mineral dispersions.

	Biocide	Refs.
general	1,2- benzisothiazolin-3-one (BIT), Bronopol (2-bromo-2-nitropropan-1,3-diol), 5-chloro-2-methyl-4-isothiazolin-3-one (CIT), glutaraldehyde, 2-methyl-4-isothiazolin-3-one (MIT), o-phenylphenol	1
calcium carbonate	(ethylenedioxy) dimethanol is ineffective because of formaldehyde-degrading microorganisms	2

Degradation of formaldehyde in calcium carbonate slurries was found to be caused by formaldehyde-resistant *Pseudomonas spp.* and methylotrophic microorganisms.[2] *Methylobacterium extorquens* seems to be responsible for the major degradation of formaldehyde in preserved calcium carbonate slurry.[2] When the formaldehyde concentration is sufficiently low, *Pseudomonas putida* takes over and suppresses the growth of *Methylobacterium extorquens*.[2]

Pseudomonas putida was found to degrade 1-naphthyl-N-methylcarbamate absorbed on montmorillonite, kaolinite, and goethite. The degradation rate depended on mineral type, with montmorillonite having the strongest and goethite the weakest effect.[3]

REFERENCES

1. Schwarzentruber, P; Gane, P A C, Application of microbiocides for the storage protection of mineral

dispersions in **Directory of Microbiocides for the Protection of Materials. A Handbook**, Paulus, W, Ed. *Springer*, Dordrecht, 2006.
2 Di Maiuta, N; Hubschmid, S; Giuliani, N; Schwarzentruber, P; Dow, C S, *Intl. Biodet. Biodeg.*, **63**, 769-77, 2009.
3 Chen, H; He, X; Rong, X; Chen, W; Cai, P; Liang, W; Li, S; Huang, Q, *Appl. Clay Sci.*, **46**, 102-108, 2009.

4.12 PETROLEUM PRODUCTS

Katarzyna Janda

4.12.1. GENERAL DATA

Microbiology of petroleum products is a very diverse subject, including microbiology of aviation, marine, and automotive fuels, transformer oils, lubricants, greases, oil emulsions, and asphalt.

First observations regarding the presence of microorganisms in petroleum products appeared in the late 19th and the early 20th century. Over the years, as industrial development became widespread and the output of the automotive industry rapidly increased, and aircraft and ships begin to use petroleum products, the problems related to microbial contamination of petroleum products become commonplace. Research conducted by several independent authors demonstrated that the microbial life can lead to malfunction of gas and diesel engines. Adverse effects related to the presence of microorganisms, such as separation of components, odors, or deterioration of performance were also observed in the fuel oils used for heating purposes, as well as in fluids used in metalworking.

Among the microbes colonizing oil products, pathogenic species can also be present, and they can become dangerous, especially when aerosols are formed; for example, in the case of metalworking fluids. Prevention and protection of petroleum products against microbial contamination are genuine issues. Often, prevention is not sufficient and protection must include either physical methods such as sedimentation, filtration, or decontamination of petroleum products by thermal or chemical method, i.e., by application of biocides. Physical methods are less detrimental for the environment, but their applicability is also limited, *inter alia*, on the ground that, for example, the entire tank cannot be disinfected by a physical method.

Hill[39] lists three categories of microbial degradation of petroleum products:

- the first concerns the distribution of microbial products used in diesel engines, where the aqueous phase volume is small compared to the oil phase. These processes were observed in the merchant marine and naval vessels, aircraft, including supersonic, and offshore drilling platforms
- the second category occurs when the aqueous phase is limited in contact with the oil phase, or where it is dispersed, and the contact is prolonged (weeks, months or even years). These processes occur in lubricating oils, hydraulic fluids, and cooling lubricants
- the third category concerns emulsions of oil in water, which are diluted 1:10 or even 1:100; these emulsions are used in steel and aluminum mills

4.12.2. MICROORGANISMS IN FUELS

Microbiological contamination of fuel causing intermittent operational problems was a known problem existing in the world for many decades, but recently the severity of these cases has increased considerably. It is worth noting that the temperature conditions prevailing during processing of oils make petroleum products to be sterile. Microbiological contamination begins to surface during distribution and storage.[31,118] This may be due to a number of factors or a combination thereof. These factors include, according to Pitcher,[76] changes in sources of fuel and its quality, increased use of additives for fuels that may contain compounds supporting the growth of microorganisms, and changes in the management and storage of fuels. Yemashova *et al.*[118] consider temperature, pH, water content, oxygen, nutrients, and chemical composition of crude oil, and products derived from it, as essential factors contributing to the biodegradation of fuel. Microorganisms can operate in a temperature range from 4 to 60°C (some even higher). pH values at which microorganisms can live range from 4 to 9, with a preference for neutral. The concentration of water sufficient for the growth of microorganisms is 1%, while the spores need only 5 to 80 ppm.[118] It is worth noting that anti-freeze, such as glycerol, reduces the availability of water for microorganisms and inhibits their growth.[118]

It is very difficult to prevent microbiological contamination of these products, because sterile conditions during transport and storage of fuel cannot be provided, considering that microorganisms capable of hydrocarbon biodegradation are common in air, water, and soil.[118] It is estimated that in 1 g of uncontaminated soil there are only 100-1,000 microorganisms capable of hydrocarbon degradation, while 1 g of oily soil contains 1×10^6 to 5×10^7, especially when the pollution is repeated periodically for a long time.[118] According to Gaylarde *et al.*,[31] there is no need to determine concentration of microorganisms in order to identify potential problems associated with their presence. Compounds synthesized by microorganisms, such as acids, polysaccharides, biosurfactants are equally important factors, and they can be detected with simple and rapid tests or their presence noted by direct observation of the fuel sample.

Fuel must be clear and transparent. Development of microorganisms usually contributes to:

- haziness and darkening
- increase in viscosity
- reduction of thermal stability
- plugging of pipes, filters, and improper collection of samples for testing
- increase in water content
- formation of sludge
- corrosion of tanks and lines
- production of particulates in fuel
- degradation of hydrocarbons
- shortening life of filters and engine parts
- increase in sulfur content of fuel
- health problems resulting from the presence of bacterial endotoxins, sulfate-reducing bacteria, and opportunistic microorganisms.[14,31,96,118]

Microorganisms often secrete biosurfactants, contributing to the emulsification of the fuel. It is probably conducive to the increased penetration by microorganisms of the hydrophobic layers of fuel. Life activity of aerobic bacteria and fungi is also associated with biosynthesis and secretion of enzymes such as peroxidase and acids.[118] The formation of visible particulates, which are a mixture of live and dead cells, and inorganic by-products (sludge formation) is the most obvious and easy to detect consequence of microbiological activity. Even a very small amount of particulate matter, i.e., 1 mg/100 ml of fuel, is a sufficient reason for problems with filtration.[118]

A classic example is the presence in fuel of fungus *Amorphoteca resinae* (previously – *Cladosporium resinae, Hormoconis resinae*), but also many species of yeast and bacteria affect the formation of these physical burdens, which can block tubes, injectors, or filters. Deposition of particulates at the bottom of tanks, pipes and other surfaces may lead to the formation of an anaerobic environment, changes in redox potential, and, thus, creation of conditions conducive to the development of sulfate-reducing bacteria (SRB – sulfate reducing bacteria). The result of their presence can be corrosion of unprotected surfaces. The presence of microorganisms and by-products synthesized by them, may contribute to adverse changes in the chemical properties of fuels. Products such as H_2S, CO_2, and SO_2 may dissolve in fuels, increasing corrosion, called the "silver strip". They are formed during the long-term storage of strategic fuel reserves in caves. These gaseous products may also constitute a serious threat to human health.[50]

State of emergency for the entire system can be defined based on the level of contamination of bottom layers by different groups of microorganisms, as shown in Table 4.57.

Table 4.57. Assessment of microbiological contamination of the tank based on the degree of contamination of water from the bottom of the tank. [According to Kwiatkowska, D; Wichary, H, *Ochrona przed korozja*, **5**, 118-121, 2000.]

Number of microorganisms at bottom layers, cm^{-3}	The degree of contamination/danger
10^3-10^4 bacteria, filamentous fungi, yeast, the lack of sulfate-reducing bacteria (SRB)	lack of pollution/low
10^5 bacteria and/or 10^3 yeast and/or 10^2 filamentous fungi and/or current SRB ($<10^2$)	growing/medium
10^6-10^8 bacteria and/or 10^4-10^7 of yeast and/or 10^3 filamentous fungi and/or $> 10^2$ SRB	very high

Petroleum and petroleum products may also be classified based on the number of microorganisms present in the layer of water from the bottom of the tank into two categories: lightly and heavily contaminated (Table 4.58).

Table 4.58. Assessment of microbiological contamination of crude oil or petroleum products based on the degree of contamination of water from the bottom of the tank, according to Yemashova *et al.*[118]

Level of contamination	Fungi count, ml^{-1}	Bacteria count, ml^{-1}
lightly contaminated	10^3-10^4	10^5
heavily contaminated	10^4-10^6	10^6-10^8

Generally, there are no existing standards for microbial content of fuels, but modern fuel, both aviation and marine, as well as automobile fuel, usually contain small amounts of microorganisms capable of survival. Fuel containing less than 50 microorganisms in dm^3 is considered to be clean. But even small amounts of water may cause growth of bacteria to large quantities (up to 10^4 per ml).[118]

If a large number of microorganisms is present in the water layer, then fuel in direct contact may also become contaminated. If long-term storage of fuels is considered, limits on the number of microorganisms should be tightened (see Table 4.59).

Table 4.59. Suggested categories of microbiological purity of fuel, according to Duda[26]

Product	Microbial count in products for 2-3 years storage, l^{-1}		Microbial count in products for long term storage (> 3 years), l^{-1}	
	filamentous fungi and yeasts	bacteria	filamentous fungi and yeasts	bacteria
aviation fuel	<100	<10	<100	<10
diesel	<1,000	<100	<100	<50
diesel (sample from water/oil interface)	<10,000		<500	<100

Typically, researchers have focused their attention on determining the presence and role of the mesophilic microorganisms. Thermophiles, that are capable of growth and development in conditions of high temperatures draw less attention, even though they may pose a potential threat. Thermophilic fungi have been isolated from many different fuels, including aviation fuel. Studies have shown that some of them are capable of growth and development in both fresh and used motor oil.[51,52] The water layer of petroleum products contained thermophilic, aerobic bacteria genus *Geobacillus* and anaerobic bacteria of the genus *Thermotoga, Thermoanaerobacter, Thermosipho, Petrotoga miotherma, Fervidobacterium, Thermococcus,* and *Archaeoglobus.*[118]

Table 4.60 contains the list of microorganisms isolated from different fuels.

Table 4.60. Microorganisms isolated from fuel. [Adapted, by permission, from Gaylarde, C; Bento, F M; Kelley, J, *Rev. Microbiol.*, **30**, 1-10, 1999.]

Bacteria	*Acinetobacter calcoaceticus, Acinetobacter cerificans, Acinetobacter sp., Actinomycetes sp., Aerobacter aerogenes, Alcaligenes, Bacillus sp., Bacillus acidocaldarius, Bacillus megatherium, Bacillus subtilis, Brevibacterium ammoniagenes, Clostridium sporogenes, Corynebacterium sp., Enterobacter cloacae, Enterobacter glomerans, Flavobacterium arborescens, Flavobacterium diffusum, Micrococcus sp., Moraxella sp., Ochrobactrum anthropii, Pasteurella sp., Pseudomonas sp., Pseudomonas aeruginosa, Pseudomonas fluorescens, Pseudomonas maliphora, Pseudomonas oleovorans, Pseudomonas putida, Serratia marcescens, Serratia odorifera,* sulfate-reducing bacteria
Filamentous fungi	*Acremonium sp., Acremonium strictum, Alternaria alternata, Alternaria sp., Aspergillus sp., Aspergillus clavatus, Aspergillus fischeri, Aspergillus flavus, Aspergillus fumigatus, Aspergillus niger, Aspergillus nidulans, Aspergillus ochraceus, Aspergillus paradoxus, Aspergillus sejunctus (A. rubrobrunneus), Aspergillus sydowi, Aspergillus versicolor, Aspergillus tamari, Botrytis cinerea, Cephalosporium sp., Chaetomium dolichotrichum, Chaetomium globosum, Cladosporium sp., Cladosporium cladosporioides, Cladosporium herbarum, Cladosporium sphaerospermum, Curvularia lunatus, Dreschlera cynodontis, Fusarium sp., Fusarium acuminatum, Fusarium moniliforme, Fusarium oxysporum, Fusarium sambucinum, Geomyces cretaceus, Geotrichum candidum, Gliomastix sp., Helminthosporium sp., Hormoconis resinae, Humicola grisea, Mucor sp., Paecilomyces sp., Paecilomyces lilacinus, Paecilomyces variotii, Paecilomyces virvus, Penicillium sp., Penicillium brevicompactum, Penicillium canescens, Penicillium citrinum, Penicillium corylophilum, Penicillium cyclopium (P. aurantiogriseum), Penicillium digitatum, Penicillium echinulatum, Penicillium expansum, Penicillium frequentans (P. glabrum), Penicillium funiculosum, Penicillium glabrum, Penicillium hirsutum, Penicillium minioluteum, Penicillium notatum (P. chrysogenum), Penicillium spinulosum, Penicillium thomii, Penicillium waksmanii, Pestalotiopsis aquatica, Phialophora sp., Phialophora richardsiae, Phoma sp., Phomopsis sp., Pseudallescheria boydii, Rhinocladiella sp., Rhizopus oryzae, Sordaria fimicola, Stemphylium botryosum, Thielavia sp., Trichoderma sp., Trichoderma harzianum, Trichoderma koningii, Trichoderma viride, Trichosporon sp., Trichothecium roseum, Tritirachium oryzae, Ulocladium sp., Ulocladium atrum, Ulocladium chartarum*
Yeasts	*Aerobasidium pullulans, Candida sp., Candida famata, Candida fluviatilis, Candida guilliermondii, Candida lipolytica, Candida rugosa, Candida tropicalis, Candida zeylanoides, Hansenula mrakii, Rhodotorula sp., Saccharomyces sp.*

4.12.2.1 Microbes in the aviation fuels and jet fuels

A condition for growth and development of microorganisms in aviation gasoline, as well as in other products, is the presence of water and nutrients. To function microorganisms must have at least 100 ppm water. In practice, water-free gasoline is difficult to achieve. Water is either supplied with fuel or comes from condensation on the walls of the tank. Nutrients (carbon, nitrogen, sulfur, oxygen, phosphorus) are components of fuels or their additives.[31,118]

In the 1950s and early 1960s, the problem of microbiological contamination of aircraft fuel was not universally recognized. It was associated with the type of fuel used. In the piston engines of aircraft, gasoline was used, which is a fuel lighter than kerosene and containing more toxic hydrocarbons, such as short-chain paraffins, as well as having a low water solubility. However, the problem of microbial contamination has been occurring, and it was generally related to the presence of anaerobic bacteria. The first publications on microbial degradation of jet kerosene appeared in the late 1950s. The fuel was less toxic to microorganisms than gasoline, and also contained more water.

Microbiological contamination of fuels is a major cause of serious problems in maintaining the quality of fuel and corrosion of metals used during the manufacturing processes and distribution, and, perhaps above all, during storage. In the recent years, the phenomenon of microbial contamination of fuels, not just aviation fuels, is growing, mostly because of the use of greater quantities of additives, which are the perfect breeding ground for microorganisms.

In the last thirty years, the most common method of storage of both crude oil and its processed products is in underground rock caverns, both natural and artificially created for this purpose.[50] This method is also used for long-term storage of fuels used for jet engines. Such storage of fuels is very attractive for reasons both economic and technical. Caves having a length from 50 to 1,000 meters can accommodate from 10 thousand m^3 to 1 million m^3 of crude oil or petroleum product. In Sweden, the fuel JP 4 is stored in caves at 6-8°C. It was observed that in some caves, fuel was subject to biocorrosion. Depending on environmental factors and the type of fuel stored, the first signs of this phenomenon were noticeable after 5 months to 5 years.[50] The effects of living organisms on aviation fuel is as follows:

- formation of sludge and slime at the bottom of storage tanks and aircraft tanks
- formation of an emulsified layer of fuel at the bottom zone of the fuel tank
- formation of mold deposits on the inner surface of tanks of ultrasonic aircraft
- changes in the structure of organic protective coatings, applied to the internal walls of both storage and aircraft tanks
- filter plugging in fuel supply lines of an aircraft
- corrosion of steel storage tanks
- corrosion of aircraft pumps, even those made out of copper and silver
- corrosion of aluminum and its alloys used in walls of aircraft tanks
- changes in fuel properties, especially composition of additives
- characteristic odors, such as hydrogen sulfide, in storage tanks.[48,50,79]

Cabral[12] has shown that samples of aviation fuel used for jets were contaminated, mainly by *Amorphoteca resinae* form *avelleneum*.

Guiamet *et al.*[34] compared the effects of pollution of aviation fuel JP_1 by *Trichosporon sp.* and *Amorphoteca resinae* and estimated the impact of their presence in fuel tanks constructed of aluminum alloy 2024 and carbon steel. They demonstrated that both strains were capable of using hydrocarbons as the sole source of carbon. The authors also speculated that *Amorphoteca resinae* is more aggressive in long-term incubation. Its adverse effects on aluminum alloys was signaled earlier by Videla.[112] The main effects caused by this species include increased acidity, caused by the production of organic acids, and local increase in redox potential.[112]

First reports on the activity of fungi in jet fuel systems appeared in 1960, when Klemme and Leonard isolated *Hormodendron sp.*, later known as *Cladosporium sp.*[32] Later tests have shown that the sludge also contained fungi *Penicillium, Aspergillus, Cladosporium*, and *Helminthosporium*. Fungi dominate contamination of aviation fuels, unlike crude oil, which is dominated by bacteria.[118]

It is known that the population of microbes colonizing aviation fuel is highly variable, but *Amorphoteca resinae* is one of the most common species.[19,31,50] This fungus has been investigated in 1969 by the Australian Parberego,[71] the best expert on *Cladosporium*. He discovered the mode of its reproduction and included it in phylum *Ascomycota*. Therefore a new definition of genera and species is currently *Amorphoteca resinae*. Itah *et al.*[48] indicated that fungus *Cladosporium resinae* has a strong ability to degrade jet fuel. Their studies have shown that over a 14-day incubation at room temperature, the fungus has caused mass loss of fuel in the range of 17.8 to 66.9%.

Amorphoteca resinae was isolated in 16 out of 24 samples of aviation fuel surveyed by Cabral.[12]

Some researchers point to the strains of the genera *Aspergillus* and *Penicillium* as frequent contaminants of aviation fuels.[36,116] Tests carried out by Solana and Gaylarde[102] have examined various fuels stored in tanks in the south of Brazil, which were contaminated by fungi and yeasts. Studies both quantitative and qualitative of aviation gasoline showed that 36% of samples contained strains belonging to *Penicillium sp.*, 18% strains of *Aspergillus sp.*, 9% *Aspergillus niger*, and 9% *Curvularis lunatus*.

Darby *et al.*[19] isolated fungi from the distribution system of military aviation fuel. Their studies showed that strains of *Cladosporium resinae f. avellenum* were the most common, followed by *Paecilomyces variotii*, and strains of *Alternaria*. *Aspergillus* and *Penicillium* were less frequent. What is clear is the overwhelming dominance of *Cladosporium*. Unlike other fungi, strains of this species produce not only the vegetative mycelium, but they are also able to produce spores *in situ* in liquids.[19]

In another study, 350 samples of aviation fuel were tested. There were fungi in all samples, and microflora was dominated by strains of *Cladosporium* and *Aspergillus*.[118] Microflora of stored fuel varied both quantitatively and qualitatively in samples of fuel stored in Swedish caves.[50] Based on the available literature, a list of microorganisms isolated from fuel and aviation gasoline is given in Table 4.61.

Table 4.61. Microorganisms isolated from fuel and aviation gasoline.

Sample	Microorganisms	Refs.
kerosine	*Actinomycetes, Aspergillus niger, Aspergillus sp., Botrytis, Cephalosporium, Chaetomium dolichotrichum, Cladosporium, Corynebacterium, Culvularis lunatus, Flavobacterium, Fusarium sambucinum, Mucor, Penicillium sp., Pestalotia aquatica, Pestalotiopsis aquatica, Phoma, Trichodema, Trichosporon, Ulocladium*	50,118

Table 4.61. Microorganisms isolated from fuel and aviation gasoline.

Sample	Microorganisms	Refs.
aviation fuel	*Aerobasidium pullulans, Acinetobacter (calcoaceticus, cerificans), Acremonium strictum, Aerobacter aerogenes, Aeromonas sp., Amorphoteca resinae (Cladosporium resinae, Hormoconis resinae), Alcaligenes sp., Alternaria sp., Arthrobacter, Aspergillus flavus, A. niger, A. fumigatus, Aspergillus sp., Bacillus brevis, B.cereus, B. licheniformis, B. subtilis, B. megaterium, Bacillus sp. (acidocaldarius* and others*)*, iron bacteria, *Brevibacterium ammoniagenes, Candida albicans, C. tropicalis, Candida (famata, lipolytica,* and others*), Cladosporium resinae f. avellaneum, Comamonas testosteroni, Enterobacter (cloacae, glomerans), Flavobacterium (arborescens, diffusum), Flavobacterium oderatum, Fusarium sp. (moniliforme* and others*), Helminthosporium sp., Micrococcus varians, Micrococcus sp., Paecilomyces varioti* (and others), *Penicillium (corylophilum, citrinum, frequentans,* and others*), Phialophora sp., Pseudomonas aeruginosa, Pseudomonas fluorescens, Rhinocladiella sp., Rhodotorula sp., Saccharomyces estuari, S. cerevisiae, Sarcina flava, Schizosaccharomyces pombe, Serratia (marcescens, odorifera), Stenotrophomonas maltophilia, Streptomyces sp., Tothersrichoderma sp. (viride* and others*), Trichosporium sp.*	50,48,81
jet fuel	*Aerobacter aerogenes, Aerobasidium pullulans, Arthrobacter sp., Aspergillus (clavatus, fumigatus, niger, fischeri, flavipes, ustus,* and others*), Bacillus subtilis, Bacillus sp. (acidocaldarius* and others*)*, sulfate-reducing bateria, *Brevibacterium ammoniagenes, Brevibacterium spp., Cephalosporium sp., Chaetomium globosum, Chaetomium sp., Cladosporium cladosporioides, C. herbarum, C. sphaerospermum, Clostridium sporogenes, Curvularia lunata, C. geniculata, Desulfibrio desulfuricans, Detzia sp., Discophaerina fagi, Epicoccum nigrum, Escherichia sp., Exophiala jeanselmei, Flavobacterium (arborescens, diffusum), Geotrichum candidum, Helminthosporium sp., Herellea spp., Hormoconis (Cladosporium) resinae, Hormodendrum (Cladosporium), spp., Humicola grisea, Isaria sp., Kocuria rhizophilia, Leucobacter komagatae, Micrococcus sp., Monocillium sp., Paecilomyces varioti* (and others), *Pantoea ananatis, Penicillium sp. (corylophilum* and others*), Pseudomonas aeruginosa, Pseudomonas fluorescens, Pseudomonas maliphora, Rhodotorula spp., Sarcina flava, Sphingomonas, Staphylococcus aureus, Staphylococcus epidermidis, Staphylococcus sp., Streptomyces spp., Trichoderma sp., Ulocladium sp.*	19,27, 50,81

It is worth noting that bacteria *Pantoea ananatis, Arthrobacter sp., Alcaligenes sp, Kocuria rhizophilia, Leucobacter komagatae, Detzia sp.,* and fungus *Discophaerina phage* had never been previously isolated from American aviation fuels. Although the presence of *Kocuria rhizophilia* and *Detzia sp.* was found earlier in the oil fields and it was proven that they were able to degrade hydrocarbons. In this light, their presence in fuels is not surprising. However, isolation of new species in this environment did not occur, which suggests that there are changes in consortia of microorganisms inhabiting fuel. This is probably a manifestation of adaptation of microorganisms to the chemical composition of fuels and biocides or simply better isolation and identification techniques.[81] It seems very likely that a similar conclusion is also true for other petroleum products.

Table 4.62. Types of microorganisms contaminating fuel during storage. [According to Yemashova, N; Murygina, V P; Zhukov, D V; Zakharyantz, A A; Glanchenko, M A; Appanna, V; Kalyuzhnyi, S V, *Rev. Environ. Sci. Biotechnol.*, **6**, 315-337, 2007.]

Source	Microorganisms
condensate from storage system of crude oil	*Actinomycetes, Aerobacter, Aeromonas, Acetobacterium, Acinetobacter, Azoarcus, Brevibacterium, Corynebacterium, Clostridium, Desulfotomaculum, Desulfovibrio, Desulfobacula, Desulfobacterium cetonicum, Desulfosarcina variabilis, Desulfococcus multivorans, Enterobacter, Flavobacterium, Pseudomonas, Rhodococcus, Thiomicrospira denitrificans, Zooglea*
condensate from storage system of fuel	*Aerobacter, Aeromonas, Acinetobacter, Bacillus, Brevibacterium, Candida, Enterobacter, Hansenula, Hormoconis resinae, Micrococcus, Pseudomonas, Rhinocladiella, Rhodococcus, Rhizopus, Saccharomyces, Serratia*
water phase	thermophile conditions: *Fervidobacterium, Petrotoga miotherma, Thermoanaerobacter, Thermococcus, Thermotoga, Thermosipho, Archaeoglobus* and sulfate-reducing bacteria

4.12.2.2 Microbes in the automotive fuel

Symptoms of microorganism presence in automotive fuels and conditions of their development are the same as in the case of aviation fuels. Based on the available literature a list of microorganisms colonizing automotive fuels is presented in Table 4.63.

Table 4.63. Microorganisms isolated from the aqueous and oil phases.[50]

Type	Species
bacteria	*Acinetobacter cerificans, Acinetobacter junii, Acinetobacter sp., Actinomycetes sp., Aeromonas hydrophila, Alcaligenes sp., Arthrobacter sp., Bacillus cereus, Bacillus firmus, Bacillus subtilis, Bacillus sp., Cochliobolus lunatus, Desulfovibrio desulfuricans, Desulfovibrio sp.,* iron bacteria, *Klebsiella sp., Leptothrix, Micrococcus sp., Mycobacterium sp., Nocardia sp., Phialophora sp., Pseudomonas aeruginosa, Pseudomonas fluorescens, Pseudomonas oleovorans, Pseudomonas sp., Sphaerotilus sp., Sphingomonas paucimobilis, Siderocapsa,* sulfate-reducing bacteria
filamentous fungi	*Alternaria alternata, Alternaria sp., Amorphoteca resinae, Aspergillus flavus, Aspergillus fumigatus, Aspergillus niger, Aspergillus ustus, Aspergillus versicolor, Aspergillus sp., Cephalosporium sp., Fusarium sp., Humicola sp., Paecilomyces variotii, Paecilomyces sp., Penicillium citrinum, Penicillium spinulosum, Penicillium sp., Rhodotorula glutinis, Rhodotorula sp., Trichoderma viride,*
yeasts	*Candida guilliermondii, Candida lipolytica, Candida rugosa, Candida silvicola, Candida tropicalis, Candida sp.*

Kwiatkowska and Wichary[62] conducted studies aiming at analysis of the risks of corrosion caused by microorganisms in the diesel tanks. In all analyzed vessels, the presence of sulfate-reducing bacteria was found, which is responsible for the progressive pitting corrosion of the internal walls of tanks.

Organic fuels (biodiesel, ecodiesel, naturdiesel, and biofuel) which are produced based on esters of rapeseed oil are becoming commonplace. The first European country producing fuel based on rapeseed oil was Austria, whose participation in European production of this fuel is currently at 60%. The United States is planning that in 40 years 50% of fuel is to be replaced by various types of biofuels, produced from vegetable, animal, and algae fats and oils.[16]

Rapeseed fuel has many advantages, which include health and environmental protection, because of reduction of the greenhouse effect, low carcinogenicity, non-toxicity, biodegradability and low smoke.[16] Currently in Poland, a mixture of methyl esters of fatty acids from rapeseed oil and mineral fuels 32:68 wt% is used. It is produced in the Czech Republic. It is designed for year-round use in all types of automatic ignition engines. This fuel can be freely used mixed with other fuels, but it should be stored in separate containers and sold with a dedicated and appropriately designated distributor. Compared with conventional gasoline, it appears that the fuel has much better exhaust gas composition, particularly with regard to the contents of toxic ingredients, such as SO_2 and CO_2. It has about half the smoke, lower sulfur and aromatic hydrocarbons content, and lower concentrations of polyaromatic components.[50] Biodiesel has many advantages, but there is no data on its vulnerability to microbiological degradation. According to Bento and Gaylarde,[7] even with the low content of methyl ester of rapeseed oil in fuel, its susceptibility to microbiological decomposition may be increased.

4.12.2.3 Microbes in marine fuels

Growth of microorganisms produces effects similar to those shown in aviation fuels, namely turbidity and corroded tanks, which are the result of the activity of sulfate-reducing bacteria.[35,76] Microbiological contamination of marine fuel is a very serious problem, especially if there are favorable conditions for their growth, such as the presence of sea water, optimum time and temperature, and mixing caused by sea waves.[96] Chung *et al.*[14] studied the effect of temperature and water on biodeterioration of marine fuels. The presence of water was found to be the most important influence. Researchers recommended storing fuel at temperatures above 65°C to prevent biodegradation.

According to Zyska,[120] in ship tanks that have the equipment to compensate for the weight of the fuel with water, sulfate-reducing bacteria may constitute up to 17% of the total microflora.

In 1990, contamination of fuel was reported in a Canadian warship. As a result of microbial activity of *Amorphotheca resinae,* damage to several gas engines was reported.[35] There were several reasons for the damage. One of them was balancing the load distribution. To achieve this, the fuel was pumped from one tank to another, resulting in contamination of clean tanks.[35,122]

In the late 1980s, an international oil company received complaints from different European countries related to an unacceptable level of turbidity and the presence of particles suspended in product. Studies have shown that the cause of these changes was the microbiological contamination of fuel. Microorganisms were isolated from water samples and from samples of fuel. In water, aerobic bacteria were present; depending on the place of collection of the sample, they amounted from 1×10^5 to 3×10^6 organisms per cm^3. The tests also identified sulfate-reducing bacteria, and in one case, yeast. The tested fuel contained aerobic bacteria in a quantity from 0 to $6x10^3$ and yeast from 0 to $4x10^2$ organisms per cm^3.[76] The following microorganisms were isolated *Penicillium spp., Aspergillus spp., Aspergillus niger, Aspergillus fumigatus, Hormoconis resinae,* and *Curvularis lunatus.*[102]

4.12.2.4 Microorganisms in gasoline

Conditions for microbial growth in gasoline, as well as the results of their presence, are similar to those for aviation fuels.

The work carried out in the Corrosion Microbiology Laboratory of Central Mining Institute shows that all test tanks, including unleaded gasoline, were contaminated by microorganisms at the bottom layers.[62] The following microorganisms were present: bacteria: *Acinetobacter sp., Alcaligenes denitrificans, Alcaligenes xylosoxidans, Bacillus sp., Citrobacter freundii, Coccobacili sp., Comamonas acidovorans, Desulfovibrio desulfuricans, Flavobacterium devorans, Klebsiella ozaenae, Micrococcus sp., Moraxella lacunata, Mycobacterium austroafricanum, Oligella sp., Pseudomonas aeruginosa, Pseudomonas alcaligenes, Pseudomonas fluorescens, Pseudomonas sp., Sphingomonas paucimobilis*, filamentous fungi: *Aspergillus unquis, Aspergillus ustus, Aspergillus versicolor, Cladosporium sp., Mycelia sterile, Paecilomyces variotii, Penicillium canescens, Penicillium sp., Penicilium versicolor*, and yeast: *Candida lipolytica* and *Candida sp.* Lead-free gasoline was mostly populated by bacteria: *Bacillus sp., Moraxella lacunata, Oligella sp., Pseudomonas aeruginosa, Pseudomonas fluorescens*, and unidentified yeasts.[49,61]

Jamison *et al.*[49] isolated 32 bacterial species from high-octane gasoline. Detailed analysis showed that they can biodegrade several components of gasoline. Lead-free gasoline and clean diesel are more susceptible to microbiological contamination than gasoline containing lead compounds.

4.12.2.5 Microorganisms in heating oil

Heating oils are used as fuel for residential heating, steam boilers, marine boilers, blast furnaces, and other industrial furnaces (e.g., in steel and glass plants). They are divided into light and heavy heating oils. Light heating oils are derived primarily from crude oil and used to heat homes. Heavy fuel oils are produced from oils and tars derived from coal, lignite, shale oil, and wood. They are used as fuel in steam boilers, marine engines, and industrial furnaces, as well as in power plants.

Fuel oils are mixtures of different hydrocarbons, which exhibit varying susceptibility to microbiological decomposition. Infection can be caused in the presence of water and air when they are mixed with fuel. Also, remnants of old oil after emptying a tank and water at the bottom of tank may cause contamination of a new oil.[119]

Sludge accumulates at the bottom of the tank and clogs filters in the supply system. Sediments are produced by microorganisms at the interface of oil and water. This may be accompanied by pH decrease, degradation of additives, unpleasant odor, and damage to the protective coating on internal surfaces of tanks. The rates of these changes depend on the chemical composition of oil and physical-chemical conditions, such as, the presence of water, pH, and temperature.

Research conducted by Roffey[82] on microbial contamination of petroleum products stored in rock caverns has shown that adverse changes in physical-chemical properties of fuels occurred. Long-term storage of fuel oils was associated with increased water content and formation of emulsion at the interface of water. In some caves, there was a significant increase in the amount of methane produced by bacteria, which could cause an explosion. Aerobic bacteria were present in the quantity of 60 per cm^3, and sulfate-reducing bacteria in an amount of 10 in 1 cm^3. Addition of 0.5 g/dm^3 of nitrate in water flowing into the cave was used to limit activity of methane-producing bacteria.

An example of the adverse effects of microorganisms on fuel oil was discussed by Barnes *et al.*[5] for Orimulsion fuel. It is a mixture of super heavy oil and water, spiked with different chemicals. It contains 70% natural bitumen, 27-30% water, and 3% additives. In 1992, one of London's power plants brought Orimulsion for trials. It turned out that the emulsion was separated. Researchers involved in evaluation noted that the emulsifier used in bituminous fuel was nonylphenol ethoxylate added in an amount of 0.2%. It was also assumed that addition of 4.74 g/dm^3 of magnesium nitrate to improve the combustion process and improve the emulsion's rheological properties of was not without significance, since it could serve as a source of nitrogen for the microorganisms. Emulsion samples taken from the tanker showed the presence of rods of Gram-negative *Pseudomonas*. Dominant species were *Pseudomonas stutzeri* and *Pseudomonas pseudomallei.* Laboratory studies showed that the emulsifier (nonylphenol ethoxylate) was very rapidly degraded by bacteria. At the same time, the content of nitrogen was rapidly reduced. Along with depletion of nitrogen source, a reduction in the number of *Pseudomonas* bacteria was noticed. If nitrogen compound was not present in emulsion, the nonylphenol ethoxylate was not undergoing microbiological degradation.

Oil degradation studies conducted by a team of Spanish researchers have shown that *Acinetobacter calcoaceticus* MM5 was able to degrade hydrocarbons, mainly n-alkanes. Rapid degradation was visible during the exponential growth of microorganisms. After five days of degradation, 31% of the total hydrocarbons and 35% of n-alkanes were degraded, and after 20 days, 60% of the total hydrocarbons and 75% of n-alkanes. Chromatographic analysis showed that these bacteria use the straight-chain paraffins, but they do not use branched hydrocarbons.[68] These results were confirmed by other literature data, since many researchers reported that n-alkanes are easily degraded by microorganisms.[25,29]

Data on microbiological contamination of the bottom layers of heating oil stored in the underground tank with a capacity of 100 m^3 are given in Table 4.64.

Table 4.64. The level of microbiological contamination in the bottom layers of oil and water in tanks of commercial chain, according to Kwiatkowska.[60]

Sample	Number of microorganism in dm^3 of sample			
	aerobic bacteria	sulfate-reducing bacteria	filamentous fungi	yeast
fuel	30	none	20	20
water	4.5×10^6	none	1×10^4	7×10^6

Significantly higher microbial contamination was found in the aqueous phase. No presence of sulfate-reducing bacteria was detected. Microflora showed the presence of *Bacillus sp., Pseudomonas fluorescens, Mycelia sterilia*, and *Candida lipolytica.*

4.12.3 MICROORGANISMS IN LUBRICATING OILS

A diverse group of products used for lubrication of surfaces in contact to reduce friction and protect metal products against corrosion is reported here. Properties and applications of these products depend on the origin of the raw material, consistency, and degree of affinity to the lubricated surface. They contain a variety of additives, such as viscosity modifiers, freezing temperature regulators, antioxidants, surfactants, and corrosion inhibitors. Due to the consistency, they are divided into liquid, lubricating oils and plastic lubricants (greases). Lubricating oils are derived from crude oil or synthetic products. Depending on the application, they serve as machine, motor, gear, hydraulic, turbine, and other lubricants. Plastic lubricants are concentrated mineral oils. Their rheological properties are modified by soap (calcium, sodium, etc.) or thickeners (e.g., ceresin or paraffin wax), and talc and graphite. They are used where the use of liquid oils is cumbersome by design, for example, in high pressure units, on low speed sliding elements, or at high temperatures. They are also used in rolling bearings operating in wet and dusty environments.

In the recent years, lubricants have become more complex and sophisticated to better fulfill their functions, but the increased number of various additives made them more susceptible to microbiological attack.[32,38] The overall share of additives may be as high as 20%.[43]

The phenomenon of microbial contamination of lubricants was given more attention at the end of the 1960s. A higher frequency of failure of diesel engines on ships was observed. In 1969, there was a failure which was caused by a malfunctioning supply of lubricating oil because of corrosion. Since then, a large number of similar cases were reported.[38] A logical explanation for the increased frequency of accidents that have occurred on ships has been searched for. According to Hill,[38] the increase in the consumption of lubricants containing additives, which were a good food source for microbes, was one of the reasons of failures. Not without significance were changes in the basic chemical composition of oils used for the manufacture of lubricants, involving a change of naphthenic to paraffinic hydrocarbons, which proved to be more susceptible to microbiological decomposition.

Hill[38] and Genner and Hill[32] found that growth of microorganisms was made possible by the presence of water. Water was present as a result of condensation due to the temperature difference between the surface and the air, and very frequent leakage and ingress of water from cooling system or from other sources. An important factor influencing the growth and development of microorganisms was the ambient temperature. Microbial attack occurs frequently in tropical climates, where the favorable thermal and moisture conditions exist. In a temperate climate there are large fluctuations in temperature, which is a factor impeding the adaptation of microorganisms. The temperature of the environment also depends on the means of transportation, in which the lubricant is used. For example, a car engine runs at about 80-90°C and lubricant probably has a similar temperature. Similar conditions prevail in high-speed marine engines, but in low-speed engines, temperature is 35-45°C. This is a temperature conducive to the growth of most microorganisms, both meso- and thermophilic.

The main problem encountered in the lubricants used in marine engines is the presence of bacteria, which prefer a slightly alkaline environment provided by modern lubricants. The main offender is *Pseudomonas*. Filamentous fungi, especially *Aspergillus fumigatus* and yeast were also isolated, but in smaller quantities. Occasionally, anaerobic bacteria are also found.[32,39]

Symptoms of microbial contamination in marine engines are water in oil, increased acidity, brown or black-brown deposits, unusual odor, honey-colored films of engine parts, rusty parts of machines, black scratches, corrosion, and clogged filters, especially during or after storm.

Microbiological contamination of lubricants has also been observed in other areas, namely in the food industry. It should be noted that lubricants used in the food industry should meet the stringent requirements for their chemical composition, due to close contact with food.[1] Lubricants used in the food industry are also produced from oil, but they always contain water.[1] According to Rossmoore,[88] air, water, and spilled foods are the major contaminants in the food industry. Phenomena which may indicate the development of microorganisms include the formation of deposits, changes in the organoleptic characteristics of the product and signs of corrosion.[65,88]

Microorganisms isolated from lubricants are shown in Table 4.65.

Table 4.65. Microorganisms isolated from the lubricants.

Type	Species	References
Bacteria	*Acinetobacter sp.*	1
	Aerobacter aerogenes	65
	Alcaligenes faecalis	65
	Alcaligenes sp.	1
	Brevibacterium lipolyticum	65
	Corynebacterium hydrocarboclastus	65
	Desulfovibrio desulfuricans	65
	Flavobacterium ferrugineum	65
	Klebsiella oxytoca	88
	Listeria monocytogenes	1, 88
	Micrococcus sp.	88
	Pseudomonas aeruginosa, Pseudomonas azatoformans, Pseudomonas desmolytica, Pseudomonas fluorescens, Pseudomonas iners, Pseudomonas maltophila, Pseudomonas nitroreducens, Pseudomonas putrefaciens, Pseudomonas schuylkiliensis, Pseudomonas stutzeri, Pseudomonas sp.	1, 57, 65, 88, 119
	Serratia marcescens	88
	Spicaria sp.	57, 65
	Staphylococcus aureus	88
	Sulfate-reducing bacteria	1

Table 4.65. Microorganisms isolated from the lubricants.

Type	Species	References
Filamentous fungi and yeasts	*Alternaria sp.*	57
	Aspergillus amstelodami, Aspergillis niger, Aspergillus flavus, Aspergillus sp.	3, 57, 65, 119
	Chaetomium globusom	3, 119
	Candida lambica, Candida lipolytica, Candida sp.	57, 88, 119
	Cladosporium resinae, Cladosporium sp.	3, 65, 119,
	Geotrichum candidum	88
	Helminthosporium sp.	65
	Paecilomyces keroti, Paecilomyces variotii, Paecilomyces sp.	3, 65, 119
	Penicillium brevicompactum, Penicillium cyclopium, Penicillium sp.	3, 57, 65, 119
	Rhodotorula	119
	Stachybotrys atra	119
	Trichoderma viride	3
	Zygosaccharomyces balii	88

More diverse composition of microorganisms has been found in the food industry (dairy products and beverages). Lubricants in these plants were a hotbed for particularly dangerous species, such as *Listeria monocytogenes*. Also, other pathogenic microorganisms, such as *Zygosaccharomyces bailii, Staphylococcus aureus, Pseudomonas fluorescens, Klebsiella oxytoca, Serratia marcescens, Micrococcus sp., Geotrichum candidum*, and *Candida lambica* were isolated.[88]

Studies have shown that both fungi and bacteria were able to hydrolyze lubricants and use them as a source of carbon.[119]

4.12.4 MICROBES IN THE HYDRAULIC OILS

Hydraulic oils are selectively refined mineral oils derived from crude oil processing. They contain corrosion inhibitors, antioxidants, antifoaming agents, and additives improving lubrication. There are several types of hydraulic oils, depending on their viscosity. Their task is to transfer energy from fluid (usually a hydraulic pump) to receiver (actuators), such as hydraulic cylinders and motors. Therefore, they must also lubricate and protect the components of the hydraulic system against corrosion in addition to heat transfer. In order to fulfill the above-mentioned functions, hydraulic fluids must have appropriate viscosity, no tendency to foam, quick air release, stability of properties, anticorrosive performance, resistance to oxidation, and high temperature performance.

Water was the first hydraulic fluid, but its presence in the systems caused corrosion and scale deposition. Water also had too low viscosity, and above all did not have lubricating properties. Currently, the vast majority of hydraulic fluids contains refined mineral oils. In view of this, like other petroleum products, hydraulic fluids are exposed to the harmful activity of microorganisms. Microbes have a tendency to locate at the bottom of reservoirs and in water droplets separated from systems. Sometimes, as a result of shocks occurring during the work of the system, the cells may be scattered throughout the hydraulic fluid and can cause malfunctioning of valves, filters, and clogging holes of narrow openings. Hydraulic fluid under elevated temperature and some amount of air are also mixed. This creates conditions conducive to the development of a broad spectrum of microorganisms. The result of this process may block the flow of small holes, tubes and filters, which causes malfunction of the entire hydraulic system.[39]

Problems with the quality of hydraulic fluids can also be caused by metabolic products of inhabiting microorganisms. The growth of microorganisms leads to a substantial depletion of additives contained in hydraulic fluids and a viscosity increase due to the selective removal of lighter hydrocarbons. Increase in the acidity of fluid, resulting from the production of acid by microorganisms, is another result of deterioration and unwelcome change.[32] According to Hill and Al-Haidary,[42] *Candida sp.* caused an increase of acidity and changes in viscosity, which accelerated corrosion of metal surfaces.

4.12.5 MICROORGANISMS IN TURBINE OILS

Turbine oil is refined lubricating oil obtained by distillation of crude oil, i.e., fractionation during which there are no chemical changes to original components. Their job is lubrication and cooling of bearings in water and steam turbines, turbine pumps, and turbochargers. They should be resistant to oxidation, emulsification, and foaming. To function properly, turbine oils must have an appropriate viscosity at a given temperature range, be stable at elevated temperatures, able to release water and air without creating foam, able to protect metal surfaces against corrosion and wear of heavily loaded gears, and have compatibility with coatings and adhesives.

There are limited reports on microbial contamination of turbine oils. Consequences of microbial growth are corrosion of metal surfaces, generation of unpleasant odors, and malfunction of various elements of control.[39] There is no available data on the qualitative composition of microorganisms inhabiting these products or biocides recommended for their protection.

4.12.6 MICROORGANISMS IN TRANSFORMER OILS

Transformer oils are refined oils, derived from crude oil. They are insulating liquid materials, having high electrical resistance and resistance to climatic factors. They are used in electrical engineering in transformers and oil circuit breakers as a means of insulating and conducting heat. Oil content depends on transformer size and it varies from 100 to 1,000 kg. Fire safety is a very important part of this application. Polish (PN-90/C-96058) and international standards list 11 characteristics of transformer oil, but none of them specifies water content, which may influence microbial growth as learned from various examples of other materials given so far.[122]

Studies on the susceptibility of this type of petroleum product to microbiological effect are scarce. Zyska[119] conducted studies using a mixed culture containing filamentous fungi, such as *Aspergillus, Paecilomyces, Penicillium, Cladosporium,* and *Chaetomium. Actinomycetes* were represented by *Mycobacterium* and *Pseudomonas*. Experiments, carried out at 28-32°C and 100% relative humidity, showed that the resistance of transformer oil to microbiological corrosion depended on the concentration of the heavy aromatic hydrocarbons, and the method and the degree of paraffins removal. There has been an increase in kinematic viscosity, acid number, saponification number, iodine number, increase of the ignition temperature, and decrease of the freezing temperature and molecular weight. Composition of hydrocarbons is also changed. Temperature conditions of the experiment show that microbial contamination of such products is possible in a temperate climate.[70] High water concentration (50 ppm, standard concentration is 20 ppm) was found. The number of bacteria reached 10^6 cells/cm^3 sludge. Also, filamentous fungi were identified. Phenyl pyrazolidone and nitrosodialkyl aniline were suggested for use to prevent microbial growth.[122]

4.12.7 MICROORGANISMS IN OIL EMULSIONS

Metalworking fluids are products that facilitate metal tooling and prolong the service life of tools. They are used for machining, metal forming, and thermal treatment. Water in oil emulsions and oil in water emulsions are used.

Water in oil emulsions are used as lubricants for agricultural machinery and mining equipment. They operate under conditions of constant exposure to water, dust, and vapors.[119] Oil in water emulsions are used as coolants or liquid cooling lubricants in machining and cold rolling of steel and nonferrous metals.

Recently, metalworking fluids have been listed in 4 classes:

- straight oils (containing the base oil without water)
- soluble oils (emulsions with high content of oil)
- semi-synthetic metalworking fluids (emulsions with a low content of oil)
- synthetic fluids (produced without petroleum)[4,111,114]

According to DIN 5138524, two types of liquid cooling lubricants are in common use, namely:

- insoluble in water (non-water-soluble MWFs) and
- soluble in water (MWFs).

Water-soluble coolants usually contain mineral oil, which, in combination with other components, are an excellent medium for growth of bacteria and fungi.[24] As in the case of other petroleum products, the presence of water is a fundamental problem. Drinking water is usually used to dilute the concentrated product. It is important that the drinking water has a bacterial count of less than 100/ml.[24,80] Moreover, what rarely calls for attention, coolants are also affected by workers themselves. The source of bacterial contamination could be food, drinks, and/or waste, but also the tools and work environment.[24,80] It is worth noting that fresh emulsions can also be easily infected by dust particles, which are reservoirs of microorganisms and aerosols from the plant.[80]

Basic functions of metalworking fluids include cooling the cutting tools, cooling the workpieces, reducing friction, improving (due to their excellent lubricating characteristics) quality of the metal finish, protecting the surface of the workpiece material from corrosion.[18,24,91,114,119]

Coolants, also known as oil emulsions, are obtained by mixing oil, and emulsifier (concentrate) with medium hard water. They are mixed with water in any ratio to form stable milky emulsions. Most commonly used emulsions contain from 2 to 8 vol% and sometimes up to 10 vol% of oil. Particle size of emulsified oil ranges from 0.1 to 20 μm. Such liquids are used in cases of high-speed machining with small loads in the contact zone. Coolants are also used in the form of a microemulsions, in which the particle size of oil is from 0.01 to 0.1 μm, which gives performance similar to the molecular mixture (solution).

Microemulsions are transparent or only slightly off-white, and sometimes are produced in different colors. During use they gradually become opaque. Microemulsions can be mixed with water in any proportion. It is assumed that microemulsion has a uniform distribution of all components. This permits maintenance of anticorrosive and biocidal properties even in very dilute solutions. Microemulsions are considered to be better than conventional coolants because of special advantages:

- they are dermatologically inert
- they have low surface tension, which ensures good wetting of tools and workpiece materials
- easy to distribute on the surface of metal, forming a thin film of lubricant, which allows to achieve correct treatment, at a very low concentration of concentrate in the coolant
- they do not emit odors and harmful gaseous substances
- they do not contain toxic components to humans and environment.[53]

These advantages of modern microemulsions may also be factors that increase their susceptibility to microbiological degradability.

Nutrient sources for microbes are hydrocarbons and additives. It is estimated that about 300 different compounds are used for production of various metalworking fluids. A single product may consist of up to 60 ingredients. Water-soluble coolant consists of 20-70% of oil (mineral oil, polyglycol, natural or synthetic esters), emulsifiers (fatty alcohols or amines), corrosion inhibitors (fatty acids or amines), foaming agents, and biocides.[80] In connection with such a rich chemical composition and the presence of water, coolant is an ideal environment for growth of microorganism.[37,80,107] Microbial degradation of coolants is a multi-stage process. Compounds which are easily degradable by microorganisms are utilized in the first place and this process occurs quickly. As a consequence, among other things, it upsets the balance between the components of coolant after their chemical composition has been changed. This leads to a reduction of quality of coolant.[80]

The most common sources of infection include coolant water used for emulsion, sediment accumulated in the dead spaces of the cooling system, and the contaminants that enter the system during operation.[3,4,18,24,77,80,109] Vedder and Rossmoore[110] have shown that the sulfate-reducing bacteria can survive on solid particles suspended in air. This gives an additional source of coolant pollution. Microorganism growth causes:

- separation of emulsion
- creation of sediments
- creation of biofilm
- clogging filters in the circulatory systems
- reduction of pH of coolant
- deterioration or loss of lubricating properties
- deterioration of the quality of surface finishes
- staining the product
- increase in corrosiveness of coolant, as a result of the presence of acidic products of microbial metabolism such as lactic, acetic, and formic acids
- damage to the reservoirs of lubricants and coolants

- leakage of liquid
- corrosion of metals
- odors
- deterioration of hygienic conditions of work and allergies and infections among workers.[18,24,69,80]

These symptoms of microorganism activity have a direct impact on economic losses. As a result, there are costs associated with the shortening of the working life of coolants, the removal of affected emulsions, and the increase in operating costs, resulting from the wear of working parts of machines, increase of their corrosion, and decreased quality of production.[80] Production downtime is associated with replacement of coolants, even simply because of their smell and potential health hazard.[30,39] These additional costs are difficult to determine, but probably are significant and should be considered.

One of the signs of microbial degradation is the separation of the emulsion that results from decomposition of emulsifier. The color change from white to gray or blue-gray is caused by secretion of hydrogen sulfide by bacteria of the genera *Desulfovibrio, Proteus, Aerobacter, Achromobacter, Escherichia,* and *Paracolobactrum*. As a result of the reaction of iron ions with hydrogen sulfide a black precipitate is formed, which causes changes in coloration.[91]

Coolants which were degraded by microorganisms can cause damage to the concrete foundations of technical equipment. Decrease in pH of the emulsion and the metabolic products of bacteria also contribute to corrosion of pipelines.

The coolant can be colonized by bacteria, both aerobic and anaerobic actinomycetes, filamentous fungi, and yeasts. Aerobic bacteria of the genus *Pseudomonas* are the predominant species. Because of their adaptive properties, they are pioneers in the ecosystem of coolants.[91,109] Construction of their cells facilitates their adhesion to the tools and biofilm formation. They are able to use all organic ingredients of coolants. The presence of these bacteria can be identified because of discoloration, separation of emulsions, and decrease of pH. As the first colonizers they prepare conditions for other microorganisms to join them in biofilm.

It was never found that anaerobic bacteria can grow in fresh coolants before the appearance of aerobic bacteria. But, as a result of their presence, the environment may become anaerobic, which may take place during work stoppages, such as during weekends. The presence of anaerobic microorganisms is noticeable because of a change in color of emulsion and release of an unpleasant odor described as "Monday morning smell". It should be noted that microorganisms are very resistant to adverse conditions of the environment.[91]

Molds and yeasts begin to grow after lowering the pH.[109] The presence of filamentous fungi causes the characteristic lesions and odors. Mycelium may block flow channels and clog showers. Due to the slimy nature of the fungi residing in the liquid cooling, systems are described as "fat Spanish moss". Odor associated with their presence is compared to the strong musty odor in a cloakroom.[91]

Lubricating-cooling fluids may also constitute a serious threat to health because they contain compounds which by themselves may be potentially harmful to workers and the environment. In addition, pollutants can be formed during operation. Some components of coolants and accidental pollutants, appearing in uncontrolled amounts, can sometimes react with each other, resulting in harmful compounds. Similar reactions may occur during mixing of liquids from different manufacturers. Harmful substances that enter the system, such as base oils, emulsifiers, and corrosion inhibitors can cause a variety of skin diseases, and affect the nervous and cardiovascular system, as well as metabolism.

High performance machining requires high turnover, which, combined with higher temperatures, leads to the formation and release of vapors and aerosols. Formation of aerosols in combination with microbiological contamination is a very serious threat to health because it promotes easy and rapid spread of microorganisms. Metal particles are also produced in the course of work. These include ions of nickel, cobalt, and chromium, classified as potential allergens. Another threat is caused by the presence of pathogenic bacteria colonizing cooling fluids. These issues have been of interest since more than half of a century.[53,122] Potential risks arising from exposure to microbiologically contaminated coolants are as follows:

- emulsion undergoes changes under the influence of the chemical activity of microbial life
- the presence of live bacteria in the air can lead to lung diseases
- toxins produced by bacteria and enzymes cause skin irritation and/or respiratory system
- emulsions could provide a source of secondary infections.

The production of ammonia and hydrogen sulfide is a manifestation of microbiological processes. The presence of hydrogen sulfide in the air, according to the limits established in the UK, should not exceed 10 mg/m^3 of air. Larger concentrations of this compound may cause eye irritation, dizziness, and even death, which may occur after 30 minutes of exposure to concentrations of 800-1000 mg/m^3 of air. Known from literature are fatal cases attributed to the presence of hydrogen sulfide in the deterioration of cargo on board of a ship and stored oil on platforms in North Sea.[39]

Aerosol generation during operation of cooling systems, containing large quantities of microorganisms, mainly bacteria, is an important hazard. If inhaled, it can lead to clinical symptoms.[4,18,24,80,95,111,114] It is estimated that around 1.2 million U.S. workers are exposed to bioaerosols, when working with coolants. Contact with them is the second most common cause of asthma related to the workplace in Michigan state.[28,114]

Contact with coolants infected with *Pseudomonas aeruginosa* may cause clinical symptoms with rashes and inflammation of skin, as well as failure of the pulmonary system.[53] According to Hodgson,[45] the most significant role of this bacterium is its ability to biosynthesis of keratinolytic enzymes that can destroy the keratin of skin and thus increase its susceptibility to irritation caused by the components of petroleum products. The cooling-lubricating fluids contained also other organisms (less pathogenic), including strains belonging to the genus *Proteus* and *Pseudomonas fluorescens*. The potential of these microorganisms to produce clinical symptoms is significantly increased if their prospective recipient is diabetic, alcoholic, or has reduced resistance for any other reason.[39]

Another important issue is the presence in aerosols of thermally stable endo- and exotoxins characteristic for Gram-negative bacteria. They are identified as causing occupational lung problems such as chronic bronchitis, abnormal cross-shift declines in pulmonary function, asthma, and other chronic diseases.[72] Presumably the level of endotoxins is highest during the decline, with the result that most skin problems prevail in heavily infected systems, which had previously been disinfected.

Microorganisms present in cooling-lubricating fluids are listed in Table 4.66.

Table 4.66. Microorganisms isolated from the cooling-lubricating fluid.[4,13,28,53,69,77,80,86,106,109,111]

Type	Species
bacteria	*Achromobacter parvulus, Achromobacter xylosoxidans, Achromobacter sp., Acinetobacter calcoaceticus, Acinetobacter genospecies, Acinetobacter haemolyticus, Acinetobacter johnsonii, Actinobacillus liginieresii, Aerobacter aerogenes, Aerobacter cloacae, Aerobacter sp., Aerococcus caviae, Aerococcus viridans, Aerococcus sp., Aeromonas caviae, Aeromonas hydrophila, Aeromonas salmonicida, Aeromonas sobria, Aeromonas sp., Agrobacterium sp., Alcaligenes faecalis, Alcaligenes faecalis subsp. faecalis, Alcaligenes metalcaligenes, Alcaligenes xylosoxydans, Alcaligenes sp., Bacillus cereus, Bacillus pumilus, Bacillus subtilis, Bacillus sp., Bordetella bronchiseptica, Brevibacterium sp., Brevundimonas diminuta, Burkholderia vietnamiensis, Chromobacterium violaceum, Chryseomonas luteola, Citrobacter amalonaticus, Citrobacter koseri, Citrobacter diversus, Citrobacter freundii, Clostridium sp., Comamonas testosteroni, Corynebacterium sp., Desulfovibrio sp., Diplococcus pneumoniae, Enterobacter aerogenes, Enterobacter agglomerans, Enterobacter cloacae, Enterobacter sakazakii, Enterococcus avium, Escherichia coli, Escherichia freundii, Exiguobacterium sp., Flavobacterium ferrugineum, Flavobacterium peregrinum, Flavobacterium sp., Klebsiella pneumoniae, Legionella bozemanii, Legionella feelii, Legionella pneumophila, Legionella sp., Leptothrix sp., Macrococcus caseolyticus, Micrococcus citreus, Micrococcus luteus, Micrococcus pyogenes var.albus, Micrococcus pyogenes var. aureus, Micrococcus varians roseus, Micrococcus sp., Moraxella catarrhalis, Moraxella sp., Morganella morganii, Mycobacteria sp., Mycobacterium chelonae, Mycobacterium immunogenum, Neisseria mucosa, Nocardia sp., Ochrobactrum anthropi, Oligella urethralis, Pantoea agglomerans, Paracolobactrum sp., Proteus mirabilis, Proteus morgani, Proteus rettgeri, Proteus vulgaris, Pseudofluorescens putida, Pseudomonas acidovorans, Pseudomonas aeruginosa, Pseudomonas alcaligenes, Pseudomonas cepatia, Pseudomonas desmolytica, Pseudomonas fluorescens, Pseudomonas maltophila, Pseudomonas marginalis, Pseudomonas mendocina, Pseudomonas oleovorans, Pseudomonas ovalis, Pseudomonas pseudoalcaligenes, Pseudomonas putida, Pseudomonas putida Biotype A, Pseudomonas putrefaciens, Pseudomonas rathonis, Pseudomonas rubescens, Pseudomonas stutzeri, Pseudomonas vesicularis, Pseudomonas fulva, Pseudomonas sp., Rhodococcus fascians, Rhodococcus sp., Rothia dentocariosa, Salmonella oranienburg, Salmonella paratyphi, Salmonella pullorum, Salmonella shottmuelleri, Salmonella typhi, Salmonella typhosa, Salmonella sp., Sarcina sp., Serratia liquefaciens, Shewanella purefacies, Shigella dysenteriae, Shigella sonnei, Shigella sp., Staphylococcus albus, Staphylococcus aureus, Staphylococcus capitis, Staphylococcus epidermis, Staphylococcus citreus, Staphylococcus xylosus, Staphylococcus sp., Stenotrophomonas maltophilia, Stenotrophomonas sp., Streptobacillus sp., Streptococcus suberis, Streptococcus pyogenes, Streptococcus sp., Vibrio cholerae, Vibrio metschnikovii, Vibrio sp.*

Table 4.66. Microorganisms isolated from the cooling-lubricating fluid.[4,13,28,53,69,77,80,86,106,109,111]

Type	Species
filamentous fungi	*Acremonium sp., Ambylosporium sp., Aspergillus candidus, Aspergillus flavus, Aspergillus fumigatus, Aspergillus oryzae, Aspergillus parasiticus, Aspergillus sp., Cephalosporium sp., Cladosporium sp., Fusarium sp., Penicillium canescens, Penicillium citrinum, Penicillium sp., Trichoderma sp.*
Yeasts	*Candida albicans, Candida humicola, Candida obtusa, Candida pseudotropicalis, Candida tropicalis, Candida sp., Rhodotorula rubra, Trichosporon sp.*

Cyprowski *et al.*[18] examined 10 samples of oil in water emulsion from three metallurgical plants in Poland. Their studies showed the presence of the following bacteria: *Shewanella putrefaciens, Micrococcus luteus, Staphylococcus warneri, Citrobacter freudi, Serratia filaria, Bacillus sp, Kocuria varians, Aeromonas salmonicida, and Brenneria nigrifluens*, and also *Actinomycetes*, *Streptomyces spp.* and fungi: *Acremonium butyri, Acremonium strictum*, and *Verticillium tenerum*. Among bacteria *Shewanella putrefaciens* was the dominant species, while among fungi it was *Acremonium butyri.*

Many microorganisms included in Table 4.66 belong to the pathogenic microorganisms. The 39 species of bacteria belong to pathogenic forms, and some species are filamentous fungi with the ability to produce mycotoxins[9,98,121] and having clinical relevance.[47,73]

A factory of diesel engines in the Ford Plant in Canada experienced problems with microbial contamination of coolants. A number of employees experienced Pontiac fever associated with infection caused by *Legionella* bacteria, manifested by fever and myalgia.[122]

In the last years, Polish industry showed interest in research of microbiological coolants. There have been many studies on this subject, but not all data are in the public domain. Table 4.67 contains some results.

Table 4.67. Microbial contamination of liquid-cooling lubricants in the Polish machine factories.[122-124]

Plant #	Machine	Number of microorganism in cm^3 of coolant.			
		aerobic	anaerobic	fungi	yeast
1	lathe, drill	5.5×10^5-7.8×10^7	present	5.0×10^2-2.4×10^3	no data
2	circular & cylindrical grinders	9.3×10^6-1.1×10^8	present	1.1×10^2	3.3×10^4
3	no data	3.2×10^4-1.1×10^8	present	0-3.0×10^3	no data
4	fresh emulsion	1.2×10^6	present	2.5×10^2	4.6×10^3
5	central supply	0-8.1×10^7	present	0-2.5×10^1	no data
6	no data	5.1×10^7	present	none	no data
7	various machines	2.0×10^3-1.5×10^8 ave: 10^7	present	1.0×10^2-3.0×10^4	no data

All analyzed samples were contaminated, with the largest contribution from aerobic bacteria (numbers exceed even 10^8 cells/cm^3). Sulfate-reducing anaerobes were present in all samples. According to Andriejuk *et al.*,[3] the number of microorganisms in coolants should not exceed 1×10^5 cells/cm^3.

In the recent years, there were reports on hypersensitivity pneumonitis (Granulomatous allergic lung disease – allergic pulmonary disorders).[28,46,95,111] So far, this was a common disease among farmers, but it also occurs among workers exposed to cooling lubricants. The disease is associated with the human immune response to the presence of microorganisms, including *Pseudomonas fluorescens* and *Mycobacterium chelonae* present in coolants. Veillette *et al.*[111] noted that the removal of coolant from a system without elimination of sources of microbiological contamination is not an effective way of dealing with hypersensitivity pneumonitis. It is important to note that *Mycobacteria* can quickly re-populate the system and the risk of occupational exposure to bioaerosols may be the same as before cleaning. This follows from the fact that these bacteria are able to metabolise a range of complex chemical compounds, including those that are in biocides, for example, morpholine. In such a situation, the disinfection of coolants eliminates a competitive microflora and permits the dominance of *Mycobacteria*, even in the case of water chlorination. In addition, the cells of these bacteria can easily form a biofilm, thus increasing their resistance to disinfection measures.[28]

In conclusion, it is worth emphasizing that according to Veillette *et al.*,[111] the key to effective control of microbial contamination of coolants is to better understand the ecology of microbes colonizing these habitats.

4.12.8. MICROBES IN THE OIL-WATER EMULSIONS USED IN HYDRAULIC SYSTEMS IN COAL MINES

Microbiological decomposition of non-flammable hydraulic fluids, powered structural supports, is one of the most burdensome in terms of corrosion problems in coal mines. Life activity under the influence of microorganisms may be seen as a change in color and consistency of the emulsions, and ultimately the loss of their functional properties.[63,119,127]

Problems associated with microbiological contamination of oil-water emulsions used in Polish coal mines began to grow rapidly in the late 1970s. This phenomenon was associated with reduced imports of biocides and emulsions. Domestic produced oils were used, which were far more vulnerable to microbes.

Currently, the adverse effect of emulsions, introduced as a response to ecological trends, is widely felt. These are biodegradable liquid emulsions. They are expected to be of high microbiological resistance during their operation and rapidly degraded by microorganisms after their useful life expired. Cieplik *et al.*[15] showed, however, that such emulsion lose their functional properties prematurely.

Microorganisms occurring in the emulsions used in hydraulic systems in coal mines may cause:

- separation of water due to the loss of emulsifiers
- form "coats" on the surface and produce suspended biomass
- loss of lubricating properties

- discoloration and gray and black coloration, formed as a result of reaction between bacteria-produced hydrogen sulfide and iron
- decrease of emulsion pH due to the release of organic acids in aerobic processes and secretion of H_2S in anaerobic processes
- corrosion of metals by lowering pH and degradation of electrochemical corrosion inhibitors
- clogging filters, which creates obstacle to circulation of fluid due to formation and accumulation of biomass and precipitation of sulfides with participation of anaerobic sulfate-reducing bacteria
- leakage of fluids due to the increased aggressiveness causing corrosion of metals and other elements of housing
- emergence of odors, such as rancid fat odors, which are volatile metabolites of aerobic bacteria and hydrogen sulfide[10,63,119]

Studies conducted over 30 years in the Microbiological Laboratory of Corrosion in the Central Mining Institute have shown that microbes grow on any kind of material used in the mine environment.[63,119,126,127] The source of infection may be coal or other fossils, mine water, and workers.[63] In coal mines, particularly conducive to the growth of microorganisms are their environmental conditions, especially high relative humidity and temperature.

Deterioration of the emulsions is caused by aerobic and anaerobic microorganisms. In addition to bacteria, yeasts, and filamentous fungi are also very active. Succession of microorganisms is similar to that which takes place in the case of other petroleum products. In well oxygenated emulsion, aerobic microorganisms are the first colonizers. As the depletion of oxygen begins, the optional aerobic microorganisms develop, which have the ability to act under both aerobic and anaerobic conditions. In the final stage, the environment is dominated by sulfate-reducing anaerobes. Measurable effect of their presence is the emission of toxic *hydrogen sulfide*. In places of stagnation, sediments and sludges are formed, which could provide a source of secondary infections of the hydraulic system. Growth of yeasts and filamentous fungi occurs when acid waste products of bacteria lower pH.[3]

Microorganisms present in the emulsions used in coal mines differ depending on type and specific conditions in mines. List of microorganisms isolated from fluids used in hydraulic systems in mines is presented in Table 4.68.

Table 4.68. Microorganisms isolated from oil-water emulsion used in coal mines.[10,63,119]

Type	Species
bacteria	*Achromobacter sp., Acinetobacter sp., Aeromonas sp., Alcaligenes sp., Citrobacter sp., Desulfotomaculum nigrificans, Desulfovibrio desulfuricans, Enterobacter cloacae var. liquiefaciens, Hafnia sp., Klebsiella pneumoniae, Moraxella sp., Proteus mirabilis, Proteus sp., Pseudomonas aeruginosa, Pseudomonas sp., Salmonella sp., Shigella sp., Streptomyces sp.*

Table 4.68. Microorganisms isolated from oil-water emulsion used in coal mines.[10,63,119]

Type	Species
filamentous fungi	*Alternaria tenuis, Aspergillus flavus, Aspergillus niger, Aspergillus ruber, Aspergillus versicolor, Botrytis cinerea, Chaetomium globosum, Cladosporium cladosporioides, Cladosporium herbarum, Fusarium chlamydosporium, Fusarium oxysporum, Fusarium solani, Fusarium sp., Penicillium citrinum, Penicillium chrysogenum, Penicillium frequentans, Penicillium notatum, Penicillium sp., Penicillium spinulosum, Penicillium variabile, Rhizopus nigricans, Trichoderma viride*
yeasts	*Candida albicans, Candida sp., Geotrichum candidum, Rhodotorula rubra*

Pathogenic microorganisms were also isolated, such as streptococci, staphylococci, and bacteria which cause food poisoning such as *Salmonella* and *Shigella*.[63]

The main sources of infection of oil emulsions are contaminated water and the remains of emulsions and sludges left in the hydraulic system. Most bacteria are attached to the surface, forming biofilm. After the breach of biofilm, bacteria grow in number to almost one hundred times of previous concentration.

The level of microbial contamination of the emulsions used in Polish coal mines is presented in Table 4.69.

Table 4.69. The degree of microbial contamination of oil-water emulsions used in Polish coal mines.[10,63,127]

Mine #	Number of microorganisms in 1 cm^3 of oil in water emulsion			
	aerobic	sulfate-reducing	filamentous fungi	yeasts
1	$3x10^7$	present	$2x10^4$	$1x10^3$
2	$2x10^7$	present	$1x10^3$	0
3	$4x10^5$	present	$5x10^2$	$2x10^2$
4	$8x10^6$	present	$8x10^3$	$1x10^5$
5	$3x10^3$	present	$2x10^4$	$1x10^1$
6	$1.0x10^6$-$2.5x10^7$	–	0-$4.0x10^3$	$1.0x10^1$-$1.6x10^4$
7	$1.3x10^5$-$5.0x10^5$	–	$4.0x10^1$-$1.0x10^3$	–
8	$6.0x10^5$-$5.0x10^6$	–	$7.5x10^1$-$9.0x10^3$	–
9	$2.0x10^5$-$7.9x10^9$	–	$1.2x10^2$-$2.8x10^3$	–

In most mines, emulsions showed a high degree of microbiological contamination. Anaerobic sulfate-reducing bacteria were omnipresent.

4.12.9 MICROBES IN ASPHALT AND INSULATING MATERIALS

Bitumens are a solid or semisolid substances with dark brown to black color. They are mixtures of saturated and naphthenic hydrocarbons with some concentrations of aromatic hydrocarbons and organic compounds of sulfur, oxygen, and nitrogen. They are divided into natural (semisolid bituminous substances occurring in natural deposits) and artificial, obtained by treatment of the specific fractions of crude oil. They are used in road construction, pavements, foundation protection, industrial paints and varnishes, in the electrical industry as insulators, in the rubber industry, plastics for the manufacture of coated fabrics, and as roofing materials, and flooring.[119]

There are few studies on microbial contamination of bitumen, its derivative products, and insulation, even though the risk of biodeterioration is real.

U.S. researchers demonstrated in 1956 that asphalt can be used as a carbon source by soil microorganisms. It was also proven that the deterioration of roofing asphalt was caused by microorganisms. It should be noted that these studies were conducted in warm climates but conclusions apply to temperate climate as well. Weight loss, increased viscosity, and the change of the structure and appearance are typical results of microbial deterioration.[119] The results of laboratory studies using some microorganisms available in soil are shown in Table 4.70.

Table 4.70. Degradation of asphalt by pure cultures of bacteria at 30°C.[20]

Species	Degradation degree, %	
	after 1 week	after 1 month
Bacillus 11-2A-W	5	–
Chromobacterium 6-4A-B	20	–
Pseudomonas 1-5A-C	3	90
Pseudomonas 7-2A-B	25	–
Pseudomonas 5-3A-B	18	–
Pseudomonas aeruginosa	20	49

The data shows that bacteria are capable of degradation. In a short time, even 90% of asphalt was degraded. The degree of degradation, however, depends on the chemical composition of asphalt and many other factors that may arise under natural conditions, such as variable temperature and humidity or the interaction with the accompanying microflora.

Kerner-Gang[54] isolated 19 strains of filamentous fungi, which were then used in the study of their ability to grow on asphalt used in flooding compositions and in the manufacture of roofing felt. Studies have shown that *Aspergillus lichtheimi, Mucor javanicus,* and *Mucor fragilis* strongly grew in the first week of experiment. In analogous studies bituminous felt was also used. Many strains were able to grow on these substrates, the strongest growth was characteristic of *Aspergillus versicolor, Chaetomium globosum,* and *Amorphoteca resinae.*[54]

Denizel *et al.*[21] also isolated *Amorphoteca resinae f. albidum* from cardboard treated with bitumen, which was found in a gold mine in South Africa. Microbial degradation of bitumen was also caused by *Pseudomonas aeruginosa, Mycobacterium ranae,* and *Nocardia coeliaca.*[119]

4.12.10 BIOCIDES IN THE PROTECTION OF PETROLEUM PRODUCTS

The issue of biocides is extremely broad. Their impact on microbial cells and resistance mechanisms is widely described in the literature.[22,23,31,40,55,59,75,90,93,94,100,104,108]

Biocides for the protection of petroleum products should:

- have anti-corrosive properties and compatibility with various petroleum systems
- be characterized by high activity and broad-spectrum of biocidal action
- have the ability to penetrate microbial consortia

- be stable over a wide range of thermal conditions
- be active for a long time
- be soluble in water and hydrocarbons
- be active at low concentrations (1-100 ppm)
- be chemically stable
- be environmentally friendly (biodegradable)
- be produced from cheap and readily available raw materials
- be safe and easy to use and store
- be cost-effective
- not contain ingredients which give harmful combustion products, such as halides, which may form dioxin
- not decompose lubricants
- not cause deterioration of products used for conservation of which they are used[31,118]

Microbiocides for coolants should also have low vapor pressure, since the temperature of the metal-working is high. They should be stable at pH values of 9 and chemically inert to all components of the emulsifying oil, and should not have unpleasant odor.[3,122] Various components of coolants, including biocides, are potentially harmful to the health of workers (they can cause irritation or skin allergies). Therefore, taking into account the specific working conditions of people in contact with cooling liquid lubricants, attention should be given to the selection of biocides, because they may pose a risk to human health. The degree of absorption through the skin of biocides could depend on the presence of other components in mixture.[113]

Selvaraju *et al.*[95] studied the impact of different biocides on *Mycobacterium immunogenum* and *Pseudomonas fluorescens* colonizing coolant. *M. immunogenum* was more resistant than *P. fluorescens* to formaldehyde-releasing biocides (Grotan and Bioban), isothiazolin (Kathon), and phenolic biocides (Preventol). Research conducted in France by Chazal[13] showed that the selection of microorganism is very important for efficacy of biocide evaluation. If one wants to determine the effectiveness of biocide used to protect the coolants in France, the strains isolated, in its climatic zone, from the cooling fluids have to be used. It was demonstrated that there are differences between strains of the same species originating from North America and France. Therefore, the results of research conducted in America cannot be directly transferred to the conditions of the French and *vice versa.*[13]

Research conducted by Dilger *et al.*[24] showed that a high number of bacteria was detected in coolants with and without biocide. This follows from the fact that bacteria, particularly those pathogenic for humans, are capable of developing biocide resistance. It should also be noted that in the case of cooling-lubricating fluid, use of biocides has a twofold significance. Firstly, they are used to protect coolant against development of microorganisms. Secondly, these products should protect workers against germs.[24]

Fuel-soluble biocides are used to protect the oil phase and water-soluble biocides are used for decontamination of storage containers.

There are four categories of biocides used to protect fuel:

- oil-dispersible products to protect water, located on the bottom of the tank

- oil soluble products, which enter the water phase for the disinfection of clean fuel, in which we expect a small amount of water
- oil-soluble blends separating into 2 components: one to protect water, and another to protect oil for the interfacial layer protection and protection of water from the bottom of the tank
- water-soluble products to add to the bottom of storage tanks[31]

Soluble biocides in the oil phase should exhibit the following characteristics:

- should be active against the microorganisms present in the oil phase, and safe at the doses used
- should not be surface-active
- should not stimulate the oxidation and corrosion
- should not contain metals and halogen

Biocides dissolved in the aqueous phase:

- should have a broad spectrum of antimicrobial activity, including the sulfate-reducing bacteria
- should not be too alkaline or too acidic
- should not be corrosion or surface-active
- should not have harmful health effects
- should not be harmful to environment when water phase is released
- should be easy to neutralize.[41]

Bento and Gaylarde[7] studied the effectiveness of Kathon FP 1.5. Its use at a dose of 0.1 ppm permits storage of a clean sample for 30 days, and at a dose of 1 or 10 ppm for up to 400 days.

Resistance against biocidal protection is widespread, especially *Pseudomonas sp*, particularly *Pseudomonas aeruginosa* are resistant. By resistance, it is understood that microorganism withstands exposure to concentration which was suggested by manufacturer for the microorganism.[104]

Biocides are poisons and may pose a potential threat to people in contact with fuels.[31]

Pitcher[76] presented results of studies of selected biocides, added to the diesel tank having a capacity of 14,000 m^3. Their impact was determined after 5 hours., 24 hrs., 5 days, and 142 days. The results are given in Table 4.71.

Table 4.71. The efficiency of biocide in the French strategic reserves of fuel. [Adapted, by permission, from Pitcher, D G, *Intl. Biodet.*, **25**, 207-18, 1989.]

Time	Sample	Number of microorganisms in 1 cm^3 of fuel.	
		oil phase	water phase
t=0 (initial state)	tank bottom	$1x10^3$ bacteria, $4x10^2$ yeast	$8x10^5$ bacteria, $4x10^3$ yeasts; sulfate-reducing bacteria present
5 hrs	tank bottom	$1x10^3$ bacteria, $4x10^2$ yeast	none
24 hrs	tank bottom, middle, top	none	none
5 days	tank bottom, middle, top	none	none

Table 4.71. The efficiency of biocide in the French strategic reserves of fuel. [Adapted, by permission, from Pitcher, D G, *Intl. Biodet.*, **25**, 207-18, 1989.]

Time	Sample	Number of microorganisms in 1 cm^3 of fuel.	
		oil phase	water phase
142 days	tank bottom, middle, top	none	none

Biocide has proven to be very effective, since after 24 hours it eliminated all microorganisms, and both water and fuel were microbiologically clean throughout the duration of the experiment.

Effectiveness of the biocides was studied also in Ireland where the bottom of the tank containing diesel was found to contain aerobic bacteria, sulfate-reducing bacteria, filamentous fungi and yeasts. Biocide efficacy testing was carried out for four months. After analyzing the data received, it was decided that tanks need decontamination every 3-4 months. Treatment fulfilled its role because the number of complaints concerning the quality of the diesel fuel was reduced by 80%.[76]

In the 1970s in England, it was calculated that the proper use of biocide in cooling-lubricating liquids brought savings of 8,000 pounds per year.[3] According to Gaylarde *et al.*,[31] decontamination cost of one ton of fuel is estimated at $3, while the continuation of prevention is $0.4-0.8/ton of fuel.

Microbial contamination of the cargo of gas oil on a tanker was also conducted in the UK. The cargo of gasoline, produced in one of Britain's refineries, was found to contain large quantities of water and microorganisms. Because the cargo was to be stored for 12 months, it was decided to treat the gasoline while still in the tanker.

The concentration of biocide was set at 2.4 ppm of the active substance. One day after biocide addition, the aqueous phase contained 150 cells/cm^3. Sulfate-reducing bacteria were absent. The oil phase also contained 150 cells/cm^3 and yeast was absent. A week after biocide addition, the aqueous phase at the bottom of the tank and the oil phase in the upper, middle, and bottom layers of the tank were microbiologically free.[76]

The presence and growth of microorganisms in cooling fluids was reduced by application of special formulas for liquid-cooling lubricants resistant to microorganisms, use of physical methods (filtration and centrifugation, radiation), and use of biocides.[69,73,89,91]

In response to the market demands, a new category of coolants was introduced. They are defined as biostable coolants. They include additives, which are non-toxic but have inhibitory effects on the development of microorganisms. These additives usually play the role of emulsifiers and corrosion inhibitors and their content in the concentrate reaches 20-30 wt%. The use of modern coolants is beneficial for improving working conditions and eliminates the cumbersome addition of biocides to the coolant during the operation. Biostatic ingredients, such as esters, salts of boric acid, alkanoamines or carboxylic acid esters often provide additional functions of emulsifiers, corrosion inhibitors and oxidation.[17,115] Interaction between the components of coolants, biocides, and corrosion inhibitors may sometimes affect properties of lubricant, and it can be avoided by the use of modern additives.[78,90]

Combination of heat treatment and biocides gives new opportunities as presented in Table 4.72.

Table 4.72. The effect of heat treatment and the biocidal activity on fluids for machining. [Adapted, by permission, from Rossmoore, H W, Biocides for metalworking lubricants and hydraulic fluids in Handbook of Biocide and Preservative Use, Blackie Academic & Professional London, Glasgow, Weinheim, New York, Tokyo, Melbourne, Madras, 1995.]

Biocide and its concentration in ppm	Number of bacterial cells in 1 cm^3 after 24 h at different temperatures	
	25°C	60°C
Control (no biocide)	$5x10^6$	10^6
Bioban P 1487 – 250 ppm	10^7	$<10^3$
Bioban P 1487 – 750	$2x10^4$	$<10^3$
Kathon 886 MW – 10	$5x10^4$	$<10^3$
Kathon 886 MW – 50	$<10^3$	$<10^3$
Uconex 345 – 250	$3x10^7$	$<10^3$
Uconex 345 – 750	10^7	$<10^3$
Triadine 10 – 280	$1.5x10^7$	$<10^3$
Triadine 10 – 750	10^6	$<10^3$

Because of the effect of biofilms, it is important to use the correct dosage of biocides. Too small amounts may help microorganisms to develop resistance.[13] Application of too high doses of biocides increases costs and may endanger workers.

Replacement of toxic biocides is one of the major directions of studies. Use of plant extracts is one of the directions. Water extracts of *Brassica nigra* and *Allium cepa* is one of the studied solutions.[33]

REFERENCES

1 Aarnisalo, K; Raaska, L; Wirtanen, G, *Food Control*, **18**, 1019-25, 2007.

2 Abu Shaqra, Q M; Hill, E C, Biodeterioration 6. The 6th International Biodeterioration Symposium, Washington, Barry, S; Houghton, D R (Eds), CAB International Mycological Institute, The Biodeterioration Society, 440-448, 1984.

3 Andriejuk, E I; Bilaj, W I; Kowal, E E; Kozlowa, U A, **Mikrobnaja korrozija i jejo wozbuditieli**, *Naukowa Dumka*, Kijew, 1980.

4 Bakalova, S; Doycheva, A; Ivanova, I; Groudeva, V; Dimkov, R, *Biotechnol. Biotechnol. EQ.*, **21**, 4, 437-441, 2007.

5 Barnes, S P; Smith, R N; Hall, S J; Craige, S F; Van Duynhovena, N, Biodeterioration and Biodegradation, 9, Bousher, A; Chandra, M; Edyvean, R, (Eds), Institution of Chemical Engineers, Rugby, UK, 629-634, 1995.

6 Bento, F M; Gaylarde, C C, *Intl. Biodet. Biodeg.*, 31-33, 1996.

7 Bento, F M; Gaylarde, C C, *Intl. Biodet. Biodeg.*, **47**, 107-112, 2001.

8 Bernstein, D I; Lummus, Z L; Santilli, G; Siskosky, J; Bernstein, I L, *Chest*, **108**, 636-641, 1995.

9 Betina, V, **Mycotoxins. Chemical biological and environmental aspects**. *Elsevier* Amsterdam-Oxford-New York-Tokyo, 1989.

10 Bohdanowicz-Murek, K J; Zyska, B J; Zankowicz, L P; Wawrzyniak, E M; Terech, I, Biodeterioration Proceedings of the Fourth International Symposium, Berlin. Oxley, T A; Becker, G; Allsopp, D, (Eds.), Pitman Publishing Ltd. London and Biodeterioration Society, 357-362, 1980.

11 Boswell, D E, Biodegradation. The Proceedings of The Third International Biodegradation Symposium.

Sharpley J M, Kaplan A M (Eds.), Applied Science Publishers Ltd, London, (1975), 215-220.
12 Cabral, P, *Intl. Biodet. Bull.*, **16**, 1, 23-27, 1980
13 Chazal, P M, *Eur. J. Epidemiology*, **11**, 1-7, 1995.
14 Chung, Y C; Chen, H C; Shyu, Y T; Hua, J, *Fuel*, **79**, 1525-1532, 2000.
15 Cieplik, Z; Kwiatkowska, D; Wichary, H; Kozlowska, R, Materialy Sympozjum-Produkty Rafinerii Trzebinia S.A. dla Górnictwa. Ustron, 1996.
16 Cieslak, J, *Zdrowa zywnosc, zdrowy styl zycia*, **3**, 53, 21-22, 2001.
17 Cunningham, J A; Lepp, N W; Saunders, V A; Stubbs, R, Biodet. Biodeg., 9. Bousher, A; Chandra, M; Edyvean, R, (Eds.), Institution of Chemical Engineers, Rugby, UK, 635-640, 1995.
18 Cyprowski, M; Piotrowska, M; Zakowska, Z; Szadkowska-Stanczyk, I, *Intl. J. Occupational Medicine Environ. Health*, **20**, 4, 365-371, 2007.
19 Darby, R T; Simmons, E G; Wiley, B J, *Intl. Biodet. Biodeg.*, **48**, 159-161, 2001.
20 Davis, J B, **Petroleum microbiology**. *Elsevier Publishing Company,* Amsterdam London New York, 1967.
21 Denizel, K; Jarvis, B; Onions, Agnes H S; Rhodes, Anette C; Samson, R A; Simmons, E G; Smith, Maud T; Hueck van der Plas, E H, *Intl. Biodet. Bull.*, **10**, 1, 3-23, 1974.
22 Denyer, S P, *Intl. Biodet.*, **26**, 89-100, 1990.
23 Denyer, S P, *Intl. Biodet. Biodeg.*, 227-245, 1995.
24 Dilger, S; Fluri, A; Sonntag, H G, *Int. J. Environ.Health*, **208**, 467-476, 2005.
25 Dott, W; Feidieker, D; Kampfer, P; Schleibinger, H; Strechel, S, *J. Ind. Microbiol.*, **4**, 365-374, 1989.
26 Duda, A, Konferencja Naukowo-Techniczna. Ekologia w przemysle rafineryjnym. Kielce, 22-27, 2001.
27 Edmonds, P; Cooney, *J J, Appl. Microbiol.*, **15**, 2, 411-416, 1967.
28 Falkinham, J O, *Appl. Environ. Microbiol.*, **75**, 7, 2057-2061, 2009.
29 Fedorak, P M; Westlake, D W S, *Can. J. Microbiol.*, **27**, 432-443, 1981.
30 Gazdzik, B, I Ogólnokrajowa Konferencja Naukowa. Rozklad i korozja mikrobiologiczna materialów technicznych. Politechnika Lódzka, Lódz, 72, 2000.
31 Gaylarde, C; Bento, F M; Kelley, J, *Rev. Microbiol.*, **30**, 1-10, 1999.
32 Genner, C; Hill, E C, Fuels and oils in **Economic Microbiology**, vol. 6, **Microbial Biodeterioration**, *Academic Press*, 1981.
33 Guiamet, P S; Gomez de Saravia, S G, *Latin Am. Appl. Res.*, **35**, 295-300, 2005.
34 Guiamet, P S; Videla, H A; Reinoso, E H, Biodeterioration and Biodegradation, 9. The 9th International Biodeterioration and Biodegradation Symposium, Leeds, UK, 615-621, 1995.
35 Haggett, R D; Morchat, R M, *Intl. Biodet.*, **29**, 1, 87-100, 1992.
36 Hettige, G E G; Sheridan, J E, *Intl. Biodet.*, **25**, 299-309, 1989.
37 Hill, E C, Proccedings of The Third International Biodegradation Symposium, Sharpley, J M; Kaplan, A M (Eds.), Applied Science Publishers Ltd., London, 243-258, 1975.
38 Hill, E C, Biodegradation of hydrocarbon-based products in industrial use in **Developments in biodegradation of hydrocarbons**, *Applied Science Publishers Ltd*, London, 1978.
39 Hill, E C, **Biodegradation of petroleum products in petroleum microbiology**, *Macmillan Publishers*, London, 1984.
40 Hill, E C, *Intl. Biodet.*, **26**, 281-285, 1990.
41 Hill, E C, Fuel biocydes in **Handbook of biocydes and preservative use**, *Blackie Academic & Professional*, London Glasgow Weinheim New York Tokyo Melbourne Madras, 1995.
42 Hill, E C; Al-Haidary, N K, Proccedings Symp. Microbial Corrosion Affecting the Petroleum Industry. Institute of Petroleum, Sunbury UK, 51-75, 1976.
43 Hill, E C; Genner, C, Biodeterioration. The Proccedings of The Fourth International Biodeterioration Symposium Berlin. Oxley, T A; Becker, G; Allsopp, D, (Eds.), Pitman Publishing Ltd London and Biodeterioration Society, 37-43, 1980.
44 Hill, E C; Hill, G C, *Trans. Inst. Marine Engrs.*, 104, 119-130, 1993.
45 Hodgson, G, *Ind. Med. Surg.*, 39, 2, 41-46, 1970.
46 Hodgson, M J; Bracker, A; Yang, C; Storey, E; Jarvis, B J; Milton, D; Lummus, Z; Bernstein, D; Cole, S, *Am. J. Ind. Med.*, 39, 6, 616-628, 2001.
47 Hoog, G S; Guarro, J; Gene, J; Figueras, M J, Atlas of clinical fungi 2nd edition. *Centraalbureau voor Schimmelcultures / Universitat Rovira i Virgili*, 2000.
48 Itah, A Y; Brooks, A A; Ogar, B O; Okure, A B, *Bull. Environ. Contam. Toxicol.*, 83, 318-327, 2009.
49 Jamison, V W; Raymond, R L; Hudson, J O, Biodegradation. The Proceedings of The Third International Biodegradation Symposium. Sharpley, J M; Kaplan, A M, (Eds.), Applied Science Publishers Ltd London, 187-196, 1975.
50 Janda, K, *Postepy Mikrobiologii*, 44, 2, 157-169, 2005.

51 Janda, K; Falkowski, J, *Postepy Mikrobiologii*, 40, 3, 287-310, 2001a.
52 Janda, K; Falkowski, J, I Ogólnokrajowa Konferencja Naukowa. Rozklad i korozja mikrobiologiczna materialów technicznych. Politechnika Lódzka, Lódz, 235-240, 2001.
53 Janda, K; Przybulewska, K, *Postepy Mikrobiologii*, **43**, 2, 167-187, 2004.
54 Kerner-Gang, W, *Mater. Organismen*, **2**, 3, 161-177, 1967.
55 Kinnimeny, S; Wimpenny, J W T, *Intl. Biodet.*, **26**, 181-194, 1990.
56 Klamer, J F, Proceedings of the 8th International Biodeterioration and Biodegradation Symposium. Rossmoore, H W, (Ed.), Windsor, Canada, Ontario, 463-464, 1991.
57 Kowalik, R, *Postepy Mikrobiologii*, **8**, 1, 105-125, 1969.
58 Kreiss, K; Cox-Ganser, J, *Am. J. Ind. Med.*, **32**, 423-432, 1997.
59 Kugler, M, II Miedzynarodowa Konferencja Naukowa. Rozklad i korozja mikrobiologiczna materialów technicznych. Politechnika Lódzka, Lódz, 353-354, 2003.
60 Kwiatkowska, D, dane nie publikowane, 2001.
61 Kwiatkowska, D, III Miedzynarodowa Konferencja Naukowa. Rozklad i korozja mikrobiologiczna materialów technicznych. Politechnika Lódzka, Lódz, 28-37, 2003.
62 Kwiatkowska, D; Wichary, H, *Ochrona przed korozja*, **5**, 118-121, 2000.
63 Kwiatkowska, D; Wichary, H; Kozlowska, R, *Wiadomosci Górnicze*, **7-8**, 299-305, 1998.
64 Lewis, M D; Janotka, E; Whitmer, M P; Bledsoe, T A, *Intl. Biodet. Biodeg.*, **47**, 89-94, 2001.
65 Lipson, D; Sadurska, I; Kowalik, R, *Postepy Mikrobiologii*, **8**, 1, 169-173, 1969.
66 Lopes, P T C; Gaylarde, C, *Intl. Biodet. Biodeg.*, 37-40, 1996.
67 Margesin, R; Schinner, F, *Appl. Microbiol. Biotechnol.*, **56**, 5/6, 650-663, 2001.
68 Marin, M; Pedregosa, A; Rios, S; Laborda, F, *Intl. Biodet. Biodeg.*, 69-75, 1996.
69 Muszynski, A; ?ebkowska, M, *Polish J. Environ. Studies*, **14**, 1, 73-79, 2005.
70 Muzik, M; Hrubant, F; Wasserbauer, R, *Elektrotechnicki obzor*, **68**, 6, 347-350, 1979.
71 Parbery, D J, *Australian J. Botany*, **17**, 331-357, 1969.
72 Park, D; Teschke, K; Bartlett, K, *Ann. Occup. Hyg.*, **45**, 7, 569-576, 2001.
73 Passman, F J, Society of Manufacturing Engineers. Conference Metal Working Fluids, Cincinanti, Ohio, Technical Paper 1-14, 1992.
74 Passman, F J; Rossmore, K; Rossmore, L, *Tribology Lubricant Technol.*, 52-55, March 2009.
75 Paulus, W, **Biodeterioration**, 7, Houghton, D R; Smith, R N; Eggins, H O, (Eds.), Elsevier Applied Science, London and New York, 1-20, 1988.
76 Pitcher, D G, *Intl. Biodet.*, **25**, 207-218, 1989.
77 Pivnick, H; Engelhard, W E; Thompson, T L, *Appl. Microbiol.*, **2**, 140-142, 1954.
78 Posenauer, B; Schweisfurth, R, *Mater. Organismen*, **19**, 3, 191-200, 1984.
79 Rabaev, M; Shapira, D; Geva, J; Fass, R; Sivan, A, *Intl. Biodet. Biodeg.*, **63**, 311-315, 2009.
80 Rabenstein, A; Koch, T; Remesch, M; Brinksmeier, E; Kuever, J, *Intl. Biodet. Biodeg.*, (2009), doi: 10.16/j.ibiod.2009.07.005.
81 Rauch, M E; Graef, H W; Rozenzhak, S M; Jones, S E; Bleckman, C A; Kruger, R L; Naik, R R; Stone, M O, *J. Ind. Microbiol. Biotechnol.*, **33**, 29-36, 2006.
82 Roffey, R, *Intl. Biodet.*, **25**, 219-236, 1989.
83 Roffey, R; Norqvist, A; Edlund, A, Biodeterioration, 6. Proceedings of the 6th International Biodeterioration Symposium Washington, DC, 476-483, 1984.
84 Roffey, R; Norqvist, A; Edlund, A, 2nd International Conference on Long-Term Storage Stabilities of Liquid Fuels. Stavinoha, L L, (Ed.), Southwest Research Institute, San Antonio, Texas, 165-279, 1986.
85 Rosenman, K D; Reilly, M J; Kalinowski, D, *Am. J. Ind. Med.*, **32**, 325-331, 1997.
86 Rossmoore, H W; Szlatky, K, *Int. Biodet. Bull.*, **13**, 4, 96-100, 1977.
87 Rossmoore, H W, **Microbial degradation of water-based metalworking fluids in comprehensive biotechnology**, *Pergamon Press*, New York, 1986.
88 Rossmoore, K, Biodeterioration, 7. Proccedings of the Seventh International Biodeterioration Symposium, Houghton, D R; Smith, R N; Eggins, H O, (Eds.), Elsevier Applied Science London and New York, 243-247, 1988.
89 Rossmoore, H W; Rossmoore, L A; Kaiser, A L, Biodeterioration, 7, Houghton, D R; Smith, R N; Eggins, H O, (Eds.), Elsevier Applied Science, New York, 517-522, 1988.
90 Rossmoore, H W, Biocides for metalworking lubricants and hydraulic fluids in **Handbook of Biocide and Preservative Use**, Blackie Academic & Professional London, Glasgow, Weinheim, New York, Tokyo, Melbourne, Madras, 1995.
91 Rossmoore, H W, II Ogólnokrajowa Konferencja Naukowa. Rozklad i korozja mikrobiologiczna materialów technicznych. Politechnika Lódzka, Lódz, 2001.

92 Rossmoore, H W; Rossmoore, L A, *Intl. Biodet.*, **27**, 145-156, 1991.
93 Russel, A D, *Intl. Biodet.*, **26**, 101-110, 1990.
94 Russel, A D, *Intl. Biodet. Biodeg.*, 247-265, 1995.
95 Selvaraju, S B; Khan, I U H; Yadav, J S, *Appl. Environ. Microbiol.*, 542-564, Jan. 2005
96 Skribachilin, V B; Mikhailova, L K; Kochanov, Yu S; Buryakovskaya, T V, *Chemmotology*, UDC 620.193.8.p. 1994. Translated from *Khimiya i Tekhnologiya Topliv i Masel*, **6**, (June 1993), 12-13.
97 Slomczynski, T; Lebkowska, M, II Ogólnokrajowa Konferencja Naukowa. Rozklad i korozja mikrobiologiczna materialów technicznych. Politechnika Lódzka, Lódz, 252-257, 2001.
98 Smith, J E; Moss, M O, Mycotoxins. Formation, analysis and significance, John Wiley & Sons Ltd, Chichester-New York-Brisbane-Toronto-Singapore, 1985.
99 Smith, R N, The 7th International Biodeterioration Symposium, Houghton, D R; Smith, R N; Eggins, H O, (Eds.),. Cambridge UK, Elsevier Applied Science Publishers Ltd, 256-262, 1988.
100 Smith, R N, *Intl. Biodet.*, **26**, 111-125, 1990.
101 Smith, R N, Proceedings of the 8th International Biodeterioration and Biodegradation Symposium. Rossmoore, H W, (Ed.), Windsor, Ontario, Canada, Elsevier Applied Science London and New York, 113-123, 1991.
102 Solana, M J V; Gaylarde, C C, Biodeterioration and Biodegradation, 9, The proceedings of the 9th International Biodeterioration and Biodegradation Symposium, Leeds, UK, 621-628, 1995.
103 Sondossi, M; Riha, V F; Rossmoore, H W, *Intl. Biodeter.*, **26**, 51-61, 1990.
104 Sondossi, M; Rossmoore, H W; Lashen, E S, *Intl. Biodet. Biodeg.*, **43**, 85-92, 1999.
105 Sprince, N L; Thorne, P S; Popendorf, W; Zwerling, C; Miller, E R; DeKoster, J A, *Am. J. Ind. Med.*, **31**, 403-413, 1997.
106 Suliman, S M A; Abukabr, M I; Mirghani, E F, *Tribology Intl.*, **30**, 10, 753-757, 1997.
107 Symes, W R; Cowap, D, Biodegradation. The Proceedings of The Third Interna-tional Biodegradation Symposium. Sharpley, J M; Kaplan, A M, (Eds.), Applied Science Publishers Ltd London, 233-242, 1975.
108 Thomas, A R; Hill, E C, *Intl. Biodet. Biodeg.*, **48**, 245-251, 2001.
109 Van der Gast, C J; Knowles, C J; Wright, M A; Thompson, I P, *Intl. Biodet. Biodeg.*, **47**, 113-123, 2001.
110 Vedder, K W; Rossmoore, H W, Biodeterioration, 6. The 6th International Biodeterioration Symposium Washington, Barry, S; Houghton, D R (Eds.), CAB International Mycological Institute, The Biodeterioration Society, 453-459, 1984.
111 Veillette, M; Thorne, P S; Gordon, T; Duchaine, C, *Ann. Occup. Hyg.*, **48**, 6, 541-546, 2004.
112 Videla, H A, The action of *Cladosporium resinae* growth on the electrochemical behaviour of aluminium in Biologically Induced Corrosion, National Assocation of Corrosion Engineers, Houston, Texas, 1986.
113 Vijay, V; Yeatts, J L; Riviere, J; Baynes, R E, *Toxicol. Lett.*, **175**, 34-43, 2007.
114 Wang, H X; Reponen, T; Martuzevicius, W Li D; Grinshpun, S A; Willeke, K, *Aerosol Sci.*, **36**, 721-734, 2005.
115 Warrilow, E O; Gillatt, J W, Biodeterioration, 6. The 6th International Biodeterioration Symposium Washington, Barry, S; Houghton, D R, (Eds.), CAB International Mycological Institute, The Biodeterioration Society, 459-465, 1984.
116 Watkinson, R J; Morgan, P, **Physiology on aliphatic hydrocarbodegrading microorganisms in physiology of biodegradative microorganisms**, *Kluwer Academic Publ.*, Dordrecht, 1990.
117 Wycislik, E T; Allsopp, D, The 5th International Biodeterioration Symposium, Oxley, T A; Barrry, S, (Eds.), *John Wiley & Sons Ltd*, Aberdeen, 453-461, 1983.
118 Yemashova, N; Murygina, V P; Zhukov, D V; Zakharyantz, A A; Glanchenko, M A; Appanna, V; Kalyuzhnyi, S V, *Rev. Environ. Sci. Biotechnol.*, **6**, 315-337, 2007.
119 Zyska, B, Mikrobiologiczna korozja materialów technicznych, Wydawnictwa Naukowo-Techniczne, Warszawa, 1977.
120 Zyska, B, Mikrobiologiczny rozklad i korozja materialów technicznych in **Mikrobiologia techniczna**, *Wydawnictwo Politechniki Lódzkiej*, Lódz, 2000.
121 Zyska, B, *Mikologia Lekarska*, 8, 127-140, 2001.
122 Zyska, B, Katastrofy, awarie i zagrozenia mikrobiologiczne w przemysle i budownictwie, *Wydawnictwo Politechniki Lódzkiej*, Lódz, 2001b.
123 Zyska, B; Bohdanowicz-Murek, K, **Chemia w walce z korozja**. *NOT Rada Wojewódzka w Czestochowie, Komitet Naukowo-Techniczny do spraw Korozji*. Czestochowa, 74-98, 1984.
124 Zyska, B; Bohdanowicz-Murek, K; Kwiatkowska, D, *Postepy Mikrobiologii*, **XVIII**, 3/4, 309-328, 1979.
125 Zyska, B; Bohdanowicz-Murek, K; Kwiatkowska, D; Cieplik Z, Proc. of the 6th Int. Symp."Enviroeffect". Praque Academy of Sciences, 1980.
126 Zyska, B; Kwiatkowska, D; Cieplik, Z, Materialy Sympozjum Naukowego. Aktualne badania i osiagniecia

w górnictwie. Główny Instytut Górnictwa, Katowice, 1985.
127 Zyska, B; Wichary, H; Cieplik, Z; Kwiatkowska, D, Proccedings of the 7th International Biodeterioration Symposium Cambridge, London a New York. Elsevier Applied Science, 263-267, 1988.

4.13 PHARMACEUTICALS, COSMETICS, AND TOILETRIES

4.13.1 RELEVANT MICROORGANISMS

Table 4.73 contains a list of microorganisms found to cause deterioration of pharmaceutical products, cosmetics, and toiletries.

Table 4.73. Microorganisms in pharmaceuticals, cosmetics and toiletries.

Material	Microorganisms	Refs.
contact lenses	*Acanthamoeba castellani, Aspergillus fumigatus, Fusarium solani, Pseudomonas aeruginosa, Staphylococcus epidermidis*	13,15
cosmetics (variety of products)	*Bacillus badius, Bacillus brevis, Bacillus circulans, Bacillus coagulans, Bacillus lentus, Bacillus licheniformis, Bacillus megaterium, Bacillus polymixa, Bacillus pumilus, Micrococcus kristinae, Micrococcus lylae, Micrococcus nishinamiyaensis, Micrococcus roseus, Micrococcus sedentarius, Staphylococcus aureus, Staphylococcus epidermidis, Staphylococcus spaprophyticus*	7
eye makeup	*Pseudomonas aeruginosa, Staphylococcus aureus, Staphylococcus epidermidis*	16
deodorants	*Campylobacter jejuni, Corynebacterium diphtheriae, Enterobacter aerogenes, Enterococcus faecalis, Escherichia coli, Klebsiella pneumoniae, Listeria monocytogenes, Mycobacterium tuberculosis var bovis, Neisseria elongata, Pseudomonas aeruginosa, Proteus vulgaris, Serratia marcescens, Shigella dysenteriae, Streptococcus pyogenes, Salmonella choleraesuis, Staphylococcus aureus*	
implants	*Staphylococcus aureus, Staphylococcus epidermidis*	5
skin antisepts	*Acinetobacter* species, *Aspergillus* species, *Bacillus cereus, Candida* species, *Citrobacter* species, *Dermatophyte, Enterobacter* species, *Enterococcus* species, *Escherichia coli, Klebsiella* species, *Morganella morganii, Proteus vulgaris, Pseudomonas aeruginosa, Salmonella/Shigella, Sarcina lutea, Serratia marcescens, Staphylococcus aureus, Streptococcus faecalis, Xanthomonas maltophilia*	12

4.13.2 ESSENTIAL PRODUCT COMPONENTS

Any implant's surface provides an ideal environment for bacterial adhesion and proliferation.[5] It has been shown that bacteria adhere preferentially to body-foreign materials as compared to bone or enamel.[5] Once a biofilm has formed, it is very difficult to treat clinically.[5] Once *Staphylococcus aureus* attaches to a surface, e.g., bone joint or metal biomaterial (implant), host cells are unable to dislodge it.[5]

4.13.3 BIOSTABILIZATION

Table 4.74 contains a list of the most suitable biocides used for protection of pharmaceuticals, cosmetics, and toiletries.

Table 4.74. Biocides used in pharmaceuticals, cosmetics, and toiletries.

Material	Biocide	Refs.
contact lenses	myristamidopropyl dimethylamine,[13] quaternary ammonium surfactants,[2] polyhexamethylene biguanide,[14] Polyquaternium-1[13]	2,13,14
cosmetics (general)	butylparaben, ethylparaben, imidazolidinylurea, methylparaben, phenoxyethanol, propylparaben, quaternary ammonium surfactants, triclosan	1,2,7
deodorants	triclosan (2,4,4'-trichloro-2'-hydroxydiphenylether)	1
eye makeups	2-bromo-2-nitropropane-1,3-diol (Bronopol), methyl paraben	9,17
hair conditioners	2-bromo-2-nitropropane-1,3-diol,[11] chloroxylenol,[11] DMDM hydantoin[9]	9,11
hand creams, lotions, soaps	2-bromo-2-nitropropane-1,3-diol, glutaraldehyde, iodopropynyl butylcarbamate, triclosan	1,8,9,10
implants	surface modification with silver	5
makeups	DMDM hydantoin, iodopropynyl butylcarbamate	8,9
mouthwash	benzalkonium chloride,[11] chlorhexidine salts,[18] surfactant-coated poly(ethylcyanoacrylate) nanoparticles,[4] triclosan	1,4,11,18
preservatives for pharmaceutical formulations	methyl parahydroxybenzoate	6
shampoo	2-bromo-2-nitropropane-1,3-diol, DMDM hydantoin, iodopropynyl butylcarbamate	8,9
shower gel	triclosan	1
skin antisepts	alcohols, chlorhexidine gluconate, iodine/iodophors, parachlorometaxylenol, quaternary ammonium compounds, triclosan, zinc pyrithione	
skin care	DMDM hydantoin	9
toothpaste	triclosan	1
wound dressing	silver nanoparticles	3

The ideal preservative should have:[11]

- broad antimicrobial activity
- effectiveness over a wide range of pH values
- effectiveness for the entire shelf life of the product
- antimicrobial activity in both oil and water vehicles
- compatibility with other ingredients
- nontoxic and nonirritating qualities
- low cost

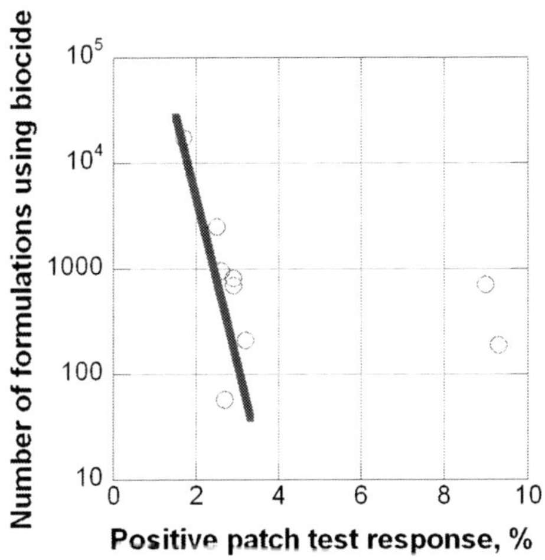

Figure 4.43. Frequency of biocide use vs. frequency of positive allergy response. [Data from Marks, J G; Elsner, P; Deleo, V A, Preservatives and Vehicles in **Contact & Occupational Dermatology**, 3rd Ed., *Elsevier*, 2002, pp 140-71.

The data presented in Figure 4.43 shows the extent to which "nontoxic and nonirritating qualities" of biocide determine its application. The data for the number of formulations come from United States Food and Drug Administration and the data on positive patch test response (allergic properties) come from a large number of tests conducted in 1996-1998. It is clear from these data that biocide that causes allergies will not survive in the market place. Parabens are least allergic for the majority of people and they dominate the market with 17,240 formulations. The next least allergic biocide (imidazolidinyl urea) is used in 2,498 formulations. The remaining biocides (DMDM hydantoin, diazolidinyl urea, methyldibromo glutaronitrile, and 2-bromo-2-nitropropane-1,3-diol) are seldom used (210-955 formulations). The two points (right side of the graph) which do not belong to a group of related results are for Quternium-15 and formaldehyde, which are no longer used in cosmetics.

Alcohol provides fast-acting, broad-spectrum antimicrobial activity.[12] However, the requirement for persistence in topical antiseptics cannot be achieved with alcohol alone.[12] Chlorhexidine gluconate, quaternary ammonium compounds, iodine, and triclosan have persistence features.[12] Zinc pyrithione was found to have a safety profile and/or antimicrobial efficacy that exceeds iodine, chlorhexidine gluconate, and triclosan.[12] It is suggested for use in skin antisepts, which are used by professional health care providers.

Triclosan acts on a defined bacterial target in the bacterial fatty acid biosynthetic pathway, NADH-dependent enoyl-[acyl carrier protein] reductase (FabI), or its homolog InhA.[1] It acts as a site-directed, picomolar inhibitor of enoyl-[acyl carrier protein] reductase by mimicking its natural substrate.[1]

Non-drug-loaded nanoparticles are being synthesized.[4] They perform not by biocidal action but by preventing adhesion of microorganisms. By adhering, microorganisms evade protective flashing actions of body fluids (e.g., saliva) or specially designed preparations (e.g., mouthwash, discussed in this source).[4]

REFERENCES

1 Schweizer, H P, *FEMS Microbiol. Lett.*, **202**, 1-7, 2001.
2 Caillier, L; Taffin de Givenchy, E; Levy, R; Vandenberghe, Y; Geribaldi, S; Guittard, F, *Eur. J. Med. Chem.*, **44**, 3201-08, 2009.
3 Jain, J; Arora, S; Rajwade, J; Khandelwal, S; Paknikar, K M, *Mol. Pharmaceutics*, Just Accepted Manuscript 10.1021/mp900056g.
4 McCarron, P A; Donnelly, R F: Marouf, W; Calvert, D E, *Intl. J. Pharm.*, **340**, 182-90, 2007.
5 Brunetto, P S; Fromm, K M, *Chimia*, **62**, 249-52, 2008.
6 Kramer, M; Suklje-Debeljak, H; Kmetec, V, *Drug Development Ind. Pharm.*, **34**, 547-57, 2008.

7 Flores, M; Morillo, M; Crespo, M L, *Intl. Biodet. Biodeg.*, **40**, 2-4, 157-60, 1997.
8 Castanedo-Tardan, M P; Zug, K A, *Dermatol. Clin.*, **27**, 265-80, 2009.
9 Biebl, K A; Warshaw, E M, *Dermatol. Clin.*, **24**, 215-32, 2006.
10 Maier, L E; Lampel, H P; Bhutani, T; Jacob, S E, *Dermatol. Clin.*, **27**, 251-64, 2009.
11 Marks, J G; Elsner, P; Deleo, V A, Preservatives and Vehicles in **Contact & Occupational Dermatology**, 3rd Ed., *Elsevier*, 2002, pp 140-71.
12 Guthery, E; Seal, L A; Anderson, E L, *Am. J. Infection Control*, **33**, 1, 15-22, 2005.
13 Codling, C E; Hann, A C; Maillard, J-E; Russell, A D, *Contact Lenses Anterior Eye*, **28**, 163-68, 2005.
14 Lucas, A D; Gordon, E A; Stratmeyer, M E, *Talanta*, **80**, 1016-19, 2009.
15 Santos, L; Rodrigues, D; Lira, M; Real Oliveira, M E S D; Oliveira, R; Yebra-Pimentel Vilar, E; Azeredo, J, *Contact Lenses Anterior Eye*, **31**, 126-30, 2008.
16 Draelos, Z D, *Clinics Dermatol.*, **19**, 4, 424-30, 2001.
17 Soni M G; Taylor, S L; Greenberg, N A; Burdock, G A, *Food Chem. Toxic.*, **40**, 1335-73, 2002.
18 Karsa, D R, Biocides in **Handbook for Cleaning/Decontamination of Surfaces**, *Elsevier*, 2007, pp 593-623.

4.14 POLYMERS

4.14.1 ACRYLICS

4.14.1.1 Relevant microorganisms

Table 4.75 contains a list of microorganisms found to cause deterioration of acrylics.

Table 4.75. Microorganisms in acrylics

Material	Microorganisms	Refs.
consolidants	*Agrobacterium, Alternaria sp., Bacillus, Brevibacillus, Cladosporium sp., Kocuria, Micrococcus, Ochrobacter, Pseudomonas, Staphylococcus*	6
dentures	*Candida albicans*	2
glass fiber composites	*Aspergillus niger, Penicillium spp., Pseudomonas aeruginosa, Pseudomonas fluorescens, Trichoderma harzianum, Trichoderma koningii*	5
paints	*Alternaria, Alternaria dianthicola, Aspergillus foetidus, Aspergillus niger, Aspergillus oryzae, Aureobasidium, Aureobasidium pullulans, Bacillariophyceae, Candida albicans, Chaetomium globosum, Chlorophyceae, Cladosporium, Cladosporium resinae, Coniophora puteana, Curvularia, Dinophyceae, Epicoccum, Gliocladium fimbriatum, Helminthosporium, Lentinus lepideus, Lenzites trabea (Gleophylum trabeum), Monascus, Mucor rouxii, Nigrospora, Penicillium funiculosum, Penicillium variabile (glaucum), Pestalotia, Pestalotiopsis, Phoma glomerata, Phoma herbarum (pigmentivora), Pullularia (Aureobasidium) pullulans, Rhizpus stolonifer, Rhodophyceae, Rhototorula rubra, Saccharomyces cerevisiae, Trichophyton mentagrophytes, Tripospermum, Xanthophyceae*	1,8
sealants	*Alternaria sp., Aspergillus fumigatis, Aspergillus niger, Aspergillus oryzae, Aspergillus versicolor, Aureobasidium pullulans, Bacillus subtilis, Chaetomium globosum, Enterobacter aerogene, Fusarium oxysporum, Myrothecium verrucaria, Penicillium citrinum, Penicillium crisogenum, Proteus vulgarus, Pseudomonas aeruginosa, Salmonella typhimurium, Stachybotrys chartarum, Staphylococcus aureus, Streptococcus faecalis*	7

4.14.1.2 Essential product components

Glass fibers in the composite material may promote fungal colonization by serving as capillaries for transporting nutrients from susceptible regions or external surfaces, stimulating extensive microbial invasion and colonization.[5]

4.14.1.3 Mechanisms of biodegradation and biodeterioration

Trichoderma's genome has an astonishing variety of secreted lytic enzymes, including cellulases, endochitinase, N-acetyl-beta-l-glucosaminidases, chitin-1,4-beta-chitobiosidases, proteases, endo and exoglucan-beta-1,3-glucosidases, endoglucan beta-1-6-glucosidases, lipases, xylanases, mannanases, pectinases, pectin lyases, amylases, phospholipases, RNases, and DNases, which can be used to digest polymeric materials.[5] In the reported case, esterase activity led to almost complete consumption of acrylic resin from filter.[5]

4.14.1.4 Results of biodeterioration

The microbial colonization of painted buildings causes aesthetic problems and can lead to degradation and spalling of the coating.[1]

ASHRAE filters are commonly used in many applications, such as residential equipment and industrial heating, ventilating, and air conditioning systems to guarantee a high level of indoor air quality. Filters are composed of glass fiber and acrylics. *Trichoderma* strains have a strong ability to colonize the glass fiber filters and penetrate progressively deeper into the substrate, leading to an effective and measurable deterioration of the composite materials, including complete decomposition of acrylic resin.[5]

4.14.1.5 Biostabilization

Table 4.76 contains a list of the most suitable biocides used in protection of acrylics.

Table 4.76. Biocides used in acrylics

Material	Biocide	Refs.
copolymers	acrylic oligomers and 2,4,40-trichloro-20-hydroxydiphenyl ether copolymer,[4] 2,4-dichlorophenyl methacrylate-vinyl acetate copolymer	3,4
dentures	apatite-coated TiO_2 photocatalyst[2]	2
paints	1,2-benzisothiazolin-3-one (in-can preservative), carbamate (carbendazin),[1] 5-chloro-2-methyl-4-isothiazolin-3-one (in-can), 2-methyl-4-isothiazolin-3-one (in-can), N-(3,4-dichlorophenyl)-N,N-dimethyl urea,[1] N-octyl-2H-isothiazolin-3-one,[1] 2-n-octyl-4-isothiazolin-3-one, sodium orthophenylphenate	1,8
sealants	1,2-benzisothiazolin-3-one (in-can preservative), diiodomethyl-p-tolylsulfone, 3-iodo-2-propynylbutylcarbamate, 2-n-octyl-4-isothiazolin-3-one	7

Prevention of biofilm formation on acrylic dentures is important for oral hygiene. It was found that use of apatite-coated titanium dioxide as photocatalyst kills microorganisms on exposure of dentures to UV.[2] Apatite coating helps to absorb bacteria and viruses on the coating surface and titanium dioxide can then kill them on exposure because microorganisms are in close contact with photocatalyst.[2] Large additions of catalyst lowers mechanical performance of denture materials but use of 5 wt% does not affect resin properties and performs catalytic functions.

2,4-dichlorophenyl methacrylate-vinyl acetate copolymer was found to have biocidal action against bacteria (*Bacillus subtilis, Escherichia coli, Staphylococcus citreus*), fungi (*Aspergillus niger, Sporotichum pulverulentum, Trichoderma lignorum*), and yeast (*Candida utilis, Saccharomyces cerevisiae, Pichia stipitis*).[3] The crosslinking by radical UV photopolymerization of mixtures of acrylic oligomers with 2,4,4'-trichloro-2'-hydroxydiphenyl ether allows to obtain polymer films which perform like hygienic coatings.[4]

4.14.1.6 Formulations

Semigloss acrylic paint

grind/pre-mix		vol%
	Water	8.50
	TAMOL 165A	0.67
	Propylene glycol	1.00
	Universal Grade TiO_2 Slurry	17.50
	KATHON LX 1.5%	0.21
letdown		
	RHOPLEX VSR-1050LOE	55.25
	Propylene glycol	0.25
	Texanol	0.60
	BYK-348	0.22
	Foamstar A-34	0.25
	Ammonia (28%)	0.05
	ACRYSOL RM-2020 NPR	3.50
	ACRYSOL SCT-275	0.50
	Water	11.50

Acrylic sealant (low modulus)[10]

PRIMAL 3362 (62%)	321.41
ACRYSOL TT-615 (30%)	1.99
TERGITOL 15-S-40 (70%)	7.97
Ethylene Glycol	9.96
OROTAN 850	1.06
KTPP	1.06
Water	4.38
Jayflex DIUP	109.60
DURCAL 2	7.97
DURCAL 5	526.09
SILANE Z-6040	7.97

References

1 Shirakawa, M A; Gaylarde, C C; Gaylarde, P M; John, V; Gambale, W, *FEMS Microbiol. Ecol.*, **39**, 165-73, 2002.

2 Shibata, T; Hamada, N; Kimoto, K; Sawada, T; Sawada, T; Kumada, H; Umemoto, T; Toyoda, M, *Dental Mater. J.*, **26**, 3, 437-44, 2007.

3 Patel, M B; Patel, S A; Ray, A; Patel, R M, *J. Appl. Polym. Sci.*, **89**, 895-900, 2003.
4 Gozzelino, G; Dell'Aquila, G A; Tobar D R, *J. Appl. Polym. Sci.*, **112**, 2334–2342, 2009.
5 Sein, T T; Spurio, R; Cecchini, C; Cresci, A, *Intl. Biodet. Biodeg.*, **63**, 901-5, 2009.
6 Cappitelli, F; Principi, P; Pedrazzani, R; Toniolo, L; Sorlini, C, *Sci. Total Environ.*, **385**, 172-81, 2007.
7 AMICAL™ Preservatives, Product information, DOW.
8 Kathon LX. Technical data sheet, Rohm and Haas (DOW).
9 Rhoplex VSR-1050LOE. Technical data sheet, Rohm and Haas.
10 Primal 3362. Acrylic emulsion for latex sealants. Rohm and Haas.

4.14.2 ACRYLONITRILE-BUTADIENE-STYRENE COPOLYMER

4.14.2.1 Relevant microorganisms

Table 4.77 contains a list of microorganisms found to cause deterioration of ABS.

Table 4.77. Organisms in ABS

Material	Microorganisms	Refs.
pipe	*Martesia striata*	1

4.14.2.2 Results of biodeterioration

ABS pipes were considered to be immune to boring organisms, until the boring bivalve *Martesia striata* was found to damage ABS pipes in a Malaysian port.[1] The holes had sizes from 3 to 19 mm^2 and they were made without any assistance of a chemical process (secreting enzymes or corrosive materials).

4.14.2.3 Biostabilization

Table 4.78 contains a list of the most suitable biocides used for protection of ABS products.

Table 4.78. Biocides used in ABS

Material	Biocide	Refs.
bathroom fixtures	Microban	
health care applications	nanosilver	2

References

1 Jennera, H A; Rajagopal, S; Van der Velde, G; Daud, M S, *Intl. Biodet. Biodeg.*, **52**, 229-32, 2003.
2 Markarian, J, *Plast. Additives. Compounding*, Jan/Feb, 18-22, 2009.

4.14.3 ALKYD RESINS

4.14.3.1 Relevant microorganisms

Table 4.79 contains a list of microorganisms found to cause deterioration of alkyd resins.

Table 4.79. Microorganisms in alkyd resins

Material	Microorganisms	Refs.
paint	*Alternaria tenuis, Aspergillus niger, Aureobasidium pullulans, Chaetomiurn globosum, Coniophora puetana, Escherichia coli, Lentinus tigrinus, Penicillium glaucum, Polyporus versicolor, Pseudomonas aeruginosa, Sclerophoma pityophila, Staphylococcus aureus, Trichoderma viride*	1,3

4.14.3.2 Essential product components

BASF has succeeded in synthesizing a binder based on emulsion polymers and water soluble alkyds which can produce the desired degree of gloss with an aqueous formulation: AQAGloss®.[5] Innovative hybrid binder COL.9®, in which nanoscale inorganic domains are permanently integrated in the organic polymer, allows paints to be formulated with a superhydrophilic, hard surface.[5] Dirt particles find it almost impossible to adhere to these surfaces and are easily washed away by dew or rainwater.[5] Since the superhydrophilic surface prevents droplets forming, the typical streak marks which detract from the appearance of buildings are prevented.[5] And the rapid drying reduces the time in which microorganisms find an optimal growth climate.[5]

4.14.3.3 Mechanisms of biodegradation and biodeterioration

Esterase action is responsible for the microbial degradation of alkyd resins.[2]

4.14.3.4 Biostabilization

Table 4.80 contains a list of the most suitable biocides used for protection of alkyd resins.

Table 4.80. Biocides used in alkyd resins.

Material	Biocide	Refs.
paint	azole+iodopropargyl butylcarbamate,[3] octylisothiazolinone, silver nanoparticles[1]	1,3

A particulate carrier material is impregnated with a biocidal formulation and serves as a vehicle for introduction of the biocide into a liquid-based media, such as a surface coating or surface cleaning compositions, in order to allow controlled release of the biocide to combat bacterial, fungal, algal or like growth for an extended period of time.[6] Amorphous silica, amorphous alumina, and zeolites are typical particulate carriers and isothiazolones are typical biocides used.[6]

4.14.3.5 Formulations

Alkyd resin formulation which does not require biocide[4]

	vol%
Long or medium oil resin	54.76
White spirits	20.24
Rheological additive (Bentone 1000)	0.41
Wetting agent (Anti-Terra U)	0.38
10% calcium octoate	0.73
Titanium dioxide	6.73
12% cobalt octoate	0.21
18% zirconium octoate	0.80
Anti-skinning agent (MEKO)	0.36
AMP (2-amino-2-methyl-1-propanol solution)	0.12
Water	15.26

References

1 Kumar, A; Vemula, P K; Ajayan, P M; John, G, *Nature Mater.*, 7, 236-41, 2008.
2 Morton, L H G;. Surman, S B, *Intl. Biodet. Biodeg.*, 203-21, 1994.

3 Uhr, H; Jaetsch, T; Kaulen, J; Spetmann, **US Patent Application** 20090192219, Lanxess Corp.
4 AMP. Angus technical bull.
5 Schuler, B, Innovative polymer technology is the key to modern high performance paints. BASF presentation during conference „Journalists and Scientists in Dialogue –Chemistry-based innovations for the construction Industry" on October 18/19, 2007 in Ludwigshafen.
6 Aldcroft, D; Jones, H; Turner, D; Edge, M; Robinson, J; Seal, K, **US Patent**, 6,905,698, Ineos Silicas Ltd., June 14, 2005.

4.14.4 EPOXY RESIN

4.14.4.1 Relevant microorganisms

Table 4.81 contains a list of microorganisms found to cause deterioration of epoxy resins.

Table 4.81. Microorganisms in epoxy resins.

Material	Microorganisms	Refs.
coatings	*Escherichia coli, Listeria monocytogenes, Pseudomonas aeruginosa, Staphylococcus aureus*	1
consolidants	*Alternaria sp., Aspergillus niger, Aureobasidium pullulans, Chaetomiumglobosum, Cladosporium spp., Gliocladium virens, Penicillium pinophilum*	3

4.14.4.2 Biostabilization

Table 4.82 contains a list of the most suitable biocides used for protection of epoxy resins.

Table 4.82. Biocides used in epoxy resins.

Material	Biocide	Refs.
coatings	silver-containing zeolite (AgIon)	1
marine coatings	coal tar,[4] ferric benzoate,[5] Sea-Nine 211N (30% solution of 4,5-dichloro-2-n-octyl-4-isothiazolin-3-one)	4,5

Silver- and zinc-containing zeolite matrix applied to stainless steel greatly enhances the antimicrobial properties of the stainless steel.[1]

The crosslinking by radical UV photopolymerization of acrylic oligomers added with 2,4,4'-trichloro-2'-hydroxydiphenyl ether permits to obtain polymer films which perform like hygienic coatings.[2] Even at low concentrations of 2,4,4'-trichloro-2'-hydroxydiphenyl ether (0.1%), epoxy coating had antibacterial activity.[2]

Maximum performance of coal tar epoxy coatings is obtained by applying them in two coats over near-white blast cleaned surfaces.[4] Typical coating thickness is 200 μm DFT per coat for a total DFT (dry film thickness) of 400 μm or more.[4] The expected service life of these coatings is seven years in seawater immersion and eight years in atmospheric marine exposure.[4] A three-coat epoxy (without coal tar) system applied over a near-white blast cleaned surface has the expected service life in saltwater immersion of six years, while the service life in the marine atmosphere is eight years.[4]

References

1 Cowan, M M; Abshire, K Z; Houk, S L; Evans, S M, *J. Ind. Microbiol. Biotechnol.*, **30**, 102-6, 2003.
2 Gozzelino, G; Dell'Aquila, G A; Tobar, D R, *J. Appl. Polym. Sci.*, **112**, 2334-42, 2009.
3 Cappitelli, F; Principi, P; Pedrazzani, R; Toniolo, L; Sorlini, C, *Sci. Total Environ.*, **385**, 172-81, 2007.
4 Bleile, H; Rodgers, S D in Marine Coatings, **Encyclopedia of Materials: Science and Technology**,

Elsevier, 2008, pp 5174-85.
5 del Amoa, B; Blusteina, G; Perez, M; Garcia, M; Deya, M; Stupak, M; Romagnoli, R, *Colloids Surf. A: Physicochem. Eng. Aspects*, **324**, 58-64, 2008.

4.14.5 ETHYLENE PROPYLENE COPOLYMERS

4.14.5.1 Relevant microorganisms

Table 4.83 contains a list of microorganisms found to cause deterioration of ethylene propylene copolymers.

Table 4.83. Microorganisms in ethylene propylene copolymers.

Material	Microorganisms	Refs.
EPDM	*Bacillus thuringiensis, Pantoea agglomerans, Pediococcus inopinatus, Pseudomonas fragi*	1

4.14.5.2 Biostabilization

Table 4.84 contains a list of the most suitable biocides used for protection of ethylene propylene copolymers.

Table 4.84. Biocides used in ethylene propylene copolymers.

Material	Biocide	Refs.
air diffuser membrane	carbolic acid	2

References

1 Storgards, E; Simola, H; Sjoberg, A-M; Wirtanen, G, *Trans. IChemE*, **77C**, 137-45, 1999.
2 Tharp, C E, **US Patent**, 6,543,753, Enviromental Dynamics Inc., April 8, 2003.

4.14.6 IONOMER

4.14.6.1 Biostabilization

Table 4.85 contains a list of the most suitable biocides used for protection of ionomers.

Table 4.85. Biocides used in ionomers

Material	Biocide	Refs.
film	benzoyl chloride, benzimidazole	1
marine coating	zinc-containing ionomer	2

Ionomer films have been used as food packaging materials because of their unique properties, including a high degree of transparency, strength, resilience, stiffness, toughness, and inertness to organic solvents and oils.[1] It was demonstrated that the incorporation of a fungicide (benomyl) onto the carboxylic groups of ionomer films gives good antimicrobial performance.[1] However, benomyl is not a legal food preservative and it is not allowed to be used in foods.[1] Benzoic acid and its salts are among the most commonly used food preservatives.[1] The incorporation of benzoic acid with an ionomer as an antimicrobial packaging film might provide several advantages.[1]

Antifouling paint has a unique zinc polymer employing the ionomer technology. Antifouling effect is due to hydrolysis of the polymer.[2] The precise control of the hydrolysis rate provides a long-term antifouling protection.[2]

References

1 Matche, R S; Kulkarni, G; Raj, B, *J. Appl. Polym. Sci.*, **100**, 3063-68, 2006.
2 Sea Granprix CF-10. Chiugoku Samhwa Paints, Ltd.

4.14.7 PHENOLIC RESINS

4.14.7.1 Relevant microorganisms

Table 4.86 contains a list of microorganisms found to cause deterioration of phenolic resins.

Table 4.86. Microorganisms in phenolic resins

Material	Microorganisms	Refs.
ceiling tiles	*Aspergillus niger, Penicillium aragonense, Penicillium chrysogenum, Penicillium glabrum*	2
composites	*Gloeophyllum trabeum, Trametes versicolor*	1
fiberboard	*Gloeophyllum trabeum, Trametes versicolor*	3

4.14.7.2 Essential product components

The fiber/matrix interface is the key area for capillary distribution of fungal metabolites (and nutrients) that may play an important role in deterioration of glass fiber/phenolic resin composites.[1]

4.14.7.3 Mechanisms of biodegradation and biodeterioration

If a ceiling tile became wet due to a leak, spill, or flood, fungal growth may be avoided by quickly and thoroughly drying the tile and keeping it dry, considering the fungal growth does not occur without moisture presence.[2]

4.14.7.4 Results of biodeterioration

No detectable weight loss of composite material, reinforced with E-glass fiber, was recorded, but there was a significant increase in the moisture content of both white and brown rot fungi exposed samples.[1]

4.14.7.5 Biostabilization

4,5-dichloro-2-n-octyl-3(2H)-isothiazolone has been encapsulated in phenolic resin to obtain a controlled-release product.[4]

References

1 Tascioglu, C; Goodell; B; Lopez-Anido, R; Peterson, M; Halteman, W; Jellison J, *Intl. Biodet. Biodeg.*, **51**, 157-65, 2003.
2 Foarde, K K; Vanosdell, D W, *Atmospheric Environ.*, **29**, 17, 2331-37, 1995.
3 Rowell, R M in **Encyclopedia of Forest Sciences**, *Elsevier*, 2004, pp 1269-74.
4 Hart, R L; Virgallito, D R; Work, D E, **US Patent** 7,550,200, Microtek Laboratories Inc., June 23, 2009.

4.14.8 POLYAMIDES

4.14.8.1 Relevant microorganisms

Table 4.87 contains a list of microorganisms found to cause deterioration of polyamides.

Table 4.87. Microorganisms in polyamides.

Material	Microorganisms	Refs.
polyamide	*Acremonium polychromum, Acremonium rutilum, Acremonium strictum, Alternaria alternata, Aspergillus amstelodami, Aspergillus niger, Aspergillus ruber, Aspergillus ustus, Aspergillus versicolor, Aureobasidium microstrictum, Aureobasidium pullulans, Chaetomium globosum, Cladosporium cladosporioides, Cladosporium herbarum, Cladosporium macrocarpum, Cladosporium sphaerospermum, Mucor racemosus, Mucor silvaticus, Oidiodendron griseum, Oidiodendron tenuissimum, Paecilomyces lilacinus, Penicillium chrysogenum, Penicillium funiculosum, Penicillium ochrochloran, Penicillium paxilli, Phoma cava, Scopulariopsis chartarum, Trichoderma koningii, Trichoderma viride, Ulocladium oudemansii*	4
polyamide-6	*Achromobacter, Aspergillus niger, Bjerkandera adusta, Corynebacterium aurantiacum, Phanerochaete chrysosporium*	1,2,5

4.14.8.2 Essential product components

Aerobic bacteria were able to grow on 6-aminocaproic acid polyamides of low molecular weight (1400, 2100, 6800) as the only carbon and energy source.[1] No growth was observed on polyamide of a molecular weight of 11,000.[1]

Polyamides have high abilities to absorb water and this is one of the most important reasons for their participation in biodegradative processes. If polyamide fiber is used in mixture with other fibers, it facilitates its biodegradation because of its ability to retain water.

The introduction of substituents such as benzyl, hydroxy and methyl greatly increases the biodegradation.[6] The higher crystallinity of polyamides due to strong interchain interactions (as compared with the more flexible polyesters with analogous structures) is behind the observed lower rates of biodegradation.[6] By comparison, the natural proteins seldom contain repeating units.[6] They have a lower tendency to pack into highly ordered morphologies.[6] For this reason, they are accessible to enzymatic attacks. But synthetic polyamides have short and regular repeating units.[6] Their higher symmetries and strong interchain hydrogen bonding result in highly ordered crystalline morphologies, which, in turn, limits the accessibility to enzyme attack.[6]

4.14.8.3 Mechanisms of biodegradation and biodeterioration

Papain, trypsin, esterase, and chymotrypsin were found to cause damage to implants based on polyamides.[3]

4.14.8.4 Results of biodeterioration

Degradation of polyamide-6 by *Phanerochaete chrysosporium* caused the decrease in the average molecular mass (50% after 3 months) and physical damage of fibers (see Figure 4.44).[2] After five months of exposure to the fungus, the damage to the fibres caused suffi-

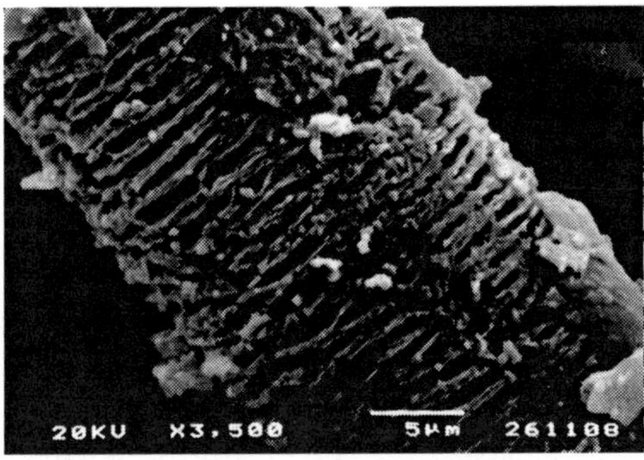

Figure 4.44. SEM micrograph of polyamide fiber biodegraded by *Phanerochaete chrysosporium* for 3 month. [Adapted, by permission, from Klun, U; Friedrich, J; Krzan, A, *Polym. Deg. Stab.*, **79**, 99-104, 2003.]

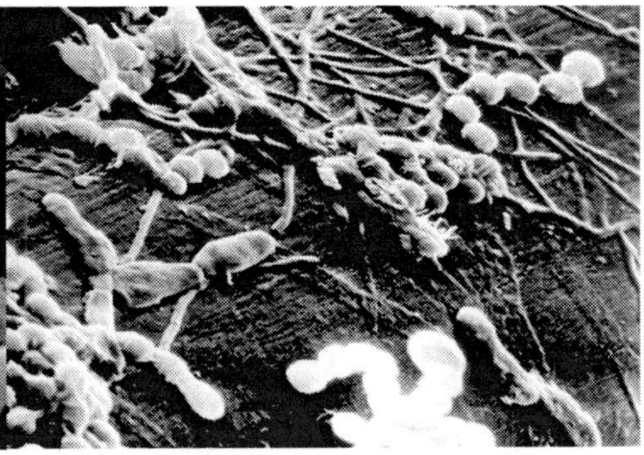

Figure 4.45. SEM micrograph of Kapron exposed to fungal deterioration for 3 years. [Adapted, by permission, from Lugauskas, A; Levinskaite, L; Peciulyte, D, *Intl. Biodet. Biodeg.*, **52**, 233-42, 2003.]

cient loss of physical strength to make them readily fragmented.[2] The enthalpies of fusion decreased in the biodegraded samples, indicating a lowering of the degree of crystallinity with further degradation.[2]

Some materials such as Kapron are colonized but do not seem to undergo changes caused by colonizing microorganisms (Figure 4.45). Aramid fiber was reported to be attacked by *Aspergillus* fungi.[6]

4.14.8.5 Biostabilization

Table 4.88 contains a list of the most suitable biocides used for protection of polyamides.

Table 4.88. Biocides used in polyamides.

Material	Biocide	Refs.
fiber	thin silver layers[7]	7
sea water reverse osmosis membrane	2,2 dibromo-3-nitrilo-propionamide,[13] chloramines, free chlorine (affects membrane by reacting with amide groups), ozone, silver nanoparticles[9]	8,9,13

Tollens' reaction has been used to apply thin silver layers onto the surface of polyamide fibers.[7] This antimicrobial finish can be used in clothing (underwear or sportwear), conveyors (transportation of food products), architectural textiles.[7]

The polyamides were synthesized by the polycondensation of diethyl-l-tartrate or chloromethylated diethyl-l-tartrate with ethylenediamine and modified to yield polymers with either quaternary ammonium or phosphonium salts.[10] They had antibacterial activity but did not show antifungal performance.[10]

Chlorination of poly(m-phenyleneisophthalamide) (Nomex) and poly(p-phenyleneterephthalamide) (Kevlar) fabrics by sodium hypochlorite aqueous solution was investigated.[11] Results indicated that stable N-halamine structures could be readily formed in Nomex, which provided potent, durable, and refreshable antibacterial functions against both Gram-negative and Gram-positive bacteria.[11] However, similar chlorination treatment of Kevlar led to serious hydrolysis of the polymer.[11]

Antimicrobial properties of polyamide-6 fabrics and increased washing durability of biofunctions were developed by grafting the fabrics with methacrylic acid to create additional carboxylic groups in polyamide-6 macromolecules, followed by subsequent reaction with dimethylalklbenzyl ammonium chloride.[12] Fabric had a strong biocidal effect on *Bacillus mycoides, Escherichia coli,* and *Candida albicans.*

2,2 dibromo-3-nitrilo-propionamide is a fast-acting, non-oxidizing biocide which is very effective at low concentrations.[13] Microorganisms that come into contact with it are rapidly killed by a mechanism that appears to involve a reaction with the protein fraction of the cell membrane and inactivation of enzyme systems.[13] The vast majority of microorganisms are killed within 5 to 10 minutes.[13] In aquatic environments, the biocide decomposes rapidly to carbon dioxide, ammonia, and bromide, and is therefore environmentally safe.[13] It can be applied either as a continuous low dosage, as an intermittent dosage while the reverse osmosis system is in operation (shock dosage, slug dose), or offline in clean-in-place mode.[13] The most frequently used mode is the intermittent dosage.[13]

References

1 Andreonif, V; Baggi, G; Guaita, C; Manfrin, P, *Intl. Biodet. Biodeg.*, **31**, 41-53, 1993.
2 Klun, U; Friedrich, J; Krzan, A, *Polym. Deg. Stab.*, **79**, 99-104, 2003.
3 Williams, D F; Zhong, S P, *Intl. Biodet. Biodeg.*, 95-130, 1994.
4 Lugauskas, A; Levinskaite, L; Peciulyte, D, *Intl. Biodet. Biodeg.*, **52**, 233-42, 2003.
5 Friedrich, J; Zalar, P; Mohorcic, M; Klun, U; Krzan, A, *Chemosphere*, **67**, 2089-95, 2007.
6 Chandra, R; Rustgi, R, *Prog. Polym. Sci.*, **23**, 1273-35, 1998.
7 Textor, T; Fouda, M M G; Mahltig, B, *Appl. Surf. Sci.*, just approved, 2009.
8 Kim, D; Jung, S; Sohn, J; Kim, H; Lee, S, *Desalination*, **238**, 43-52, 2009.
9 Yang, H-L; Lin, J C-T; Huang, C, *Water Res.*, **43**, 3777-86, 2009.
10 Kenawy, E; Abdel-Hay, F I; Shahada, L; El-Shanshoury, A E R; El-Newehy, M H, I, **102**, 4780-90, 2006.
11 Sun, Y; Sun, G, *Ind. Eng. Chem. Res.*, **43**, 5015-20, 2004.
12 Shalaby, S E; Al-Balakocy, N G; Abdel-Fatah, O M; Elshafei, A M, *J. Appl. Polym. Sci.*, **110**, 738-46, 2008.
13 Filmtec Membranes. DOW Form No. 609-00423-0705.

4.14.9 POLYCARBONATE

4.14.9.1 Relevant microorganisms

Table 4.89 contains a list of microorganisms found to cause deterioration of polycarbonate.

Table 4.89. Microorganisms in polycarbonate.

Material	Microorganisms	Refs.
bisphenol A	*Sphingomonas bisphenolicum*	1
compact disks (CD)	*Geotrichum*	2
phenol	*Ochrobactrum sp.*	7
polycarbonate	*Arthrobacter sp., Enterobacter sp.*	4

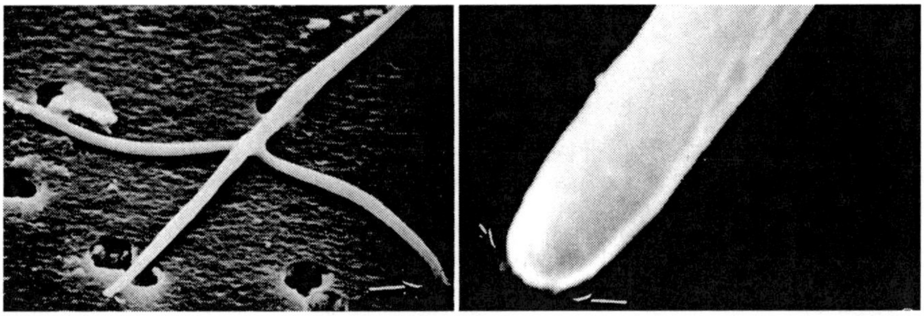

Figure 4.46. *Alatospora acuminata*. Conidia attached to the polycarbonate surface by secreted mucilage (arrowed) at the tip of the conidial arms. [Adapted, by permission, from Jones, E B G, *Mycol. Res.*, **98**, 9, 961-81, 1994.]

4.14.9.2 Essential product components

Polycarbonate is an amorphous polymer which has very low moisture absorption. Low moisture absorption reduces risk of biodegradation. Plasma treatment in the presence of CO_2 changes the character of the polycarbonate surface from hydrophobic to hydrophilic, which renders it to become biodegradable.[5]

4.14.9.3 Mechanisms of biodegradation and biodeterioration

Arthrobacter sp. and *Enterobacter sp.* have the ability to produce esterase needed for degradation of polycarbonates.[4]

4.14.9.4 Results of biodeterioration

Geotrichum was able to grow on a CD under tropical conditions (about 30°C and 90% RH).[2] It caused a fractal structure and the destruction of the pits, followed by the loss of the information stored on the disk.[2]

Figure 4.46 illustrates the tips of *Alatospora acuminata* attached to a polycarbonate membrane by mucilage, but there is no sign of the spore germination.[6] Zoosporic fungi have been shown to attach by the release of fibrilar or electron-opaque adhesive materials.[6]

4.14.9.5 Biostabilization

Table 4.90 contains a list of the most suitable biocides used for protection of polycarbonate.

Table 4.90. Biocides used in polycarbonate.

Material	Biocide	Refs.
medical devices	silver-based	8

References

1 Oshiman, K; Tsutsumi, Y; Nishida, T; Matsumura, Y, *Biodeg.*, **18**, 247-55, 2007.
2 Ranalli, G; Zanardini, E; Sorlini, D in Biodeterioration - including cultural heritage, **Encyclopedia of Microbiology**, *Elsevier*, 2009, pp 191-205.
3 Martina, M; Hutmacher, D W, *Polym. Int.*, **56**, 145-57, 2007.
4 Goela, R; Zaidi, M G H; Soni, R; Lata, K; Shouche, Y S, *Intl. Biodet. Biodeg.*, **61**, 167-72, 2008.
5 Kim, M C; Masuoka, T, *Appl. Surf. Sci.*, **255**, 4684-88, 2009.

6 Jones, E B G, *Mycol. Res.*, **98**, 9, 961-81, 1994.
7 Kilic, N K, *Intl. Biodet. Biodeg.*, **63**, 778-81, 2009.
8 Markarian, J, *Plast. Additives Compounding*, Jan/Feb, 18-22, 2009.

4.14.10 POLYETHERETHERKETONE

4.14.10.1 Relevant microorganisms

Table 4.91 contains a list of microorganisms found to cause deterioration of polyetheretherketone.

Table 4.91. Microorganisms in polyetheretherketone.

Material	Microorganisms	Refs.
polyetheretherketone	*Aspergillus versicolor, Cladosporium cladosporioides, Chaetomium spp.*	1

4.14.10.2 Biostabilization

Table 4.92 contains a list of the most suitable biocides used for protection of polyetheretherketone.

Table 4.92. Biocides used in polyetheretherketone.

Material	Biocide	Refs.
impregnation of graphite fibers	72% sodium o-phenylphenate	2

4.14.10.3 Formulations

Slurry for impregnation of graphite fibers

	Parts
Polyox SR 301	6
Distilled water	594
Fluorad FC 430	1
Dowcide A	0.01
Propylene glycol	0.05
Antifoam A	0.01
Triton X-400	0.05
Polyetheretherketone powder 150-300 mesh	10-100

References

1 Gu, J-D; Ford, T; Thorp, K; Mitchell, R, *Intl. Biodet. Biodeg.*, 197-204, 1996.
2 Reinhart, T J, **US Patent** 5,236,972, US Airforce, August 17, 1993.

4.14.11 POLYETHYLENE

4.14.11.1 Relevant microorganisms

Table 4.93 contains a list of microorganisms found to cause deterioration of polyethylene.

Table 4.93. Microorganisms in polyethylene.

Material	Microorganisms	Refs.
bags for oranges	*Botrytis cinerea, Botrydiplodia lecanidion, Penicillium digitatum, Penicillium italicum*	2

Table 4.93. Microorganisms in polyethylene.

Material	Microorganisms	Refs.
oxidized LDPE	*Acinetobacter sp., Delftia acidovorans, Delftia sp., Flavobacterium sp., Pseudomonas sp., Ralstonia sp., Rhodococcus erythropolis, Rhodococcus sp., Stenotrophomonas sp.*	7
polyethylene	*Aspergillus terreus, Aureobasidium pullulans, Brevibacillus borstelensis, Fusarium sp., Penicillium simplicissimum, Pseudomonas aeruginosa, Rhodococcus rubber*	6,9
polyethylene containing starch	*Aspergillus flavus, Mucor rouxii, Streptomyces sp.*	9
wood flour-PE film	*Aspergillus sp., Aureobasidium pullulans, Gliocladium virens, Penicillium pinophilum*	5

4.14.11.2 Essential product components

Adding a paraffin ($C_{32}H_{66}$) to an HDPE does not initially increase its degradation rate. This is because the paraffin is preferentially degraded.[6] Oxidated HDPE is degraded to about one-third of that of the HDPE without oxidant.[6]

The microorganisms capable of adhering to a preoxidized PE surface and metabolizing some of its oxidation products are not rare in soils.[7]

4.14.11.3 Mechanisms of biodegradation and biodeterioration

Microbial PE degradation is a two-step process involving an initial abiotic photooxidation, followed by a cleavage of the polymer carbon backbone.[8] However, the mechanism of the second step needs extensive analysis before plausible conclusions can be drawn confidently.[8] Lower molecular weight PEs, including paraffins, can be biodegraded and paraffin undergoes oxidative hydroxylation to form an alcohol group, followed by formation of carboxylic acid.[8]

At higher temperatures, ketones, alcohols, aldehydes, lactones, and carboxylic acids are formed abiotically in 6 weeks.[8]

4.14.11.4 Results of biodeterioration

Buried conduits are ubiquitously used for carrying various materials, such as water, natural gas, oil, and sewage.[10] A major problem with buried conduits comprised of metal or concrete with metal reinforcements is corrosion.[10] A multi-layered material was developed which contains biocide and it is used to encase buried conduits to prevent corrosion.[10]

4.14.11.5 Biostabilization

Table 4.94 contains a list of the most suitable biocides used for protection of polyethylene.

Table 4.94. Biocides used in polyethylene.

Material	Biocide	Refs.
bag for oranges	thiabendazole	2
biocide concentrate	2,4,4'-trichloro-2'-hydroxydiphenyl ether	11
food film	benzoic anhydride, nisin	3

Table 4.94. Biocides used in polyethylene.

Material	Biocide	Refs.
multilayer film	silver-polyethylene nanocomposite	4
pipe wrap	2,4,4'-trichloro-2'-hydroxy diphenyl ether	10
shrink-wrap film for pepper and cheese	1-(2-(2,4-dichlorophenyl)-2-(2-propenyloxy)ethyl)-1H-imidazole (Imazalil)[1]	1
wood flour-PE film	silver, zinc	5

References

1 Matche, R S; Kulkarni, G; Raj, B, *J. Appl. Polym. Sci.*, **100**, 3063-68, 2006.
2 Chien, P-J; Sheu, F; Lin, H-R, *Food Chem.*, **100**, 1160-64, 2007.
3 Quintavalla, S; Vicini, L, *Meat Sci.*, **62**, 373-80, 2002.
4 Sanchez-Valdes, S; Ortega-Ortiz, H; Ramos-de Valle, L F; Medellin-Rodriguez, F J; Guedea-Miranda, R, *J. Appl. Polym. Sci.*, **111**, 953-62, 2009.
5 Dawson-Andoh, B; Matuana, L, *Holz Roh. Werkst.*, **65**, 331-34, 2007.
6 Albertsson, A-C; Karlsson, S, *Intl. Biodet. Biodeg.*, **31**, 161-70, 1993.
7 Koutny, M; Amato, P; Muchova, M; Ruzicka, J; Delort, A-M, *Intl. Biodet. Biodeg.*, **63**, 354-57, 2009.
8 Gu, J-D, *Intl. Biodet. Biodeg.*, **52**, 69-91, 2003.
9 Shah, A A; Hasan, F; Hameed, A; Ahmed, S, *Biotechnol. Adv.*, **26**, 246-65, 2008.
10 Crook, J A, **US Patent** 6,488,998, Fulton Enterpises, Inc., December 3, 2002.
11 Stadler, U, **US Patent** 6,475,505, Ciba Specialty Chemical Corporation, November 5, 2002.

4.14.12 POLY(ETHYLENE OXIDE)

4.14.12.1 Relevant microorganisms

Table 4.95 contains a list of microorganisms found to cause deterioration of poly(ethylene oxide).

Table 4.95. Microorganisms in poly(ethylene oxide).

Material	Microorganisms	Refs.
poly(ethylene oxide)	*Gloeophyllum trabeum, Pseudomonas aeruginosa*	1,2,3

4.14.12.2 Essential product components

Poly(ethylene oxide) chains are highly mobile and have extremely large exclusion volumes, they make the surface difficult to approach for incoming particles such as proteins.[1] It was demonstrated that PEO-brushes on a glass surface inhibited the adhesion of bacteria and yeast by decreasing the Lifshitz-van der Waals attraction between the cells and the glass surface.[1] The PEO-brush-coated glass surface reduced bacterial adhesion with optimized chain length and density by more than 98%.[1] However, *P. aeruginosa* strains that are hydrophobic and release surfactants adhered to the PEO-brush coating due to acid-base interactions.[1]

4.14.12.3 Mechanisms of biodegradation and biodeterioration

The molecular weight strongly affects the biodegradation propensity of PEO.[2] In most cases microbial species are able to assimilate only PEO with molecular weights lower than 1,000, whereas high molecular weight (10 kD) PEO samples were shown to be substantially recalcitrant to biological attack.[2] However, significant biodegradation of samples with molecular weights up to 20 kD have also been reported.[2]

The cleavage of PEO ether links occurs through oxidative or non-oxidative pathways.[2] Three different membrane-bound enzymes were recognized as PEO-dehydrogenase, PEO-aldehyde-dehydrogenase and PEO-carboxylate-dehydrogenase, respectively.[2] It was suggested that these enzymes act sequentially to produce terminal carbonyl and carboxyl groups from the terminal units of poly(ether) chains, followed by the release of C_2 units as glioxylic acid.[2]

References

1 Hori, K; Matsumoto, S, *Biochem. Eng. J.*, just accepted doi:10.1016/j.bej.2009.11.014.
2 Chiellini, E; Corti, A; D'Antone, S; Solaro, R, *Prog. Polym. Sci.*, **28**, 963-1014, 2003.
3 Fahr, K; Wetzsteinb, H-G; Grey, R; Schlosser, D, *FEMS Microbiol. Lett.*, **175**, 127-32, 1999.

4.14.13 POLY(ETHYLENE TEREPHTHALATE)

4.14.13.1 Relevant microorganisms

Table 4.96 contains a list of microorganisms found to cause deterioration of poly(ethylene terephthalate).

Table 4.96. Microorganisms in poly(ethylene terephthalate).

Material	Microorganisms	Refs.
PET	*Aspergillus spp., Fusarium solani pisi, Pseudomonas funiculosum, Pseudomonas sp.*	2,3,4

4.14.13.2 Essential product components

Modified poly(ethylene terephthalate) contains co-monomers, such as ether, amide or aliphatic monomers that provide "weak" linkages that are susceptible to biodegradation through hydrolysis.[1] The mechanism involves a combination of hydrolysis of the ester linkages and enzymatic attack on ether and amide bonds.[1] With modified PET it is possible to adjust and control degradation rates by varying the co-monomers used.[1] Depending on the application, up to three aliphatic monomers are incorporated into the PET structure.[1]

4.14.13.3 Mechanisms of biodegradation and biodeterioration

Abiotic hydrolysis is the most important reaction for initiating the environmental degradation of synthetic polymers such as poly(ethylene terephthalate).[1]

Cutinases and carboxylesterases have both shown the potential to hydrolyze polyester bonds similarly to lipases.[3]

4.14.13.4 Biostabilization

Table 4.97 contains a list of the most suitable biocides used for protection of poly(ethylene terephthalate).

Table 4.97 Biocides used in poly(ethylene terephthalate).

Material	Biocide	Refs.
concentrate	2,4,4'-trichloro-2'-hydroxydiphenyl ether	6
surface grafting	N-hexyl-N'-(4-vinylbenzyl)-4,4'-bipyridinium bromide chloride	5

A surface N^+ concentration of 25 $nmol/cm^2$ on PET has been shown to be highly effective. At this concentration, more than 99% of *Escherichia coli* bacteria cells in a suspension of $1x10^7$ cells/ml are killed when the suspension is placed in contact with the substrate.[5]

References

1 Shah, A A; Hasan, F; Hameed, A; Ahmed, S, *Biotechnol. Adv.*, **26**, 246-65, 2008.
2 Mueller, R_J; Kleeberg, I, Deckwer, W-D, *J. Biotechnol.*, **86**, 87-95, 2001.
3 Vertommen, M A M E; Nierstrasz, V A; van der Veer, M; Warmoeskerken, M M C G, *J. Biotechnol.*, **120**, 376-386, 2005.
4 Lubertozzi, D; Keasling, J D, *Biotechnol. Adv.*, **27**, 53-75, 2009.
5 Shi, Z; Neoh, K G; Kang, E T, *Biomaterials*, **26**, 501-8, 2005.
6 Stadler, U, US Patent 7,041,307, Ciba Specialty Chemicals Corporation, May 9, 2006.

4.14.14 POLYIMIDE

4.14.14.1 Relevant microorganisms

Table 4.98 contains a list of microorganisms found to cause deterioration of polyimide.

Table 4.98. Microorganisms in polyimide

Material	Microorganisms	Refs.
polyimide	*Alcaligenes denitrificans, Aspergillus amstelodami, Aspergillus niger, Aspergillus ustus, Aspergillus versicolor, Aureobasidium microstrictum, Cladosporium cladosporioides, Cladosporium macrocarpum, Chaetomium sp., Mucor alboater, Ochrobactrum anthropi, Paecilomyces variotii, Penicillium brevicompactum, Penicillium chermesinum, Penicillium chrysogenum, Penicillium cyaneum, Penicillium decumbens, Penicillium nalgiovense, Penicillium ochrochloron, Pseudomonas aeruginosa, Scopulariopsis chartarum, Trichoderma koningii, Trichoderma viride, Vibrio harveyi, Xanthomonas maltophila*	1,4

4.14.14.2 Results of biodeterioration

Two steps are involved in degradation: an initial decline of resistance related to the partial ingress of water and ionic species into the polymer matrix. This is followed by further deterioration of the polymer by activity of the fungi, resulting in a large decrease in resistivity.[2]

At the end of the incubation experiments, physically damaged but intact polyimide film showed a severe decrease of capacitance and increase of conductance, further supporting the suggestion that degradation of polymer dielectric properties were solely due to damage of polymer by fungal growth.[3]

References

1 Gu, J-D; Roman, M; Esselman, T; Mitchell, R, *Intl. Biodet. Biodeg.*, **41**, 25-33, 1998.
2 Gu, J-D, *Intl. Biodet. Biodeg.*, **52**, 69-91, 2003.
3 Gu, J-D, *Intl. Biodet. Biodeg.*, **59**, 179-0-79, 2007.
4 Lugauskas, A; Levinskaite, L; Peciulyte, D, *Intl. Biodet. Biodeg.*, **52**, 233-42, 2003.

4.14.15 POLYMETHYLMETHACRYLATE

4.14.15.1 Relevant microorganisms

Table 4.99 contains a list of microorganisms found to cause deterioration of polymethylmethacrylate.

Table 4.99. Microorganisms in polymethylmethacrylate.

Material	Microorganisms	Refs.
PMMA	*Amphibalanus reticulatus, Aspergillus flavus, Bacillus subtilis*	5,6

4.14.15.2 Results of biodeterioration

The majority of self-polishing paints are based on polymethylmethacrylate which undergoes a controlled hydrolysis and is completely released into sea water.[1] The rate of hydrolysis is designed to achieve the polishing of the coating in 36 or 60 months.[1] New formulations based on PMMA without biocides are designed with an elevated hydrolysis rate for 12 months.[1] Thus the complete paint with a dry film thickness of 150-200 μm is released into the water.[1]

4.14.15.3 Biostabilization

Table 4.100 contains a list of the most suitable biocides used for protection of polymethylmethacrylate.

Table 4.100. Biocides used in polymethylmethacrylate.

Material	Biocide	Refs.
PMMA	chitosan, copper thiocyanate, cuprous oxide, quaternary ammonium compound, tolylfluanid	2,3,4,5

References

1 Watermann, B T; Daehne, B; Sievers, S; Dannenberg, R; Overbeke, J C; Klijnstra, J W; Heemken, O, *Chemosphere*, **60**, 1530-41, 2005.
2 Thouvenin, M; Peron, J J; Langlois, V; Guerin, P; Langlois, J-Y; Vallee-Rehel, K, *Prog. Org. Coat.*, **44**, 85-92, 2002.
3 Fay, F; Linossier, I; Peron, J J; Langlois, V; Vallee-Rehel, K, *Prog. Org. Coat.*, **60**, 194-206, 2007.
4 Markarian, J, *Plast. Additives Compounding*, Jan/Feb, 18-22, 2009.
5 Prashantha, K V H; Lakshman, L; Shamala, T R; Tharanathan, R N, *Intl. Biodet. Biodeg.*, **56**, 115-20, 2005.
6 Sangeetha, R; Kumar, R Venkatesan, R; Doble, M; Vedaprakash, L; Lakshmi, K, *Mater. Sci. Eng. C*, **30**, 112-19, 2010.

4.14.16 POLYOXYMETHYLENE

Polyoxymethylene provides continuous release of formaldehyde biocide by long-term decomposition. The polyoxymethylenes used are water-insoluble materials with their decomposition rates controlled by pH, temperature, and certain decomposition catalysts.[1]

References

1 Fenton, J T; Miller, J F, **US Patent** 4,911,923, Conoco, Inc., March 27, 1990.

4.14.17 POLYPROPYLENE

4.14.17.1 Relevant microorganisms

Table 4.101 contains a list of microorganisms found to cause deterioration of polypropylene.

Table 4.101. Microorganisms in polypropylene.

Material	Microorganisms	Refs.
construction materials	*Aspergillus flavus, Aspergillus niger, Aspergillus terreus, Aspergillus versicolor, Aureobasidium pullulans, Candida albicans, Chaetomium globisum, Citrobacter freundii, Escherichia coli, Enterobacter aerogenes, Geotrichum candidum, Klebsiella pneumoniae, Penicillium funiculosum, Penicillium sp., Pseudomonas aeruginosa, Pseudomonas oleovorans, Saccharomyces cerevisiae, Salmonella choleraesuis, Staphylococcus aureus*	13
polypropylene	*Aspergillus niger, Aspergillus terreus, Aureobasidium pullulans, Bacillus flexus, Chaetomium globosum, Chlorella saccharophila, Chlorella vulgaris, Escherichia coli, Gloeophyllum trabeum, Paecilomyces variotii, Phanerochaete chrysosporium, Postia placenta, Pseudomonasaeruginosa, Scopulariopsis brevicaulis, Staphylococcus aureus, Trametes versicolor, Ulocladium atrum*	1,2,4-7

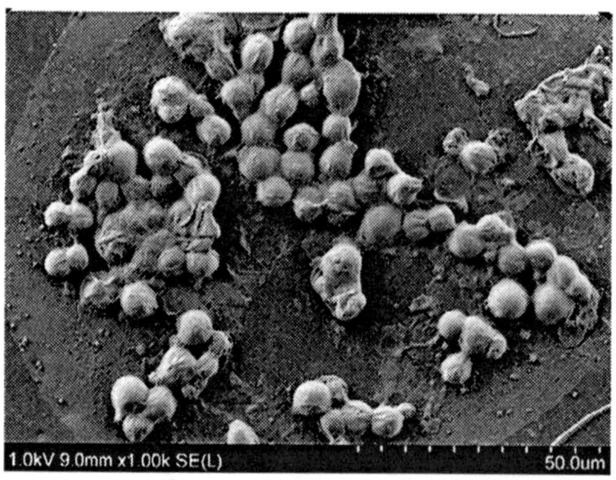

Figure 4.47. Biofilm formation on the surface of polypropylene. [Adapted, by permission, from Stroemberg, E; Karlsson, S, *Intl. Biodet. Biodeg.*, **63**, 1045-53, 2009.]

4.14.17.2 Essential product components

Figure 4.47 shows surface colonization of polypropylene. Different compositions had different rates of colonization (polypropylene containing wood flour underwent the most rapid colonization). Algae were responsible for the initial colonization of the materials, followed by fungi that use algae as a nutrition source at first and later attack the material by both chemical and mechanical degradation mechanisms.[1] *Staphylococcus aureus* adhered better to the polypropylene surface at 12°C than at 30°C, while the reverse was true for *Escherichia coli*.[4] Fungi usually develop as a result of the presence of dust and debris containing nutrients.[5]

4.14.17.3 Mechanisms of biodegradation and biodeterioration

The mechanism may involve the formation of hydroperoxides which destabilize the polymeric carbon chain to form a carbonyl group.[3]

4.14.17.4 Results of biodeterioration

Degradation of PP results in a decrease of its tensile strength and molecular weight.[3]

4.14.17.5 Biostabilization

Table 4.102 contains a list of the most suitable biocides used for protection of polypropylene.

Table 4.102. Biocides used in polypropylene.

Material	Biocide	Refs.
construction materials	1, 2-benzisothiazolin-3-one	13
membrane	poly(ethylene glycol) monomethacrylate brush	10
polypropylene	N-halamine precursor, silver nanoparticles, silver powder	8,9,11
thin films	TiO_2-anatase	12

The surface-functionalized membranes with molecular brushes exhibit long-term durability and antibacterial efficacy in repeated applications.[10] The bacteria adsorption, deposition, growth and colony forming are affected by the surface properties of the membrane and conditions of operation.[10]

References

1 Stroemberg, E; Karlsson, S, *Intl. Biodet. Biodeg.*, **63**, 1045-53, 2009.
2 Arkatkar, A; Arutchelvi, J; Bhaduri, S; Uppara, P V; Doble, M, *Intl. Biodet. Biodeg.*, **63**, 106-11, 2009.
3 Gu, J-D, *Intl. Biodet. Biodeg.*, **52**, 69-91, 2003.
4 Pompermayer, D M C; Gaylarde, C C, *Food Microbiol.*, **17**, 361-65, 2000.
5 Lugauskas, A; Levinskaite, L; Peciulyte, D, *Intl. Biodet. Biodeg.*, **52**, 233-42, 2003.
6 Lomeli-Ramirez, M G; Ochoa-Ruiz, H G; Fuentes-Talavera, F J; Garcia-Enriquez, S; Cerpa-Gallegos, M A; Silva-Guzman, J A, *Intl. Biodet. Biodeg.*, **63**, 1030-35, 2009.
7 Oh, Y J; Lee, N R, Jo, W; Jung, W K; Lim, J S, *Ultramicroscopy*, **109**, 874-80, 2009.
8 Badrossamay, M R, Sun, G, *J. Biomed. Mater. Res. Part B: Appl. Biomater.*, **89B**, 93–101, 2009.
9 Kumar, A; Vemula, P K; Ajayan, P M, John, G, *Nature Mater.*, **7**, 236-41, 2008.
10 Yao, F; Fu, G-D; Zhao, J; Kang, E-T; Neoh, K G, *J. Membrane Sci.*, **319**, 149-57, 2008.
11 Radheshkumar, C; Muenstedt, H, *Reactive Functional Polym.*, **66**, 780-88, 2006.
12 Kubacka, A; Ferrer, M; Cerrada, M L; Serrano, C; Sanchez-Chaves, M; Fernandez-Garcia, M; de Andres, A; Jimenez Rioboo, R J; Fernandez-Martin, F; Ferna ndez-Garcia, M, *Appl. Catalysis B: Environ.*, **89**, 441-47, 2009.
13 Bioban Ultra BIT 20 Le Antimicrobial. DOW. Product information.

4.14.18 POLYSTYRENE

4.14.18.1 Relevant microorganisms

Table 4.103 contains a list of microorganisms found to cause deterioration of polystyrene.

Table 4.103. Microorganisms in polystyrene.

Material	Microorganisms	Refs.
PS	*Aspergillus flavus, Aspergillus niger, Aspergillus terreus, Aspergillus versicolor, Aureobasidium pullulans, Candida albicans, Chaetomium globisum, Citrobacter freundii, Colletotrichum lindemuthianum, Escherichia coli, Enterobacter aerogenes, Geotrichum candidum, Klebsiella pneumoniae, Penicillium funiculosum, Penicillium sp., Pseudomonas aeruginosa, Pseudomonas oleovorans, Pseudomonas putida, Saccharomyces cerevisiae, Salmonella choleraesuis, Staphylococcus aureus*	1,2,3

4.14.18.2 Biostabilization

Table 4.104 contains the most suitable biocide used for protection of polystyrene.

Table 4.104. Biocides used in polystyrene.

Material	Biocide	Refs.
polystyrene	1,2-benzisothiazolin-3-one, medetomidine	3,5

Ozone treated PS surfaces are able to inactivate various species of vegetative and sporulated bacteria in a relatively short contact time.[4] Denaturation of proteins seems likely based on a significant loss of enzymatic activity of the lysosome protein.[4]

References

1 Jones, E B G, *Mycol. Res.*, **98**, 9, 961-81, 1994.
2 Ko, C-H; Hunag, C-Y; Hsieh, S-P, Kuo, P-A, *Intl. Biodet. Biodeg.*, **60**, 250-57, 2007.
3 Bioban Ultra BIT 20 Le Antimicrobial. DOW. Product information.
4 Mahfoudh, A; Barbeau, J; Moisan, M; Leduc, A; Seguin, J, *Appl. Surface Sci.*, just accepted, doi:10.1016/j.apsusc.2009.11.074.
5 Nyden, M; Fant, C; Holmber, K; Swanson, L, **US Patent** 7,531,581, I-Tech AB, May 12, 2009.

4.14.19 POLYSULFONE

4.14.19.1 Relevant microorganisms

Table 4.105 contains a list of microorganisms found to cause deterioration of polysulfone.

Table 4.105. Microorganisms in polysulfone.

Material	Microorganisms	Refs.
membrane	*Escherichia coli, Pseudomonas mendocina*	2

4.14.19.2 Biostabilization

Table 4.106 contains a list of the most suitable biocides used for protection of polysulfone.

Table 4.106. Biocides used in polysulfone.

Material	Biocide	Refs.
membrane	silver nanoparticles	1,2
reverse osmosis	silver coating	3

Bactericidal action of silver ions increases with increasing temperature and pH.[2]

References

1 Taurozzi, J S; Arul, T; Bosak, V Z; Burban, A F; Voice, T C; Bruening, M L; Tarabara, V V, *J. Membrane Sci.*, **325**, 58-68, 2008.
2 Zodrow,K; Brunet, L; Mahendra, S; Li, D; Zhang, A; Li, Q; Alvarez, P J J, *Water Res.*, **43**, 715-23, 2009.
3 Yang, H-L; Lin, J C-T; Huang, C, *Water Res.*, **43**, 3777-86, 2009.

4.14.20 POLYTETRAFLUOROETHYELENE

4.14.20.1 Relevant microorganisms

Table 4.107 contains a list of microorganisms found to cause deterioration of polytetrafluoroethylene.

Table 4.107. Microorganisms in polytetrafluoroethylene.

Material	Microorganisms	Refs.
PTFE	*Listeria monocytogenes, Pseudomonas fluorescens, Salmonella enteritidis*	1,2

References

1 Austin, J W; Sanders, G; Kay, W W; Collinson, S K, *FEMS Microbiol. Lett.*, **162**, 295-301, 1998.
2 Meylheuc, T; Renault, M; Bellon-Fontaine, M N, *Intl. J. Food Microbiol.*, **109**, 71-78, 2006.

4.14.21 POLYURETHANES

4.14.21.1 Relevant microorganisms

Table 4.108 contains a list of microorganisms found to cause deterioration of polyurethanes.

Table 4.108. Microorganisms in polyurethanes.

Material	Microorganisms	Refs.
aliphatic PU	*Rhodococcus globerulus*	12
plasticizer degradation	*Sphingobium chungbukense*	4
polyester polyurethane	*Aureobasidium pullulans, Cladosporium sp., Comamonas acidovorans, Corynebacterium sp., Curvularia senegalensis, Fusarium solani, Pseudomonas aeruginosa*	1,10
polyurethane	*Acinetobacter calcoaceticus, Anaerovibrio lipolytica, Arthrobacter globiformis, Arthrobacter sp. , Aspergillus terreus, Aureobasidium pullulans, Bacillus sp., Bacillus subtilis, Butyrivibrio fibrisolvens, Candida albicans, Chaetomium globosum, Cladosporium sp., Clostridium perfringens, Comamonas acidovorans, Corynebacterium sp., Curvularia senegalensis, Escherichia coli, Fusarium solani, Micrococcus sp., Prevotella ruminicola, Pseudomonas aeruginosa, Pseudomonas cepacia, Pseudomonas chlororaphis, Pseudomonas fluorescens, Pseudomonas putida, Pseudomonas sp., Selenomonas ruminantium, Serratia marcescens, Torpedo californica*	1,2,7,8,10

4.14.21.2 Essential product components

The regularity in synthetic polymers allows the polymer chain to pack easily, resulting in the formation of crystalline regions.[1] This limits accessibility of the polymer chains to degradative agents.[1]

It was observed that sulfur-cured polyester and polyether PUs had some fungal inertness.[1]

Phthalate esters are frequently used as plasticizers in polyurethane formulations.[4]

4.14.21.3 Mechanisms of biodegradation and biodeterioration

Esterase activity (both membrane-bound and extracellular) has been noted in microbes which allow them to utilize polyurethane.[1] Microbial degradation of polyester polyurethane is hypothesized to be mainly due to the hydrolysis of ester bonds by esterase enzymes.[1]

Polyurethane degradation is the result of synergistic activity between endopolyurethanases and exopolyurethanases.[1] Endoenzymes hydrolyze the PU molecule at random locations throughout the polymer chain, leading to loss of tensile strength.[1] Exoenzymes remove successive monomer units from the chain ends and show little loss of tensile strength.[1]

Most species are capable of utilizing polyurethane as a sole carbon and energy source.[1]

Bacteria produce several classes of lipase:

- carboxylesterases (hydrolyze small ester-containing molecules at least partly soluble in water)
- true lipases (display maximum activity towards water-insoluble long-chain triglycerides
- phospholipases

Enzymes involved in biodeterioration require the presence of cofactors, such as cations present in the material matrix and coenzymes synthesized by microorganisms to help in the breakdown of specific bonds.[9] The biodeterioration may proceed by two mechanisms: bulk and surface erosion.[9]

In bulk erosion, fragments are lost from the entire polymer mass and the molecular weight changes due to bond cleavage.[9] This lysis is provoked by chemicals (e.g., H_2O, acids, bases, transition metals and radicals) or by radiation, but not by enzymes.[9] Enzymes are too large to penetrate throughout the matrix framework.[9]

In surface erosion, matter is lost but there is no detectable change in the molecular weight of polymers of the matrix.[9] If the diffusion of chemicals throughout the material is faster than the cleavage of polymer bonds, the polymer undergoes bulk erosion.[9] If the cleavage of bonds is faster than the diffusion of chemicals, the process occurs mainly at the surface of the matrix.[9]

Comamonas acidovorans grew on polyurethane in a medium in which polyurethane was a sole source of carbon and nitrogen (48% was degraded after seven days).[11] Diethylene glycol and adipic acid were detected as degradation products, but no metabolites were found to indicate cleavage of the urethane linkage. The enzyme was isolated from *Comamonas acidovorans*.[11] The enzyme is a cell-bound esterase that can hydrolyze polyester chains in polyurethane to diethylene glycol and adipic acid.[11]

4.14.21.4 Results of biodeterioration

Polyester-based polyurethanes are more susceptible to biodeterioration than polyether-based polyurethanes.[6]

4.14.21.5 Biostabilization

Table 4.109 contains a list of the most suitable biocides used for protection of polyurethanes.

Table 4.109. Biocides used in polyurethanes.

Material	Biocide	Refs.
catheters	chitosan	2
coatings	2,4-di-tert-butylphenol, nonylphenol disulfide	5
fibrous membrane	quaternization of the grafted pyridine groups with hexylbromide	13
Foley catheters	gentamicin and fluoroquinolones	17
microporous PU	grafting of 2,2,5,5-tetramethyl-imidozalidin-4-one	16
military applications	quaternary ammonium compound (Biosafe)	18
paint	chloromethyl-/methylisothiazolone and a glycol-based formaldehyde releaser	20
plastic articles	4,5,-dichloro-2-n-octyl-4-isothiazolone-3-one biocide, 2-n-octyl-4-isothiazolin-3-one	19
polyether PU	gold nanoparticles, silver nanoparticles	3
polyurethanes	glucose oxidase and/or horseradish peroxidase (use hydrogen peroxide and chloride ion to generate hypochlorite which kills pathogens),[14] hydantoin-containing polyurethane surface modifier obtained by ring-opening copolymerization,[15] 10,10'-oxybisphenoxyarsine	14,15

References

1 Howard, G T, *Intl. Biodet. Biodeg.*, **49**, 245-52, 2002.
2 Zilberman, M; Navon, A; Sandovsky-Losica, H; Segal, E, *J. Biomed. Mater. Res.*, **81A**, 392–398, 2007.
3 Chou, C-W; Hsu, S-H; Wang, P-H, *J. Biomed. Mater. Res.*, **84A**, 785–794, 2008.
4 Park, J-M; Jeon, M; Lim, E-S; Um, H-J; Kim, Y-C; Min, J; Kim, Y-H, *Curr. Microbiol.*, **57**, 515-18, 2008.
5 Wynne, J H; Pant, R R; Jones-Meehan, J M; Paige Phillips, J, *J. Appl. Polym. Sci.*, **107**, 2089–2094, 2008.
6 Warscheid, T; Braams, J, *Intl. Biodet. Biodeg.*, **46**, 343-68, 2000.
7 Gautam, R; Bassi, A S; Yanful, E K; Cullen, E, *Intl. Biodet. Biodeg.*, **60**, 4, 245-49, 2007.
8 Rowe, L; Howard, G T, *Intl. Biodet. Biodeg.*, **50**, 33-40, 2002.
9 Lucas, N; Bienaime, C; Belloy, C; Queneudec, M; Silvestre, F; Nava-Saucedo, J-E, *Chemosphere*, **73**, 429-42, 2008.
10 Shah, A A; Hasan, F; Hameed, A; Ahmed, S, *Biotechnol. Adv.*, **26**, 246-65, 2008.
11 Shimao, M, **Current Opinion in Biotechnology**, *Elsevier*, Vol. 12, Issue 3, pp 242-247.
12 Gu, J-D; **Encyclopedia of Microbiology**, *Elsevier*, 2009, pp 259-269.
13 Yao, C; Li, X; Neoh, K G; Shi, Z; Kang, E T, *J. Membrane Sci.*, **320**, 259-67, 2008.
14 Amitai, G; Andersen, J; Wargo, S; Asche, S; Chir, J; Koepsel, R; Russell, A J, *Biomaterials*, **30**, 6522-29, 2009.
15 Grunzinger, S J; Kurt, P; Brunson, K M; Wood, L; Ohman, D E; Wynne, K J, *Polymer*, **48**, 4653-62, 2007.
16 Tan, K; Obendorf, S K, *J. Membrane Sci.*, **289**, 199-209, 2007.
17 Williams, G J; Stickler, D J, *J. Urology*, **178**, 697-701, 2007.
18 Markarian, J, *Plast. Additives Compounding*, Jan/Feb, 18-22, 2009.
19 Redlich, G H; Hamel, R G; Rein, N M, **US Patent** 6,875,797, Rohm and Haas, April 5, 2005.
20 Rocima 625. Rohm and Haas, Technical Data Sheet. April 2009.

4.14.22 POLY(VINYL ALCOHOL)

4.14.22.1 Relevant microorganisms

Table 4.110 contains a list of microorganisms found to cause deterioration of poly(vinyl alcohol).

Table 4.110. Microorganisms in poly(vinyl alcohol).

Material	Microorganisms	Refs.
PVA	*Bacillus megaterium, Endomyces fibuliger, Fusarium lini, Geotrichum fermentans, Lipomyces, Nadsonia, Phanerochaete chrysoporium, Pichia, Pseudomonas putida, Pseudomonas sp., Pseudomonas vesicularis, Rhodotorula spp., Saccharomyces rouxii, Zigosaccharomyces*	5

4.14.22.2 Essential product components

The long-term acclimation of the microbial populations is a stringent parameter for an efficient removal of poly(vinyl alcohol).[5]

4.14.22.3 Mechanisms of biodegradation and biodeterioration

The initial attack on PVA chains occurred extracellularly in water solution.[5] The first microorganisms capable of assimilating PVA as their sole carbon source were isolated from soil samples and identified as *Pseudomonas* species.[5] Further studies showed that a similar endocleavage mechanism was active in the degradation of PVA by other bacterial species.[5]

The major biodegradation mechanism of PVA is represented by the random endo-cleavage of the polymer chains.[5] The initial step is the specific oxidation of 1,3-hydroxyl groups, mediated by oxidase and dehydrogenase type enzymes, to give β-hydroxylketone as well as 1,3-diketone moieties.[5] The latter groups are susceptible to carbon-carbon bond cleavage promoted by specific β-diketone hydrolase, giving rise to the formation of carboxyl and methyl ketone end groups.[5]

4.14.22.4 Biostabilization

Table 4.111 contains a list of the most suitable biocides used for protection of poly(vinyl alcohol).

Table 4.111 Biocides used in poly(vinyl alcohol).

Material	Biocide	Refs.
hydrogel	paraquat dichloride	3
nanofibers	quaternized chitosan	2
PVA	4-vinylpyridine grafting	4

Electrospinning of an aqueous solution of poly(vinyl alcohol) with chitosan derivative such as N-[(2-hydroxy-3-trimethylammonium)propyl] chitosan chloride has been prepared.[1] Mats produced had a good antibacterial activity against Gram-positive bacteria, *Staphylococcus aureus*, and Gram-negative bacteria, *Escherichia coli*.[1]

References

1 Alipour, S M; Nouri, M; Mokhtari, J; Bahrami, S H, *Carbohydrate Res.*, **344**, 2496-2501, 2009.
2 Ignatova, M; Starbova, K; Markova, N; Manolova, N; Rashkov, I, *Carbohydrate Res.*, **341**, 2098-2107, 2006.
3 Alemzadeh, I; Vossoughi, M, *Chem. Eng. Process.*, **41**, 707-10, 2002.
4 Wang, C-C; Yang, F-L; Liu, L-F; Fu, Z-M; Xue, Y, *J. Membrane Sci.*, **345**, 223-32, 2009.
5 Chiellini, E; Corti, A; D'Antone, S; Solaro, R, *Prog. Polym. Sci.*, **28**, 963-1014, 2003.

4.14.23 POLY(VINYL CHLORIDE)

4.14.23.1 Relevant microorganisms

Table 4.112 contains a list of microorganisms found to cause deterioration of poly(vinyl chloride).

Table 4.112. Microorganisms in poly(vinyl chloride).

Material	Microorganisms	Refs.
mattresses	*Scopulariopsis brevicaulisa*	17
plasticizers	*Bacillus subtilis, Bacillus sp., Rhodococcus rhodochrous, Sphingobium chungbukense*	6,8,9
PVC	*Aureobasidium microstrictum, Aspergillus amstelodami, Aspergillus niger, Aspergillus ruber, Aspergillus ustus, Aspergillus versicolor, Aureobasidium pullulans, Bacillus circulans Bacillus cereus, Bacillus subtilis, Chaetomium globosum, Chaetomium sphaerospermum, Fusarium sp., Mortierella isabellina, Mucor racemosus, Neurospora crassa, Penicillium lanosum, Penicillium nalgiovense, Penicillium ochrochloron, Penicillium variotii, Penicillium funiculosum, Pseudomonas aeruginosa, Pseudomonas resinovorus, Pseudomonas veronii, Scopulariopsis chartarum, Scopulariopsis sp., Serratia marcescens, Trichoderma sp., Trichoderma viride, Ulocladium oudemansii, Verticillium sp.*	5,10,11, 13

4.14.23.2 Essential product components

Plasticizers and some costabilizers (epoxidized oils) are the most essential components for the prediction of biostability. Their presence is sufficient for biodegradation to occur. In absence of these two components, PVC is a biostable material and in most cases does not require addition of biocides. Although rigid PVC does not contain high plasticizer levels that serve as a nutrient source for microbes, outdoor applications can pick up organic matter from the environment or develop an organic film with age that may be attacked by fungus or algae.[18] Fungal staining is an occasional problem for siding.[18] The wood flour in PVC is a nutrient source for microbes that can stain surfaces.[18]

4.14.23.3 Mechanisms of biodegradation and biodeterioration

Numerous studies indicate that phthalate esters are degraded by a wide range of bacteria and actinomycetes under both aerobic and anaerobic conditions.[15] Primary degradation half-lives in surface and marine waters range from <1 day to 2 weeks and in soils from <1 week to several months.[15] Longer half-lives may occur in anaerobic, oligotrophic, or cold environments.[15] Numerous experiments have shown that the bioaccumulation of phthalate esters in the aquatic and terrestrial food chain is limited by biotransformation, which increases with increasing trophic level.[15] Consequently, models that ignore biotransformation grossly exaggerate bioaccumulation potential of higher molecular weight phthalate esters.[15]

4.14.23.4 Results of biodeterioration

During vinyl flooring life service, attack by fungi and microorganisms leads to the degradation of DOP and the release of some volatile organic compounds.[3]

Increases in the strength of PVC films as the result of microbial attack have been observed and are considered to be caused by degradation of the plasticizer.[11] Products of plasticizer biodegradation are well established. For example, di-ethylhexyl adipate is biodegraded with formation of adipic acid, 2-ethylhexanol, 2-ethylhexanal, and 2-ethanoic acid.[16] Products of biodegradation of di-ethylhexyl phthalate are the same with exception of acid, which is phthalic acid.[16]

Biodeterioration is demonstrated by stains, change of color, odor, and changes in mechanical properties.

Arsine, phosphine, and stibine are toxic gases that are produced after the breakdown of poly(vinyl chloride) mattress coverings by fungi or bacteria.[17] In-vitro incubation of *Scopulariopsis brevicaulisa* common domestic fungus with samples of PVC cot mattresses leads to stibine and phosphine production from mattress fire retardants, such as antimony trioxide and phosphate plasticizers.[17]

4.14.23.5 Biostabilization

Table 4.113 contains a list of the most suitable biocides used for protection of poly(vinyl chloride).

Table 4.113. Biocides used in poly(vinyl chloride).

Material	Biocide	Refs.
PVC	copper nanoparticle,[7] 4,5-dichloro-2-n-octylisothiazolin-3-one,[2] 2-n-octyl-isothiazolin-3-one,[14] 10,10'-oxybisphenoxarsine,[2] surface azidation,[4] tebuconazole,[1] 2,3,5,6-tetrachloro-4-(methylsulphonyl)pyridine,[5] zeolite encapsulated 2-n-octyl-4-isothiazolin-3-one,[10] zinc pyrithione[2]	1,2,4,5,7, 10,14

Encapsulation provides a method for reducing the concentration of antimicrobial used in products and its leaching into the environment.[10]

References

1 Liu, Y; Laks, P; Heiden, P, *J. Appl. Polym. Sci.*, **86**, 608–614, 2002.
2 Bessems, E, *J. Ind. Textiles*, **30**, 3, 185-200, 2001.
3 El-Aghoury, A; Vasudeva, R K; Banu, D; Elektorowicz, M; Feldman, D, *J. Polym. Environ.*, **14**, 135-47, 2006.
4 Lakshmi, S; Pradeep Kumar, S S; Jayakrishnan, A, *J. Biomed. Mater. Res.*, **61**, 26–32, 2002.
5 Jones, C; Robson, G D; Greenhaulgh, M; Eastwood, I; Handley, P S, *Biofouling*, **18**, 1, 21-27, 2002.
6 Grochowalski, A R; Cooper, D G; Nicell, J A, *Biodegradation*, **18**, 283-93, 2007.
7 Cioffi, N; Torsi, L; Ditaranto, N; Sabbatini, L; Zambonin, P G; Tantillo, G; Ghibelli, L; D'Alessio, M; Bleve-Zacheo, T; Traversa, E, *Appl. Polym. Lett.*, **85**, 12, 2417-19, 2004.
8 Pour, A K; Mamer, O A; Cooper, D G; Maric, M; Nicell, J A, *J. Mass Spectrometry*, **44**, 662-71, 2009.
9 Park, J-M; Jeon, M; Lim, E-S; Um, H-J; Kim, Y-C; Min, J; Kim, Y-H, *Curr. Microbiol.*, **57**, 515-18, 2008.
10 Coulthwaitea, L; Bayley, K; Liauw, C; Craig, G; Verran, J, *Intl. Biodet. Biodeg.*, **56**, 86-93, 2005.
11 Whitney, P J, *Intl. Biodet. Biodeg.*, **37**, 3-4, 205-13, 1996.

12 Pintado, J L; Montero, F, *Intl. Biodet. Biodeg.*, **29**, 3-4, 357-65, 1992.
13 Lugauskas, A; Levinskaite, L; Peciulyte, D, *Intl. Biodet. Biodeg.*, **52**, 233-42, 2003.
14 Hamel, R G; McEntee, T C, *Plast. Additives Compounding*, Oct., 24-27, 2000.
15 Staples, C A; Peterson, D R; Parkerton, T F; Adams, W J, *Chemosphere*, **35**, 4, 667-749, 1997.
16 Nalli, S; Cooper, D G; Nicell, J A, *Sci. Total Environ.*, **366**, 286-94, 2006.
17 **Mattresses and cot death** in Noticeboard of *The Lancet*, **337**, 969, 1991.
18 Makarian, J, *Plast. Addivies Compounding*, May/June, 34-37, 2005.

4.14.24 SILICONE

4.14.24.1 Relevant microorganisms

Table 4.114 contains a list of microorganisms found to cause deterioration of silicone.

Table 4.114. Microorganisms in silicones.

Material	Microorganisms	Refs.
insulating rubber	*Chlorella saccharophila, Chlorella vulgaris, Cladosporium cladosporides, Cladosporium tenuissimum, Curvularia lunata, Epicoccum nigrum, Fusarium semitecum, Microsphaeriopsis, Pestalitiopsis guepinii, Polyscytalum fecundissimum, Stagonospora*	12
medical devices	*Candida tropicalis*	9
stone protective coating	*Aspergillus fumigatus, Aureobasidium pullulans, Cladosporium cladosporioides, Exophiala jeanselmei, Paecilomyces farinosus, Penicillium aurantiogriseum, Phoma leveillei, Trichoderma koningii, Verticillium lecanii*	10

4.14.24.2 Essential product components

Silicone materials support fungal growth when an external carbon source is available. Extra nutrition is always present in an outdoor environment, thus none of the formulations can be called bioresistant, but silicone rubbers are not biodegradable either.[11]

4.14.24.3 Results of biodeterioration

The significant decrease in hydrophobicity of silicone-treated samples during massive fungal growth is a result of the degradation of hydrophobic groups of the thin resin films by organic acids and other compounds as extracellular polymeric substances.[10]

4.14.24.4 Biostabilization

Table 4.115 contains a list of the most suitable biocides used for protection of silicones.

Table 4.115. Biocides used in silicones.

Material	Biocide	Refs.
coatings	sodium benzoate, zosteric acid	5,8
dental applications	N-chloramine	13
non-eroding coatings	silicone backbone (biocide-free antifouling coatings)	4
general applications	silanols	3
medical devices	triclosan	7
mortar protection	2,3,5,6-tetrachloro-4-methylsulfonyl-pyridine, 2,3,5,6-tetrachloro-4-methylsulfonylpyridine	6

Table 4.115. Biocides used in silicones.

Material	Biocide	Refs.
sealants	4,5-dichloro-2-n-octyl-4-isothiazol-3-one, 4,5-dichloro-2-cyclohexyl-3-isothiazolinone, 2-n-octyl-4-isothiazol-3-one	14

5-chloro-2-(2,4-dichlorophenoxy) phenol was modified with alkenyl moieties and incorporated into a silicone backbone through covalent bonds for use in antifouling coating.[1]

Polymeric materials may be protected against biological attack by lowering probability of colonization due to either low adhesion or toxicity or both. Silicones have low surface energy (24 dyne/cm as compared with PTFE 18.5 dyne/cm). Their adhesion property is $3.6x10^6$ dyne/cm^2, which means it is 26.6 times lower than for PTFE ($3.3x10^9$).[2] This low adhesion makes them excellent candidates for antifouling coatings and reduces the probability of colonization.

References

1 Thomes, J; Choi, S-B; Fjeldheim, R; Boudjoul, P, *Biofouling*, **20**, 4/5, 227-36, 2004.
2 Omae, I, *Chem. Rev.*, **103**, 3431-48, 2003.
3 Kim, Y-M; Farrah, S; Baney, R H, *Electronic J. Biotechnol.*, **9**, 2, 176-80, 2006.
4 Watermann, B T; Daehne, B; Sievers, S; Dannenberg, R; Overbeke, J C; Klijnstra, J W; Heemken, O, *Chemosphere*, **60**, 1530-41, 2005.
5 Al-Juhni, A A; Newby, B-m Z, *Prog. Org. Coat.*, **56**, 135-45, 2006.
6 Urzi, C; De Leo, F, *Intl. Biodet. Biodeg.*, **60**, 25-34, 2007.
7 McBride, M C; Malcolm, R K; Woolfson, A D; Gorman, S P, *Biomaterials*, **30**, 6739-47, 2009.
8 Barriosa, C A; Xu, Q; Cutright, T; Newby, B-m Z, *Coll. Surf. B: Biointerfaces*, **41**, 83-93, 2005.
9 Busscher, H J; de Boer, C E; Verkerke, G J; Kalicharan, R; Schutte, H K; van der Mei, H C, *Intl. Biodet. Biodeg.*, **33**, 4, 363-90, 1994.
10 Leznicka, S; Kuroczkin, J; Krumbein, W E; Strzelczyk, A B; Petersen, K, *Intl. Biodet.*, **28**, 91-111, 1991.
11 Wallstrom, S; Dowling, K; Karlsson, S, *Polym. Deg. Stab.*, **78**, 257-62, 2002.
12 Wallstrom, S; Karlsson, S, *Polym. Deg. Stab.*, **85**, 841-46, 2004.
13 Pusineri, C; Del Torto, M, **US Patent** 6,559,199, Rhodia Chimie, May 6, 2003.
14 McEntee, T C, **US Patent** 6,280,657, Rohm and Haas, August 28, 2001.

4.15 PULP AND PAPER

4.15.1 RELEVANT MICROORGANISMS

Table 4.116 contains a list of microorganisms found to cause deterioration of pulp and paper.

Table 4.116. Microorganisms in pulp and paper.

Material	Microorganisms	Refs.
aspen wood extractives	*Aspergillus awamori, Aspergillus flavus, Aspergillus luchuensis, Aureobasidium pullulans, Bjerkandera adjusta, Cunninghamella elegans, Gliocladium roseum, Ophiostoma ainoae, Ophiostoma piceae, Ophiostoma piliferum, Peniophora polygonia, Phanerochaete chrysosporium, Trametes versicolor*	11

Table 4.116. Microorganisms in pulp and paper.

Material	Microorganisms	Refs.
historical paper	*Alternaria alternata, Aspergillus hollandicus, Aspergillus nidulans, Aspergillus niger, Aspergillus terreus, Aspergillus versicolor, Botryotinia fuckeliana, Botrytis cinerea, Byssoclamys fulva, Chaetomium globosum, Chromelosporium carneum, Cladosporium cladosporioides, Cladosporium herbarum, Coprinus sp., Debaryomyces hansenii, Epicoccum nigrum, Eurotium amstelodami, Eurotium chevalieri, Liposcelis bostrychophila, Mucor, Myrothecium verrucaria, Paecilomyces variotii, Penicillium commune, Penicillium chrysogenum, Penicillium rubrum, Penicillium sp., Phlebia subserialis, Phlebiopsis gigantea, Penicillium helicum, Rhizopus oryzae, Skeletocutis sp., Stachybotrys atra, Thanatephorus cucumeris, Toxicocladosporium irritans, Trichoderma, Ulocladium botrytis*	6,8,10,12
human safety risk	*Bacillus cereus, Klebsiella pneumoniae, Staphylococcus aureus*	15
industrial paper	*Alternaria, Aspergillus, Chaetomium, Cladosporium, Eladia sacculum, Paecilomyces, Penicillium, Stachybotrys*	3
kraft-lignin	*Bacillus cereus, Bacillus pumilus, Bacillus sphaericus, Bacillus subtilis, Bacillus sp., Bacillus thuringiensis*	1
odors and tastes	*Clostridium spp., Desulfovibrio spp., actinobacteria*	15
paper	*Aspergillus terreus, Chaetomium elatum, Penicillium chrysogenum, Stachybotrys atra*	2
paper mill slime	*Bacillus psychrophilus, Bacillus sp., Beggiatoa, Chryseobacterium, Cytophaga sp., Flavobacterium columnare, Flectobacillus sp., Herpetosiphon, Norcardiopsis alba, Paenibacillus sp., Sphaerotilus sp., Streptomyces albidoflavus, Streptomyces limosus, Streptomyces odorifer*	9
paper quality reduction	*Bacillus cereus, Staphylococcus aureus*, coliforms, molds	15
pitch deposit formation	24 fungal and bacterial strains caused precipitation of resin	4
production of slime	*Bacillus spp., Burkholderia cepacia, Deinococcus spp.*	15
recycled liquid containers	*Acremonium, Aspergillus, Bacillus subtilis, Oidiodendron, Pantoea agglomerans, Penicillium, Rhizopus, Scopulariopsis, Trichoderma, Verticillium*	7,15

4.15.2 ESSENTIAL PRODUCT COMPONENTS

Paper consists mainly of cellulose, plus minor components that could be the targets of degradation by accessory enzyme activities, e.g., proteases and amylases acting on glues or additives.[3] Celullolytic, proteolytic, and amylolytic activities of species determine their potential for paper biodeterioration.[3]

Deacidification is a chemical process involving the neutralization of the acids present in the paper and the deposition of an alkaline buffer to prevent, or at least retard, further acidification.[13] 3-aminopropylmethyldiethoxysilane is the most effective deacidifying agent, showing the highest fungi inhibitory activity.[13]

Uncooked starch suspensions can easily spoil.[16] If suspensions are to be held for more than a few hours, a preservative (biocide) needs to be added. Some commonly used reagents, such as chlorine or sodium hypochlorite, will attack (oxidize) starch.[16] This reaction has to be considered when choosing a preservative.[16]

4.15.3 RESULTS OF BIODETERIORATION

The activity of fungi on paper can produce several types of damage, such as different colored stains, discoloration of pigments and inks, and changes in the chemical and physical properties of the paper.[5]

Conversion of cellulose into humic acids was detected in recycled containers stored for 6-18 months.[7] Concentration of CO_2 inside the bale of recycled board is an indicator of the activity of aerobic microorganisms and might be suitable for identifying deteriorated bales and removing them from the production line.[7] The concentration of the humic acid fraction varied in the range 0.3 - 0.6% of the organic matter in fresh bales and 2.2% in one old bale.[7]

Accumulation of pigments that may stain their support causes a phenomenon referred to as foxing.[8] Some foxing stains occur as reddish and/or brownish marks on the paper support, and are thought to have both biotic and abiotic origins, from the metabolic activity of microorganisms, to metal or ink oxidation.[8] However, the majority of foxing stains are supposedly due to the effect of microorganisms, or their metabolites.[8] Foxing may occur on other non-cellulosic materials, as long as an optimum environment for "foxing-causing fungi" is present, because cellulose is not always an essential nutrient for fungal growth.[8]

Biocides are added to control:

- formation of deposits on paper
- brightness
- strength of fibers
- holes in paper
- pH
- odors
- corrosion of machinery
- degradation of additives such as starch, polymers, and fillers
- biodeterioration of paper.

4.15.4 BIOSTABILIZATION

Table 4.117 contains a list of the most suitable biocides used for protection of pulp and paper.

Table 4.117. Biocides used in pulp and paper.

Material	Biocide	Refs.
books and paper	aminoalkylalkoxysilanes	13
construction felt paper	zinc pyrithione	16

Table 4.117. Biocides used in pulp and paper.

Material	Biocide	Refs.
fresh and circuit water control	2,2-dibromo-3-nitrilo-propanamide, glutaraldehyde, chlorine dioxide, 2-mercapto-benzothiazole, peracetic acid, quaternary ammonium compounds, sodium hypochlorite, sodium tosyl-chloramide	14
historical paper	0.5% methyl paraben and 1% propyl paraben in 85% ethanolic solution[5]	5
pitch deposit prevention	chlorine- or oxygen-releasing bleaching agents and elevated temperature	4
slimicides	1, 2-benzisothiazolin-one, bronopol, 2,2-dibromo-3-nitrilo-propanamide, didecyl ammonium chloride, glutaraldehyde, 2-methyl-4-isothiazolin-3-one, peracetic acid, sodium dimethyl-thiocarbamate	14

Oxidizing biocides are inexpensive and effective in use but they may cause corrosion, destroy felt, and affect dyes and other additives.

REFERENCES

1 Raj, A; Reddy, M M K; Chandra, R; Purohit, H; Kapley, A, *Biodegradation*, **18**, 783-92, 2007.
2 Fabbri, A A; Ricelli, A; Brasini, S; Fanelli, C, *Intl. Biodet Biodeg.*, **39**, 1, 61-65, 1997.
3 Rojas, J A;Cruz, C; Mikan, J F; Villalba, L S; de Garcia, M C C; Restrepo, S, *Intl. Biodet. Biodeg.*, **63**, 169-75, 2009.
4 Stranger-Johannessen, *Biotech. Adv.*, **2**, 319-27, 1984.
5 Neves, E R; Schaefer, S; Phillips, A; Canejo, J; Macedo, M F, *Intl. Biodet. Biodeg.*, **63**, 267-72, 2009.
6 Michaelsen, A; Pinar, G; Montanari, M; Pinzari, F, *Intl. Biodet. Biodeg.*, **63**, 161-68, 2009.
7 Koivula, N; Hanninen, K, *Chemosphere*, **38**, 8, 1873-87, 1999.
8 Mesquita, N; Portugal, A; Videira, S; Rodriguez-Echeverria, S; Bandeira, A M L; Santos, M J A; Freitas, H, *Intl. Biodet. Biodeg.*, **63**, 626-29, 2009.
9 Oppong, D; King, V M; Bowen, J A, *Intl. Biodet. Biodeg.*, **52**, 53-62, 2003.
10 Michaelsen, A; Pinzari, F; Ripka, K; Lubitz, W; Pinar, G, *Intl. Biodet. Biodeg.*, **58**, 133-41, 2006.
11 Leone, R; Breuil, C, *Intl. Biodet. Biodeg.*, **41**, 133-37, 1998.
12 Green, P W C, *J. Stored Prod. Res.*, **44**, 258-63, 2008.
13 Rakotonirainy, M S; Dupont, A L; Lavedrine, B; Ipert, S; Cheradame, H, *J. Cultural Heritage*, **9**, 54-59, 2008.
14 European Commission. Article 5(3) of Regulation. Status on June 3, 2009.
15 Viikari, L; Suurnäkki, A; Grönqvist, S; Raaska, L; Ragauskas, A, **Encyclopedia of Microbiology**, *Elsevier*, 2009.
16 Maurer, H W, **Starch**, *Elsevier*, 2009, pp 657-713.
17 Blanpied, R H; Murphy, F L, **US Patent** 7,410,553, Atlas Roofing Corporation, August 12, 2008.

4.16 ROOFING MATERIALS

4.16.1 RELEVANT MICROORGANISMS

Table 4.118 contains a list of microorganisms found to cause deterioration of roofing.

Table 4.118. Microorganisms in roofing.

Material	Microorganisms	Refs.
algae-resistant roofing	*Gloeocapsa magma*	8

Table 4.118. Microorganisms in roofing.

Material	Microorganisms	Refs.
asbestos cement roofing	*Acarospora cervina, Acarospora sp., Aspicilia sp, Buellia subalbula, Caloplaca cirrochroa, Caloplaca citrina, Caloplaca holocarpa, Candelariella vitellina, Lecanora dispersa, Leparia sp.* and blue-green algae, *Nitrobacter sp., Nitrosomonas sp., Pbyscia adscendens, Thiobacillus thioparus, Toninia caeruleonigricans, Xanthoria parietina*	1,3,5
flagstone (cathedral roofing)	*Aspicilia hoffmannii, Aspicilia intermutans, Lecanora murali*	4
polypropylene wood composite	*Gloeophyllum trabeum, Phanerochaete chrysosporium, Postia placenta, Trametes versicolor, Xylophagus fungi*	6
roofing granules	*Gloeocapsa magma*	2
roofing slates	*Leptospirillum ferrooxidans, Thiobacillus ferrooxidans*	7

4.16.2 ESSENTIAL PRODUCT COMPONENTS

Many species can be present on roofing, including cyanobacteria, algae, lichens, mildew, moss, etc.[2] A lichen is a symbiotic combination of algae with a fungus; a fungus is a primitive plant-like organism that lacks chlorophyll.[2] Mildew (e.g., on wood) is due to various species of fungi.[2] In the roofing industry, it is often not known exactly what species make up the biomass deposits on roofing.[2] In fact, the terms algae and fungi are sometimes used interchangeably.[2]

4.16.3 MECHANISMS OF BIODEGRADATION AND BIODETERIORATION

During the metabolic activity of nitrifying bacteria, HNO_2 and HNO_3 are produced, which with ammonia can produce NH_4NO_2 and/or NH_4NO_3.[3] The reaction of metabolites with lime components of asbestos-cement stone produces highly soluble calcium salts, which weakens material and makes it porous.[3]

It has been reported that many lichen species growing on ancient monuments create microclimatic effects at the thallus/substratum interface, which undoubtedly lead to mechanical damage to stonework.[4] In addition, organic acids secreted by the fungal hyphae etch the minerals of the substratum.[4] These effects have been observed in crustose lichens.[4]

Iron-oxidizing bacteria such as *Thiobacillus ferrooxidans* and *Leptospirillum ferrooxidans,* under highly acidic conditions, can catalyze the oxidation of Fe^{2+} to Fe^{3+}, and oxidation of pyrite by Fe^{3+} then becomes the main mechanism whereby acid is produced.[7] This mechanism is considered as the reason for corrosion of roofing slates.[7]

4.16.4 RESULTS OF BIODETERIORATION

It is well known that in humid climates, such as the southeastern United States, the growth of biomass on roofing leads to visible stains.[2] In some cases, it has been noted that staining does not occur near copper and galvanized (zinc) flashings.[2]

A number of light-colored PVC roof membranes that had been exposed for a number of years had obvious biological growth, presumably fungal in nature.[2]

Asbestos–cement roofs, the most widespread sources of airborne, toxic and carcinogenic asbestos fibres, are often colonized by lichens.[5] Fibre loss is significantly lower (<30%) where lichens develop and offer a physical barrier to the fibre detachment.[5] Below the most covering lichens (*Acarospora cervina, Candelariella sp.*), chrysotile and crocidolite undergo a partial incongruent dissolution.[5] Lichens act as effective spontaneous bioattenuation agents.[5]

4.16.5 BIOSTABILIZATION

Table 4.119 contains a list of the most suitable biocides used for protection of roofing.

Table 4.119. Biocides used in roofing.

Material	Biocide	Refs.
algae-resistant roofing	silver-coated zinc oxide particles	8
asbestos cement roofing	cupric ethanolamine, cupric sulphate, barium metaborate	1
PVC membrane	2-n-octyl-4-isothiazolin-3-one	2

REFERENCES

1 Martin, A; Johnson, G C; McCarthy, D F; Filson, R B, *Intl. Biodet. Biodeg.*, **30**, 261-71, 1992.
2 Berdahl, P; Akbari, H; Levinson, R; Miller, W A, *Construction Building Mater.*, **22**, 423-33, 2008.
3 Wasserbauer, R; Zadak, Z; Novotny, J, *Intl. Biodet.*, **24**, 153-65, 1988.
4 Garcia-RoweJ; Saiz-Jimenez, C, *Intl. Biodet.*, **28**, 151-63, 1991.
5 Favero-Longoa, S E; Castelli, D; Fubini, B; Piervittori, R, *J. Hazardous Mater.*, **162**, 1300-8, 2009.
6 Lomeli-Ramirez, M G; Ochoa-Ruiz, H G; Fuentes-Talavera, F J; Garcia-Enriquez, S; Cerpa-Gallegos, M A; Silva-Guzman, J A, *Intl. Biodet. Biodeg.*, **63**, 1030-35, 2009.
7 Van den Eynde, V C; Paradelo, R; Monterroso, C, *Corrosion Sci.*, **51**, 2387-92, 2009.
8 Kalkanoglu, H M; Hong, K C; Shiao, M L; Jacobs, G F, **US Patent** 7,595,107, CertainTeed Corporation, Septemebr 29, 2009.

4.17 RUBBER

4.17.1 RELEVANT MICROORGANISMS

Table 4.120 contains a list of microorganisms found to cause deterioration of rubber.

Table 4.120. Microorganisms in rubber.

Material	Microorganisms	Refs.
natural latex	*Aerobacter, Azotobacter, Bacillus, Clostridium, Escherichia, Lactobacillus, Micrococcus, Propionibacterium, Pseudomonas, Sarcina, Serratia, Streptomyces*	11
natural rubber	*Bacillus subtilis, Gordonia sp., Gordonia polyisoprenivorans, Micromonospora aurantiaca, Mycobacterium fortuitum, Pseudomonas aeruginosa*	4,5

Table 4.120. Microorganisms in rubber.

Material	Microorganisms	Refs.
rubber (general)	*Acremonium polychromum, Acremonium rutilum, Acremonium strictum, Aspergillus amstelodami, Aspergillus niger, Aspergillus ruber, Aspergillus ustus, Aspergillus versicolor, Aureobasidium pullulans, Cladosporium sphaerospermum, Mortierella isabellina, Mucor alboater, Oidiodendron echinulatum, Paecilomyces variotii, Penicillium lanosum, Penicillium meleagrinum, Penicillium nalgiovense, Penicillium ochrochloron, Phoma cava, Phoma eupyrena, Scopulariopsis brevicaulis, Ulocladium atrum, Ulocladium oudemansii*	3
rubber straps in Russian space station	*Aspergillus sp., Fusarium sp., Penecillium sp.*	9
tire rubber	*Nocardia sp.*	2

4.17.2 ESSENTIAL PRODUCT COMPONENTS

A decrease of 60.88% MPa and 41.66% MPa was observed for uncrosslinked and dicumyl peroxide-crosslinked natural rubber, respectively, which was exposed to *Pseudomonas sp.*[7]

4.17.3 MECHANISMS OF BIODEGRADATION AND BIODETERIORATION

Biochemical oxidation of natural rubber was reported.[8] The infrared studies show an increase in the concentration of carbonyl groups during exposure. *Nocardia sp.* was able to cleave double bonds in natural rubber.[2,8] Degrading enzymes are secreted by microorganisms.

4.17.4 RESULTS OF BIODETERIORATION

Various microbiological process were implicated in biodegradation of latex, smoke sheet, and manufactured goods. Loss of weight up to 43%, reduction of molecular weight, degradation of double bonds, changes in color and formation of color spots were the most frequently observed results of biodeterioration.[11]

4.17.5 BIOSTABILIZATION

Table 4.121 contains a list of the most suitable biocides used for protection of rubber.

Table 4.121. Biocides used in rubber.

Material	Biocide	Refs.
natural rubber	boric acid esters	1
rubber (general)	1, 2-benzisothiazolin-3-one,[12] titanium oxide nanoparticles[10]	10,12

The proliferation of *Legionella pneumophila* in plumbing systems can be stimulated or inhibited by rubber materials used in the construction of building water services.[6] Accelerators such as tetramethyl thiuram disulfide and monosulfide (TMTS) may inhibit the microbial degradation.[6]

REFERENCES

1 Grachek, V I; Lukashik, A N, *Russ. J. Appl. Chem.*, **79**, 5, 819-22, 2006.
2 Tsuchii, A; Tokiwa, Y, J. *Biosci. Bioeng.*, **87**, 4, 542-44, 1999.
3 Lugauskas, A; Levinskaite, L; Peciulyte, D, *Intl. Biodet. Biodeg.*, **52**, 233-42, 2003.
4 Berekaa, M M; Linos, A; Reichelt, R; Keller, U; Steinbuechel, A, *FEMS Microbiol. Lett.*, **184**, 199-206, 2000.
5 Linos, A; Reichelt, R; Keller, U; Steinbuechel, A, *FEMS Microbiol. Lett.*, **182**, 155-61, 2000.
6 Colbourne, J S; Ashworth, J, *The Lancet*, **328**, 8506, 583, 1986.
7 Roy, R V; Das, M; Banerjee, R; Bhowmick, A K, *Bioresource Technol.*, **97**, 2485-88, 2006.
8 Tsuchii, A, *Prog. Ind. Microbiol.*, **32**, 177-87, 1995.
9 Gu, J-D, *Intl. Biodet. Biodeg.*, **59**, 170-79, 2007.
10 Goebbert, C, *Handbook of Cleaning/Decontamination of Surfaces*, *Elsevier*, 2007, pp 813-29.
11 Zyska, B, in Kauczuk i Guma, **Mikrobiologia Materialow**, Zyska, B; Zakowska, Z, Ed., *Politechnika Lodzka Press*, Lodz, 2005, 231-67.
12 Bioban Ultra BIT 20, Product information. DOW.

4.18 SEALANTS AND ADHESIVES

4.18.1 RELEVANT MICROORGANISMS

Table 4.122 contains a list of microorganisms found to cause deterioration of adhesives and sealants.

Table 4.122. Microorganisms in adhesives and sealants.

Material	Microorganisms	Refs.
acetoxy silicone sealants	*Anabaena flos-aquae, Aspergillus niger, Aureobasidium pullulans, Bacillus subtilis, Candida albicans, Chlorella pyrenoidosa, Chlorococcum oleofaciens, Cladosporium cladosporoides, Escherichia coli, Microcystis aeruginosa, Nostoc commune, Oscillatoria prolifera, Penicillium purpurogenum, Pseudomonas aeruginosa, Pseudomonas cepacia, Scenedesmus quadricauda, Scytonema hofmanni, Synechococcus leopoliensis, Staphylococcus aureus, Trichoderma viride, Ulocladium atrum, Ulothrix acuminata*	3
acrylic sealants	*Alternaria dianthicola, Ankistrodesmus falcatus, Aspergillus niger, Aspergillus oryzae, Aspergillus repens, Aureobasidium pullulans, Bacillus subtilis, Brevibacterium ammoniagenes, Candida albicans, Chaetomium globisum, Chlorella pyrenoldosa, Cladosporium resinae, Coccomyxa elongata, Enterobacter aerogenes, Escherichia coli, Lenzites lepideus, Lenzites trabea, Nostoc sp., Penicillium funiculosum, Phoma glomerata, Phoma pigmentivora, Phormidium sp., Phormidium retzil, Proteus mirabilis, Proteus vulgaris, Pseudomonas aeruginosa, Rhizopus stolonifer, Rhodotorula rubra, Saccharomyces cerevisiae, Salmonella typhosa, Scenedesmus obliquus, Staphylococcus aureus, Staphylococcus epidermidis, Trichophyton interdigitale*	6

4.18.2 BIOSTABILIZATION

Table 4.123 contains a list of the most suitable biocides for adhesives and sealants.

Table 4.123. Biocides used in adhesives and sealants.

Material	Biocide	Refs.
acetoxy silicone sealants	4.5-dichloro-2-n-octyl-4-isothiazolin-3-one	1
acrylic sealants	5-chloro-2-methyl-4-isothiazolin-3-one,[7] 2-methyl-4-isothiazolin-3-one,[7] 2-n-octyl-4-isothiazolin-3-one[6]	6,7
water-based sealants and adhesives	2-methyl-4-isothiazolin-3-one and 1,2-benzisothiazolin-3-one at a ratio of 1:1	2

4.18.3 FORMULATIONS

Low modulus acrylic sealant[4]

	wt%
Primal 3362 (62%)	32.1
Acrysol TT-615 (30%)	0.2
Tergitol 15-S-40 (70%)	0.8
Ethylene Glycol	0.1
Orotan 850	0.1
KTPP	0.1
Water	1.2
Jayflex DIUP	11.0
Durcal 2	0.8
Durcal 5	52.6
Silane Z-6040	0.8
Kathon LXE	0.1
Skane M-8	0.1

High performance acrylic sealant[5]

	weight parts
Rhoplex A-920 (62%)	468.5
Water	1.5
Octyphenoxy polyethoxy ethanol, (70%)	11.4
Kathon™ LX, 1.5%	1.5
Ethylene glycol	14.3
Tamol™ 850	1.4
Skane™ M-8	1.1
Defoamer (Nopco NXZ)	2.0
Plasticizer (Santicizer S-160)	144.9
Thickener (Natrosol 250HR)	3.4
Mineral spirits (Varsol #1)	11.4
Adhesion promoter (Silane Z-6040)	0.7
Dispersant (KTPP)	1.4
Calcium carbonate (Drikalite)	512.2

Pigment (TiPure R-901)	11.4
Ammonium hydroxide (28% NH_3)	2.3

REFERENCES

1 Kathon 910 SB Biocide. Technical data sheet. Rohm and Haas. April 2009.
2 Rocima MBX. Technical data sheet. Rohm and Haas. March 2009.
3 Kathon 287 PXE. Technical data sheet. Rohm and Haas. November 2005.
4 Primal 3362. Technical data sheet. Rohm and Haas. January 2002.
5 Rhoplex A-920. Technical data sheet. Rohm and Haas. October 1997.
6 Skane M-8. Mildewcide. Technical data sheet. Rohm and Haas. September 2004.
7 Kathon LX 1.5% Microbiocide. Technical data sheet. Rohm and Haas. July 2007.

4.19 STONES INCLUDING CULTURAL HERITAGE

4.19.1 RELEVANT MICROORGANISMS

Table 4.124 contains a list of microorganisms found to cause deterioration of stones.

Table 4.124. Microorganisms in stones.

Material	Microorganisms	Refs.
black crust	*Aspergillus, Bacillus sp., Botriotrichum piluliferum, Engyodontium album, Fusurium, Gloeothece sp., Gloeocupsa, Mucor circinelloides, Nocardia sp., Perticillium, Phormidium sp., Pseudomonas sp., Ulocludium atrum*	6
concrete	*Fusarium sp., Thiobacillus intermedius, Thiobacillus neopolitabus, Thiobacillus sp., Thiobacillus thiooxidans*	5
Italy (Roman catacombs – heterotrophic microorganisms)	*Acinetobacter radioresistens, Amycolatopsis azurea, Amycolatopsis coloradensis, Arthroderma curreyi, Acremonium bacillisporum, Acremonium sp., Aspergillus ustus, Aspergillus sp., Bacillus mycoides, Bacillus pumilus, Bacillus thuringiensis, Beauveria alba, Buttiauxella izardii, Cladosporium sp., Cladosporium cladosporioides, Crossiella cryophila, Cryptococcus albidus, Cryptococcus curvatus, Debaryomyces hansenii, Dermacoccus nishinomiyaensis, Dietzia maris, Enterobacter intermedius, Fusarium sp., Janibacter limosus, Kluyvera cochleae, Lecanicillium aranearum, Lecanicillium psalliotae, Microbacterium liquefaciens, Microbacterium maritypicum, Microbacterium resistens, Micrococcus luteus, Mycobacterium fortuitum, Nocardioides albus, Nocardioides fulvus, Nocardioides flavus, Nocardia asteroides, Nocardia ignorata, Oxalobacter sp., Paenibacillus polymyxa, Paenibacillus burgondia, Pectobacterium cypripedii, Penicillium brevicompactum, Penicillium chrysogenum, Penicillium glabrum, Penicillium minioluteum, Penicillium pinophilum, Penicillium sp., Phialophora cinerescens, Promicromonospora aerolata, Pseudomonas alcaligenes, Pseudomonas asplenii, Pseudomonas fuscovaginae, Pseudomonas gingeri, Pseudomonas jessenii, Pseudomonas putida, Rhodotorula glutinis, Rhodotorula minuta, Sinorhizobium morelense, Sporobolomyces salmonicolor, Staphylococcus capitis, Staphylococcus epidermidis, Staphylococcus hominis, Staphylococcus warneri, Stenotrophomonas maltophilia, Stenotrophomonas nitritireducens, Streptomyces avendulae, Streptomyces exfoliatus, Streptomyces griseomycini, Streptomyces griseus, Streptomyces venezuelae, Streptomyces virginiae, Streptomyces vulgaris, Streptomyces zaomyceticus, Torrubiella sp.*	14

Table 4.124. Microorganisms in stones.

Material	Microorganisms	Refs.
Italy (most common Cyanobacteria, Chlorophyta, and Bacillariophyta)	*Apatococcus lobatus, Chlorella fusca, Chlorella homosphaera, Chroococcus minor, Gloeocapsa biformis, Klebsormidium accidum, Microcoleus vaginatus, Muriella terrestris, Myxosarcina concinna, Navicula mutica, Nitzschia communis, Nostoc punctiforme, Phormidium autumnale, Phormidium tenue, Plectonema boryanum, Stichococcus bacillaris*	7
main groups of microorganisms	Actinomycetes, Algae, Archaea (halophilic bacteria), Chemolithotrophic bacteria (nitrogen cycle and sulfur cycle), Chemoorganitrophic bacteria (acidobacteria), Chloriflexi, Cyanobacteria, Fungi, Lichens	8
mortars (original Roman and repair in Spain)	*Aspicilia contorta, Aspicilia hoffmanii, Bryum bicolor, Bryum radiculosum, Caloplaca aurantia, Caloplaca chalybaea, Caloplaca citrina, Caloplaca erythrocarpa, Caloplaca flavescens, Caloplaca lactea, Caloplaca teicholyta, Caloplaca variabilis, Caloplaca velana, Candelariella aurella, Catapyrenium sp., Catapyrenium crispum, Catapyrenium squamulosum, Collema crispum, Clauzadea immersa, Collema auriforme, Collema crispum, Collema tenax, Dirina massiliensis f. sorediata, Endocarpon pusillum, Fulgensia sp., Lecania turicencis, Lecanora albescens, Lecanora dispersa, Lecanora pruinosa, Lepraria lesdainii, Lepraria nivalis, Lepraria sp., Lepraria lesdainii, Leproplaca xantholyta, Opegrapha calcarea, Pseudosagedia linearis, Rinodina bischoffii, Sarcogyne regularis, Squamarina concrescens, Toninia aromatica, Tortula brevissima, Verrucaria hochstetteri, Verrucaria macrostoma, Verrucaria muralis, Verrucaria nigrescens, Verrucaria viri*	13,16
Spain (heterotrophic bacteria)	*Bacillus cereus, Bacillus cereus, Bacillus circulans, Bacillus licheniformis, Bacillus polymyxa, Microciccus agilis, Micrococcus halobius, Microciccus luteus, Microciccus roseus, Micrococcus varians, Thiobacillus albertis, Thiobacillus denitrifikans, Thiobacillus neapolitanus, Thiobacillus thioparus, Thiobacillus thioxidans, Thiobacillus versutus*	12
Turkey (granite, limestone, murduge, tufa)	*Alternaria alternata, Arconobacterium haemolyticum, Bacillus agri, Bacillus cereus, Bacillus coagulans, Bacillus lentimorbus, Bacillus licheniformis, Bacillus mycoides, Bacillus subtilis, Bacillus thuringiensis kenyae, Bacillus thuringiensis sotto, Brevibacillus laterosporus, Cladosporium orchidis, Embellicia hyacinthi, Paecilomyces lilacinus, Paenibacillus apiarius, Paenibacillus larvae-pulvifaciens, Penicillium frequentans, Phoma sp., Streptoverticillium reticulum, Virgibacillus pantothenticus*	2

4.19.2 ESSENTIAL PRODUCT COMPONENTS

Biological infections and the intensity of the biodeterioration processes are strongly influenced by water availability.[1] This is determined by both material-specific parameters, such as porosity and permeability and by environmental conditions of the site and exposure of the object.[1] The presence of an extensive inner pore surface (>5m^2/g as determined by BET method) resulting from a high porosity or the presence of clay minerals, facilitates the spreading of microflora within the pore system.[1] On the other hand, large-pore sandstones, due to their short water retention, promote microbial contamination only temporarily, small-pore stones, with longer water retention time, offer more suitable conditions for the settlement of stone-colonizing microorganisms.[1]

Stones containing significant amounts of weathering-prone minerals (i.e., feldspars, clays, and ferruginous minerals, etc.) are particularly susceptible to hosting microorganisms.[1]

Cement-based materials now contain more organic species used as adjuvants in order to enhance materials properties, but they bring organic matter to the matrix, and thus could enhance its primary bioreceptivity.[3]

Photosynthetic micro/macro-organisms such as mosses and lichens were not dominant in Turkish monuments because the extreme cold weather in winter time restricted photosynthesis in Erzurum.[2] Heterotrophic microflora, composed of fungi, actinomycetes and bacteria dominate the stone surfaces in all of the stone monuments during the winter months.[2]

Comparison of climatic conditions in Rome shows that pollution has increased and rainfall substantially decreased in the 20th century as compared with the 19th century.[4] This has implications for biodeterioration. Increased air pollution caused the reduction of biodeterioration phenomena during the 20th century.[4] Air pollutants caused the disappearance of the most sensitive species, such as many lichens, and caused a reduction of the floristic diversity. However, resistant species among algae, cyanobacteria, fungi and also lichens (e.g., *Lecanora muralis* and *Lecanora dispersa*) can still survive in favorable microclimatic conditions forming large populations.[4] The substantial reduction of rainfall and humidity reduces the rate and extent of biodeterioration.[4]

Populations of chemolithotrophic and heterotrophic bacteria and fungi were suppressed in the polluted habitat comparing with the unpolluted location.[9] The diversity of the microflora was also reduced in the surface biofilms on gravestones in the city contaminated by air pollution.[9]

4.19.3 MECHANISMS OF BIODEGRADATION AND BIODETERIORATION

Microbial colonization is most frequently initiated with phototrophic microorganisms such as cyanobacteria and algae.[8] Lichens follow. The accumulation of photosynthetic biomass provides a nutrient base for the subsequent heterotrophic microflora.[8] Colonization by heterotrophic communities is also possible without the pioneering participation of phototrophic organisms.[8] This may facilitate the subsequent growth of photosynthetic populations.[8]

Chemolithotrophic microorganisms use, as an energy source, reduced sulphur and ammonium compounds that are oxidized to sulfates and nitrates. The degradative action is due to the release of their autotrophic metabolism end-products as sulfuric, nitric, and nitrous acids.[11] The degradation to the stone caused by these bacteria is similar to that caused by atmospheric chemical pollutants in urban areas.[11]

4.19.4 RESULTS OF BIODETERIORATION

It has been calculated that, for limestone, an average of 1.5–3 mm of rock will erode away in 100 years in temperate climates, leading to the disappearance of inscriptions on tombstones in the United Kingdom within 300 years.[8] Part of it is due to the action of microorganisms.

Microorganisms are able to remove some of the most abundant components of black crusts, such as gypsum and polycyclic aromatic hydrocarbons.[6] This shows that nature develops strategies for biodegradation of pollutants in urban environments.[6]

Mechanical damage by lichens is due to penetration of the substrate by the hyphae.[15] The depth to which the thallus penetrates depends on the lichen species and the nature of the substrate.[15]

4.19.5 BIOSTABILIZATION

Table 4.125 contains a list of the most suitable biocides used for protection or remediation condition of stones.

Table 4.125. Biocides used in stones

Material	Biocide	Refs.
algal antifouling	2,3,5,6-tetrachloro-4-methylsulfonyl-pyridine and silane-siloxane mixture as water repellent (autoclaved aerated concrete), 2,3,5,6-tetrachloro-4-methylsulfonyl-pyridine and alkyl alkoxy silane (white architectural concrete)	19
concrete mix	addition of photocatalytic titanium dioxide	20
lichens	alkyleneoxide (97.6%), alkylaminotriazine (0.98%), N'-(3,4-dichlorophenyl)-N,N-dimethyl urea (0.98%), and denaturated alkyltrihydroxybenzene polyoxide (0.48%) (Koretrel),[10] 2,3,5,6-tetrachloro-4- methylsulfonyl-pyridine[18]	10,18

It is well-known that applying a biocide on stone before a water-repellent leads to a diminishment of the hydrophobic effect of the treatment. The application of biocide after a water-repellent treatment affects the stone's hydrophobicity.[17]

REFERENCES

1 Warscheid, T; Braams, J, *Intl. Biodet. Biodeg.*, **46**, 343-68, 2000.
2 Nuhoglu, Y; Oguz, E; Uslu, H; Ozbek, A; Ipekoglu, B; Ocak, I; Hasenekoglu, I, *Sci. Total Environ.*, **364**, 272-83, 2006.
3 Wiktor, V; De Leo, F; Urzi, C; Guyonnet, R; Grosseau, P; Garcia-Diaz, E, *Intl. Biodet. Biodeg.*, **63**, 1061-65, 2009.
4 Caneva, G; Gori, E; Montefinale, T, *Sci. Total Environ.*, **167**, 205-14, 1995.
5 Gu, J-D; Ford, T E; Berke, N S; Mitchell, R, *Intl. Biodet. Biodeg.*, **41**, 101-9, 1998.
6 Saiz-Jimenez, C, *Intl. Biodet. Biodeg.*, **40**, 2-4, 225-32, 1997.
7 Tomaselli, L; Lamenti, G; Bosco, M; Tiano, P, *Intl. Biodet. Biodeg.*, **46**, 251-58, 2000.
8 Scheerer, S; Ortega-Morales, O; Christine Gaylarde, C, in **Microbial Deterioration of Stone Monuments—An Updated Overview, Advances in Applied Microbiology**, Vol. 66, *Elsevier*, 2009, pp 97-139.
9 Mitchell, R; Gu, J-D, *Intl. Biodet. Biodeg.*, **46**, 299-303, 2000.
10 Tretiach, M; Crisafulli, P; Imai, N; Kashiwadani, H; Moon, K H; Wada, H; Salvadori, O, *Intl. Biodet. Biodeg.*, **59**, 44-54, 2007.
11 Bartolini, M; Monte, M, Proceedings of the 9th International Congress on Deterioration and Conservation of Stone, 2000, pp 453-460.
12 Flores, M; Lorenzo, J; Gomez-Alarcon, G, *Intl. Biodet. Biodeg.*, **40**, 2-4, 241-46, 1997.
13 Puertas, F; Blanco-Varela, M T; Palomo, A; Ortega-Calvo, J J; Arino, X; Saiz-Jimenez, C, *Sci. Total Environ.*, **153**, 123-31, 1994.
14 Saarela, M; Alakomi, H-L; Suihko, M-L; Maunuksela, L; Raaska, L; Mattila-Sandholm, T,

Intl. Biodet. Biodeg., **54**, 27-37, 2004.
15 Lisci, M; Monte, M; Pacini, E, *Intl. Biodet. Biodeg.*, **51**, 1-17, 2003.
16 Arin, X; Gomez-Bolea, A; Saiz-Jimenez, C, *Intl. Biodet. Biodeg.*, **40**, 2-4, 217-24, 1997.
17 Moreau, C; Verges-Belmin, V; Leroux, L; Orial, G; Fronteau, G; Barbin, V, *J. Cultural Heritage*, **9**, 394-400, 2008.
18 Urzi, C; De Leo, F, *Intl. Biodet. Biodeg.*, **60**, 25-34, 2007.
19 De Muynck, W; Ramirez, A M; De Belie, N; Verstraete, W, *Intl. Biodet. Biodeg.*, **63**, 679-89, 2009.
20 Giannantonio, D J; Kurth, J C; Kurtis, K E; Sobecky, P A, *Intl. Biodet. Biodeg.*, **63**, 252-59, 2009.

4.20 SWIMMING POOLS

4.20.1 RELEVANT MICROORGANISMS

Table 4.126 contains a list of microorganisms found in swimming pools, spas, etc.

Table 4.126. Microorganisms in swimming pools.

Material	Microorganisms	Refs.
bacteria	*Escherichia coli, Mycobacterium marinum, Pseudomonas aeruginosa, Staphylococcus aureus, Staphylococcus epidermidis, Streptococcus faecalis, Streptococcus salivarius*	3
fungi	*Trichophyton mentagrophytes var. granulosum*	3
protozoa	*Naegleria fowleri*	3
swimming pools in Grece	*Aeromonas hydrophila, Alternaria spp., Aspergillus spp., Candida albicans, Chryseobacterium indologenes, Enterobacter cloacae, Klebsiella pneumoniae, Leuconostoc, Mucor spp., Ochrobactrum anthropi, Penicillium spp., Pseudomonas alcaligenes, Pseudomonas aeruginosa, Pseudomonas fluorescens, Rhizopus spp., Staphylococcus aureus, Staphylococcus wernerii, Trichophyton spp.*	1
test organisms	*Candida albicans, Escherichia coli, Legionella pneumophila, Pseudomonas aeruginosa, Staphylococcus aureus*	4
viruses	Adenovirus, Coxsackie, hepatitis B	3
water-borne pathogens	*Acanthamoeba, Aeromonas, Ancylostoma duodenale, Ascaris lumbricoides, Balantidium coli, Blastocystis hominis, Campylobacter* species, *Clonorchis sinensis, Cryptosporidium parvum, Cyclospora, Diphyllobothrium latum, Dracunculus medinensis, Echinococcus granulosus, Entamoeba histolytica, Fasciola hepatica, Giardia intestinalis,* Hepatitis E, *Isospora belli,* Miscellaneous enteric viruses (>100 types), Norovirus, *Paragonimus westermani,* Poliovirus, *Salmonella* (primarily *typhi*) species, *Shigella species, Strongyloides stercoralis, Trichuris trichiura, Vibrio cholerae, Yersinia enterocolitica*	7

4.20.2 ESSENTIAL PRODUCT COMPONENTS

At the onset of the outdoor swimming season, a new, freshly-filled pool of balanced water is an essentially oligotrophic environment providing relatively low microbial growth potential.[3] As the season progresses and the level of nutrients in the pool rises, the environment becomes increasingly more conducive to microbial growth.[3]

4.20.3 MECHANISMS OF BIODEGRADATION AND BIODETERIORATION

The following features are expected from biocide:[3]

- broad-spectrum (antibacterial, antialgal, antifungal, antiviral, antiprotozoal)
- activity at low concentration
- rapid rate of kill
- persistence and retention of activity in presence of organic and inorganic loading
- minimal development of microbial resistance in the presence of use levels of the agent
- non-volatile
- stable in the presence of light
- compatible with chlorine and other oxidizing agents
- no complex formation with copper ions
- low haze-forming potential
- no smell or taste imparted to the water
- quantitatively determined with the aid of a simple-to-use poolside test kit
- safe to handle in concentrated form
- presents no hazard to the environment when the pool water is discharged into rivers, draining systems, or on to the soil
- biodegradable

Chlorine reacts rapidly with water to form hypochlorous acid which is considered to be the most potent biocide in the aqueous mixture, which also contains inactive chloride and hypochlorite ions.[5] The rate of microbial kill depends on pH. Usually slightly acidic or neutral pH gives sufficiently fast kill rates.[5]

4.20.4 RESULTS OF BIODETERIORATION

Antimicrobial agents must be added to pool water to eliminate pathogens and control non-pathogenic organisms (e.g., algae), which create undesirable visible effects.[3]

4.20.5 BIOSTABILIZATION

Table 4.127 contains a list of the most suitable biocides used for protection in swimming pools.

Table 4.127. Biocides used in swimming pools.

Material	Biocide	Refs.
biofilm removal	polyhexamethylene biguanide, super chlorination (information below the table)	2
private swimming pools (most common)	calcium hypochlorite, chloro isocyanurate, sodium hypochlorite, hydrogen peroxide, silver ions	4
public swimming pools (most common)	chlorine, chloro isocyanurate, ozone, sodium hypochlorite, UV	4
swimming pools in Greece	chlorine+cyanurated chlorine, coagulation–filtration– ozonization–activated carbon filtration–chlorination (hydrotherapy pool)	1

Shock treatments (10 mg/l polyhexamethylene biguanide) were effective at minimizing biofilm accumulation.[2] 16 h super-chlorination (free chlorine concentration was maintained between 1 and 3 ppm. Every Wednesday, the system was dosed with 10 ppm free chlorine) was as effective as a 64 h super-chlorination against biofilm accumulation.[2]

All cells were completely killed by sodium hypochlorite after 10, 20 and 30 minutes.[4] No significant microbicidal effect was found with regard to the products based on hydrogen peroxide, which were tested up to a concentration of 150 ppm hydrogen peroxide.[4] It can be expected that hydrogen peroxide has quite a good biostatic activity in the tested concentration range.[4] However, a quick biocidal (killing) effect is required in order to avoid spreading of infections from one swimmer to the next.[4]

It was reported that the inactivation efficiencies of *Escherichia coli* and MS-2 phage by silver ions were significantly enhanced by UV-A or visible light irradiation.[6] The photochemical reaction of silver–cysteine complex was suggested as a possible mechanism for this enhancement.[6]

4.20.6 FORMULATIONS

Thickened bleach formulation

Cocoalkyl betaine (or cocodimethyl amine oxide)	0.72%
Sodium laurate	0.28%
Perfume	0.08%
Sodium hydroxide	0.05%
Sodium hypochlorite (15% chlorine)	66.6%
Water	to 100%

REFERENCES

1 Papadopoulou, C; Economou, V; Sakkas, H; Gousia, P; Giannakopoulos, X; Dontorou, C; Filioussis, G; Gessouli, H; Karanis, P; Leveidiotou, S, *Int. J. Hyg. Environ.-Health*, **211**, 385-97, 2008.
2 Goeres, D M; Palys, T; Sandel, B B; Geiger, J, *Water Res.*, **38**, 3103-9, 2004.
3 Singer, M, *Intl. Biodet.*, **26**, 159-68, 1990.
4 Borgmann-Strahsen, R, *Intl. Biodet. Biodeg.*, **51**, 291-7, 2003.
5 Karsa, D R in Biocides, **Handbook for Cleaning/Decontamination of Surfaces**, *Elsevier*, 2007, pp 593-623.
6 Kim, J Y; Lee, C; Cho, M; Yoon, J, *Water Res.*, **42**, 356-62, 2008.
7 Backer, H, Water Disinfection for International Travelers, **Travel Medicine**, 2nd Ed., *Elsevier*, 2008, pp 47-58.

4.21 WATER

This section contains discussion on treatment in the prevention of biofouling.

4.21.1 RELEVANT MICROORGANISMS

Table 4.128 contains a list of microorganisms found in water biofouling.

Table 4.128. Microorganisms in water biofouling.

Material	Microorganisms	Refs.
copper pipe corrosion	*Acidovorax delafieldii, Acidovorax sp., Cytophaga johnsonae, Flavobacterium sp., Micrococcus kristinae, Pseudomonas fluorescens, Pseudomonas paucimobilis, Pseudomonas solanacearum, Rhodotorula sp., Sphingomonas sp.*	10

Table 4.128. Microorganisms in water biofouling.

Material	Microorganisms	Refs.
dental water	*Aeromonas salmonicida, Acidovorax temperans, Comamonas acidovorans, Novosphingobium subarctica, Sphingomonas paucimobilis, Sphingomonas aerolata, Sphingomonas trueperi, Porphyrobacter tepidarius, Pseudomonas fluorescens, Arthrobacter agilis, Kocuria palustris, Microcococcus luteus, Rhodococcus fascians, Variovorax paradoxus*	4
drinking water (*Mycobacteria* only)	*Mycobacterium abscessus, Mycobacterium aurum, Mycobacterium avium, Mycobacterium brumae, Mycobacterium chelonae, Mycobacterium chlorophenolicus, Mycobacterium flavescens, Mycobacterium fortuitum, Mycobacterium gadium, Mycobacterium gastri, Mycobacterium genavense, Mycobacterium gilvum, Mycobacterium gordonae, Mycobacterium haemophilum, Mycobacterium intracellulare, Mycobacterium kansasii, Mycobacterium lentiflavum, Mycobacterium malmoense, Mycobacterium marinum, Mycobacterium mucogenicum, Mycobacterium nonchromogenicum, Mycobacterium peregrinum, Mycobacterium phlei, Mycobacterium scrofulaceum, Mycobacterium septicum, Mycobacterium shimoidei, Mycobacterium simiae, Mycobacterium smegmatis, Mycobacterium szulgai, Mycobacterium terrae, Mycobacterium tusciae, Mycobacterium vaccae, Mycobacterium xenopi*	11
spring water	*Brucella melitensis, Escherichia coli, Leptospira icterohaemorrhagiae, Pasteurella tularensis, Pseudomonas pseudomallei, Salmonella hirschfeldii, Salmonella paratyphi, Salmonella schottimuelleri, Salmonella typhosa, Shigella dysenteriae, Shigella flexneri, Shigella sonnei, Shigella paradysenteriae, Vibrio cholerae, Vibrio comma*	9

4.21.2 ESSENTIAL PRODUCT COMPONENTS

Most of the bacteria in a cooling water system are likely to be in biofilm.[5] Almost immediately after water enters a pipe, organic molecules adhere to the surface.[5] In the initial stage, pioneer bacteria attach themselves by electrostatic attraction and physical forces.[5] The daughter cells produce a polysaccharide matrix which facilitates maturation of the biofilm.[5]

4.21.3 RESULTS OF BIODETERIORATION

Biofouling causes the following problems in water cooling systems:

- reduced heat transfer
- plugging pipes and reduction of flow rate
- corrosion.

4.21.4 BIOSTABILIZATION

Table 4.129 contains a list of the most suitable biocides used in water.

Table 4.129. Biocides used in water.

Material	Biocide	Refs.
cooling water	calcium or sodium hypochlorite, chlorine, chlorine dioxide, glutaraldehyde, ozone, peracetic acid, quats, thiocyanates	

Table 4.129. Biocides used in water.

Material	Biocide	Refs.
dental water	Ecasol (combination of chlorine-containing and ozone disinfectants), hydrogen peroxide	4,6
seawater cooling	chlorine	7
seawater desalination	nanosilver, hypochlorite, hypobromite	1, 8
spiral wound membrane	copper sulfate daily dosing	2
ultrafiltration membrane	silver nanoparticles	3

REFERENCES

1 Yang, H-L; Lin, J C-T; Huang, C, *Water Res.*, **43**, 3777-86, 2009.
2 Cornelissen, E R; Vrouwenvelder, J S; Heijman, S G J; Viallefont, X D; Van Der Kooij, D; Wessels, L P, *J. Membrane Sci.*, **287**, 94-101, 2007.
3 Zodrow, K; Brunet, L; Mahendra, S; Li, D; Zhang, A; Li, Q; Alvarez, P J J, *Water Res.*, **43**, 715-23, 2009.
4 O'Donnell, M J; Boyle, M; Swan, J; Russell, R J; Coleman, D C, *J. Dentistry*, **37**, 748-62, 2009.
5 Tanji, Y; Nishihara, T; Miyanaga, K, *Biochem. Eng. J.*, **35**, 81-86, 2007.
6 Walker, J T; Marsh, PD, *J. Dentistry*, **35**, 721-31, 2007.
7 Nebot, E; Casanueva, J F; Casanueva, T; Fernandez-Baston, M M; Sales, D, *Appl. Thermal Eng.*, **26**, 1893-1900, 2006.
8 Lattemann, S; Höpner, T, *Desalination*, **220**, 1-15, 2008.
9 Fotouhi, F; Kresic, N in Springwater treatment, **Groundwater Hydrology of Springs**, Elsevier, 2010, pp 269-304.
10 Reyes, A; Letelier, M V; De la Iglesia, R; Gonzalez, B; Lagos, G, *Intl. Biodet. Biodeg.*, **61**, 135-41, 2008.
11 Vaerewijck, M J M; Huys, G; Palomino, J C; Swings, J; Portaels, F, *FEMS Microbiol. Rev.*, **29**, 911-34, 2005.

4.22 WOOD

4.22.1 RELEVANT MICROORGANISMS

Table 4.130 contains a list of microorganisms found to cause deterioration of wood.

Table 4.130. Microorganisms in wood.

Material	Microorganisms	Refs.
Antarctica wood samples	*Alternaria sp., Ascomycete sp., Cadophora fastigiata, Cadophora luteo-olivacea, Cadophora malorum, Cadophora sp., Chaetomium funicola, Cladosporium cladosporioides, Cosmospora vilior, Cryptococcus carnescens, Cryptococcus sp., Cryptococcus victoriae, Dactylella lobata, Eurotium sp., Exophiala spinifera, Fusarium oxysporum, Geomyces sp., Geomyces pannorum, Hormonema dematioides, Monodictys castaneae, Nectria sp. olrim, Penicillium echinulatum, Pseudeurotium desertorum, Rhodoturula laryngis, Sarea difformis, Sporidiobolus salmonicolor, Sporobolomyces symmetricus*	25

Table 4.130. Microorganisms in wood.

Material	Microorganisms	Refs.
archeological waterlogged wood from 19 European sites	*Aequorivita sublithincola, Algoriphagus antarcticus, Algoriphagus ratkowskyi, Blattabacterium clevelandi, Blattabacterium relictus, Capnocytophaga canimorsus, Capnocytophaga cynodegmi, Cellulophaga (Cytophaga) fucicola, Chitinophaga pinensis, Chryseobacterium formosense, Chryseobacterium indologensis, Chryseobacterium jll, Chryseobacterium scophthaimum, Cyclobacterium marinum, Cytophaga arvensicola, Cytophaga aquatilis, Cytophaga aurantiaca, Cytophaga sp., Cytophaga succinicans, Flammeovirga (Cytophaga) aprica, Flavobacterium columnare, Flavobacterium ferrugineum, Flavobacterium hydatis, Flavobacterium sp., Flectobacillus major, Flectobacillus speluncae, Flexibacter aggregans, Flexibacter aurantiacuc, Flexibacter filiformis, Flexibacter litoralis, Flexibacteraceae bacterium, Flexithrix dorotheae, Haliscomenobacter hydrossis, Hymenobacter actinoscierus, Hymenobacter roseosalivarius, Sporocytophaga cauliformis, Sporocytohaga myxococcoides*	12
brazilwood	*Anadenanthera macrocarpa, Chaetomium globosum, Eucalyptus grandis, Gloeophyllum trabeum, Pinus elliottii, Pycnoporus sanguineus*	8
copper-tolerant microorganisms	*Antrodia vaillantii, Antrodia radiculosa, Gloeophyllum trabeum, Meruliporia incrassata, Meruliporia incrassata, Postia placenta, Serpula lacrymans, Tyromyces palustris, Wolfiporia cocos*	21,22
cork test species	*Chrysonilia sitophila, Mucor plumbeus, Penicillium glabrum, Penicillium olsonii, Trichoderma longibrachiatum*	24
Norway spruce	*Antrodia vaillantii, Gloeophyllum trabeum*	11
prehistoric wood	*Acremonium kiliense, Acremonium sp., Aspergillus sp., Chaetomium sp., Cladosporium cladosporioides, Fusarium decemcellulare, Fusarium gibbosum, Fusarium merismoides, Fusarium oxysporum, Fusarium solani, Mucor sp., Penicillium sp., Stachybotrys chartarum, Stemphylium sp., Trichoderma koningii*	9
rubber wood	*Ganoderma applanatum, Lenzites palisotii*	7
Southern pine	*Gloeophyllum trabeum, Trametes versicolor*	1
wood decay fungi	brown-rot: *Coniophora puteana, Dacrymyces stillatus, Gleophyllum trabeum, Fistulina hepatica, Rigidoporus crocatus, Rigidoporus lineatus* white rot: *Armillaria borealis, Armillaria cepistipes, Armillaria gallica, Armillaria mellea, Auricularia auricularia, Ganoderma adspersum, Heterobasidion annosum, Inonotous hispidus, Meripilus giganteus, Oudemansiella mucida, Phellinus contiguus*	23

4.22.2 ESSENTIAL PRODUCT COMPONENTS

Fungi (*Gloeophyllum trabeum, Coniophora puteana*, and *Serpula lacrymans*) increased the apparent percent crystallinity early in the decay process while decreasing the total amounts of both crystalline and amorphous material.[15] Data also showed an apparent decrease of approximately 0.05 Å in the average spacing of the crystal planes in all degraded samples after roughly 20% weight loss, as well as a decrease in the average observed relative peak width at $2\theta=22.2^{o}$.[15] These results may indicate a disruption of the outer most semi-crystalline cellulose chains comprising the wood microfibril.[15]

The wood cell wall is organized in layers of different thicknesses and different ratios of cellulose, hemicellulose and the matrix material lignin.[23] The cell wall proper consists of a thin primary wall, to which a much thicker secondary wall, consisting of three layers (S1, S2 and S3), is added after initial formation of the cell.[23] The walls of adjacent cells are bonded together by the middle lamella (Figures 4.48.1 to 4.48.3), consisting of lignin, calcium and pectic substances (Figure 4.48.4).[23] Cellulose is the main structural component of the walls of young wood.[23] Cellulose exists as a system of fibrils of 3–4 nm diameter aggregated in larger structural units.[23] The cellulose microfibrils are helically wound at different angles in the various layers of the cell wall.[23] These different helical windings are thought to contribute to the mechanical resilience of the wood.[23]

4.22.3 MECHANISMS OF BIODEGRADATION AND BIODETERIORATION

During brown-rot decay, the carbohydrate fraction of the holocellulose is actively metabolized, leaving behind a chemically modified lignin residue, which is typically brown and crumbly.[6] The depolymerization of the cellulosic fraction of the cell wall dramatically and rapidly reduces the strength of the wood.[6] Most enzymes are too large to penetrate intact cell wall layers, therefore it is hypothesized that nonenzymatic low-molecular mediators or chelators diffuse into the cell wall, producing localized hydroxyl free radicals *via* modified Fenton reactions.[6]

It has been proposed that low molecular mediators are involved in the process of initiation of depolymerization of wood cell wall constituents.[13] These small molecular weight agents in conjunction with metals mediate oxidative reactions.[13] Brown-rot fungi such as *Gloeophyllum trabeum* or *Poria placenta* synthesize extracellular catecholates, which are involved in the reduction of Fe(III), allowing the initiation of the Fenton reaction, and also production of hydroxyl radicals.[13] In the case of white-rot fungi, low-molecular weight compounds with chelating properties and iron reductive activity are probably involved in the wood degradation process in addition to peroxidases and laccases.[13] For this reason, the addition of chelators enhances the action of biocides on white-rot fungi wood degradation.[13]

Brown-rot fungi preferentially depolymerize and degrade structural carbohydrates (i.e. cellulose and hemicelluloses) in the cell wall, leaving behind the modified lignin.[14] White-rot fungi can progressively utilize all major cell wall components, including both carbohydrates and lignin.[14]

4.22.4 RESULTS OF BIODETERIORATION

Wood decay is caused by enzymes and low-molecular-weight mediators secreted by fungi in their efforts to obtain a carbon source for their survival.[6] The resulting wood decay is classified into three types: soft rot, brown rot, and white rot.[6] Basidiomycota members typically cause the more active and destructive brown- or white-rot decay.[6] Brown-rot fungi and brown-rot decay predominate on pines and pine products such as lumber.[6]

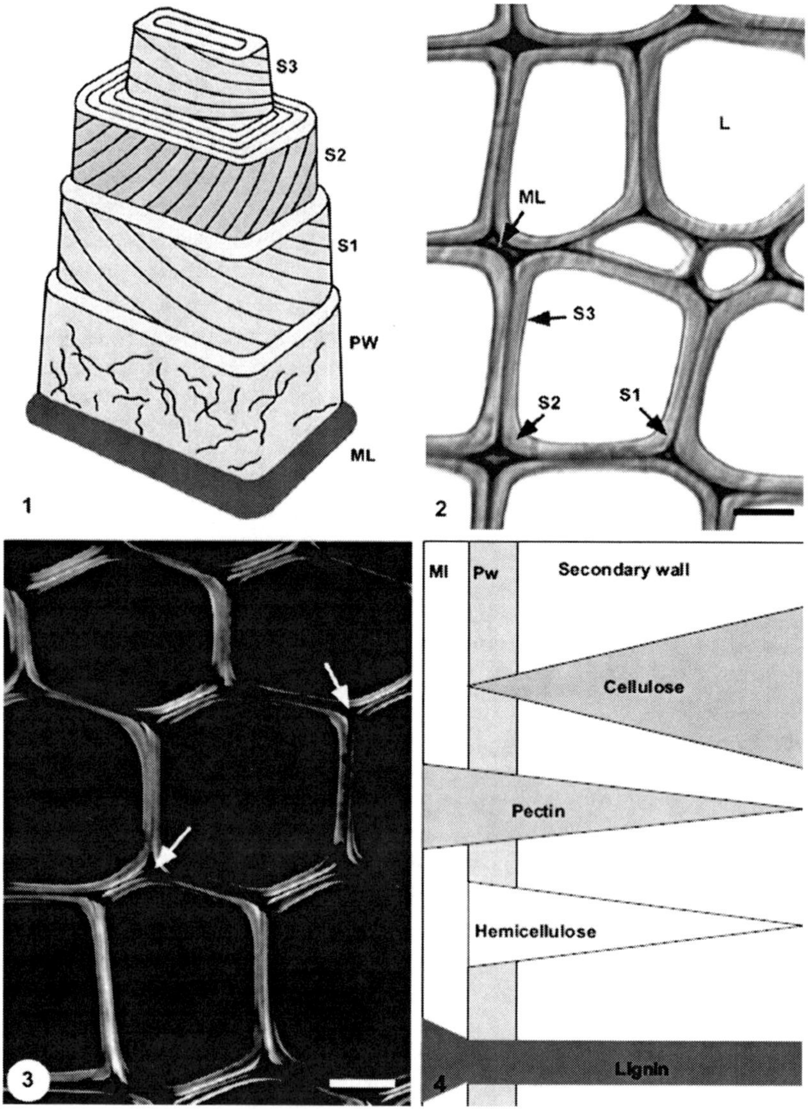

Figure 4.48. Woody cell wall. (1) Conventional cell-wall model which distinguishes five cell-wall layers. These are the middle lamella (ML), the primary wall (PW), and the three-layer secondary wall (S): outer (S1), middle (S2) and inner secondary wall layer (S3). (2) Transverse section (T.S.) of earlywood tracheids. ML, middle lamella; S1, S2, S3, outer, middle and inner layers of secondary wall. L, lumen. Bar, 10 μm. (3) T.S. of earlywood viewed between crossed Nicols. Note secondary walls appear bright, whereas the amorphous compounds in the middle lamella are non-birefringent (arrows). Bar, 10 μm. (4) Diagrammatic representation of the relative distribution of the main cell wall constituents within the different layers of the cell wall. [Adapted, by permission, from Schwarze, F W M R, *Fungal Biol. Rev.*, **21**, 133-70, 2007.]

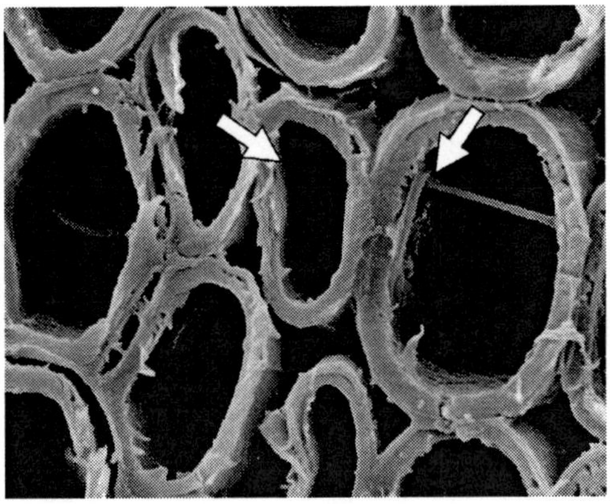

Figure 4.49. Cell walls penetrated by hyphae. [Adapted, by permission, from Pournou, A; Bogomolova, E, *Intl. Biodet. Biodeg.*, **63**, 371-78, 2009.]

Cellulolytic bacteria are quite common and have the capacity to break down lignin and other wood components such as resins, gums, dye, tannic acid, waxes, and fats.[8]

In an outdoor environment, when the moisture content of the wood is above 20%, it becomes susceptible to microbial attack by microfungi and bacteria.[8] The predominant fungal infections, by exoenzyme production and subsequent destruction of natural polymers, are white rot (*Fomes sp.*, and *Pleurotus sp.*), brown rot (*Poria spp., Merulis sp.*, and *Coniophora sp.*), and soft rot (*Chaetomium sp., Alternaria sp.*, and *Humicola sp.*). Because both bacteria (*Pseudomonas sp.* and *Achromobacter sp.*) and Actinomycetes (*Micromonospora sp.*) require a higher water content, their role is important in outdoor and marine environments.[8]

The Neolithic settlement of Dispilio, Greece, revealed significant amounts of waterlogged wood. Seventeen species of fungus were identified.[8] Figure 4.49 shows the extent of biodeterioration in which wood is penetrated by hyphae.[8]

With increasing decay, the relative lignin concentration increases, as does the concentration of nitrogen and phosphorus.[18] Due to the ability of the microorganisms to degrade mainly polysaccharides, the hemicellulose and cellulose content decrease with increasing decay time.[18] The amount of phenolic compounds in waterlogged wood was considerably lowered, whereas the ash content increased due to the very long water storage conditions.[18]

Very old dry wood is relatively rare and mainly comes from tombs in Egypt.[26] One of the most spectacular examples of ancient wood was the Solar Boat of Cheops discovered in 1954.[26] Wood over 4000 years old taken from several Egyptian pyramids still had the outward appearance of new wood except for color changes.[26] Electron microscopy revealed cracks and fissures not found in recent wood, indicating weakening of the ultrastructure, particularly in the middle lamella region.[26]

4.22.5 BIOSTABILIZATION

Table 4.131 contains a list of the most suitable biocides used for protection wood.

Table 4.131. Biocides used in protection of wood.

Material	Biocide	Refs.
antibacterial, antimold, and antitermite protection	borate base supplemented with either 0.1% azole or 0.5% thujaplicin	20

Table 4.131. Biocides used in protection of wood.

Material	Biocide	Refs.
beech	2-hydroxypyridine-N-oxide+ethylenediaminetetraacetic acid+phenolic antioxidant (Irganox 1076) (synergistic combination)	13
HDPE/wood composite	zinc borate	2
Japanese beech & cedar	ammonium borate oleate (biocidal effect and water repellency)	4
maple	silver and zinc biocides	5
Norway spruce	boric acid+montan wax,[17] copper–ethanolamine-based aqueous solution[11]	11,17
Scots pine	chitosan,[3] 3-iodo-2-propynyl-butyl carbamate and silane[10]	3,10
Southern pine	polymeric nanoparticles containing the fungicides tebuconazole and chlorothalonil	1
sugi wood	silane+boric acid	16
wood flooring	colloidal silver treatment (antibacterial performance)	19
wood (injection with particles)	particles of different chemical composition having diameter between 0.2 and 0.7 mm, and density greater than 5.5 g/cc for direct injection to wood	28

The most important drawback of boron is prominent leaching in a wet environment. In order to improve its fixation and performance against wood decay fungi, boric acid was combined with montan wax emulsion.[17]

A method for preserving wood has been patented which comprises contacting the wood with a complex of a cationic monomeric or polymeric biocide and an anionic monomeric or polymeric biocide. Suitable cationic biocides are those that contain functionalities such as amidine, guanidine, biguanide, quaternary, phosphonium, sulfonium and gemini quaternary functionalities. Suitable anionic biocides include: phenolics; saturated, unsaturated or substituted carboxylates; organomercaptide; tetrathiocarbonates; cyanodithioimidocarbamates; dithiodialkylcarbamates; anionic oxides of transition metals; aminocarboxylic acids; aminoorganophosphonic acids; monoalkyl phosphates; dialkylphosphates; and substituted or unsubstituted 2-hydroxy-2,4,6-cycloheptatrianone.[29]

4.22.6 FORMULATIONS

Rotten wood stabilizer[27]

PVC emulsion (Vinac DPN890	82-88%
Water	12-15
Bactericide (Acticide LA)	0.01-0.15
Film forming agent (Texanol)	1-2
Deformer	0.01-0.1
Mildewcide (Skane M8)	0.01-0.1

REFERENCES

1 Liu, Y; Laks, P; Heiden, P, *J. Appl. Polym. Sci.*, **86**, 615–621, 2002.

2 Pendleton, D E; Hoffard, T A; Adcock, T; Woodward, B; Wolcott, *Forest Prod. J.*, **52**, 6, 21-27, 2002.
3 Eikenes, M; Alfredsen, G; Christensen, B E; Militz, H; Solheim, H, *J Wood Sci.*, **51**, 387-94, 2005.
4 Lyon, F; Pizzi, A; Imamura, Y; Thevenon, M ; Kartal, S N; Gril, J, *Holz Roh Werkst*, **65**, 359-66, 2007.
5 Dawson-Andoh, B; Matuana, L, *Holz Roh Werkst*, **65**, 331-34, 2007.
6 Kang, Y-M; Prewitt, M L; Diehl, S V, *Intl. Biodet. Biodeg.*, **63**, 1036-44, 2009.
7 Edwin, E; Ashraf, P M, *Intl. Biodet. Biodeg.*, **57**, 31-36, 2006.
8 Alves da Silva, C; Bacellar Monteiro, M B; Brazolin, S; Carballeira Lopez, G A; Richter, A; Braga, M R, *Intl. Biodet. Biodeg.*, **60**, 285-92, 2007.
9 Pournou, A; Bogomolova, E, *Intl. Biodet. Biodeg.*, **63**, 371-78, 2009.
10 De Vetter, L; Van den Bulcke, J; De Windt, I; Stevens, M; Van Acker, J, *Intl. Biodet. Biodeg.*, **63**, 1093-1101, 2009.
11 Humar, M; Bucar, B; Pohleven, F, *Intl. Biodet. Biodeg.*, **58**, 9-14, 2006.
12 Landy, E T; Mitchell, J I; Hotchkiss, S; Eaton, R A, *Intl. Biodet. Biodeg.*, **61**, 106-16, 2008.
13 Mabicka, A; Dumarcay, S; Rouhier, N; Linder, M; Jacquot, J P; Gerardin, P; Gelhay, E, *Intl. Biodet. Biodeg.*, **55**, 203-11, 2005.
14 Oses, R; Valenzuela, S; Freer, J; Baeza, J; Rodriguez, J, *Intl. Biodet. Biodeg.*, **57**, 129-35, 2006.
15 Howell, C; Steenkjær Hastrup, A C; Goodell, B; Jellison, J, *Intl. Biodet. Biodeg.*, **63**, 414-19, 2009.
16 Kartal, S N; Yoshimura, T; Imamura, Y, *Intl. Biodet. Biodeg.*, **63**, 187-190, 2009.
17 Lesar, B; Kralj, P; Humar, M, *Intl. Biodet. Biodeg.*, **63**, 306-10, 2009.
18 Gelbrich, J; Mai, C; Militz, H, *Intl. Biodet. Biodeg.*, **61**, 24-32, 2008.
19 Kim, S; Kim, H-J, *Intl. Biodet. Biodeg.*, **57**, 155-62, 2006.
20 Clausen, C A; Yang, V, *Intl. Biodet. Biodeg.*, **59**, 20-24, 2007.
21 Steenkjaer Hastrup, A C; Green, F; Clausen, C A; Jensen, B, *Intl. Biodet. Biodeg.*, **56**, 173-77, 2005.
22 Green, F; Clausen, C A, *Intl. Biodet. Biodeg.*, 56, 75-79, 2005.
23 Schwarze, F W M R, *Fungal Biol. Rev.*, **21**, 133-70, 2007.
24 Silva Pereira, C; Soares, G A M; Oliveira, A C; Rosa, M E; Pereira, H; Moreno, N; San Romao, M V, *Intl. Biodet. Biodeg.*, **57**, 244-50, 2006.
25 Arenza, B E; Held, B W; Jurgens, J A; Farrell, R L; Blanchette, R A, *Soil Biol. Biochem.*, **38**, 3057-64, 2006.
26 Schniewind, A P in Wood Products: Aging of Wood, **Encyclopedia of Materials: Science and Technology**, *Elsevier* 2008, pp 9688-90.
27 Van Foster, R, **US Patent** 7,569,157, Hunt Holdings Inc., August 4, 2009.
28 Richardson, H W; Hodge, R L, **US Patent** 7,426,948, PhibroWood, LLC, September 23, 2008.
29 Stockel, R F, **US Patent** 7,074,459, July 11, 2006.

5

ANALYTICAL METHODS IN BIODEGRADATION, BIODETERIORATION, AND BIOSTABILIZATION

5.1 STANDARDS

5.1.1 ADHESIVES AND SEALANTS

ASTM standard[1] is designed to determine the susceptibility of adhesive film to biodegradation and to establish if antifungal properties are capable to prevent growth at bond line. Potato dextrose agar (PDA) provides a complete medium for the growth of fungi, while mineral salts agar (MSA) lacks a carbohydrate source and provides a less favorable medium.[1] PDA tests are used to determine ability of adhesive film to resist the growth of fungi on its surface as well as its ability to repel a growth of fungi on the adjacent agar surface.[1] MSA tests the ability of adhesive film to resist the growth of fungi on its surface.[1]

Another ASTM standard[2] is used to determine if adhesive formulation is sufficiently protected by biocide to resist attack by bacteria, yeast, and fungi during adhesive storage.[2] The results obtained when using the procedures given in these methods apply only to the species which are used for the testing.[2] The standard contains a list of species typically used for such studies but the study can also be conducted with any other selected species.[2] Letheen agar and broth are recommended by the procedures described.

5.1.2 ANTIFOULING COATINGS

The erosion rates of ablative antifouling paint systems are measured by exposing them to flowing water at velocities designed to subject the paint system to shear stresses experienced in service.[3] The test data are intended to serve as a guide for predicting the service life of ablative antifouling paints and calculation of the necessary paint thickness.[3] Erosion rates of antifouling paints in service will vary depending on such factors as: berthing location, geographic area of operation, salinity, pH, and temperature of seawater.[3] It is also recognized that different areas of the ship are subject to different erosion rates.[3] Another test leads to similar data but after exposure of antifouling coating to seawater.[4]

Antifouling panels are also studied in shallow submergence.[5] This method is used as a screening test.[5] Barnacle adhesion strength in shear is based on US Military Specification: MIL-P-24441/1.[6] Standard practices of inspection of the coating systems of a ship are broadly discussed in another standard.[7] Relevant standards and publications for commercial shipbuilding are included in the standard guide.[8]

Several standards address properties of metal additives, their amount, and methods of exposure.[9-13] One standard gives specification for metallic copper powder,[9] another gives the test method required to determine total copper in antifouling paints,[10] copper[11] and organotin[12] release rates are measured by specialized techniques, and methods of exposing and evaluation of different metals to seawater are given in still another standard.[13]

British standards have several relevant publications. These include methods of determination of release rate of biocide from antifouling paint,[14] determination of pyridine-triphenylborane concentration in the extract and calculation of the release rate of biocide from antifouling paint,[15] calculation of the zinc ethylene-bis (dithiocarbamate) (zineb) release rate by determination of the concentration of ethylenethiourea in the extract,[16] calculation of the tolylfluanid and dichlofluanid release rate by determination of the concentration of dimethyltolylsulfamide and dimethylphenylsulfamide in the extract,[17] and determination of copper ion concentration in the extract, and calculation of the release rate.[18]

International standards, European standards, and other countries' standards also contain information on testing and modelling properties of antifouling coatings. These include: modelling of biocide release rate from antifouling paints by mass-balance calculation,[19] extraction of biocides to determine their release rate, [20] determination of copper ion to determine its release rate from antifouling paint,[21] calculation of zinc ethylene-bis(dithiocarbamate),[22] determination of pyridine-triphenylborane,[23] determination of tralopyril release rate by quantification of its degradation product in the extract,[24] calculation of the tolylfluanid and dichlorofluanid release rates by determination of the concentration of dimethyltolylsulfamide and dimethylphenylsulfamide in the extract,[25] marine environmental risk assessment method on active substances used for antifouling systems on ships,[26] and general methods of biocide extraction.[27]

5.1.3 ANTISEPTIC DRUGS AND HANDWASH

The antimicrobial activity of drug products applied topically to the skin that are intended to help in preventing infection in minor cuts, scrapes and burns can be determined by a standardized procedure.[28] Liquids, ointments, powders, films, or dressing containing or impregnated with an antimicrobial are tested for their effect to reduce an enhanced skin microflora or their effect to suppress the growth of the skin flora.[28]

The effectiveness of an antibacterial handwash for reducing the level of aerobic bacterial flora on the hands, after an extended period of use can be determined.[29]

Formulations containing ingredients intended to inhibit growth of bacteria on intact skin are evaluated for the difference, post-product-exposure, between numbers of bacterial colonies in active test formulation plates and numbers on control plates, expressed as percent inhibition after studying by agar patch technique.[30]

Aseptic processing areas, which include readily cleanable floors, walls, and ceilings that have smooth, non-porous surfaces; particulate, temperature, and humidity controls; and cleaning and disinfecting procedures to produce and maintain aseptic conditions, require certain effectiveness of disinfectants and procedures of disinfection.[31] These can be measured by their ability to ensure that potential contaminants are adequately removed from surfaces.[31]

There are available procedures to assess effectiveness of healthcare personnel handwash formulations,[32] surgical handscrub,[33] and handscrub agents.[34,35]

German standards evaluate basic bacterial activity of chemical disinfectants and antiseptics[36] and their fungicidal and yeasticidal activity.[37] A special method exists for testing disinfectants and antiseptics used in food, industrial, domestic, and institutional areas,[38] and veterinary[39,40] and medical areas.[41,42] Sporicidal activity, after use of antiseptics and disinfectants, is determined by still another method.[43]

5.1.4 CHEMICAL MATERIALS IN GENERAL

Many test methods are used for studies of biodegradation of chemical materials. Some of these methods are listed below:

- die-away biodegradation method conducted in a natural water source to follow the disappearance of an added amount of the test substance resulting from the activity of microorganisms in the water sample[44]
- anaerobic biodegradation potential of any given chemical[45]
- anaerobic bioremediation[46]
- biodegradability of organic chemicals by semi-continuous activated sludge method[47]
- determination of ready, ultimate biodegradability of organic chemicals (sealed vessel CO_2 production test)[48]
- determination of biodegradability of organic chemicals during secondary waste water treatment (activated sludge)[49]
- determination of chemical fate of a substance using site-specific microcosms (sediment/water conditions in particular site)[50]
- microbial detoxification of chemically contaminated water and soil[51]
- sample collection of soils for environmental studies[52]
- sample collection of soils containing volatile organic compounds[53]
- guide to soil and rock collecting devices[54]
- determination of adverse effects on chemicals on *Daphnia magna* Straus[55]
- EcoRBCA process of a streamlined decision-making process for implementing corrective action protective of ecological resources at chemical release sites in a consistent manner[56]

5.1.5 COATINGS AND PAINTS

ASTM test method of determination of the relative resistance of emulsion paints to attack in the container by microorganisms.[57] The microbiological condition of paint raw materials and plant area can be examined using another method.[58] This method provides a simple screening test to discover the source of contamination and the level of contamination.[58]

Defacement of paint and coating films by algal growth are controlled by differences in the environment, lighting, temperature, substrate, and other factors in addition to the coating composition.[59] The test method[59] attempts to provide a means to comparatively evaluate different coating formulations for their relative performance under a given set of conditions.[59]

The German standard[60] gives a method of laboratory evaluation of fungal efficiency of film preservatives. Efficiency of film preservatives against algae is given by another method.[61]

5.1.6 COOLING WATER SYSTEMS

The test method[62] determines potentially effective microbiocides for use in cooling water systems using cooling water and deposits/biofilm obtained from the field. The addition of deposits/biofilms addresses the need to include the major source of microorganisms in cooling water systems.[62] Even with this addition, laboratory results may not be totally predictive of microbicidal effectiveness in the field.[62] This is because conditions in the field affecting microbiocide effectiveness are difficult to mimic in the laboratory.[62] These include blow down rate, addition of makeup water, water hardness, hydrocarbon leaks, pH, sediment loading, dissolved solids, microbes in slime, and deposits (biofilms) on surfaces.[62] An additional factor is the difficulty in enumerating all microbes in the water due to the lack of adequate recovery media.[62]

Current tests for monitoring the biomass concentration in cooling towers require at least 36 h for growth of the microorganisms on a solid agar surface for counting.[63] Replication of microorganisms over the 36-h period before results are available creates an aqueous environment which is no longer represented by the data generated.[63] Timely test results can assist in minimizing biocide addition to control biomass concentrations.[63] Kool Kount provides data within hours to allow for more precise control of active biomass concentrations in the waters.[63]

The German standard provides methods for the examination of water, waste water, and sludge, and determination of selected plant treatment agents, biocides and breakdown products. Method uses gas chromatography (GC-MS) after solid-phase microextraction.[64] The same method is included in the ISO standard.[65]

5.1.7 DETERGENTS

If the surfactant reduction in the presumptive test equals or exceeds 90%, the material is considered to be adequately biodegradable without further testing.[66] The test method is capable of detecting small concentrations of detergents so that they can be controlled.[67-69]

5.1.8 FUELS AND FUEL SYSTEMS

Although the numbers of viable bacteria and fungi recovered from fuel-phase samples are likely to be several orders of magnitude smaller than those found in water-phase samples, fuel-phase organisms are often the most readily available indicators of fuel and fuel system microbial contamination.[70] This practice covers a membrane filter procedure for the detection and enumeration of heterotrophic bacteria and fungi in liquid fuels.[70]

A guide provides information addressing the conditions that lead to fuel microbial contamination and biodegradation and the general characteristics of and strategies for controlling microbial contamination. It compliments and amplifies information provided in Practice ASTM D 4418 for handling gas-turbine fuels.[71]

Fuels containing aromatic compounds, such as diesel fuel and gasoline, as well as other aromatic-containing hydrocarbon materials, such as crude oil, coal oil, and motor oil, can be determined by the test method.[72]

Manual sampling procedure for microbiological testing is included in standardized procedure.[73]

The effect of environmental conditions, such as a variety of fuel additives, metal surfaces, and climatology, are variables that can be included in specific tests.[74] Aviation turbine fuel is specified in the ASTM standard.[75]

5.1.9 GEOMEMBRANES AND GEOTEXTILES

A guide provides recommendations for the selection of appropriate test methods (including monitoring biodegradation) for prefabricated bituminous sheet used in geomembrane applications to provide consistency in data reporting.[76]

The test method is used to determine the potential for, and relative degree of, biological growth which can accumulate on geotextile or geotextile/soil filters.[77] This test method uses the measurement of flow rates over an extended period of time to determine the amount of clogging.[77]

5.1.10 HYDRAULIC FLUIDS

The effectiveness of antimicrobial agents intended for microbial control in invert emulsions and other water containing hydraulic fluids is given in the ASTM standard.[78]

A guide discusses ways to assess the likelihood that a hydraulic fluid will undergo biodegradation if it enters an environment that is known to support biodegradation of some substances; for example, the material used as the positive control in the test.[79] The information can be used in making or assessing claims of biodegradability of a fluid formula.[79] Another standard[80] contains classification of some aspects of environmental impact related to environmental persistence of which biodegradability is one component and acute ecotoxicity is the other. Although environmental persistence is discussed first, this classification does not imply that considerations of environmental persistence should take precedence over concerns for ecotoxicity.[80] The performance classification for biodegradable fire-resistant hydraulic fluids that are used in the industrial/mobile and mining industries are included in a standard.[81]

5.1.11 LUBRICANTS

A test method concerns the determination of the degree of aerobic aquatic biodegradation of fully formulated lubricants or their components on exposure to an inoculum under laboratory conditions.[82] Another procedure[83] covers determination of carbon dioxide evolution resulting from biodegradation of lubricant or its component. This test method is designed to be applicable to all non-volatile lubricants or lubricant components that are not toxic and not inhibitory at the test concentration to the organisms present in the inoculum.[83]

The degree of biodegradability of lubricants or their components in an aerobic aqueous medium on exposure to an inoculum under controlled laboratory conditions is measured by closed respirometer method.[84] This test method is suitable for evaluating the biodegradation of volatile as well as nonvolatile lubricants or lubricant components.[84]

A special standard covers definitions relating to biodegradability and ecotoxicity of lubricants.[85] Still another standard contains terminology used in petroleum, petroleum products, and lubricants.[86]

5.1.12 LUMBER, PALLETS, AND WOOD BOXES

Standard specifications for wirebound pallet-type wood boxes[87] and general purpose wirebound shipping boxes,[88] and wood-cleated shipping boxes[89] include information about fungicidal protection and biodegradability.

Standard practice[90] includes information on treatment and marking of wood packaging materials.

5.1.13 METALWORKING FLUIDS

Laboratory procedures are provided for rating the relative bioresistance of metalworking fluid formulations, for determining the need for microbiocide addition prior to or during fluid use in metalworking systems and for evaluating microbiocide performance.[91] The factors affecting challenge population numbers, taxonomic diversity, physiological state, inoculation frequency and biodeterioration effects in recirculating metalworking fluid systems are varied and only partially understood. Consequently, the results of tests completed in accordance with this practice should be used only to compare the relative performance of products or microbiocide treatments included in a test series.

General considerations for microbiocide selection are provided in another standard.[92] This practice summarizes the steps in the antimicrobial pesticide selection process, reviewing technical and regulatory considerations inherent in the process.[92] Steps in the antimicrobial selection process include: needs identification, use strategy selection, efficacy testing, chemical compatibility testing, regulatory consideration review, handling and disposal issue review.[92]

Standard terminology for metalworking fluids and operations is included in another ASTM standard.[93]

The determination of endotoxin concentrations in metalworking fluids is a parameter that can be used in decision-making for prudent fluid management practices (fluid draining, cleaning, recharging, or biocide dosages).[94] This test provides a method for analysts who perform quantitative endotoxin analyses of water-miscible metalworking fluids.[94]

Another standard method[95] measures the concentration of adenosine triphosphate, ATP, present in the sample. ATP is a constituent of all living cells, including bacteria and fungi.[95] Consequently, the presence of ATP is an indicator of total microbial contamination in metalworking fluids.[95] ATP is not associated with matter of non-biological origin.[95]

5.1.14 OILFIELD AND REFINERY

A standard for evaluating and qualifying oilfield and refinery corrosion inhibitors in the laboratory.[96]

5.1.15 OIL SPILL RESPONSE

The guide provides remediation managers and spill response teams with guidance on an alternate means (called bioremediation) of treating oil spills safely and effectively on and below terrestrial surfaces.[97]

Gas chromatography coupled with positive ion electron impact low resolution mass spectrometry provides analytical characterization of petroleum oils for comparison purposes.[98] Certain classes of source-specific chemical compounds are targeted in this qualitative comparison; these target compounds are both unique descriptors of an oil and materials which are chemically resistant to environmental degradation.[98] Spilled oil can be assessed in this way as being similar or different from potential source samples by the direct visual comparison of specific extracted ion chromatograms.[98] In addition, other, more weathering-sensitive chemical compound classes can also be examined in order to crudely assess the degree of weathering undergone by an oil spill sample.[98]

The use of oil spill dispersants to assist in the control of oil spills is described for different scenarios and geographic locations.[99-102]

5.1.16 PACKAGING

Several international standards exist which discuss packaging materials suitable for composting and biodegradation:

- requirements for compostable and biodegradable packaging[103,104]
- ultimate aerobic biodegradability[105]

5.1.17 PAPER

The test method is used to determine if a slime control agent has application in the paper industry for control of bacterial or fungal slime.[106] This test method is run in acid, alkaline, or acid and alkaline conditions to determine the efficacy of the slime control agent.[106] The test conditions may be modified to reflect intended use patterns in typical paper mill systems, including use of actual paper mill furnish.[106]

Paper industry sludges are industrial by-products derived from wastewater treatment operations at plants associated with the manufacturing of paper.[107] These sludges typically consist of clay and organic matter.[107] They may also contain low levels of inorganic and organic contaminants and can be rich in microbes.[107]

A standard method is used to determine if an antimicrobial preserves pigment suspensions, dye solutions, pulp slurries, starch solutions, polymers, sizing agents, latex emulsions, and other aqueous-based materials used in the paper industry from bacterial spoilage.[108]

A specification covers biodegradable plastics and products (including packaging), where plastic film or sheet is attached (either through lamination or extrusion directly onto the paper) to substrates and the entire product or package is designed to be composted in municipal and industrial aerobic composting facilities.[109]

5.1.18 PLASTICS AND POLYMERS

Many different standards discuss plastics recycling, composting, biodegradation, and other related similar subjects, forming broad analytical background for decision and selection processes. The existing standards address the following areas of expertise:

- biodegradability in seawater of radiolabelled plastics[110]
- waste reduction, resource recovery, and reuse[111]
- antimicrobials preventing staining[112]
- activated-sludge biodegradation[113]

- degradation by combination of oxidation and biodegradation[114]
- composting conditions[115]
- biodegradation in the presence of municipal sewage sludge[116]
- accelerated landfill conditions[117]
- aerobic degradation after composting[118]
- high-solids anaerobic-digestion conditions[119]
- aerobic biodegradation in marine environment[120]
- aerobic biodegradation of radiolabelled plastics[121]
- toxicity and compost quality testing[122]
- specification for compostable plastics[123]
- non-floating biodegradable plastics in marine environment[124]
- standard guide to compostable plastics[125]
- laboratory-scale simulated landfill environment[126]
- compost stability by measuring oxygen consumption[127]
- sample preparation[128,138]
- biogas production method[129,131,132,138-142]
- carbon dioxide measurement[130,134,135,136]
- closed respirometer for measurement of oxygen demand[133]
- degree of disintegration after composting[137]

5.1.19 STONE CONSOLIDANTS

A standard guide provides information on methods that can assist the design and conservation professional in the selection and use of consolidants for stone.[143] Use of a consolidant should be considered only after developing a thorough understanding of all factors contributing to the deterioration of the stone.[143] Such an understanding is best reached through a comprehensive survey of existing conditions, environment, and a definition of stone performance requirements, followed by laboratory analysis of the stone and its deterioration products.[143] With this information it should be possible to identify the decay mechanisms and to develop a conservation plan for the stone.[143] However, careful preliminary investigation may indicate that consolidation is an inappropriate treatment.[143]

5.1.20 SURGICAL IMPLANTS AND MEDICAL DEVICES

A specification covers material requirements for chemical composition of inorganic xenogeneic or allogeneic bone (apatite) intended for surgical implants.[144] Another test method covers *in vitro* degradation of hydrolytically degradable polymers intended for use in surgical implants.[145]

A classification outlines aspects of Tissue Engineered Medical Products which includes their individual components.[146] The categories outlined in this classification are intended to list, identify, and group the areas pertinent to Tissue Engineered Medical Products.[146]

Quantification of potential degradation products in medical devices can be conducted using an international standard.[147]

5.1.21 WATER SYSTEMS

A guide for determining the appropriateness of remediation by natural attenuation and implementing remediation by natural attenuation at a given petroleum release site, either as a stand alone remedial action or in combination with other remedial actions.[148] Natural attenuation is a potential remediation alternative for containment and reduction of the mass and concentration of petroleum hydrocarbons in the environment to protect human health and the environment.[148] Remediation by natural attenuation depends upon natural processes such as biodegradation, dispersion, dilution, volatilization, hydrolysis, and sorption to attenuate petroleum constituents of concern to achieve remedial goals.[148]

Potential biodegradability of aqueous waste is assessed by a toxicity test with a luminescent marine bacterium.[149] Oil from one crude oil field is readily distinguishable from another, and differences in the makeup of oils from the same crude oil field can often be observed as well.[150] A standard guide helps in their identification.[150]

The use of oil spill dispersants to assist in the control of oil spills is discussed in the ASTM standard.[151] This guide is written with the goal of minimizing the environmental impacts of oil spills.[151]

Procedures for obtaining laboratory data concerning bioconcentration of a test material added to dilution water—but not to food—by freshwater and saltwater fishes and saltwater bivalve mollusks using the flow-through technique are given in the ASTM guide.[152]

An international standard gives a selection of tests for biodegradability of water contaminants.[153]

5.2 OTHER ANALYTICAL METHODS

A large number of different directions of studies require a variety of testing methods. The main subjects include:

- identification of microorganisms involved in biodegradation
- studies of biofilms
- biomass growth rate
- selection of biocides
- determination of minimal concentration of biocide required for protection
- resistance to biocides
- determination of biodegradation rate
- products of biodegradation
- morphological effects of biodegradation
- chemical changes of materials caused by biodegradation
- preservation of cultural heritage
- restoration and protection methods
- bioremediation

General guidance regarding testing methods and their applications in degradation studies can be taken from the monographic source on degradation of various materials.[154] The data in the monograph[154] contain specific information for degradation studies, especially useful in spectroscopic methods of analysis.

Because studies on biodegradation are frequently conducted in solution and suspension, high pressure liquid chromatography and gas chromatography are more frequently used than in other degradation studies but generally all basic analytical techniques are employed and their choice depends on particular requirements of conducted studies. For this reason, it is not useful to discuss analytic methods here because the type of studies imposes a selection rather than the general area of research.

There are a few types of methods which require to be mentioned because of their importance and less usual applications in general subjects of analytical chemistry.

DNA-based molecular techniques are frequently used for detection and accurate identification of microorganisms involved in biodegradative processes.[155,156] The proper identification is a prerequisite for understanding biodeterioration processes and finding remedies for restoration and prevention strategies.[155] Microorganism morphology cannot be used as a reliable criterion for identification purposes.[155] Classical methods for the detection and identification of microorganisms consist in attempting to culture the organisms.[155] These techniques present severe limitations.[155] It is now accepted that 99% of the microorganisms present in an environmental sample cannot be cultivated.[155] DNA-based molecular techniques do not require culturing procedures.[155]

DNA techniques are based on the analysis of ribosomal DNA sequences that are present in the genome of all living organisms.[155] The sequences can be used as biomarkers for the accurate identification of microorganisms.[155] Ribosomal genes contain both highly conserved and highly variable regions in which "signature sequences" have been identified for each domain of life (e.g., Bacteria, Eucarya and Archaea) as well as for microbial groups, families, genera and species.[155] By knowing their ribosomal sequences, the microorganisms can be classified and identified.[155]

Methods based on measurements of chemical components which form mycelium cells (ergosterol, chitin) become important in determination of biomass production during processes of biodeterioration. Chemical determination of ergosterol permits determination of both active and inactive mycelium as required in most studies.[157] Ergosterol is a basic sterol of cellular membranes in filamentous fungi and yeast.[157] In plant cells it is absent or occurs in small quantities.[157] Considering low ergosterol content in the yeast cells and the contamination of building materials mainly by mould, it can be assumed that ergosterol plays a significant role in the quantitative evaluation of mould contamination.[157]

FTIR has the ability to record spectra from a very wide range of different compounds and compositions. In organic and polymer chemistries it is frequently used for the so-called "fingerprinting" of various compounds or their mixtures. The capacity to express a specific "fingerprint" enables FTIR spectroscopy to be used for identification of unknown microbial strains using spectral data libraries.[158] The FTIR technique has great sensitivity, the microbial sample preparation is simple, no reagent is needed, and data acquisition is faster than with other physical-chemical techniques.[158] It has been used as a powerful tool for differentiating fungi.[158] IR spectra are highly specific for each strain and species, representing the total cell chemical composition such as lipids, proteins, nucleic acids, and polysaccharides.[158]

The major advantages of the adoption of Raman spectroscopy for the analysis of artefacts and works of art are twofold: the technique is nondestructive and requires little or no chemical and mechanical pretreatment of the specimen, and the molecular signatures from both the organic and inorganic components are obtained in the same spectrum.[159] The effects of deterioration with time and storage or of degradation in changing environments are often extremely difficult to quantify, yet this information is critical for conservators and restorers.[159] Such information can be acquired using Raman spectroscopy.

Raman spectroscopy is very convenient to use because portable fiber optics microprobes provide the possibility of acquiring spectra directly from the object of study without need to damage objects of study to obtain samples for analysis.[160]

Raman spectroscopy can be also used to study specific links, such as, for example, disulfide links present in cysteine, which is part of keratin. Using these methods, biodeterioration of skin, hair, nail, feathers, fur, claws and horn can be selectively studied.[161]

Radiolabelling is another useful technique in studies of biodegradation. Materials containing a randomly distributed ^{14}C marker can be exposed to selected microbial environments.[162] The amount of ^{14}C carbon dioxide evolved is estimated using a scintillation counter.[162] This method is not subject to interference by biodegradable impurities or additives in the polymer.[162] Biodegradability investigations using this technique for polymeric materials in different microbial environments show a high degree of precision and consistency.[162] Kinetics of biodegradation can be followed without interfering with biochemical processes.

The above methods addressed mostly needs in identification of microorganism content during biodegradative processes. In order to assess the extend and type of damage, morphological studies are very important. They should be designed in such a manner that they will permit analysis of integrity of degraded objects as well as changes to the original crystalline structure of material under the study.

Various microscopic methods are in common use, including the most frequently used optical microscopy, SEM, AFM, and TEM. SEM offers a high resolution (approximately 100 Å) and high field depth giving a 3D effect to the images.[155] For this reason it is the most frequently used in degradation studies, both for research of degraded materials and biofilms. The biggest disadvantage in the case of non-conductive samples (most studied) is that they must be coated with a conductive metal film before the sample can be imaged to avoid the build up of local electron concentrations that impede the formation of clean images.[155] Contrary to SEM, ESEM (so-called environmental SEM) does not operate at high vacuum and provides fast, accurate images of biological deposits (biofilms), showing their spatial relationship and surface chemistry without needing extensive manipulation of the samples.[155] This instrument has a unique secondary electron detector capable of forming high resolution images at pressures in the range of 0.1–20 torr.[155] At these relatively high pressures, non-conductive biological specimens are directly observed without metallic coating and any special preparation.[155]

The polarizing microscope (also called a petrographic microscope) is often used to examine thin sections of constructional materials and it is very useful in the study of rock transformation due to weathering effects.[155]

Several X-ray techniques are especially useful in studies of biodeterioration. In the energy dispersion X-ray analysis, EDX, technique, a solid-state detector is placed close to the sample to emit an electric pulse when the interaction with the X-rays is produced.[155] The intensity of the electric pulse is proportional to the energy of the X-ray beam.[155] If the number of pulses and the intensity of the X-rays are measured, the chemical composition of the sample can be determined.[155] X-ray diffraction, XRD, is a convenient and useful tool for the identification of crystalline compounds as well as for assessing the structure of complex natural compounds. The theoretical basis of the method is the fact that X-ray diffraction patterns are unique for each crystalline structure.[155] Grazing incident diffraction, GID, is a valuable technique that provides very precise information on surface and interface of atomic arrangement in crystalline structures.[155] By varying the angle of incidence, the penetration of the X-ray beam into the material can be controlled allowing the analysis of surfaces and buried interfaces.[155] GID is a powerful tool for surface structural characterization of properties such as atomic structure and roughness.[155] Atomic force microscopy is also useful in these studies.

REFERENCES

1 ASTM D4300-01(2008) Standard Test Methods for Ability of Adhesive Films to Support or Resist the Growth of Fungi
2 ASTM D4783 - 01(2008) Standard Test Methods for Resistance of Adhesive Preparations in Container to Attack by Bacteria, Yeast, and Fungi
3 ASTM D4938 - 89(2007) Standard Test Method for Erosion Testing of Antifouling Paints Using High Velocity Water
4 ASTM D4939 - 89(2007) Standard Test Method for Subjecting Marine Antifouling Coating to Biofouling and Fluid Shear Forces in Natural Seawater
5 ASTM D3623 - 78a(2004) Standard Test Method for Testing Antifouling Panels in Shallow Submergence
6 ASTM D5618-94(2005) Standard Test Method for Measurement of Barnacle Adhesion Strength in Shear
7 ASTM F1130-99(2009) Standard Practice for Inspecting the Coating System of a Ship
8 ASTM F1547-09 Standard Guide Listing Relevant Standards and Publications for Commercial Shipbuilding
9 ASTM D964-03(2008)e1 Standard Specification for Metallic Copper Powder for Use in Antifouling Paints
10 ASTM D6632-01 Standard Test Method for Total Copper in Antifouling Paints
11 ASTM D6442-06 Standard Test Method for Determination of Copper Release Rate From Antifouling Coatings in Substitute Ocean Water
12 ASTM D5108-90(2007) Standard Test Method for Organotin Release Rates of Antifouling Coating Systems in Sea Water
13 ASTM G52-00(2006) Standard Practice for Exposing and Evaluating Metals and Alloys in Surface Seawater
14 BS EN ISO 15181-1:2007 Paints and varnishes. Determination of release rate of biocides from antifouling paints. General method for extraction of biocides
15 BS EN ISO 15181-4:2008Paints and varnishes. Determination of release rate of biocides from antifouling paints. Determination of pyridine-triphenylborane (PTPB) concentration in the extract and calculation of the release rate
16 BS EN ISO 15181-3:2007Paints and varnishes. Determination of release rate of biocides from antifouling paints. Calculation of the zinc ethylene-bis (dithiocarbamate)(zineb) release rate by determination of the concentration of ethylenethiourea in the extract
17 BS EN ISO 15181-5:2008 Paints and varnishes. Determination of release rate of biocides from antifouling paints. Calculation of the tolylfluanid and dichlofluanid release rate by determination of the concentration of dimethyltolylsulfamide (DMST) and dimethylphenylsulfamide (DMSA) in the extract
18 BS EN ISO 15181-2:2007 Paints and varnishes. Determination of release rate of biocides from antifouling paints. Determination of copper-ion concentration in the extract and calculation of the release rate
19 Paints and varnishes - Modelling of biocide release rate from antifouling paints by mass-balance

calculation
20 DIN EN ISO 15181-1 Paints and varnishes - Determination of release rate of biocides from antifouling paints - Part 1: General method for extraction of biocides (ISO 15181-1:2007)
21 DIN EN ISO 15181-2 Paints and varnishes - Determination of release rate of biocides from antifouling paints - Part 2: Determination of copper-ion concentration in the extract and calculation of the release rate (ISO 15181-2:2007)
22 DIN EN ISO 15181-3 Paints and varnishes - Determination of release rate of biocides from antifouling paints - Part 3: Calculation of the zinc ethylene-bis(dithiocarbamate) (zineb) release rate by determination of the concentration of ethylenethiourea in the extract (ISO 15181-3:2007)
23 DIN EN ISO 15181-4 Paints and varnishes - Determination of release rate of biocides from antifouling paints - Part 4: Determination of pyridine-triphenylborane (PTPB) concentration in the extract and calculation of the release rate (ISO 15181-4:2008)
24 Project. Paints and varnishes - Determination of release rate of biocides from antifouling paints - Part 6: Determination of tralopyril release rate by quantification of its degradation product in the extract
25 DIN EN ISO 15181-5 Paints and varnishes - Determination of release rate of biocides from antifouling paints - Part 5: Calculation of the tolylfluanid and dichlofluanid release rate by determination of the concentration of dimethyltolylsulfamide (DMST) and dimethylphenylsulfamide (DMSA) in the extract (ISO 15181-5:2008)
26 Project. Ships and marine technology - Risk assessment on anti-fouling systems on ships - Part 1: Marine environmental risk assessment method on active substances used for anti-fouling systems on ships
27 BS EN ISO 15181-1 Paints and varnishes - Determination of release rate of biocides from antifouling paints - General method for extraction of biocides
28 ASTM E1589 - 05 Standard Test Method for Evaluation of First Aid Antiseptic Drug Products
29 ASTM E1883 - 02(2007) Standard Test Method for Assessment of an Antibacterial Handwash Product by Multiple Basin Wash Technique
30 ASTM E1882 - 05 Standard Test Method for Evaluation of Antimicrobial Formulations by the Agar Patch Technique
31 ASTM E2614 - 08 Standard Guide for Evaluation of Cleanroom Disinfectants
32 ASTM E1174-06 Standard Test Method for Evaluation of the Effectiveness of Health Care Personnel Handwash Formulations
33 ASTM E1115-02 Standard Test Method for Evaluation of Surgical Hand Scrub Formulations
34 ASTM E2613-08 Standard Test Method for Determining Fungus-Eliminating Effectiveness of Hygienic Handwash and Handrub Agents Using Fingerpads of Adults
35 ASTM E2276-03e1 Standard Test Method for Determining the Bacteria-Eliminating Effectiveness of Hygienic Handwash and Handrub Agents Using the Fingerpads of Adult Subjects
36 DIN EN 1040 Chemical disinfectants and antiseptics - Quantitative suspension test for the evaluation of basic bactericidal activity of chemical disinfectants and antiseptics - Test method and requirements
37 DIN EN 1275 Chemical disinfectants and antiseptics - Quantitative suspension test for the evaluation of basic fungicidal or basic yeasticidal activity of chemical disinfectants and antiseptics - Test method and requirements
38 DIN EN 1276 Chemical disinfectants and antiseptics - Quantitative suspension test for the evaluation of bactericidal activity of chemical disinfectants and antiseptics used in food, industrial, domestic and institutional areas - Test method and requirements
39 DIN EN 1656 Chemical disinfectants and antiseptics - Quantitative suspension test for the evaluation of bactericidal activity of chemical disinfectants and antiseptics used in veterinary area - Test method and requirements
40 DIN EN 1657 Chemical disinfectants and antiseptics - Quantitative suspension test for the evaluation of fungicidal or yeasticidal activity of chemical disinfectants and antiseptics used in the veterinary area - Test method and requirements
41 DIN EN 13624 Chemical disinfectants and antiseptics - Quantitative suspension test for the evaluation of fungicidal and yeasticidal activity in the medical area - Test method and requirements (phase2, step 1)
42 DIN EN 13727 Chemical disinfectants and antiseptics - Quantitative suspension test for the evaluation of bactericidal activity in the medical area - Test method and requirements (phase 2, step 1)
43 DIN EN 14347 Chemical disinfectants and antiseptics - Basic sporicidal activity - Test method and requirements (phase 1)
44 ASTM E1279 - 89(2008) Standard Test Method for Biodegradation By a Shake-Flask Die-Away Method
45 ASTM E2170 - 01(2008) Standard Test Method for Determining Anaerobic Biodegradation Potential of Organic Chemicals Under Methanogenic Conditions

46 ASTM F1834 - 98(2004) Standard Guide for Consideration of Anaerobic Bioremediation as a Chemical Pollutant Mitigation Method on Land
47 ASTM E1625 - 94(2008) Standard Test Method for Determining Biodegradability of Organic Chemicals in Semi-Continuous Activated Sludge
48 ASTM E1720 - 01(2008) Standard Test Method for Determining Ready, Ultimate, Biodegradability of Organic Chemicals in a Sealed Vessel CO_2 Production Test
49 ASTM E1798 - 96(2008) Standard Test Method for Assessing Treatability or Biodegradability, or Both, of Organic Chemicals in Porous Pots
50 ASTM E1624 - 94(2008) Standard Guide for Chemical Fate in Site-Specific Sediment/Water Microcosms
51 ASTM D5660-96(2009) Standard Test Method for Assessing the Microbial Detoxification of Chemically Contaminated Water and Soil Using a Toxicity Test with a Luminescent Marine Bacterium
52 ASTM D6640-01(2005) Standard Practice for Collection and Handling of Soils Obtained in Core Barrel Samplers for Environmental Investigations
53 ASTM D4547-09 Standard Guide for Sampling Waste and Soils for Volatile Organic Compounds
54 ASTM D6169-98(2005) Standard Guide for Selection of Soil and Rock Sampling Devices Used With Drill Rigs for Environmental Investigations
55 ASTM E1193 - 97(2004) Standard Guide for Conducting Daphnia magna Life-Cycle Toxicity Tests
56 ASTM E2205 / E2205M - 02(2009)e1 Standard Guide for Risk-Based Corrective Action for Protection of Ecological Resources
57 ASTM D2574 - 06 Standard Test Method for Resistance of Emulsion Paints in the Container to Attack by Microorganisms
58 ASTM D5588 - 97(2008) Standard Test Method for Determination of the Microbial Condition of Paint, Paint Raw Materials, and Plant Areas
59 ASTM D5589 - 09 Standard Test Method for Determining the Resistance of Paint Films and Related Coatings to Algal Defacement
60 DIN EN 15457 Paints and varnishes - Laboratory method for testing the efficacy of film preservatives in a coating against fungi
61 DIN EN 15458 Paints and varnishes - Laboratory method for testing the efficacy of film preservatives in a coating against algae; German version EN 15458:2007
62 ASTM E645 - 07 Standard Test Method for Efficacy of Microbiocides Used in Cooling Water Systems
63 ASTM D6530 - 00(2006) Standard Test Method for Total Active Biomass in Cooling Tower Waters (Kool Kount Assay; KKA)
64 DIN 38407-34 German standard methods for the examination of water, waste water and sludge - Jointly determinable substances (group F) - Part 34: Determination of selected plant treatment agents, biocides and break-down products; Method using gas chromatography (GC-MS) after solid-phase micro extraction
65 ISO/FDIS 27108 Water quality - Determination of selected plant treatment agents and biocide products - Method using solid-phase microextraction (SPME) followed by gas chromatography-mass spectrometry (GC-MS)
66 ASTM D2667 - 95(2008) Standard Test Method for Biodegradability of Alkylbenzene Sulfonates
67 ASTM D2330 - 02 Standard Test Method for Methylene Blue Active Substances
68 DIN 38409-20 German standard methods for the examination of water, waste water and sludge; parameters characterizing effects and substances (group H); determination of substances that react with disulfine blue (H20)
69 German standard methods for the examination of water, waste water and sludge; general measures of effects and substances (group H), determination of methylene blue active and bismuth active substances (H23)
70 ASTM D6974 - 09 Standard Practice for Enumeration of Viable Bacteria and Fungi in Liquid Fuels—Filtration and Culture Procedures
71 ASTM D6469 - 08 Standard Guide for Microbial Contamination in Fuels and Fuel Systems
72 ASTM D5831 - 09 Standard Test Method for Screening Fuels in Soils
73 ASTM D7464 - 08 Standard Practice for Manual Sampling of Liquid Fuels, Associated Materials and Fuel System Components for Microbiological Testing
74 ASTM E1259 - 05 Standard Practice for Evaluation of Antimicrobials in Liquid Fuels Boiling Below 390°C
75 ASTM D7223 - 05 Standard Specification for Aviation Certification Turbine Fuel
76 ASTM D6455 - 05 Standard Guide for the Selection of Test Methods for Prefabricated Bituminous Geomembranes
77 ASTM D1987 - 07 Standard Test Method for Biological Clogging of Geotextile or Soil/Geotextile Filters

78 ASTM E979 - 09 Standard Test Method for Evaluation of Antimicrobial Agents as Preservatives for Invert Emulsion and Other Water Containing Hydraulic Fluids
79 ASTM D6006 - 97a(2008) Standard Guide for Assessing Biodegradability of Hydraulic Fluids
80 ASTM D6046 - 02(2006) Standard Classification of Hydraulic Fluids for Environmental Impact
81 ASTM D7044 - 04a Standard Specification for Biodegradable Fire Resistant Hydraulic Fluids
82 ASTM D5864 - 05 Standard Test Method for Determining Aerobic Aquatic Biodegradation of Lubricants or Their Components
83 ASTM D6139 - 00(2005) Standard Test Method for Determining the Aerobic Aquatic Biodegradation of Lubricants or Their Components Using the Gledhill Shake Flask
84 ASTM D6731 - 01(2005) Standard Test Method for Determining the Aerobic, Aquatic Biodegradability of Lubricants or Lubricant Components in a Closed Respirometer
85 ASTM D6384 - 99a(2005) Standard Terminology Relating to Biodegradability and Ecotoxicity of Lubricants
86 ASTM D4175 - 09a Standard Terminology Relating to Petroleum, Petroleum Products, and Lubricants
87 ASTM D6254 / D6254M - 07 Standard Specification for Wirebound Pallet-Type Wood Boxes
88 ASTM D6573 / D6573M - 01(2007) Standard Specification for General Purpose Wirebound Shipping Boxes
89 ASTM D6251 / D6251M - 07e2 Standard Specification for Wood-Cleated Panelboard Shipping Boxes
90 ASTM D6253 - 10 Standard Practice for Treatment and/or Marking of Wood Packaging Materials
91 ASTM E2275 - 03(2008) Standard Practice for Evaluating Water-Miscible Metalworking Fluid Bioresistance and Antimicrobial Pesticide Performance
92 ASTM E2169 - 01(2007) Standard Practice for Selecting Antimicrobial Pesticides for Use in Water-Miscible Metalworking Fluids
93 ASTM E2523 - 07 Standard Terminology for Metalworking Fluids and Operations
94 ASTM E2657 - 09 Standard Test Method for Determination of Endotoxin Concentrations in Water-Miscible Metalworking Fluids
95 ASTM E2694 - 09 Standard Test Method for Measurement of Adenosine Triphosphate in Water-Miscible Metalworking Fluids
96 ASTM G170 - 06 Standard Guide for Evaluating and Qualifying Oilfield and Refinery Corrosion Inhibitors in the Laboratory
97 ASTM F1693 - 96(2003) Standard Guide for Consideration of Bioremediation as an Oil Spill Response Method on Land
98 ASTM D5739 - 06 Standard Practice for Oil Spill Source Identification by Gas Chromatography and Positive Ion Electron Impact Low Resolution Mass Spectrometry
99 ASTM F1209 - 08 Standard Guide for Ecological Considerations for the Use of Oil Spill Dispersants in Freshwater and Other Inland Environments, Ponds and Sloughs
100 ASTM F1210 - 08 Standard Guide for Ecological Considerations for the Use of Oil Spill Dispersants in Freshwater and Other Inland Environments, Lakes and Large Water Bodies
101 ASTM F1279 - 08 Standard Guide for Ecological Considerations for the Restriction of the Use of Surface Washing Agents: Permeable Land Surfaces
102 ASTM F2205 - 07 Standard Guide for Ecological Considerations for the Use of Chemical Dispersants in Oil Spill Response: Tropical Environments
103 DIN EN 13432 Packaging - Requirements for packaging recoverable through composting and biodegradation - Test scheme and evaluation criteria for the final acceptance of packaging
104 DIN EN 13432 Berichtigung 2 Packaging - Requirements for packaging recoverable through composting and biodegradation - Test scheme and evaluation criteria for the final acceptance of packaging
105 DIN EN 14046 Packaging - Evaluation of the ultimate aerobic biodegradability of packaging materials under controlled composting conditions - Method by analysis of released carbon dioxide
106 ASTM E1839 - 07 Standard Test Method for Efficacy of Slimicides for the Paper Industry—Bacterial and Fungal Slime
107 ASTM D7243 - 06a Standard Guide for Measuring the Saturated Hydraulic Conductivity of Paper Industry Sludges
108 ASTM E723 - 07 Standard Test Method for Efficacy of Antimicrobials as Preservatives for Aqueous-Based Products Used in the Paper Industry (Bacterial Spoilage)
109 ASTM D6868 - 03 Standard Specification for Biodegradable Plastics Used as Coatings on Paper and Other Compostable Substrates
110 ASTM D6692-01 Standard Test Method for Determining the Biodegradability of Radiolabelled Polymeric Plastic Materials in Seawater (Withdrawn 2009)

111 ASTM D7209 - 06 Standard Guide for Waste Reduction, Resource Recovery, and Use of Recycled Polymeric Materials and Products
112 ASTM E1428 - 99(2009) Standard Test Method for Evaluating the Performance of Antimicrobials in or on Polymeric Solids Against Staining by Streptoverticillium reticulum (A Pink Stain Organism)
113 D5271-02 Standard Test Method for Determining the Aerobic Biodegradation of Plastic Materials in an Activated-Sludge-Wastewater-Treatment System
114 ASTM D6954 - 04 Standard Guide for Exposing and Testing Plastics that Degrade in the Environment by a Combination of Oxidation and Biodegradation
115 ASTM D5338 - 98(2003) Standard Test Method for Determining Aerobic Biodegradation of Plastic Materials Under Controlled Composting Conditions
116 ASTM D5210 - 92(2007) Standard Test Method for Determining the Anaerobic Biodegradation of Plastic Materials in the Presence of Municipal Sewage Sludge
117 ASTM D5526 - 94(2002) Standard Test Method for Determining Anaerobic Biodegradation of Plastic Materials Under Accelerated Landfill Conditions
118 ASTM D5988 - 03 Standard Test Method for Determining Aerobic Biodegradation in Soil of Plastic Materials or Residual Plastic Materials After Composting
119 ASTM D5511 - 02 Standard Test Method for Determining Anaerobic Biodegradation of Plastic Materials Under High-Solids Anaerobic-Digestion Conditions
120 ASTM D6691 - 09 Standard Test Method for Determining Aerobic Biodegradation of Plastic Materials in the Marine Environment by a Defined Microbial Consortium or Natural Sea Water Inoculum
121 ASTM D6340 - 98(2007) Standard Test Methods for Determining Aerobic Biodegradation of Radiolabelled Plastic Materials in an Aqueous or Compost Environment
122 ASTM D5951 - 96(2002) Standard Practice for Preparing Residual Solids Obtained After Biodegradability Standard Methods for Plastics in Solid Waste for Toxicity and Compost Quality Testing
123 ASTM D6400 - 04 Standard Specification for Compostable Plastics
124 ASTM D7081 - 05 Standard Specification for Non-Floating Biodegradable Plastics in the Marine Environment
125 ASTM D6002 - 96(2002)e1 Standard Guide for Assessing the Compostability of Environmentally Degradable Plastics
126 ASTM D6776-02 Standard Test Method for Determining Anaerobic Biodegradability of Radiolabelled Plastic Materials in a Laboratory-Scale Simulated Landfill Environment
127 ASTM D5975-96(2010) Standard Test Method for Determining the Stability of Compost by Measuring Oxygen Consumption
128 Projects Plastics - Preparation of samples for biodegradation testing Planned document number: ISO/DIS 10210
129 Projects Plastics - Determination of the ultimate anaerobic biodegradation of plastic materials in controlled slurry digestion systems - Method by measurement of biogas production. Planned document number: ISO/WD 13975
130 DIN EN ISO 14855-2 Determination of the ultimate aerobic biodegradability of plastic materials under controlled composting conditions - Method by analysis of evolved carbon dioxide - Part 2: Gravimetric measurement of carbon dioxide evolved in a laboratory-scale test (ISO 14855-2:2007)
131 BS ISO 14853 Plastics - Determination of the ultimate anaerobic biodegradation of plastic materials in an aqueous system - Method by measurement of biogas production
132 BS ISO 15985 Plastics - Determination of the ultimate anaerobic biodegradation and disintegration under high-solids anaerobic-digestion conditions - Method by analysis of released biogas
133 NF T51-801; NF EN ISO 14851:2004-12-01 Determination of the ultimate aerobic biodegradability of plastic materials in an aqueous medium - Method by measuring the oxygen demand in a closed respirometer.
134 NF T51-802; NF EN ISO 14852:2004-12-01 Determination of the ultimate aerobic biodegradability of plastic materials in an aqueous medium - Method by analysis of evolved carbon dioxide.
135 NF T51-803-1; NF EN ISO 14855-1:2008-07-01 Determination of the ultimate aerobic biodegradability of plastic materials under controlled composting conditions - Method by analysis of evolved carbon dioxide - Part 1: general method.
136 NF T51-803-2; NF EN ISO 14855-2:2009-10-01 Determination of the ultimate aerobic biodegradability of plastic materials under controlled composting conditions - Method by analysis of evolved carbon dioxide - Part 2: gravimetric measurement of carbon dioxide evolved in a laboratory-scale test.
137 NF T51-806; NF EN ISO 20200:2005-10-01 Plastics - Determination of the degree of disintegration of plastic materials under simulated composting conditions in a laboratory-scale test.

138 ISO/DIS 10210 Plastics - Preparation of samples for biodegradation testing
139 ISO 14853 Plastics - Determination of the ultimate anaerobic biodegradation of plastic materials in an aqueous system - Method by measurement of biogas production
140 ISO 14853 Technical Corrigendum 1 Plastics - Determination of the ultimate anaerobic biodegradation of plastic materials in an aqueous system - Method by measurement of biogas production; Technical Corrigendum 1
141 ISO 15985 Plastics - Determination of the ultimate anaerobic biodegradation and disintegration under high-solids anaerobic-digestion conditions - Method by analysis of released biogas
142 ISO 15985 Technical Corrigendum 1 Plastics - Determination of the ultimate anaerobic biodegradation and disintegration under high-solids anaerobic-digestion conditions - Method by analysis of released biogas; Technical Corrigendum 1
143 ASTM E2167 - 01(2008) Standard Guide for Selection and Use of Stone Consolidants
144 ASTM F1581 - 08e1 Standard Specification for Composition of Inorganic Bone for Surgical Implants
145 ASTM F1635 - 04a Standard Test Method for in Vitro Degradation Testing of Hydrolytically Degradable Polymer Resins and Fabricated Forms for Surgical Implants
146 ASTM F2211 - 04 Standard Classification for Tissue Engineered Medical Products
147 DIN EN ISO 10993-9 Biological evaluation of medical devices - Part 9: Framework for identification and quantification of potential degradation products (ISO 10993-9:1999)
148 ASTM E1943 - 98(2004) Standard Guide for Remediation of Ground Water by Natural Attenuation at Petroleum Release Sites
149 ASTM D5660 - 96(2009) Standard Test Method for Assessing the Microbial Detoxification of Chemically Contaminated Water and Soil Using a Toxicity Test with a Luminescent Marine Bacterium
150 ASTM D3415 - 98(2004) Standard Practice for Identification of Waterborne Oils
151 ASTM F1231 - 08 Standard Guide for Ecological Considerations for the Use of Oil Spill Dispersants in Freshwater and Other Inland Environments, Rivers and Creeks
152 ASTM E1022 - 94(2007) Standard Guide for Conducting Bioconcentration Tests with Fishes and Saltwater Bivalve Mollusks
153 FD T90-315; FD ISO/TR 15462:2006-08-01 Water quality - Selection of tests for biodegradability.
154 Wypych, G, **Handbook of Material Weathering**, 4th Ed., *ChemTec Publishing*, Toronto, 2008.
155 Herrera, L K; Videla, H A, *Intl. Biodet. Biodeg.*, **63**, 813-22, 2009.
156 Oshiman, K; Tsutsumi, Y; Nishida, T; Matsumura, Y, *Biodeg.*, **18**, 247-55, 2007.
157 Gutarowska, B; Zakowska, Z, *Intl. Biodet. Biodeg.*, **49**, 299-305, 2002.
158 Santos, C; Fraga, M E; Kozakiewicz, Z; Lima, N, *Res. Microbiol.*, in press, 2010.
159 Edwards. H G M, **Frontiers of Molecular Spectroscopy**, pp 133-73, *Elsevier*, 2009.
160 Perez-Alonso, M; Castro, K; Alvarez, K; Madariaga, J M, *Anal. Chim. Acta*, **524**, 379-89, 2004.
161 Edwards, H G M; Gniadecka, M; Petersen, S; Hansen, J P H; Nielsen, O F; Christensen, D H; Wulf, H C, *Vibrational Spectroscopy*, **28**, 3-15, 2002.
162 Shah, A A; Hasan, F; Hameed, A; Ahmed, S, *Biotechnol. Adv.*, **26**, 246-65, 2008.

6

Biostabilizers – Health & Safety

This chapter contains information on health and safety related to biocide use. It discusses only general health and safety acts. Legislation specific to biocides is reported in Chapter 8. Information is supported with data on:

- toxic substance control
- carcinogenic effect
- workplace exposure limits
- food regulatory acts

6.1 TOXIC SUBSTANCE CONTROL

The United States Congress enacted the Toxic Substances Control Act.[1] The following are the premises of the Act: "Human beings and the environment are being exposed each year to a large number of chemical substances and mixtures; among the many chemical substances and mixtures which are constantly being developed and produced, there are some whose manufacture, processing, distribution in commerce, use, or disposal may present an unreasonable risk of injury to health or the environment".

The chemical substances in the Chemical Substances Inventory contain chemical compounds which are permitted for use by industry. New compounds, not available on this list, must undergo a process established by the Environmental Protection Agency, EPA.

The Inventory is created for the purpose stated in Section 2624: "The Council on Environmental Quality, in consultation with the Administrator, the Secretary of Health and Human Services, the Secretary of Commerce, and the heads of other appropriate Federal departments or agencies, shall coordinate a study of the feasibility of establishing a standard classification system for chemical substances and related substances, and a standard means for storing and for obtaining rapid access to information respecting such substances."

This very useful regulation, after more than a quarter century of practice, still does not give sufficient information. The EPA has selected a group of High Production Volume, HPV, chemicals, which constitutes 3,300 organic chemicals produced or imported at or above 1 million pounds per year, in the United States. The Organization for Economic Co-operation and Development (OECD) also maintains a list of 5,235 HPV chemicals, compiled in 2000. The OECD list includes chemicals which have annual production volumes greater than 1 thousand metric tones (2.2 million pounds) in more than one economically developed country.

These chemicals have a large impact on health and safety because of their widespread application. The EPA's analysis found that no basic toxicity information, i.e., neither human health nor environmental toxicity, is publicly available for 43% of the high volume chemicals manufactured in the US, and that a full set of basic toxicity information is available for only 7% of these chemicals. More information is available elsewhere.[2-4] Since then, information is gradually being improved.

The lack of this basic toxicity information, on many high volume chemicals, is a serious issue for risk assessment, safeguarding children's health, expanding the public's right-to-know, and promoting the pollution prevention ethic, which are important EPA initiatives.

Table 6.1 lists biocides which are HPV chemicals, and information on data availability on acute toxicity, AT, chronic toxicity, CT, teratogenicity, T, or developmental and reproductive toxicity, mutagenicity, M, ecotoxicity, E.

Table 6.1 Biocides on HPV chemicals list and data availability[5]

CAS #	Additive name	Data available (Y – yes; N – no)				
		AT	CT	T	M	E
65-85-0	Benzoic acid	Y	Y	Y	Y	Y
77-92-9	Citric acid	Y	Y	Y	Y	Y
334-48-5	Decanoic acid	Y	N	Y	Y	N
64-18-6	Formic acid	Y	Y	Y	Y	Y
143-07-7	Lauric acid	Y	N	N	Y	N
112-05-0	Nonanoic acid	Y	Y	Y	Y	Y
124-07-2	Octanoic acid	Y	N	Y	Y	N
79-21-0	Peroxyacetic acid	Y	N	N	Y	N
69-72-7	Salicylic acid	Y	N	Y	Y	Y
110-44-1	Sorbic acid	Y	N	Y	Y	N
90-43-7	2-Biphenylol	Y	Y	Y	Y	Y
64-17-5	Ethanol	Y	Y	Y	Y	Y
122-99-6	Ethanol, 2-phenoxy-	Y	N	Y	Y	Y
71-23-8	Propyl alcohol	Y	Y	Y	Y	Y
67-63-0	Isopropyl alcohol	Y	Y	Y	Y	Y
107-02-8	Acrolein	Y	Y	Y	Y	Y
104-55-2	Cinnamaldehyde	Y	N	Y	Y	Y
50-00-0	Formaldehyde	Y	Y	Y	Y	Y

Table 6.1 Biocides on HPV chemicals list and data availability[5]

CAS #	Additive name	Data available (Y – yes; N – no)				
		AT	CT	T	M	E
111-30-8	Glutaraldehyde	Y	N	Y	Y	Y
107-22-2	Glyoxal	Y	N	Y	Y	Y
330-54-1	Diuron	Y	Y	Y	Y	Y
137-26-8	Thioperoxydicarbonic diamide	Y	Y	Y	Y	Y
533-74-4	1,3,5-Tetrahydro-3,5-dimethyl-2H-1,3,5-thiadiazine-2-thione	Y	N	N	Y	Y
126-11-4	Tris(hydroxymethyl)nitromethane	Y	N	N	N	N
87-90-1	Trichloroisocyanuric acid	Y	N	N	Y	Y
80-46-6	4-Tert-amylphenol	Y	N	N	Y	Y

The above table shows that only some biocides (the simplest chemical compounds) are included on the list and these are quite well covered by testing. Many generic compounds present in biocides in use today are not included in the HPV list of chemicals and basic information on their effects is scarce.

6.2 CARCINOGENIC EFFECT

To regulate carcinogenic substances, federal agencies follow rulemaking procedures. Rulemaking procedures may be:

- risk-based (for example, the Delaney clause that bans a food additive shown to cause cancer in humans or in animal tests)
- technology-based, which might require the use of "best available technology" or "best practical technology" to control emissions from a particular source
- risk-benefit or cost-benefit balancing, which permits the consideration of competing health risks and benefits (for example, cancer-causing drugs treat fatal illnesses but may affect other organs).[6]

The Occupational Safety and Health Act of 1970 established the Occupational Safety and Health Administration, OSHA, and the National Institute for Occupational Safety and Health, NIOSH. OSHA is a regulatory agency which, among its other duties, issues and enforces regulations that limit exposure to carcinogens in the workplace. NIOSH is a research agency that has supported epidemiological and toxicological research, and makes recommendations to OSHA concerning changes in occupational health standards.[6]

In its rulemaking procedure, OSHA has used two different approaches for limiting exposures: setting permissible exposure limits, and requiring specific process technology and procedures. In addition to the data from the National Toxicology Program, NTP, OSHA also considers the classification of the International

Agency for Research on Cancer, IARC. Based on this data, OSHA makes its own decisions. Table 6.2 contains information on substances used as biocides and included on lists of carcinogens.

Table 6.2. Biocides on carcinogen lists[7]

Chemical substance	IARC	NTP	OSHA-Z	Prop. 65
Arsenic and arsenic compounds	1	K		no
Biphenyl				yes
Creosote	2A	K		no
Diuron				yes
Fenoxycarb				yes
Folpet				yes
Formaldehyde	1	P	1910 Z	yes
Mancozeb				yes

2A the chemical is probably carcinogenic to humans
2B the chemical is possibly carcinogenic to humans
P the chemical may reasonably be anticipated to be carcinogenic
Prop. 65 Proposition 65 List of Carcinogens (California) updated Feb. 5, 2010

Table 6.2 shows that some biocidal compounds are found on carcinogen lists.

6.3 WORKPLACE EXPOSURE LIMITS

There are numerous limiting values set to protect employees against workplace exposures. The definitions of these values are given below:

1. *Threshold Limit Value*, TLV, is an exposure standard set by a committee of the American Conference of Governmental Industrial Hygienists, ACGIH. The TLVs are published annually. The guidelines are based on available animal and human exposure studies. The rationale for setting the TLVs is given in an ACGIH publication.[8] The ACGIH committee is independent and flexible, incorporates new data rapidly, and is relatively free of bureaucratic constrains of official government agencies. The TLV's are recommended values, not legal limits. They do not guarantee protection to all workers, and they are not intended to be used for community exposure. TLVs are values that should not be exceeded. The goal is to minimize workers' exposure to hazardous concentrations as much as possible.[9]
2. *Recommended Exposure Limits*, REL, are set by the National Institute for Occupational Safety and Health, NIOSH, which is a part of the department of Health and Human Services. NIOSH scientists recommend exposure limits to OSHA, based on animal and human studies. NIOSH RELs are often more conservative than the TLV, and NIOSH's consideration of available research and studies is regarded as thorough.[9]
3. *Permissible Exposure Limits*, PEL, are set by the Occupational Safety and Health Administration, OSHA, and are the law of the land in the United

States. Workers' exposure may not exceed these standards, and OSHA has the power to warn, cite, and fine violators. The Act required OSHA to set standards that will provide safe working conditions by negotiation and consensus. As a result, only about 25 permanent standards have been set since 1973. To protect workers in the meantime, OSHA was allowed to adopt existing standards or develop Emergency Temporary Standards (ETS). OSHA adopted the ACGIH TLV as the interim standard, giving legal status to what was meant to serve as recommended limits. The permanent standards are thorough and reflect the extensive effort invested in their preparation. They include action levels that are typically half the TWA exposure limits. When the action limit is exceeded, several steps must be taken such as medical monitoring, air sampling, and control measures. Each permanent standard includes recommendations for air sampling procedures, regulations for record keeping, engineering control methods, labeling and warning, and other pertinent regulations.[9] The PELs are published in 29 CFR 1910.1000.[10] Time-Weighted Average, TWA, is the concentration of contaminants over an 8-hour period. It is determined by sampling the breathing zone of the worker for 8 hours. It is expressed as follows, when a series of successive samples have been taken:

$$TWA = \frac{\sum_i C_i t_i}{\sum_i t_i} \qquad [6.1]$$

where:

t_i the period of time during which one sample is taken
C_i the average concentration over time period t_i.

To determine the level of exposure, the TWA reading is compared with a standard such as the threshold limit value, TLV, or permissible exposure limit, PEL. If the reading exceeded the standard, overexposure occurred.[9]

4. *Short-Term Exposure Limit*, STEL, is a 15-minute TWA concentration that may not be exceeded, even if the 8-hour TWA is within the standards. TWA-STEL are given for contaminants for which short-term hazards are known. For the rest, an excursion factor of 3 has been often used: STEL should not exceed 3 times the TWA limit.[9]
5. *Ceiling* is a concentration that should not be exceeded at any time. Note that both TWA and STEL permit limited excursion if, in the end, the average is below the exposure limit. The ceiling value, however, may not be exceeded.[9]
6. *Immediately Dangerous to Life and Health*, IDLH, is defined as conditions that pose immediate danger to life or health, or conditions that pose a threat of severe exposure. IDLH limits are created mainly to assist in making decisions regarding respirator use: above the IDLH only supplied air respirators should be used, below the IDLH, air purifying respirators may be used, if appropriate. Two factors were considered when establishing the IDLH limits:
   - workers must be able to escape such an environment without suffering permanent health damage

- workers must be able to escape without severe eye or respiratory tract irritation or other conditions that might impair their escape

The current definition has no exposure duration associated with it. Workers should not be in an IDLH environment for any length of time, unless they are equipped and protected to be in that environment. IDLH values were determined based on animal and human data. They may be found in the NIOSH Pocket Guide to Chemical Hazards.[11]

Occupational Exposure Limits, OELs, are set by the European Union. In 1990, at the request of the Council, the European Commission had set up an informal group of scientists, known as the Scientific Expert Group, SEG, to give advice on setting limit values, after having reviewed the different approaches in the Member States. To encourage the Occupational Exposure Limits, OELs, the European Commission decided to formalize the SEG, and with its Decision 95/320/EC of 12 July 1995 has set up a formal base for the work on the scientific evaluation of risk at the workplace related to chemical substances.[12] The proposed values include:

- the eight-hour time weighted average, TWA
- short-term/excursion limits, STEL

Table 6.3 gives limits values for biocides available from OSHA[10] and NIOSH.[11]

Table 6.3. Limits for air contaminants from OSHA[10] and NIOSH[11]

Additive	PEL, ppm	REL, ppm	IDLH, ppm
Acrolein	0.1	0.1	2
Arsenic (organ8ic compound, as As)	0.5 mg/m^3		
Chlorine	1	0.5	10
Copper dust	1 mg/m^3	1 mg/m^3	100 mg/m^3
Ethanol	1000	1000	3300
Formaldehyde	0.75	0.1	20
Glutaraldehyde		0.2	
Hydrogen peroxide	1	1	75
Iodine	0.1	0.1	2
Isopropanol	400	400	2000
Propanol	200	200	800
Silver	0.01 mg/m^3	0.01 mg/m^3	10 mg/m^3

(r) respiratory
(t) total
* carbon black in the presence of polycyclic aromatic hydrocarbons

6.4 FOOD REGULATORY ACTS

In the United States, the Food and Drug Administration is entrusted by a Code of Federal Regulations[13] to control use of different materials in the production of food and its packaging. Two types of additives are used in connection with food: additives permitted to be added to food directly, and additives which are used in materials in contact with food products.

Table 6.4 shows the relevant parts of the Regulation[13] and biocides which may be directly added to food.

Table 6.4. Biocides permitted as direct additives to food[13]

Part	
Part 172. Food Additives Permitted for Direct Addition to Food for Human Consumption	
172.167	silver nitrate and hydrogen peroxide solution
Part 173. Secondary Direct Food Additives Permitted in Food for Human Consumption	
173.300	chlorine dioxide, sodium hypochlorite
173.320	disodium ethylene bis(dithiocarbamate), sodium dimethyldithiocarbamate, glutaraldehyde
173.370	peroxyacetic acid, octanoic acid, acetic acid, hydrogen peroxide

Table 6.4 shows the relevant parts of the Regulation,[13] and additives sanctioned as indirect food additives.

Table 6.5 shows the relevant parts of the Regulation[13] and biocides sanctioned as indirect food additives.

Table 6.5. Biocides permitted as indirect additives to food[13]

Part	Description	Additives
Part 175 - Indirect food additives: Adhesives and Components of Coatings		
175.105	Adhesives	boric acid, formaldehyde, glutaraldehyde, hydrogen peroxide, thiram
Part 176. Indirect Food Additives: Paper and Paperboard Components		
176.300	Slimicides	acrolein, alkenyl (C_{16}-C_{18}) dimethylethyl-ammonium bromide, n-alkyl (C_{12}-C_{18}) dimethyl benzyl ammonium chloride, 1,2-benzisothiazolin-3-one, bis(1,4-bromoacetoxy)-2-butene, 5,5-bis(bromoacetoxymethyl)dioxane, 2,6-bis(dimethylaminomethyl) cyclohexanone, 1,2-bis(monobromoacetoxy) ethane, bis(trichloromethyl)sulfone, 4-bromoacetoxymethyl-m-dioxolane, 2-bromo-4'-hydroxyacetophenone, 2-bromo-2-nitropropane-1,3-diol, β-bromo-β-nitrostyrene, chloroethylenebisthiocyanate, 5-chloro-2-methyl-4-isothiazolin-3-one, chloromethyl butanethiolsulfonate, cupric nitrate, n-dialkyl (C_{12}-C_{18}) benzylmethylammonium chloride, 1,2-dibromo-2,4-dicyanobutane, 2,2-dibromo-3-nitrilopropionamide, 2,3-dibromopropionaldehyde, 4,5-dichloro-1, 2-dithiol-3-one, 1,3-dihalo-5,5-dimethylhydantoin, 4-(diiodomethylsulfonyl) toluene, 3,5-dimethyl 1,3,5,2H-tetrahydrothiadiazine-2-thione, dipotassium and disodium ethylenebis(dithiocarbamate), disodium cyanodithioimidocarbonate, n-dodecylguanidine hydrochloride, glutaraldehyde, 2-(p-hydroxyphenyl) glyoxylohydroximoyl chloride, 2-hydroxypropyl methanethiol sulfonate, 2-mercaptobenzothiazole, methylenebisbutanethiolsulfonate, methylenebisthiocyanate, 2-nitrobutyl bromoacetate, potassium 2-mercaptobenzothiazole, potassium N-hydroxymethyl-N-methyldithiocarbamate, potassium N-methyldithiocarbamate, potassium pentachlorophenate, potassium trichlorophenate, silver fluoride, silver nitrate, sodium dimethyldithiocarbamate, sodium 2-mercaptobenzothiazole, sodium pentachlorophenate, sodium trichlorophenate, 1,3,6,8-tetraazatricyclododecane,

Table 6.5. Biocides permitted as indirect additives to food[13]

Part	Description	Additives
176.300	Slimicides	3,3,4,4-tetrachlorotetrahydrothiophene-1,1-dioxide, tetrakis(hydroxy-methyl)phosphonium sulfate, 2-(thiocyanomethylthio) benzothiazole, vinylene bisthiocyanate
Part 178. Indirect Food Additives: Adjuvants, Production Aids, and Sanitizers		
178.1005	H_2O_2 solution	<35% hydrogen peroxide
178.1010	Sanitizing solutions	decanoic acid, dichloroisocyanuric acid, hydrogen peroxide, iodine, isopropanol, acetic acid, octanoic acid, peroxyacetic acid, potassium iodide, quaternary ammonium compounds, sodium hypochlorite, sodium p-toluenesulfonchloroamide, trichloroisocyanuric acid

REFERENCES

1 Title 15 - Commerce and Trade. Chapter 53. Toxic Substances Control. Subchapter I - Control of Toxic Substances.
2 Wypych, G., **Handbook of Plasticizers**, Wypych G., Ed., *ChemTec Publishing*, Toronto 2004.
3 Chemical Hazard Data Availability Study. What Do We Really Know About the Safety of High Production Volume Chemicals? EPA's 1998 Baseline of Hazard Information that is Readily Available to the Public Prepared by EPA's Office of Pollution Prevention and Toxics, April 1998.
4 EPA information on 2863 U.S. HPV Chemicals from 1990 IUR Update.
5 Master Summary for the Chemical Hazard Data Availability Table, EPA.
6 Identifying and Regulating Carcinogens, NTIS, November 1987.
7 2001 Toxic Release Inventory. Public Data Release. Appendix C. Basis of OSHA Carcinogen Listing for Individual Chemicals.
8 Documentation of the Threshold Limit Values and Biological Exposure Indices, 7th Ed. ACGIH, 2001.
9 Office of response and Restoration, National Ocean Service, National Oceanic and Atmospheric Administration. Occupational Exposure Limits, March 2002.
10 US Department of Labor. Occupational Safety & Health Administration. Standards - 29 CFR. Table Z-1. Limits for Air Contaminants. - 1910.1000.
11 NIOSH Pocket Guide to Chemical Hazards. NIOSH Publication No. 2005-149. September 2005.
12 Occupational Exposure Limits, OELs, Official J., L188, p.14, (1995).
13 Code of Federal Regulations. Title 21. Food and Drugs. Chapter 1. Food and Drug Administration, Department of Health and Human Services.

7

ENVIRONMENTAL FATE OF BIOSTABILIZERS

Biocides must be available on the surface of a material in order to perform their functions and in many cases they have to produce a certain concentration in liquid surrounding the perimeters of the protected product. This mechanism of action does not leave any doubt that biocides are present in the surrounding media (air, water, and/or soil). Whether this presence is harmful to the environment depends on concentration, toxicity, rate of their degradation, and products of degradation. The above subjects are the essence of this chapter.

7.1 CONCENTRATION

Table 7.1 contains a summary of biocide concentration found in the water of various geographic locations indicated in the table.

Table 7.1. Concentration of biocides in water from various locations as indicated.

Biocide	Concentration, ng/l	Location (year of testing)	Ref.
Acrylated glyphosate	none	experimental study of coating	6
Chlorothalonil	none to 1380	marinas & ports, Europe (98-00)	1
Dichlofluanid	none to 600 1-3	marinas & ports, Europe (98-00) marinas, Greece (2003)	1 8
Diuron	1-768 <1 to 6742 90-1130 5-2000 none to 3054 <1 to 1379	various marinas in UK (1998-1999) various estuaries in UK (1998-2000) marinas, Holland (1998-2000) Mediterranean, Spain (97-00) ports & marinas, Japan (1999) Gulf of Naples, Italy (05-06)	1 1 1 1 1 7
DMSA (dichlofluanid metabolite)	1-28	marinas, Greece (2003)	8
Irgarol 1051	<1 to 1421 <1 to 680 <2-14 22-1700 <5 to 280 negl to 1 <10 to 1000 negl. to 90 10-24	various marinas in UK (1993-2001) various estuaries in UK (1993-2000) ports, UK coast (1993) marinas, France (1992-1995) ports, France (1992-1995) beaches, France (1992-1995) marinas, Spain (1996-2000) marinas, Greece (1999-2000) ports, Greece (1999-2000)	1 1 1 1 1 1 1 1 1

Table 7.1. Concentration of biocides in water from various locations as indicated.

Biocide	Concentration, ng/l	Location (year of testing)	Ref.
Irgarol 1051	8-90 11-440 <1 to 144.2 <0.2 to 173 <1 to 960 8-27	marinas, coast, Holland (2000) marinas, North & Baltic Sea, Germany (1993-1996) Florida, US Gulf of Naples, Italy (05-06) Brighton Marina UK (1999-2001) S. England (2004-2005)	1 1 1 7 11 13
Mecoprop	40,000-160,000	sewage (from bitumen & plastic sheets) Switzerland, 2006	5
Sea Nine 211	none to 3300	marinas, Europe (1998-2000)	1
TCMTB-TCMS	none	marinas, UK (1998)	1
Zinc pyrithione	<20	marinas, UK (1998)	1

Table 7.2 contains summary of biocide concentration found in sediment in various geographic locations indicated in the table.

Table 7.2. Concentration of biocides in sediments from various locations as indicated.

Biocide	Concentration, ng/l	Location (year of testing)	Ref.
Chlorothalonil	none to 165	marinas & ports, Europe (98-00)	1
DCOIT	2	in-service leach rate, Denmark	9
Dichlorofluanid	none to 688.2 5-100	marinas & ports, Europe (98-00) marinas, Greece (2003)	1 8
Diuron	<12 to 395 <12 to 1420 <100 <0.4-136 0.4-6.2	various marinas in UK (1998-1999) various estuaries in UK (1998-2000) marinas, Holland (1998-2000) Mediterranean, Spain (97-00) Southampton, UK	1 1 1 1 12
DMSA (dichlorofluanid metabolite)	3-10	marinas, Greece (2003)	8
Irgarol 1051	<1 to 1011 <1 to 222 <0.2 to 88 75 to690 none to 19 <1 3-220 0.388-21.26 <1 to 77 0.3-1.2 9-27	various marinas in UK (1999-2001) various estuaries in UK (1993-2000) marinas, Spain (1996-2000) marinas, Greece (1999-2000) ports, Greece (1999-2000) marinas, coast, Holland (2000) marinas, North & Baltic Sea, Germany (1993-1996) costal sediments, Hong Kong (2006) Brighton Marina UK (1999-2001) Southampton, UK S. England (2004-2005)	1 1 1 1 1 1 1 10 11 12 13
Sea Nine 211	<0.4-4	marinas, Europe (1998-2000)	1

Substantial concentrations were detected between 1993 and 2000 in the areas of high yachting activity, particularly in marinas and harbours, as a consequence of their increased use in antifouling paints.[1]

In antifouling paints, the release rate is important to calculate the thickness of paint required for a certain service life of paint. It is also important here since it determines the amount of active components present in the environment. The release rate depends more on the cleaning schedule and its severity than it does on paint formulation.[2] It also depends on the medium in which studies are conducted. In artificial seawater the removal rate was up to 39 times faster than in natural seawater.[2] Typical rates of copper leaching were in the range of 20-25 μg cm^{-2} d^{-1}.[2] Table 7.3 gives release rates of common biocides as measured according to the ISO standard.[3]

Table 7.3. Release rates of common biocides.[2,4]

Biocide	Trade name	Release rate, μg cm^{-2} day^{-1}	Refs.
Copper		20-25	2
Cuprous oxide		25-40	4
Dichlofluanid	Preventol A4, Euparen	0.6	4
Diuron		3.3 0.6	4 12
Irgarol 1051		5 2.6	4 12
Sea Nine 211	DCOI	2.9	4
TBT		1.5-4.0	
TCMS pyridine	Densil S	0.6	4
TCMTB	Busan		4
Zinc pyrithione	Zinc Omadine	3.3	4

The highest concentrations of Irgarol 1051 were recorded in Brighton Marina, UK, in November through January.[11] They were probably associated with maintenance of vessels in an adjacent boatyard.[11] Elevated levels were also encountered at the beginning of the season, coinciding with the introduction of newly antifouled vessels.[11] Increased concentrations also followed dredging, possibly through re-mobilisation of Irgarol 1051.[11]

The sorptive behavior of Irgarol, dichlofluanid, chlorothalonil and diuron is dictated by a number of physiochemical parameters in the aquatic environment.[14] Diuron will be present mainly in the dissolved phase, whereas chlorothalonil and Irgarol will be present in significant amounts in both the adsorbed phase and in the particulate matter.[14] Dichlofluanid exhibits stronger adsorption characteristics and will bind more strongly to sediments.[14] Sorption depends on sediment type. Clay and silt type sediments are more receptive to adsorption than larger claysands.[14]

7.2 TOXICITY

Table 7.4 gives environmental toxicity data for various biocides.

Table 7.4. Environmental toxicity data for various biocides.

Biocide	Class	Test organism	Data	Refs.
Chlorothalonil	Ascidiacea	*Botryllus schlosseri*	LC50: 13,827	21
	Crustacean	*Artemia salina*	24-h LC50: 1,000	15
	Mollusc	*Mytilus edulis*	EC50: 0.125	16
	Sea squirt	*Ciona intestinalis*	EC50: 0.466	16
	Sea urchin	*Paracentrotus lividus*	EC50: 0.095	16
Cu pyrithione	Amphipod	*Elasmopus rapax*	96 h-LC50: 78	19
	Crustacean	*Artemia salina*	24-h LC50: 830	15
	Diatom	*Thalassiosira pseudonana*	96 h-EC50: 0.97	19
	Polychaeta	*Hydroides elegans*	48 h-LC50: 120	19
Dichlofluanid	Algae	*Scedenesmus subspicatus*	96h-EC50: 1	8
	Crustacean	*Daphnia magna*	48h-EC50: 0.42	8
	Mollusc	*Mytilus edulis*	EC50: 0.732	16
	Rainbow trout	*Oncorhynchus mykiss*	96h-LC50: 10	8
	Sea squirt	*Ciona intestinalis*	EC50: 2.538	16
	Sea urchin	*Paracentrotus lividus*	EC50: 5.643	16
DMSA dichlo-fluanid metabolite	Algae	*Scedenesmus subspicatus*	96h-EC50: 97.7	8
	Crustacean	*Daphnia magna*	48h-EC50: >95.6	8
	Rainbow trout	*Oncorhynchus mykiss*	96h-LC50: >100,000	8
Diuron	Algae	*Selenastrum capricornotum*	3-day EC50: 45	1
	Chinook salmon	*Oncorhynchus tshawytscha*	24-h EC50: 52,000	1
	Crustacean	*Artemia salina*	24-h LC50: 12,010	15
		Daphnia magna	48-h EC50: 8,600	1
		various species	EC50: 15,000-25,000	7
	Phytoplankton	various species	EC50: 5.8-500	7
	Rainbow trout	*Oncorhynchus mykiss*	7-day LC50: 74,000	1
Glutaraldehyde	Barnacle	*Balanus improvisus*	10-d LC50: 7,500	20
	Bivalvia	*Crassostrea virginica*	48-h LC50: 550	20
	Bluegill sunfish	*Lepomis macrochirus*	96-h LC50: 9,400-11,000	20
	Branchiopoda	*Ceriodaphnia dubia*	8-d LC50: 4,700	24
	Chlorophyceae	*Pseudokirchneriella sub.*	96-h LC50: 1000-1800	24
	Coho salmon	*Oncorhynchus kisutch*	96-h LC50: 3,000	20
	Crustacean	*Daphnia magna*	48-h LC50: 5,000	20
	Fathead minnow	*Pimephales promelas*	96-h LC50: 5,400-6,000	20
	Mussel	*Mytilus edulis*	5-d LC50: 200	20
	Rainbow trout	*Salmo gairdneri*	96-h LC50: 11,000	20
		Oncorhynchus mykiss	62-d LC50: 1,820	24

Table 7.4. Environmental toxicity data for various biocides.

Biocide	Class	Test organism	Data	Refs.
Irgarol 1051	Algae	*Closterium ehrenbergii*	5-day EC50: 2.5	1
		Selenastrum capricornotum	3-day EC50: 10.8-1.7	1
	Bacteria	*Vibrio fischeri*	15-min EC50: 50,800	1
	Chinook salmon	*Oncorhynchus tshawytscha*	24-h EC50: 10^5 <	1
	Crustacean	*Daphnia magna*	48-h EC50: 7,300	1
	Cyanobacteria	*Chroococcus minor*	96h-EC50: 0.771	17
	Diatom	*Skeletonema costatum*	96h-EC50: 0.00029	17
		Thalassiosira pseudonana	96h-EC50: 0.00041	17
	Duckweed	*Lemna minor*	7-day EC50: 7.3-8.9	1
	Macrophytes	*Elodea canadensis*	EC20: 17-52	1
		various species	EC50: 1,922-8,100	7
	Mollusc	*Mytilus edulis*	EC50: 24	16
	Phytoplankton	various species	EC50: 0.441-0.647	7
		various species	EC50: 117-3,110	1
	Rainbow trout	*Oncorhynchus mykiss*	7-day LC50: 25,000	1
	Seagrass	*Zostera marina*	10-day EC50: 2.5	
	Sea squirt	*Ciona intestinalis*	EC50: 32.97	16
	Seaweed	*Porphyra yezoensis*	4-day EC50: 0.60	1
		Eisenia bicyclis	4-day EC50: 2.6-7.4	1
	Sea urchin	*Paracentrotus lividus*	EC50: 93.9	16
Sea-nine 211	Algae	*Enteromorpha intestinalis*	LD50: 2	1
		Selenastrum capricornotum	3-day EC50: 3	1
	Ascidiacea	*Botryllus schlosseri*	LC50: 7,055	21
	Bacteria	*Vibrio fischeri*	15-min EC50: 12	1
	Barnacles	*Balanus amphitrite larvae*	LD50: 340	1
	Chinook salmon	*Oncorhynchus tshawytscha*	24-h EC50: 2,600	1
	Crustacean	*Daphnia magna*	48-h EC50: 4	1
	Diatom	*Amphora coffeaeformis*	LD50: 3	1
	Macroalga	*Fucus serratus*	72-h EC50: 0.049-2.05	23
	Mollusc	*Mytilus edulis*	EC50: 0.135	16
	Rainbow trout	*Oncorhynchus mykiss*	7-day LC50: 14	1
	Sea squirt	*Ciona intestinalis*	EC50: 1.322	16
	Sea urchin	*Paracentrotus lividus*	EC50: 0.153	16
Slimicide C-30	Mussel	*Mytilus galloprovincialis*	96-h EC50: 0.07	18
	Mycid	*Leptomysis mediterranea*	96-h LC50: 0.11-0.30	18
Tolylfluanid	Mollusc	*Mytilus edulis*	EC50: 0.612	16
	Sea squirt	*Ciona intestinalis*	EC50: 0.792	16
	Sea urchin	*Paracentrotus lividus*	EC50: 3.35	16
Triphenylborane pyridine	Crustacean	*Artemia salina*	48-h LC50: 130	21
	Diatom	*Skeletonema costatum*	48-h LC50: 2.2	21
Zn pyrithione	Amphipod	*Elasmopus rapax*	96 h-LC50: 29	19
	Crustacean	*Artemia salina*	24-h LC50: 3,170	15
	Diatom	*Thalassiosira pseudonana*	96 h-EC50: 0.0019	19
	Polychaeta	*Hydroides elegans*	48 h-LC50: 4.4	19

EC50 - effect concentration (mg/l), LC50, lethal concentration (μg/l)
Extremely toxic: EC50: <1 mg/l, LC50 <10 μg/l
Highly toxic: EC50: 1-50 mg/l, LC50 10-100 μg/l
Moderately toxic: EC50: 50-500 mg/l, LC50 100-1000 μg/l

Irgarol 1051 causes the inhibition of photosynthetic electron transport in chloroplasts that can occur at concentrations lower than 1 mg/l.[1] A maximum permissible concentration in water of 24 ng/l for Irgarol 1051 is proposed by the Dutch National Institute of Public Health and the Environment.[14]

Diuron inhibits photosynthesis and was found to have similar toxicity with Irgarol 1051.[1] For diuron, a maximum permissible concentration of 430 ng/l has been derived by the appropriate Dutch authorities.[14]

Sea-nine 211 was also found to be more toxic than Irgarol and Diuron, using *Daphnia magna* and *Vibrio fischeri* and suspension cultured fish cells in toxicity bioassays.[1]

After extensive studies conducted in Gulf of Naples, it was concluded that the levels of Irgarol 1051 and diuron detected were not expected to induce serious damage to the local marine ecosystem.[7]

The toxicity of several binary, ternary and quaternary mixtures of biocides were evaluated using the brine shrimp *Artemia salina* as the test organism.[15] The zinc pyrithione/copper pyrithione combination was found to have synergistic toxic effect which requires attention, because of frequent coexistence of these two biocides in the marine environment.[15]

It is noticeable that results are not always consistent between different experiments. It is also noticeable that many biocides are toxic at low concentrations (some can be classified as having high toxicity). At the same time, it is important to compare information in all three sections of this chapter to make an adequate judgement, because, for example, some of the biocides may be toxic at low concentrations but they are not present at such concentration either because they are not so frequently used to leach sufficient quantities or they are rapidly degraded in the natural environment.

7.3 DECAY

Efficiency, durability, and their biodegradation of biocides depend on their physical-chemical properties. Table 7.5 gives some properties of major biocides.

Table 7.5. Properties of major biocides.[1,4,5,8,30,41]

Biocide	Henry's law constant, atm m^3 mol^{-1}	Vapor pressure, atm @25°C	Water solubility, mg/l	log Kow
BIT			1000	
Carbendazim			8	
Chlorothalonil	2x10^{-7}		0.9	4.0
CMIT			5000	
Dichlofluanil	9.5x10^{-8}	3.7x10^{-11}	1.3	3.7
Diuron	2x10^{-6}	3.6x10^{-7}	35-42	2.8
DMSA, dichlofluanid metabolite		2.47x10^{-9}	1300	1.59
Irgarol 1051	2.5x10^{-8}	8.7x10^{-10}	7	3.95

Table 7.5. Properties of major biocides.[1,4,5,8,30,41]

Biocide	Henry's law constant, atm m^3 mol^{-1}	Vapor pressure, atm @25°C	Water solubility, mg/l	log Kow
Mecoprop			700	
MIT			>10,000	
OIT			480-500	
Sea-nine 211	$5.1x10^{-7}$	$9.7x10^{-9}$	14	2.8
TCMTB	$2.7x10^{-7}$	$1.17x10^{-8}$	10.4	3.3
Terbutryn			25	
Thiram	$2.7\text{-}10^{-7}$	$2.3x10^{-8}$		
Triclocarban		$4.75x10^{-12}$	0.6479	4.9
Triclosan		$6.12x10^{-9}$	4.621	4.7
Zn pyrithione	non-volatile		8	0.9
Zineb			0.07-10	0.8

There are many indicators of biocide biodegradation rate used to express research data, some of them are included in Table 7.6.

Table 7.6. Biocide degradation indicators.[1,4,27-37,40,41]

Biocide	Degradation rate constant		Half-life		Residence time
	water	soil	data	environment	
Carbendazim			35-175 d	surface water	
Chlorothalonil			43 h 4-150 h 4 w 7-47.8 h	biotic freshwater seawater photolysis, sun	
Cu pyrithione			12.9 d 37 h to 4 d 7.1-29.1 min 2-22 h 0.5 h	art. seawater nat. seawater photolysis aerobic anaerobic	42,700 y (Cu)
Dichlofluanid			18 h 53 h 43-83	bioassay seawater photolysis, sun	
Diuron			14-35 d	anaerobic	
Glutaraldehyde			10-35 h	lake water	
Irgarol 1051	0.0054 d^{-1} 0.0004 h^{-1}	0.086 y^{-1}	100-200 d	sea/fresh H_2O	10.2 y
MIT			>30 d	distilled water	
OIT			stable	water	

Table 7.6. Biocide degradation indicators.[1,4,27-37,40,41]

Biocide	Degradation rate constant		Half-life		Residence time
	water	soil	data	environment	
Sea-nine 211	0.44 d^{-1}	6100 d^{-1}	<0.5 d <24 h to 8.4 d <1 h to 24 h 8-433 h >720 h 60 h	anaerobic seawater seawater+biodegrad photolysis hydrolysis pH=9	4.76-5 d
TBT	0.061 d^{-1}	0.24 y^{-1}	w to m 6-9 m	seawater seawater+bio	16.9 d
TCMTB			740	hydrolysis	
Terbutryn			stable	water	
Triclocarban			0.5 d 60 d 60 d 240 d	air water soil sediment	
Triclosan			14.6-25.7 min 0.66 60 60 240	phototransformation air water soil sediment	
Zn pyrithione			8.3 min to 1 h <2 h 24 h 96-120 d >90 d 2-4 d 2-22 h 0.5 h	photolysis biodegradation water column hydrolysis artif. seawater nat. seawater aerobic anaerobic	
Zineb			96 h	hydrolysis	

min – minute, h – hour, d – day, w – week, y – year, artif. – artificial, nat. – natural

Table 7.7 contains information on the modes of degradation of some biocides and the metabolites formed.

Table 7.7. Modes of degradation of biocides and metabolites.

Biocide	Mode of degradation	Metabolites	Refs.
Chlorothalonil	biodegradation, photodegradation	chloro-1,3-dicyanobenzene, dichloro-1,3-dicyanobenzene, trichloro-1,3-dicyanobenzene, benzamide, 4-hydroxy-2,5,6-trichloroisophthalonitrile, 5-cyano-4,6,7-trichloro-2H-1,2-benzisothiazol-3-one	1,4,34
Dichlofluanid	photodegradation, hydrolysis, anaerobic degradation	N,N-dimethyl-N'-phenyl-sulfamide, n-dichlorofluoromethylthio-aniline, aniline, dichlorofluoromethane	1,35

Table 7.7. Modes of degradation of biocides and metabolites.

Biocide	Mode of degradation	Metabolites	Refs.
Diuron	aerobic, anaerobic, photodegradation, hydrolysis	1-(3,4-dichlorophenyl)-3,1-dimethylurea, 1-(3,4-dichlorophenyl)-3-methylurea, 1-(3,4-dichlorophenyl)urea, dimethyldiuron, 3,4-dichloroaniline	1,42
Glutaraldehyde	anaerobic	1,5-pentanediol, 3-formyl-6-hydroxy-2-cyclohexene-1-propanal	20
Irgarol 1051	biological, chemical, and photodegradation	2-methylthio-4-tert-butylamino-6-amino-s-triazine, 3-[4-tert-butylamino-6-methylthiol-s-triazin-2-amino]-propionaldehyde	1,36
Sea-nine 211	photolysis	N-octyl acetamide, N-octyl-hydroxypropionamide, N-n-octyl malonamic acid	39
TCMB	hydrolysis	2-mercaptobenzothiazole, 2-(methylthio) benzothiazole	4
Thiram	hydrolysis, photodegradation		4
Triclosan	phototransformation		31
Zn pyrithione	biodegradation, hydrolysis, transformation	2-pyridine sulfonic acid, zinc chelate of 2-pyridinethiol-l-oxide	1,4
Zineb	hydrolysis	5,6-dihydro-3H-imidazo (2,l-c)-l,2,4-dithiazole-3-thione, ethylene diisothiocyanate, ethylenethiourea	4

2-methylthio-4-tert-butylamino-6-amino-s-triazine (product of degradation of Irgarol 1051, which is called GS26575) was determined at lower concentrations than Irgarol 1051,[25] which suggests that the environmental transformation rate of Irgarol 1051 is relatively slow and that the rate of GS26575 degradation is greater than that of its formation.[1] However, these data conflict with other studies,[26] which have shown Irgarol 1051 and GS26575 to have a similar seawater $t_{1/2}$ and proposed that GS26575 has a greater environmental persistence than Irgarol 1051.[1]

Under sunlight conditions, Irgarol underwent rapid degradation to produce GS26575, which was more persistent than Irgarol.[38] GS26575 was stable to hydrolysis in aqueous solutions and was heat resistant under the conditions tested.[38] A long-term field experiment of GS26575 dissolved in five aqueous solutions indicated that 78–93% of GS26575 was degraded after 10 months exposure to sunlight, although 75–88% of GS26575 remained under dark conditions.[38]

Diuron is reported to be relatively persistent in seawater, and considerably stable to hydrolysis and sunlight irradiation.[1] Diuron is predominantly found in the dissolved phase and only weakly sorbed to sediments.[1] Diuron's metabolites have lower or equal half-life than diuron.[1]

Sea-nine 211 degrades rapidly, both biologically and chemically, in natural seawater and sediment.[1] Biological degradation of Sea-nine 211 is considered to be over 200 times faster than hydrolysis and photolysis.[1] Its metabolites are ring-opened structures and their toxicity is reduced by four to five orders of magnitude.[1]

Permethrin and carbendazim were efficiently removed in the conventional activated sludge process.[32] The mechanism of breakdown is not known.[32] High degradation rates (60-90%) were recorded.[32]

Glutaraldehyde meets the OECD criteria to be classified as readily biodegradable in the freshwater environment.[20] Pretreatment with sodium bisulfite, in a molar ratio of 2 to 3 parts per part of glutaraldehyde, is the best method to deactivate aqueous solutions of glutaraldehyde prior to disposal to treatment systems.[20]

Degradation of pyrithione is mainly caused by the UV-light component of the sunlight spectrum, with wavelengths of 320–355 nm being the most effective in producing photodegradation.[40] Photodegradation is affected by factors controlling the penetration of near ultraviolet radiation in natural waters, like: depth of mixing, turbidity, presence of dissolved material absorbing in near UV, cloud cover, etc.[40] In moderately turbid coastal waters, incident light with wavelengths of 380 nm or less is almost completely attenuated at depths of 1–2 m but, in very clear parts of the ocean, 20 m may be required to remove 90% of the radiation entering the surface.[40] The amount of Zn and Cu pyrithione that will photodegrade is, therefore, dependent on the depth to which the 320–355 nm wavelengths can penetrate in the water column, together with the photon flux density.[40] If Zn and Cu pyrithiones do not degrade in the watercolumn, the compounds may accumulate in the sediment.[40] The sedimentation rates of Zn and Cu pyrithiones depend on the release rate, the degree of photo- and biodegradation of the compounds, and the amount of suspended material.[40] Biodegradation without the influence of sunlight was negligible over the time-period investigated.[40]

REFERENCES

1 Konstantinou, I K; Albanis, T A, *Environ. Intl.*, **30**, 235-48, 2004.
2 Yebra, D M; Kiil, S; Weinell, C E; Dam-Johansen, K, *Biofouling*, **22**, 1, 33 – 41, 2006.
3 DIN EN ISO 15181 Paints and varnishes - Determination of release rate of biocides from antifouling paints.
4 Thomas, K V, *Biofouling*, **17**, 1, 73-86, 2001.
5 Burkhardt, M; Kupper, T; Hean, S; Haag, R; Schmid, P; Kohler, M; Boller, M, *Water Sci. Techn.*, **56**, 12, 63–67, 2007.
6 Piunova, V; Berger, D; Neckers, D C; Bullerjahn, G S; McKay, R M; Fedorov, A V, *Photochem. Photobiol. Sci.*, **7**, 1565–1569, 2008.
7 Di Landa, G; Parrella, L; Avagliano, S; Ansanelli, G; Maiello, E; Cremisini, C, *Water Air Soil Pollut.*, **200**, 305–21, 2009.
8 Hamwijk, C; Schouten, A; Foekema, E M; Ravensberg, J C; Collombon, M T; Schmidt, K; Kugler, M, *Chemosphere*, **60**, 1316-24, 2005.
9 Steen, R J C A; Ariese, F; van Hattum, B; Jacobsen, J; Jacobson, A, *Chemosphere*, **57**, 513-21, 2004.
10 Tsang, V W-H; Lei, N-Y; Lam, M H-W, *Marine Pollution Bull.*, **58**, 1462-71, 2009.
11 Bowman, J C; Readman, J W; Zhou, J L, *Marine Pollution Bull.*, **46**, 444-51, 2003.
12 Thomas, K V; McHugh, M; Waldock, M, *Sci. Total Environ.*, **293**, 117-27, 2002.
13 Zhou, J L, *Sci. Total Environ.*, **406**, 239-46, 2008.
14 Lamoree, M H; Swart, S P; van der Horst, A; van Hattum, B, *J. Chromatogr. A*, **970**, 183– 90, 2002.
15 Koutsaftis, A; Aoyama, I, *Sci. Total Environ.*, **387**, 166-74, 2007.
16 Bellas, J, *Sci. Total Environ.*, **367**, 573-85, 2006.
17 Zhang, A G; Leung, K M Y; Kwok, K W H; Bao, V W W; Lam, M H W, *Marine Pollution Bull.*, **57**, 575-86, 2008.
18 Lucu, C; Pavicic, J; Skreblin, M; Mastrovic, M, *Marine Pollution Bull.*, **11**, 294-96, 1980.
19 Bao, V W W; Leung, K M Y; Kwok, K W H; Zhang, A Q; Lui, G C S, *Marine Pollution Bull.*, **57**,

616-23, 2008.
20 Leung, H-W, *Ecotoxicol. Environ. Safety*, **49**, 26-39, 2001.
21 Cima, F; Bragadin, M; Ballarin, L, *Aquatic Toxicol.*, **86**, 299-312, 2008.
22 Okamura, H; Kitano, S; Toyota, S; Harino, H; Thomas, K V, *Chemosphere*, **74**, 1275-78, 2009.
23 Braithwaite, R A; Fletcher, R L, *J. Experim. Marine Biol. Ecol.*, **322**, 111-21, 2005.
24 Sano, L L; Krueger, A M; Landrum, P F, *Aquatic Toxicol.*, **71**, 283-96, 2005.
25 Thomas, K V; McHugh, M; Hilton, M; Waldock, M, *Sci. Total Environ.*, **293**, 117–27, 2002.
26 Hall, L W; Giddings, J M; Solomon, K R; Balcomb, R, *Crit. Rev. Toxicol.*, **29**, 367–437, 1999.
27 Ranke, J, *Environ. Sci. Technol.*, **36**, 1539-1545, 2002.
28 Turley, P A; Fenn, R J; Ritter, J C, *Biofouling*, **15**, 1-3, 175-82, 2000.
29 Jacobson, A H; Wiligham, G L, *Sci. Total Environ.*, **258**, 103-110, 2000.
30 Jungnickel, C; Stock, F; Brandsch, T; Ranke, J, *Environ. Sci. Pollut. Res.*, **15**, 3, 258-265, 2008.
31 Tixier, C; Singer, H P; Canonica, S; Mueller, S R, *Environ. Sci. Technol.*, **36**, 3482-89, 2002.
32 Kupper, T; Plagellat, C; Brandli, A C; de Alencastro, L F; Grandjean, D; Tarradellas, J, *Water Res.*, **40**, 2603-12, 2006.
33 Sano, L L; Bartell, S M; Landrum, P F, *Marine Pollution Bull.*, **50**, 1050-1060, 2005.
34 Sakkas, V A; Lambropoulou, D A; Albanis, T A, *Chemosphere*, **48**, 939-45, 2002.
35 Sakkas, V A; Konstantinou, I K; Albanis, T A, *J. Chromat. A*, **930**, 135-44, 2001.
36 Lam, K-H; Lei, N-Y; Tsang, V W-H; Cai, Z; Leung, K M Y; Lam, M H-W, *Marine Pollution Bull.*, **58**, 272-79, 2009.
37 Steen, R J C A; Ariese, F; van Hattum, B; Jacobsen, J; Jacobson, A, *Chemosphere*, **57**, 513-21, 2004.
38 Okamura, H, *Chemosphere*, **48**, 43-50, 2002.
39 Sakkas, V A; Konstantinou, I K; Albanis, T A, *J. Chromatog. A*, **959**, 215-27, 2002.
40 Maraldo, K; Dahlloef, I, *Marine Pollution Bull.*, **48**, 894-901.
41 Ying, G-G; Yu, X-Y; Kookana, R S, *Environ Pollution*, **150**, 300-5, 2007.
42 Giacomazzi, S; Cochet, N, *Chemosphere*, **56**, 1021-32, 2004.

8

LEGISLATION

8.1 EUROPEAN UNION

In 1998, The European Commission and Parliament adopted Biocidal Product Directive 98/8/EC, which determines steps required from manufacturers to introduce new biocides to the market. The aim of the legislation was to remove trade barriers between member states and to ensure a harmonized high degree of protection for man and the environment.[1]

One of the important achievements of this regulation was systematic classification of biocidal products, which were divided into four main groups and 23 product groups (product types or PTs). The full classification was adapted in this book and it is given in Chapter 1.

Three steps were envisioned in the application of the legislation:[1]

1. provisional listing of active substances in the market
2. the actual Review Regulation programme (approved by the Standing Committee for Biocides in June 2000)
3. initiation of the actual evaluation of the active substances

By February 2000, the European Chemical Industry Council submitted a list containing 915 entries and the list was then circulated to the 15 member states of the EU. The most recent list of participants/applicants to the Review Programme of existing active substances used in biocidal products has been last updated on July 3, 2009 and it is considerably smaller. In the most important sections for this book, it has the following number of substances:

- 6. In-can preservatives: 57 chemicals
- 7. Film preservatives: 30 chemicals
- 8. Wood preservatives: 40 chemicals
- 9. Fibre, leather, rubber and polymerized materials preservatives: 41 chemicals
- 10. Masonry preservatives: 28 chemicals
- 11. Preservatives for liquid-cooling and processing systems: 53 chemicals
- 12. Slimicides: 43 chemicals
- 13. Metalworking-fluid preservatives: 36 chemicals

Considering that many of the same chemical compounds are used in each group, the initial number of entries was drastically reduced. All biocides which are found on this list are included in tables of Chapter 3, and their properties are discussed in groups generated according to their chemical type. The tables in Chapter 3 also contain biocides which are not permitted for use in the European Union but are used in the USA.

The Directive distinguishes between "new active substances" (those placed in the market after 14 May 2000) which must be approved before being marketed for biocidal purposes; and "existing active substances" (listed prior to 14 May 2000).[1] New active sub-

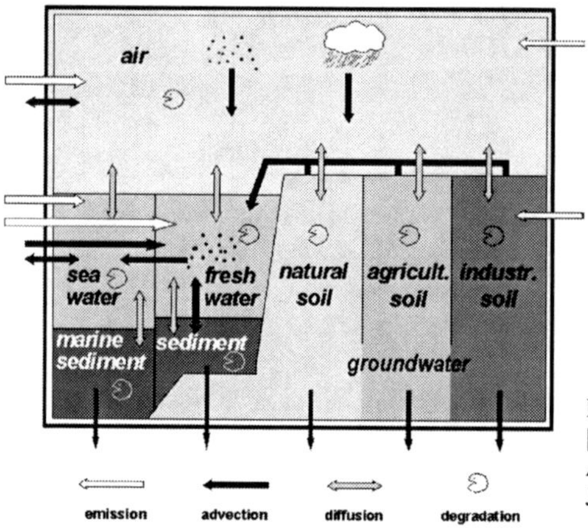

Figure 7.1. Schematic representation of the regional model in EUSES 2.0. [Adapted, by permission, from Vermeire, T; Rikken, M; Attias, L; Boccardi, P; Boeije, G; Brooke, D; de Bruijn, J; Comber, M; Dolan, B; Fischer, S; Heinemeyer, G; Koch, V; Lijzen, J; Mueller, B; Murray-Smith, R; Tadeo, J, *Chemosphere*, **59**, 473-85, 2005.]

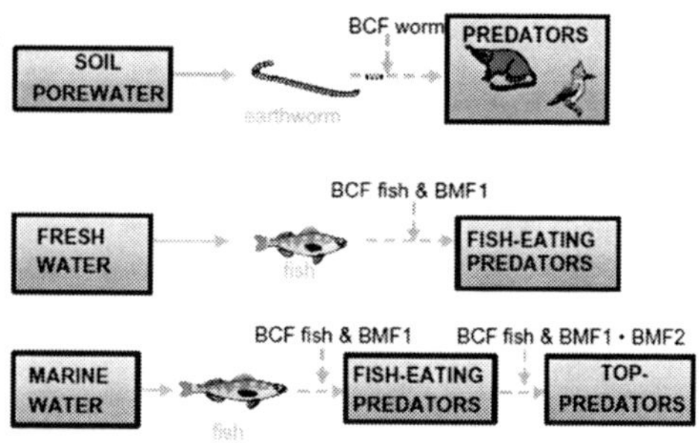

Figure 7.2. Indicative food chains of EUSES 2.0. [Adapted, by permission, from Vermeire, T; Rikken, M; Attias, L; Boccardi, P; Boeije, G; Brooke, D; de Bruijn, J; Comber, M; Dolan, B; Fischer, S; Heinemeyer, G; Koch, V; Lijzen, J; Mueller, B; Murray-Smith, R; Tadeo, J, *Chemosphere*, **59**, 473-85, 2005.]

stances cannot be placed in the market before they have been evaluated.[1]

EUSES, the European Union System for the Evaluation of Substances, is a harmonized quantitative risk assessment tool in the form of the PC-implementation of the technical guidelines developed within the framework of EU chemical legislation for industrial chemicals and biocides.[2] As such, it is designed to support decision-making by risk managers in government and industry and to assist scientific institutions in the risk assessment for these substances.[2] The EU risk assessment includes risks to the marine environment. A simplified representation of the regional model, including a marine compartment, is presented in Figure 7.1. The marine local assessment is practically identical to the inland freshwater assessment, except for the use of a sewage treatment plant and the dilution factor.[2] The food chain for the marine environment is additionally modelled with a top-predator preying on fish-eating predators (Figure 7.2).[2] Measured biomagnification data are at present very limited and therefore default values are used.[2] EUSES is a complex system, since it covers the whole risk assessment chain from emissions and exposure to effects and risks for both humans and the environment in an integrated way.[2]

Biocidal Product Directive 98/8/EC is the specific legislature regulating use and development of biocides. There are also other regulations which govern use of biocidal products in various sections of industry. For example, water protection and management – marine pollution is the object of Regulation EC 782/2003. It aims at elimination of adverse effects on the marine environment and human health caused by organotin compounds, which act as active biocides in antifouling systems used on ships flying the flag of, or operating under the authority of, a Member State or sailing to or from ports of the Member States.[3] Composition of cosmetics and their safety are regulated by Directive 76/768/EEC.[4] According to this regulation, industry is obliged to provide the following information:[4]

- the qualitative and quantitative composition of the product
- physico-chemistry, microbiology and purity of the ingredients and the cosmetic product
- the manufacturing method
- safety assessment of the finished cosmetic product
- name and address of the safety assessor
- existing data on undesirable effects on human health
- proof of the effects claimed
- data on animal testing (animal testing was banned in EU since September 11, 2004)

Annexes of Directive 76/768/EEC contain information on permitted components of cosmetics.[4]

Registration, Evaluation, Authorization and Restriction of Chemicals, commonly referred to as REACh, intends to bring the data requirements for existing and new chemical substances to the same level and therefore introduces tonnage-linked data requirements for existing substances.[5] IUCLID 5 is the database, which contains data for submission to REACh and other programs.[6]

There are also local developments in EU member states. Information about these developments in UK,[7] Denmark,[8] Finland,[9] Italy,[10] Spain,[11] Poland,[12] Germany,[13] India,[14] and Czech Republic[15] can be found in the review chapters.[7-15]

8.2 INTERNATIONAL

The World Health Organization has published the WHO recommended classification of pesticides by hazard and the guidelines to classification. The document is arranged as follows:

- Part I: The Classification as recommended by the World Health Assembly. This part is not subject to periodic review and the classification table and text can only be changed by resolution of the World Health Assembly
- Part II: Guidelines to Classification. Individual products are classified in a series of tables, according to the oral or dermal toxicity of the technical products, and their physical states. The tables are subject to review periodically.

Pesticides are classified according to the guidelines included in Table 8.1.

Table 8.1. WHO classification of toxicity of pesticides

Class	Description	LD50 for rat, mg/kg body weight			
		Oral		Dermal	
		Solids	Liquids	Solids	Liquids
Ia	Extremely hazardous	5 or less 20	20 or less	10 or less	40 or less
Ib	Highly hazardous	5-50	20-200	10-100	40-400
II	Moderately hazardous	50-500	200-2000	100-1000	400-4000
III	Slightly hazardous	over 50	over 2000	over 1000	over 4000

Pesticides are also classified in tables by their main uses, including bacteriostats, fungicide, insecticide, molluscicide, and rodenticide.

Many other world activities are discussed in a recent monograph.[16] Items relevant to our topic include:[16]

- The Strategic Approach to International Chemicals Management (SAICM)
- The Classification and Labelling of Chemicals (Globally Harmonized System, GHS)
- The International Programme on Chemical Safety (IPCS)
- The World Library of Toxicology
- The Organization for Economic Co-operation and Development (OECD).

8.3 USA

In the USA, biocides are regulated together with pesticides. The Environmental Protection Agency, EPA, and the states (usually a state's agriculture office) register or license pesticides use. States may have more restrictive requirements, but pesticides must be registered both by the EPA and the state before distribution. Some key EPA and state pesticide regulatory activities include:

- evaluation of new pesticides and their uses
- new approaches to minor uses
- providing for special local needs and emergency situations
- review of registered pesticides
- registration of producing establishments
- enforcing pesticide requirements

The EPA also regulates pesticides imported to the US and participates in a wide variety of international activities, such as regulatory agreements and coordination activities.

There are numerous acts involved in biocide legislation. These include both biocide specific acts:

- Federal Insecticide, Fungicide, and Rodenticide Act (FIFRA) (the basis for regulation, sale, distribution and use of pesticides in the USA)
- Pesticide Registration Improvement Act (PRIA) (regulates registration service fees)

and final product group-specific regulations, such as:

- Federal Food, Drug, and Cosmetic Act (FFDCA) (restricts maximum residual levels in food products)
- Food Quality Protection Act of 1996 (FQPA) (safety standards for new and old pesticides in processed and unprocessed foods)
- Endangered Species Act (ESA) (prohibits any action that can adversely affect an endangered or threatened species or its habitat)

Information about biocides can be found in a comprehensive database – The National Pesticide Information Retrieval System (NPIRS), which is a collection of pesticide-related databases available by subscription. NPIRS is under the administration of the Center for Environmental and Regulatory Information Systems, CERIS, at Purdue University in West Lafayette, Indiana.[17] The database has been available since 1984. There are several subset databases, including:[18]

- Pesticide Product
- Pesticide Document Management System
- Tolerance Index
- Data Submitters' List
- Federal Register
- NPIRS State Pesticide Information Retrieval System

The collection of information in these databases is the best available anywhere in the world.

REFERENCES

1 Rasmussen, K; MacLellan, M A, *Environ. Sci. Policy*, **4**, 137-46, 2001.
2 Vermeire, T; Rikken, M; Attias, L; Boccardi, P; Boeije, G; Brooke, D; de Bruijn, J; Comber, M; Dolan, B; Fischer, S; Heinemeyer, G; Koch, V; Lijzen, J; Mueller, B; Murray-Smith, R; Tadeo, J, *Chemosphere*, **59**, 473-85, 2005.
3 Arvanitoyannis, I S; Tserkezou, P; Choreftaki, S, **Waste Management Food Ind.**, 135-288, *Elsevier*, 2008.
4 Pauwels, M; Rogiers, V, *Toxicol. Appl. Pharmacol.*, **243**, 260-74, 2010.
5 Pauwels, M; Rogiers, V, *Regulatory Toxicol. Pharmacol.*, **49**, 308-15, 2007.
6 http://iuclid.eu/
7 Ashdown, L; Harrison, P T C; Holmes, P; Takamiya, M, **Information Resources in Toxicology**, 4th Ed, *Elsevier*, 2009, pp 1177-1211.
8 Bailling, H, **Information Resources in Toxicology**, 4th Ed, *Elsevier*, 2009, pp 853-66.
9 Komulainen, H, **Information Resources in Toxicology**, 4th Ed, *Elsevier*, 2009, pp 875-89.
10 Marinovich, M; Della Seta, M; Preziosi, P; Marcello, I; Binetti, R, **Information Resources in Toxicology**, 4th Ed, *Elsevier*, 2009, pp 967-89.
11 Repetto, G; Repetto, M, **Information Resources in Toxicology**, 4th Ed, *Elsevier*, 2009, pp 1125-45.
12 Banasik, M; Stedeford, T, **Information Resources in Toxicology**, 4th Ed, *Elsevier*, 2009, pp 1061-80.
13 Desel, H; Kahl, R, **Information Resources in Toxicology**, 4th Ed, *Elsevier*, 2009, pp 899-911.
14 Patel, S; Dewan, A, **Information Resources in Toxicology**, 4th Ed, *Elsevier*, 2009, pp 913-44.
15 Hornychova, M; Veseal, J, **Information Resources in Toxicology**, 4th Ed, *Elsevier*, 2009, pp 835-52.
16 Kuellmer, J; Schlottmann, U, **Information Resources in Toxicology**, 4th Ed, *Elsevier*, 2009, pp 718-36.
17 http://ppis.ceris.purdue.edu/htbin/epachem.com
18 http://ppis.ceris.purdue.edu/htm/databases.htm

9

Personal Protection

The following personal safety equipment is covered in this chapter:

- clothing
- gloves
- eye protection
- respiratory protection

9.1 CLOTHING

Chemicals can cause adverse effects on unprotected skin, ranging from contact dermatitis to permeation of the skin and systemic toxic effects. Chemical protective clothing, comprising gloves, boots, suits, and other related components, can prevent direct skin contact and contamination.

According to the Occupational Safety and Health Administration, OSHA, Personal Protective Equipment Standard,[2] selection of protective clothing follows a hazard assessment, which determines a need for required protections.

The OSHA regulation,[2] Appendix B, provides the following, non-mandatory, guidelines for selection of protective equipment:

- become familiar with the potential hazards, the type of protective equipment that is available, and its protective features (e.g., splash protection, impact protection, etc.)
- compare the hazards associated with the environment (i.e., impact velocities, masses, projectile shape, radiation intensities, etc.) with the capabilities of the available protective equipment
- select the protective equipment which ensures a level of protection greater than the minimum required to protect employees from hazards
- fit the user with a protective device, and give instructions regarding its care and use. It is very important that employees be made aware of all warning labels for and limitations of their protective equipment.

Careful consideration must be given to comfort and fit. Protective equipment that fits poorly will not afford the necessary protection. Continued wearing of the device is more likely if it fits the wearer comfortably. Protective devices are generally available in a variety of sizes. Care should be taken to ensure that the right size is selected.[2]

The National Institute for Occupational Safety and Health believes that protective clothing should be considered as the last line of defense to protect against accidental contact (e.g., spills, splashes). This is in conformance with the generally accepted industrial hygiene strategy for controlling workplace exposures to chemical hazards, which recommends, beginning with the highest preference, the following hierarchy of primary controls:[3]

(1) substitution or elimination
(2) process change
(3) isolation/enclosure
(4) ventilation
(5) good housekeeping
(6) personal protection.

Protective clothing must provide resistance to permeation, penetration, and degradation for chemicals in use. A standard test method for permeation resistance has been developed by the American Society for Testing and Materials.[4] The test determines both the breakthrough time, and the steady state permeation rate of chemicals through a sample of the protective barrier. Breakthrough (permeation) resistance depends on temperature of the challenge material, the environment, and thickness of the barrier. Evaluation methods are discussed in detail in a special monograph[5] on this subject, and further information is provided in the next section of this chapter.

The selected clothing depends on the severity of exposure (duration and concentration). The chemical resistance of various materials is discussed in the next section of this chapter. Selection should be made by a professionally trained person.

9.2 GLOVES

Many types of gloves are manufactured for different purposes. Proper selection of gloves is important to ensure that the solvent, or other chemical compound, cannot penetrate through the glove, where its subsequent evaporation would be prevented by the glove itself. Exposure to the additive entrapped between the glove and the skin is more severe than the exposure to the additive without the glove.

The first requirement in the selection of a glove is that the materials from which gloves are produced are not dissolved or swollen by additives. This alone is not sufficient. The other important requirement in glove selection is permeability of all of the materials in its structure. Permeability is measured according to ASTM Standard F 739,[4] or European Standard EN 374.[6] Permeability results are usually expressed in mg of permeate which penetrates 1 m^2 of material per minute.

Good materials for gloves should have permeability below 1 mg m^{-2} min^{-1}. Materials having permeability in the range from 1 to 10 mg m^{-2} min^{-1} may still be considered as suitable materials.

Breakthrough time is another measure of glove fitness for this purpose. This parameter measures the time in minutes during which a measurable amount of a particular chemical compound penetrated through a particular garment. Specialized monographs give data on breakthrough times for a large number of gloves, and other protective clothing and numerous penetrants.[5,7] These sources[5,7] are the best collections of data which should be considered in glove selection. There is a relationship between permeation rate and breakthrough time. For practical purposes, gloves are selected based on the assumption that they should resist penetrant breakthrough for more than eight hours.

The European standard[6] classifies gloves according to their breakthrough times as follows:

Breakthrough time	*Performance level*
> 10 min	class 1
> 30 min	class 2
> 60 min	class 3
> 120 min	class 4
> 240 min	class 5
> 480 min	class 6

The American National Standard[8] provides guidance for selecting suitable gloves that will protect workers and assist employers in compliance with OSHA Regulation 29 CFR 1910.[2] The standard covers glove selection criteria in specific categories related to testing, and performance properties in chemical and industrial applications, such as:

- cut method
- puncture resistance
- abrasion
- protection from cold
- protection from heat
- chemical resistance (chemical permeation and chemical degradation)
- viral penetration
- dexterity ratings
- liquid tight integrity
- flame resistance
- heat resistance

Monitoring the quality of protection during use is no less important. Materials are not permanent. Frequent inspection of gloves is therefore very critical. Any mechanical damage will compromise protection properties. The presence of cuts, cracks, or holes immediately makes the gloves useless. Gloves should be inspected for changes of color, texture, or hardening, which are signs of material degradation.

It should be borne in mind that protective clothing is not universal for all chemical materials. For this reason, protection against one penetrant does not suggest that the glove has protecting qualities against other materials. Mixtures of additives behave in a different manner than do their pure components. The temperature at which exposure occurs is also an important factor. Generally, as temperature increases, permeability increases.

Note that glove materials are not equal (even if they are made from the same generic polymer). They vary with the formulation used by a specific manufacturer and the thickness of the protective layer. For each selection, the manufacturer's specification should be consulted to estimate previously discussed parameters. The best practice requires that gloves are tested for particular conditions of their use. The other good option is to consult results of measurements included in the specialized monographs and manufacturer's literature.[5,7]

Some gloves are manufactured for specific applications, and some are for a broad spectrum of chemicals. Gloves are manufactured as supported (e.g., cotton flannel lining) and unsupported. Unsupported gloves are usually made of thicker film (30 mils) (supported gloves have film coated on the support material). There are some gloves from very thin laminate such as Safety4 4H (2.7 mils thick, 5 ply laminate).

A broad range of glove products is suggested by manufacturers of biocides, such as PVC, neoprene, butyl, and nitrile rubber, but also canvas and cotton gloves. It is very likely that glove selection may be determined by simultaneous use of some other chemical products which are more corrosive/toxic than biocides.

9.3 EYE PROTECTION

The employer shall ensure that each employee uses appropriate eye or face protection, when exposed to eye or face hazards from flying particles, molten metal, liquid chemicals, acids or caustic liquids, chemical gases or vapors, or potentially injurious light radiation.[10] Protective eye and face devices purchased after July 5, 1994 shall comply with the ANSI Standard,[11] which is incorporated by reference as specified in Sec. 1910.6.

Industrial-thickness glass, plastic and polycarbonate lenses meet or exceed the requirements of the eye protection standard. In general, each type of lens does offer certain advantages and disadvantages. Glass lenses provide good scratch resistance and withstand chemical exposure. But glass is heavy. Plastic and polycarbonate lenses are lighter and are not likely to fog. Unless specially coated, these lenses are not as scratch resistant as glass.

Indirect-vent, impact, and splash resistant goggles are suggested for use with liquid and solid chemical products such as biocides. Face shield should additionally be used when working with corrosive, highly irritating, or toxic substances in combination.

9.4 RESPIRATORY PROTECTION

The selection of a respirator, general suggestions for its use, the adsorption capacity of filters, and breakthrough time are reviewed below.

The selection of a respirator is based on the nature of respiratory hazards, such as permissible exposure limits, established concentration immediately dangerous to life, actual concentration of hazardous substances, and amount of oxygen. When using filters, the ambient air must contain at least 20% oxygen at sea level. Filter protection should not be used for unknown contaminants and contaminants that arc immediately dangerous to life and health.

Permissible exposure limits and actual concentration (among other factors) determine the breakthrough time as discussed below.

Only approved respirators should be used. Each country has a regulatory body which can approve respirators for use. In the USA, respirators are approved jointly by the National Institute for Occupational Safety and Health (US Department of Health and Human Services), and the Mine Safety and Health Administration (Department of Labor).

In addition to the above parameters determining respirator selection, the following factors are also considered: nature of the operation process, location of the hazardous area (especially in respect to supply of respirable air), employee activity, and intensity of effort required to perform the work (determines the amount of air which must be supplied to lungs), physical characteristics, and limitations of various respirators.

The selection of a respirator must be done with full consideration of these many factors, which require specialized knowledge going beyond this discussion. Further information on this subject can be found in the specialized monograph.[12]

3M offers free Online Respirator Selection Software.[13] This software helps make the process of selecting a respirator easier and faster.[13] It can analyze more than 700 different chemicals and provide an audit trail of respirator decision logic.[13] Knowing the contaminants and their concentrations, the program can then lead to an appropriate respirator recommendation.[13] The software combines contaminant reference information from a variety of resources.[13] This includes immediately dangerous to life and health (IDLH), exposure limit, odor threshold, molecular weight, and more.[13]

Respirators can be generally divided into two main groups: air-purifying respirators, and atmosphere-supplying respirators. The first group is more common, and for this reason will be discussed in more detail below. Air-purifying respirators are still divided into non-powered and powered, and these are further divided into gas/vapor and particle removing. Our main interest here is given to vapor-removing, non-powered respirators, which are the most common in industry and laboratories in solvent applications.

Two additional subjects are an integral part of good protection: respirator fit testing, and training of users. Non-powered, vapor-removing respirators are also termed as negative pressure respirators. This means that the respirator is under positive pressure during exhalation, and under negative pressure during inhalation. If the respirator does not fit the user properly, some air from the surroundings will be drawn into the respirator during inhalation because of leaks. This will result in a diminishing efficiency of protection. For this reason, each user should be given help from the employer in selecting the most suitable type and size of respirator for the particular individual. In addition, the reasons for fitting should be clearly explained so that they are fully understood by the employee. It is only the employee who may later assure that the respirator is used properly, and this will depend on his/her full understanding of the principles.

The first matter of importance is the rate of breathing under various conditions. The amount of air we breathe depends on our energy requirements, or more precisely, on the intensity with which we expend energy. If no work is performed, the average human being requires about 10 l of air per minute. The rate of breathing increases with the intensity of work (light work 20-30, heavy work 70-100 l/min). The use of a filter reduces air flow rate due to the pressure drop, and many respirators cannot cope with high rates of flow, which becomes one important limitation of negative pressure respirators that must be evaluated during the selection process.

The pressure drop in the filter depends on its design. Chemical cartridges, used for chemical material adsorption, are filled with activated carbon or activated charcoal, which are amorphous forms of carbon characterized by their ability to rapidly absorb many gases and vapors. The carbon is obtained by pyrolysis of wood, nutshells, animal bones, petroleum products, and other carbonaceous materials. Activated carbon for respirators usually comes from coconut shells or coal, after activation at temperatures in the range from 800 to 900°C. It has a porous structure with a surface area of 10,000 m^2/g. On the one hand, a maximizing surface contributes to increased capacity; on the other, it may lead to an increased pressure drop during breathing (depending on particles shapes and sizes). It is thus important to consider both the capacity of the filter and its pressure drop.

The adsorption of vapor is a physical process which can be reversed. Desorption of vapor does not require a high energy and is equilibrium driven (equilibrium relative to the concentration of vapor in surrounding atmosphere, and distribution of adsorbed vapor in activated carbon). Desorption during storage or non-use time can result in migration of previously adsorbed molecules. This phenomenon is very important for effective use of respirators. On the one hand, it precludes the possibility of cartridge recycling. On the other hand, it imposes restriction on cartridge use. For good performance, the mask should be worn 100% of the time. It is estimated that removal of the respirator for 5 min in an 8-hour day causes a loss of 50% of the expected protection. It is also important to limit filter use to one shift, even if it was not fully exhausted in that time.

The absorption capacity is calculated from the following equation:

$$W = \rho W_o \exp\left[-\frac{BT^2}{\beta}\left\{\log\frac{p_s}{p}\right\}^2\right] \qquad [9.1]$$

where:

W	adsorption capacity per unit weight of carbon
ρ	density
W_o	total volume of adsorption space
B	microporosity constant of carbon
T	temperature
β	affinity coefficient of additive vapor to carbon
p_s	saturated vapor pressure of additive at temperature T
p	equilibrium partial pressure of the additive.

This equation shows that both additive type and type of carbon affect performance. On the additive side, its density, affinity, saturated vapor pressure, and its actual concentration determine adsorption. On the carbon side, its porous structure and surface area available for adsorption determine the capacity of a particular filter. The data for the parameters of the equation can be found in chemical handbooks, and therefore, the equation can be used for predicting the adsorption capacity of a filter. The capacity of commercial filters is normally rated by an experimental method, in which, the amount of the additive adsorbed by filter is determined under conditions which specify concentration of the additive, rate of flow, and time. Vapor pressures of additives are very low; therefore, calculated service life of filters is usually very long.

Under conditions of use it is important to predict how long a particular cartridge will last under real conditions. This is done by estimation of the breakthrough time from the following equation:

$$t = \frac{W\rho An}{QC_o}\left[z + \frac{1}{a_c\rho_c}\left(\frac{dG}{\eta}\right)^{0.41}\left(\frac{\eta}{\rho_a D}\right)^{0.67}\ln\left(\frac{C_b}{C_o}\right)\right] \qquad [9.2]$$

where:

W	adsorption capacity per unit weight of carbon
ρ_c	carbon density

A	cross-sectional area of adsorbent bed
n	number of cartridges tested
Q	flow rate
C_o	concentration of solvent
z	V/A where V is carbon volume
a_c	specific surface area
d	diameter of granule
G	mass velocity through the cartridge
η	viscosity of air-vapor stream
ρ_a	density of air-vapor stream
D	diffusion coefficient
C_b	breakthrough concentration

This equation gives information on the relevance of major parameters of performance of filters. Breakthrough time increases with the increase of the following parameters: adsorption capacity of carbon, its density, its volume, and the cross-sectional area of the adsorbent bed, and size of the granules. It decreases with increased flow rate, concentration of the additive, and an increase in the value of the diffusion coefficient. The toxicity of material plays a role here as well, because with the increase of material toxicity, the breakthrough concentration is decreased, which decreases breakthrough time. Breakthrough time for 3M cartridges can be calculated using available software by providing the type of material, its concentration, and type of work (light, medium, heavy).[13] In addition, relative humidity is accounted for in the calculations. The adsorption of vapors is not affected by humidity below 50% but decreases rapidly as the relative humidity increases above 50%. Adsorption temperature is also an essential factor.

The above discussion shows that, although there is generally one type of cartridge used for organic vapor, all cartridges are not the same in terms of performance. The conditions of use of these respirators determine if they can perform specific protective functions. Considering that one cartridge, typically disposed after one day of work, may contain 40 g of adsorbed material, which would otherwise be inhaled, the selection and use of respirators is not a trivial matter, and should be given serious attention.

Most organic chemical materials used in biocides have low volatility, with the exception of some products, such as formaldehyde, chlorine, creosote, and iodine, which are not only volatile but also very toxic.

REFERENCES

1 Recommendations for Chemical Protective Clothing. A Companion to the NIOSH Pocket Guide to Chemical Hazards. NTIS No. PB98-137730. February 1998.

2 Occupational Safety and Health Administration. 29 CFR Part 1910. Subpart I. Personal Protective Equipment for General Industry.

3 Birmingham D J. Occupational dermatoses in Clayton G D, Clayton F E, Eds. **Patty's industrial hygiene and toxicology**. 4th rev. ed., Vol. 1, part A. New York: *Wiley Interscience Publishers*, 1991.

4 ASTM F739-07. Standard Test Method for Permeation of Liquids and Gases through Protective Clothing

Materials under Conditions of Continuous Contact
5 Forberg K, Keith L H, **Chemical Protective Clothing. Performance Index**. Second Edition. *John Wiley & Sons, Inc., New York 1999.*
6 BSI BS EN 374-3 Protective gloves against chemicals and micro-organisms Part 3: Determination of resistance to permeation by chemicals.
7 K Forsberg, S Z Mansdorf, **Quick Selection Guide to Chemical Protective Clothing**. 5rd Ed., *John Wiley & Sons*, 2007.
8 ANSI/ISEA 105-2000. Hand Protection Selection Criteria.
9 Chemrest. Glove selection software from Best Manufacturing.
W H Revoir, C-T Bien, **Respiratory Protection Handbook**, *Lewis Publishers*, Boca Raton, 1997.
10 Occupational Safety and Health Administration. 29 CFR Part 1910.133 Eye and face protection.
11 ANSI Z87.1-2003. American National Standard Practice for Occupational and Educational Eye and Face Protection.
12 Revoir W H, Bien C-T, **Respiratory Protection Handbook**, *Lewis Publishers*, Boca Raton, 1997.
13 http://solutions.3m.com/wps/portal/3M/en_US/Health/Safety/Resources/Four/
14 NIOSH Pocket Guide to Chemical Hazards. NIOSH Publication No. 2005-151, September 2005.

INDEX

Numerics

1,2-benzotiazolinion-3-one 159
1,4-glycosidic bond 107
1-naphthyl-N-methylcarbamate 201
2,4,6-trichlorophenol 167
2-(thiocyanmethylthio)benzothiazole 159
2-benzimidazole-methylcarbamate 172
2-benzimidazolyl-methylcarbamate 167
2-bromo-2-nitropropane-1,3-diol 237
2-mercaptobenzothiazole 167
2-n-octylisothiazolin-3-one 165
2-n-octylisothiazoline-3-one 165
5-chloro-2-methyl-4-isothiazolin-3-one/2-methyl-4-isothiazolin-3-one 200
5-methyl-2-isopropylphenol 178

A

abietane 99
abiotic 99
  factors 32, 146
  hydrolysis 253
  photooxidation 251
  processes 31
ablative antifouling 291
abrasion 337
absorption capacity 341
acclimation 262
acetal aldehyde-releasing compounds 48
acetoxy silicone sealant 273
acid 50, 204
  -base interaction 252
  esters 49
Acidobacteria 14
acidophilic character 132
acquired resistance 46
  mechanisms 46
acrylic 239
  resin 103
  sealants 273
acrylonitrile-butadiene-styrene copolymer 241
Actinobacteria 8, 9, 36
Actinomycetes 142, 277
activated
  carbon 340
  charcoal 340
  sludge
    biodegradation 297
    method 293
    process 326
active halogen products 51
acute toxicity 310
additives mixture 338
adenosine triphosphate 45, 296
  synthesis 45
adherence 199
adhesion 33, 266
adhesive 93, 273, 291
  film biodegradation 291
  patch 102, 103
adjuvants 162, 277
adsorption
  capacity 341
  temperature 342
aerobic 200
  biodegradability 297
  biodegradation 298
  corrosion 197, 198
  microorganisms 268
affinity coefficient 341
aggregation 33
air
  contaminants 314
    limits 314
  diffuser membrane 244
  flow rate 340
  -purifying respirators 339
albumin 119, 132

alcoholic 221
alcohols 53
aldehydes 54
Alexander the Great 1
algae 35, 39, 42, 43, 89, 270, 277, 280, 294
  nutrition source 256
  -resistant roofing 269, 271
algal
  antifouling 278
  growth 293
alkaline phosphatase 123
alkalinity 83
alkyd 92, 242
  resin 241
alteration 4
alternating corrosion 198
alum 146
  tanned leather 179
aluminum 128, 197
American Conference of Governmental Industrial Hygienists 312
amides 55
aminoacid 114
  biosynthesis 45
  composition 148
aminoalkyloxysilane 100
ammonia 221
amorphous
  alumina 242
  region 108
  silica 242
amylase 173, 267
anaerobic 200
  conditions 119, 143
  corrosion 197, 198
analytical methods 291
anatase 103
animal glue 146
Animalia 7, 8
anode 198
anodic reaction 104
Antarctica wood 283
anthrax 116
antifouling
  boats 1
  coatings 291
  paint 76, 188
  system 189
antimicrobial 4
  activity 236
  pesticides 4
antioxidants 214
antiphonal parchment 151
antiseptic 4, 293
  drugs 292
antistatic agents manufacturers 338
apatite 103
  -coated titanium dioxide 239
apical pore 187
Apollo spacecraft 1
appendages 34
aquaculture equipment 186
Aramid 247
archaeological leather 142, 176
Archea 7
Archeabacteria 7
archeological waterlogged wood 284
architectural textiles 247
Aristotle 1
asbestos 271
  cement roofing 270, 271
Ascomycota 23, 24
ascospore 34
ascus 24
ash 287
aspen wood 266
asphalt 202, 226, 227
assimilation 30, 31
assimilatory biodeterioration 38
asthma 221
athlete's foot 135
atmosphere-supplying respirators 339
atomic force microscopy 301, 302
attraction forces 197
autolysis 117
automotive fuel 202, 210
autotrophic metabolism 277
aviation 202
  gasoline 206, 208
  turbine fuel 295
azoles 57

B

Bacteria 8

bacteria
  adhesion 235, 252
    force 198
  electrokinetic potential 106
  Gram-positive 41
  iron-oxidizing 270
  primary attachment 33
  proteolitic 130
  survival rate 140
bacterial 42
  activity 293
  cell intrinsic resistance 42
  count 136, 137
  quinone 198
  spore 42
  surface 198
bacteriostatic effect 162
Bacteroidetes 11
barnacle adhesion strength 291
basic toxicity information 310
Basidiomycota 23, 24, 27
bathroom 85
  fixture 241
bating 127, 128
bedroom 85
beech 288
benomyl 244
benzalkonium chloride 46
benzisothiazolone 46
benzoic acid 46, 244
β-glucosidase 108
binding 142, 144
bioaerosol 221, 224
bioattenuation agent 271
bioavailability 73, 74
biochemical process 111
biocidal 73
  efficiency 73, 74
  Product Directive 2, 329
  textiles 109
biocide 41, 73, 84, 190
  action 105
  activity 44, 73
  alteration 169
  carrier 94
  classification 2
  concentrate 251
  concentration 73, 75, 317, 318
  controlled release 242
  decay 322
  degradation mode 324
  diffusion 188
    barriers 41
  efficiency 229
  environmental
    fate 317
    toxicity 320
  features 279
  health effect 170
  immobilization 174
  industrial 41
  lethal action 45
  mechanisms of action 35
  metabolites 324
  microencapsulation 93
  migration rate 77
  new introduction 329
  performance 296
  quantity 170
  release rate 319
  risk assessment 330
  sorption 319
  sublethal dose 174
  systematic classification 329
  toxicity data 320
biocorrosion 186
biodegradability 170
biodegradable 265, 326
  materials 31
  packaging 297
biodegradation 4, 5, 81
  anaerobic 293
  die-away method 293
  mechanism 30, 81
biodeterioration 4, 5, 81
  mechanism 30, 81
  processes 38
  result 81
biofilm 42, 45, 98, 104, 195, 198, 199, 235, 239, 282, 294, 299
  formation 35, 42
  matrix 43, 190
  reduced diffusion 43
  thickness 38

biofouling 38, 187, 281
biofragmentation 30, 31
biogas 298
biological marker 85
biomagnification data 330
biomarker 85
biomass
  concentration 294
  growth rate 299
  production 300
bioreceptivity 4, 35, 277
bioremediation 296, 299
  anaerobic 293
bioresistant 265
biostabilization 41, 81
  mechanism 41
biostabilizer selection 73
biostatic properties 174
biosurfactants 204
biotic 32, 99
birch 77
bisphenol A 248
BIT 162
bitumen 226
black crust 275
Blastocladiomycota 28
blistering 90
"blue water" phenomenon 199
BMC 172
book cover 145
bookbinding 142, 144, 179
  biological damage 145
  chemical damage 145
  white leather 146
books 268
boots 177, 335
boric acid 162
boring organisms 241
bovine 113
brazilwood 284
breakdown mechanism 326
breakthrough
  concentration 342
  time 336, 337, 339, 341
breathing rate 340
brick 82, 84, 85
brightness 268
bromelain 119
Bronopol 170
brown rot fungi 245, 285
building materials 82, 83, 85
bulk erosion 76, 260
buoys 186
burns 110, 195

C

cable 104, 105
cages 185, 187
calcite
  crystals 36
  precipitation 36
calcium
  carbonate 201
    nucleation points 36
    slurry 201
  hydroxide 127
  hydroxyapatite 193
calf 120, 122
cancer 311
  -causing drugs 311
capacitance 254
capillary distribution 245
carapace 190
carbamates 58
carbendazim 172
carbolic acid 244
carbon
  dioxide measurement 298
  source 265
  steel 197
carbonation 84
carbonyl group 256
carboxylesterase 253, 260
carcinogen list 312
carcinogenic
  effect 309
  substances 311
carotenoid biosynthesis 45
casein degradation 121
catacombs 35
catheters 261
cathode 198
cathodic
  depolarization 199
    theory 106

polarization 4, 199
reaction 104
cationic membrane 44
cattle 120
Cavalier-Smith 7
ceiling 313
tiles 245
cell
lysis 44
membrane 44, 108
cellobiose 107
cellular
membrane 89
protein denaturation 180
cellulase 89, 99, 173
complex 107
cellulolytic
bacteria 287
fungi 99
cellulose 107, 174, 267, 285, 287
acetate 99
biodegradation 107, 108
crystallinity 108
fiber protection 110
molecular weight 108
cementitious matrix 82, 85
ceramic tiles 82, 85
ceresin 214
channels 35
chemical
cartridge 340
changes 108
gases 338
protective clothing 335
resistance 337
Substances Inventory 309
chemolithotrophic 277
Cheops 287
chitin 24, 300
chitosan 103
Chlamydiae 12
chloramine 46
chlorhexidine 44, 46
gluconate 237
chlorination reaction 109
chlorine dioxide 47
p-chloro-m-cresol 164, 165, 180
chloroallyltriazine-azoniadamantane 46
Chlorobi 11
Chloroflexi 13
chlorophenol 46
chlorophyll 24
a 85
chloroplasts 322
chondroitin sulfate 113
Chromista 8
chromium 128
chronic
bronchitis 222
toxicity 310
chymotrypsin 246
Chytridiomycota 28
cinematographic film 35, 37, 99
classification 7, 8
clays 277
cleaning 196
climate
control 110
cycle 32
condition 32
closed respirometer 298
clothing 247, 335
CMK 166
coal
mine 224, 225
oil 295
tar epoxy 243
coating 87, 93, 243, 261, 265
cockle 115
codon-anticodon recognition 45
cold rolling 218
collagen 112, 119, 127, 132, 142, 144
catalytic decomposition 144
composition 114
damage 132
degradation 147
fiber 121, 127, 133, 145
crack 152
flattened 147
relaxation 148
spongy structure 152
split 151
fibril 146
hydration 148

partial depolymerization 148
collagenase 119, 132, 152
activity 119
collagenic activity 152
collagenolytic enzyme 147
colonization 145
sequence 83, 186
colophony 99
color 338
spots 272
staining 149
comfort 335
compact disks 248
composites 238, 245
compostable plastics 298
composting 297, 298
computer motherboard 105
concrete 84, 275
mix 278
conductance 254
conductive substance 105
conduits 251
conidia 34, 37
conidial arms 34
connective tissue 118
consolidants 238, 243
construction
felt 268
materials 256
consumers 32, 33
contact
dermatitis 335
lenses 235, 236
time 45, 73, 258
contamination 335
screening test 293
controlled release 103
conveyors 247
coolants 218-220, 221
cooling
-lubricating fluid 222
techniques 154
tower 200
water 200, 282
system 294
copolymers 239
copper 77, 85, 100, 197
ion 188, 292
copper
pipe 85
corrosion 281
powder 292
sheet 188
tubing 199
-tolerant microorganisms 284
cork test species 284
corrosion 104, 197, 268
inhibitor 200, 214, 296
cortex 42
cosmetics 235, 236, 331
industry 2
cost 75
cotton 106, 109
Council on Environmental Quality 309
crackling 99
cresylic acid 2
crude oil 205, 295
crystalline
morphology 246
region 259
structures 302
crystallinity 77, 284
cultural heritage 97, 100, 175, 275, 299
cuprous oxide 188
cut method 337
cutinase 253
Cyanobacteria 13, 14, 35, 89, 277
sheaths 36
cytochrome 45
cytoplasm residues 108

D

Daphnia magna 293
darkening 203
data 309
DCOIT 191
Dead Sea 148
deamination 108
Debye forces 33
decay 4
decomposers 24
degradation rates control 253
dehairing 147
dehydrogenase 253, 262
delamination 90

Delaney clause 311
deliming 128, 147
dental
  application 265
  biofilm 102
  materials 101, 102
  water 282, 283
    system 102
dentures 101, 102, 238, 239
  stomatitis 102
deodorants 235, 236
depilating action 127
depilation 127, 147
depolymerization 108
dermatophytes 135, 175
dermis 113, 118, 121, 144, 156
desalination 185, 188
desaturase 45
desiccation 35
destroyers 32, 33
detergents 294
developmental and reproductive toxicity 310
dexterity rating 337
dextrose agar 291
diabetic 221
diatoms 36
diazolidinyl urea 237
dibromodicyanobutane 46
dichlofluanid 188, 292
dielectric properties 105, 254
diesel fuel 295
diffusion 41, 260
  coefficient 342
digestive enzymes 24
Dikarya 23
dimethoxydimethylhydantoin 46
dimethyldithiocarbamate 46
dimethyloxazolidine 46
dimethylphenylsulfamide 292
dimethyltolylsulfamide 292
DIMTS 172
direct
  additives to food 315
  skin contact 335
discoloration 90, 99, 268
disfigurement 90
disinfect 5
disinfectant 293
disinfection 105
dispersants 297, 299
dissimilatory biodeterioration 38
disulfide bond 108, 198
diuron 47
dizziness 221
DMDM hydantoin 237
DNA 45, 86
  synthesis 45
Domain 7
door profile 84
doors 82, 83, 85
dosing frequency 73
double helix 86
dressing 292
drinking water 282
durable goods 31
dust 144, 176
  particles 218
dyeing 170

E

ecosystem 322
ecotoxicity 295, 310
ectoparasites 114
efflorescence 5
Egypt 287
Egyptian pyramid 287
elastase 119
elastin 119
  fiber 150
electric
  current 105
  potential 104
  resistance 105
electrical 104
  equipment 105
electrochemical
  cell 104
  potential 199
electron
  acceptor 198
  beam 155
  microscopy 121
  transfer 198
  transferring molecule 198
  transport 86

electronic 104, 105
electrostatic
  attraction force 197, 198
  interaction 198
elongase 45
Emergency Temporary Standards 313
emission control 311
Empire 8
emulsifier decomposition 220
emulsion 218, 221
  paint 85
encapsulation 77
Endangered Species Act 332
endocleavage 262
endocuticle 108
endoglucanase 89, 107
endopolyurethanase 260
endoscope tubing 194
endotoxins 203
energy dispersion X-ray analysis 198, 302
enhanced efflux 46
entrapped chemicals 336
environmental
  effect 75
  factors 35
  impact 170
  persistence 325
  Protection Agency 309, 332
enzymatic
  degradation 150
  detoxification 46
  digestion 128
  oxidation 105
  trapping 46
enzyme 127
  activity 119
  mediated resistance 43
EPDM 244
epidermis 113, 118, 121, 135
epoxy 189
  coating service life 243
  resin 243
equilibrium
  partial pressure 341
  relative humidity 83
ergosterol 85, 300
eroding coating 189
erosion rate 291
ESEM 301
esterase 99, 242, 246, 249, 260
ethylene propylene copolymer 244
ethylene-diamino-tetra-sodium acetate 160
ethylenethiourea 292
etiological factors 135
ettringite 85
Eubacteria 7
Eukaryota 8
eukaryotic 99
European
  Standard 336
  Union 314, 329
    classification 47
EUSES 330
evaporation 336
exclusion volume 252
exoenzyme production 287
exoglutanases 107
exopolymer 198
exopolymeric substance 35, 186
exopolysaccharides 42
exopolyurethanase 260
exotoxin 222
exposure
  limits 339
  temperature 338
external conditions 118
extracellular
  enzymes 89
  polymer 35, 42
eye
  irritation 221
  makeup 235, 236
  protection 335, 338
    standard 338

F

face
  protection 338
  side 127
Family 7
fat 132, 135
feather 144
Federal Food, Drug, and Cosmetic Act 332
feldspars 277
Fenton reaction 285

fermentative microorganism 198
fiber 106, 108, 109, 247
  coat 189
  structure 113
  /matrix interface 245
  board 245
fibril 112
Fibrobacteres 14
fibroin 107
fibrous membrane 261
ficain 119
FIFRA 332
fig leaves 100
filamentous
  bacteria 199
  fungi 150, 164, 173, 180
fillers 77, 201
film preservation 87, 88, 90
  fungal efficiency 294
filter 239, 339, 340
  capacity 340
  life 341
fingerprinting 300
finishing 170
firing temperature 84
Firmicutes 14
first invaders 88
fish farming 185, 187
flagella 29
flagstone 270
flame resistance 337
flax 107
flooring 264
flux rate 77
flying particles 338
fogging resistance 338
Foley catheters 261
Food and Drug Administration 315
food
  film 251
  packaging material 244
  Quality Protection Act 332
  regulatory acts 309, 315
footwear 140, 173, 174
  disinfection 174
  hygienic finishes 175
  production 112
  service life 140
formaldehyde 2, 46, 237, 255
  -degrading microorganism 201
  -releasing compounds 60
formulation 338
fouled hull 187
foxing 5, 99, 268
free
  chlorine 47
  radical oxidation 148
  radicals 285
freeze-drying 157
fresco 97, 100
freshwater assessment 330
friction 214
frustule 187
FTIR 300
fuel 294
  consumption 187
  microbiological purity 205
  microorganisms 206
  storage 210
  strategic reserve 229
  suspended particles 211
fungal
  adhesion 35
  filaments 39
  growth 84
  slime 297
  spores 42
Fungi 7, 8, 23
fungi 39, 42, 43, 145
  arbuscular mycorrhizal 42
  hypha 137
  keratinophilic 143
  occurrence 138
  pathogenic 139
  saprophytic 136
fungicide distribution 165, 166
fusion enthalpy 247
Fusobacteria 17

G

gadfly 114
gamma radiation 155
gas
  chromatography 294, 297
  formation 90

transmission pipeline 199
gasoline 212, 295
gas-turbine fuel 294
gelatin hydrolysis 121
genetic adaptation 43
genome 300
Genus 7
geomembrane 295
geotextile 295
germ cell wall 42
germination 117
gingivitis 103
glass 97, 100
fiber 239
surface 252
wool 85
glioxylic acid 253
globulin 119, 132
glomalin 42
Glomeromycota 29
gloss 242
glove 174, 335, 336
fitness 337
selection 336
glucose 107
glutaraldehyde 46, 128
glycocalyx 43, 46
formation 33
glycogen 113, 117
glycolipids 43
glycoprotein 33
goat skin 113
goethite 201
goggles 339
good housekeeping 336
Gram
-negative 11, 12, 17, 41, 222
-positive 9, 14
granite 276
graphite 214
fibers 250
grazing incident diffraction 302
greases 202
Greece 287
green
algae 189
biofilm 36
gypsum 85
board 82, 83, 85
gyrase 45

H

hair conditioners 236
haloalkylthio compounds 61
halophile 120, 156, 162
bacteria 162
hand
creams 236
scrub 293
wash 292, 293
hazard assessment 335
hazardous substances 339
hazards 313
haziness 203
health
and safety 310
problems 203
care application 241
heat
exchanger 188
resistance 337
heating oil 212, 213
heavy metals 100
helical windings 285
hemicellulose 99, 287
heritage leather 175, 178
heterocyclic N,S compounds 63
heterotrophic 277
bacteria 276, 294
microflora 277
microorganisms 275
heterotrophs 33
hexahydrotriethyltriazine 46
High Production Volume chemicals 309
historical
note 1
paper 267, 269
holocellulose 285
horse 122
house paint 94
hull cleaning 187
human safety risk 267
humic acid 268
hyaluronic acid 113
hybrid binder 242

hydraulic
fluids 295
oils 216
system 224
hydrodynamic drag 187
hydrogen 199
embrittlement 5, 199
peroxide 46, 281
sulfide 84, 220, 221
hydrolysis 325
hydrolytic enzymes 173
hydroperoxides 256
hydrophilic microorganisms 83
hydrophobicity 197, 198, 265, 278
hydrophobins 42
hydroxyethylcellulose 103
hydroxyl radical 285
hydroxyproline 147
hypersensitivity pneumonitis 224
hyphae 37, 278, 287
hyphal
cell 195
penetration 39
hypochlorite ion 280
hypochlorous acid 280
hypodermatosis 177

I

ice-wedging action 39
imidazolidinyl urea 46, 237
Immediately Dangerous to Life and Health 313, 339
impact
protection 335
velocities 335
impaired 46
uptake 46
imperfect fungi 142
implants 101, 102, 235, 236, 246
impression material 101, 102
inactivation 43, 46
in-can preservation 87, 90
indirect additives to food 315
indole 143
industrial
paper 267
products 81
indwelling devices 195
infections 194
inhalation 340
initiating 41
step 41
injury 309
inner spore coat 42
insoles 136
insulating
material 226
rubber 265
insulation 105
intensity of effort 339
intercellular communication 190
interchain interaction 246
interior paint 92
International Association for Research on Cancer 311
intracellular 44
free radicals 44
intravascular catheters 195
intrinsic resistance 45
mechanisms 46
iodine 46, 237
ionic
interactions 43
strength 35, 198
ionomer 244
iron
salt 142
-reducing microorganisms 197
irradiation 153
isoelectric point 198
Italy 276
IUCLID 331

J

Japanese beech 288
jet kerosene 207

K

kaolinite 201
Kapron 247
keratin 28, 108, 119, 132, 221
keratinase 108
keratinolysis process 108
keratinolytic enzyme 221
kerosene 207
Kevlar 247
Kingdom 7

Kool Kount 294

L

landfill 298
Lascaux Cave 13
latent period 118
latex paint 93
leachable biocides 76
lead 100
leaks 340
leather 81, 97, 100, 110
  conservation methods 176
  fungal species 133
  gloss 133
  import 112
  industry water consumption 172
  market share 113
  microbiocides 163
  pH 126
  processes involved 126
  producing countries 111
  product 135, 142
  production 111
  technology 125
  vegetable tanned 144
  water sensitivity 146
  white 146
legislation 170, 329
lethal 45
letheen agar 291
library material 142
lice 114
lichen 39, 84, 270, 277, 278
Lifshitz-van der Waals attraction 252
lignin 99, 107, 285, 287
  concentration 287
limestone 1, 276
liming 127, 147
linen 107
Linneaeus 7
lipase 99, 133, 173, 253, 260
lipid 45, 108
  biosynthesis 45
lipolytic taste 142
lipophilic properties 41
lipopolysaccharide 41, 43
liquid chemicals 338
lobster pots 186
longevity 76
lotions 236
lotus leaf 190
louse 114
low carbon steel 200
lubricants 202, 214, 215, 295
lubricating oils 214
lumber 296
lung diseases 221
lymph node 118
lyophilization 153
lysis 260

M

machining 218
maintenance 319
makeups 236
malodor 90
manganese-oxidizing bacteria 198
manufacturing method 331
maple 288
marine 202
  coatings 243
  environment 298, 330
  fuels 211
  installations 184
  sediments 11
  sensors 186
  transport 184
mass
  loss 149
  spectrometry 297
mattress 264
maximum permissible concentration 322
MBT 164
mechanical
  damage 337
  deterioration 38
  properties 90
medetomidine 188
medical
  devices 249, 265, 298
  equipment 194
  instrument 196
  materials 33
medieval paintings 100
membrane 257
  disintegration 45

function 45
mercuric salts 46
mercury 100
mesophilic aerobe 152
metabolic
activity 43
mechanism 140
metal 104, 197
additives 292
-containing products 64
corrosion 2, 187
mechanism 198
structure weakening 199
surface 198
metalworking fluids 202, 218, 296
methanol 143
methyl paraben 46
methyldibromo glutaronitrile 237
methylene bis(thiocyanate) 159
methylenebischlorophenol 46
methylisothiazolone 46
micafungin 195
microbial
adhesion 35
cluster 35
detoxification 293
ecosystem 198
-induced corrosion 5
microbiocide performance 296
microbiological
contamination 204
corrosion 144
microemulsions 219
microfractures 99
micronodules 198, 199
microorganism 44
growth climate 242
inactivation mechanism 44
killing ability 73
morphology 300
salt-tolerant 120
microporosity 341
microporous PU 261
Microsporidia 29
Middle Ages 147, 148
middle lamella 285, 286, 287
migration 77
mild steel 197, 200
mildew 270
military applications 261
Mine Safety and Health Administration 339
mineral
dispersion 200
salts 132
agar 291
minimum inhibition concentration 74
MIR station 1
mites 115
moisture 173
content 245
mold
attack 192
colonization 85
count 85
growth 105
odor 108
spores 179
molecular
sieving 43
target site 45
techniques DNA-based 300
weight 246, 260
Monera 7
monochloramine 47
monofilaments 112
montmorillonite 201
monuments 1
mortar 276
protection 265
mosses 277
motor oil 295
mouthwash 236
mucilage 34, 249
mucopolysaccharides 113, 152
multilayer film 252
municipal sewage sludge 298
mural 97, 100
paintings 99
murduge 276
museum
collection 142
objects 175
mussel 190
mutagenicity 310

myalgia 223
mycelium 133, 145
  cells 300
  conductivity 105
mycobacteria 42
mycotoxin 85, 223
N
nail stratification 135
nanoparticles 77
National Institute for Occupational Safety and Health 311, 312, 336, 339
National Toxicology Program 311
natural
  latex 271
  rubber 271
Neocallimastigomycota 29
netting 185, 187
neutralization 179
N-halamine 109, 247
nitrifying bacteria 270
nitrosodialkyl aniline 218
Nitrospirae 17
Nocardia sp. 272
Nomex 106, 108, 109, 247
non-eroding coating 189, 265
nonylphenol ethoxylate 213
Norway spruce 284, 288
nosocomial infection 109, 195
NPIRS 332
nucleic remnants 108
nucleotide biosynthesis 45
nutrient 88
  starvation 35
nutritional
  requirement 173
  value 145
O
Occupational Exposure Limits 314
Occupational Safety and Health Act 311
Occupational Safety and Health Administration 311, 312, 335
octenidine 110
odor 90, 267
offshore constructions 186
oil
  drilling 200
  emulsions 202, 218
  field 296, 299
  oxidation 146
  spill 296, 297, 299
ointments 292
OITZ 166
old dry wood 287
oligotrophic character 173
onychomycosis 140
o-phenylphenol 165, 178
OPP 166
oral care 101, 102
orange bag 250, 251
Order 7
organic fuels 211
organosilicone 192
organotin compounds 184
ornamental embossing 146
osmotic pressure 156, 161
outdoor environment 265
outer
  membrane structure 43
  spore coat 42
oxidants 43
oxidase 262
oxidation 298
oxidative reactions 285
oxidizing
  agents 66
  stress response 43
oxygen
  concentration 88
  demand 298
P
packaging materials 297
paint 84, 87, 192, 193, 238, 239, 241, 242, 261
  defacement 293
  self-polishing 255
  thickness 291, 319
  thinning 89
painting 98, 100
pallets 296
papain 119, 246
paper 98, 100, 266, 267, 268, 297, 315
  insulation 104
  mill slime 267
  quality 267
paperboard 315

papillae 113
papillary layer 113
parabens 44, 237
para-chloro-meta-cresol 165
paraffin 251
  wax 214
parchment 97, 142, 146, 147, 149, 151, 179
  manufacture 148
particulate carrier 242
pathogenic
  fungi 174
  microorganism 174
  species 202
pathway compensation 46
patina 5
pearl culture 185, 188
pectic substances 107, 285
penetrants 337
penetration depth 39
peptidoglycan 42
perforation 127
permeability 336
permeation resistance 336
permissible 73
  concentration 73
  Exposure Limits 312
peroxidase 204
persistence 170
Personal Protective Equipment Standard 335
pest 5
pesticide 5
petrographic microscope 301
petroleum products 81, 202, 205
pH 35, 73, 75, 77, 88, 104, 198, 203, 236, 258, 268
  change 90
pharmaceuticals 235
phenol 2, 248
phenolic 67
  compounds 287
  resin 245
phenols 44
phenoxyethanol 46
phenylethyl alcohol 46
phenylmercuric acetate 46
o-phenylphenol 165
phosphatase 123
  inhibitor 123
phospholipase 260
phospholipids 43
photographic material 98, 100
photon flux density 326
photosynthesis inhibition 322
photosynthetic electron transport 45, 322
phototrophic microorganisms 277
phthalate esters 263
phylogenetic 8
Phylum 7
physiological needs 140
pickling 128, 153, 170
pig 120, 122
pigments 120, 201
pimarane 99
pink staining 152
pipe 82, 83, 84, 86, 186, 241
  wrap 252
pitch deposit 267, 269
pitting corrosion 199
Planctomycetes 17
plant hygiene 87, 90
Plantae 7, 8
plaque 102
plasma cell membrane 42
plasmid 46
  -borne mechanism 46
plasterboard 84
plastic 297
  article 261
  profiles 82
plasticizer 105, 263
  degradation 259
plumbing 272
polarizing microscope 301
pollutants 32
pollution prevention 310
poly(ethylene oxide) 252
  brush 252
  molecular weight 252
poly(ethylene terephthalate) 253
  co-monomers 253
poly(vinyl alcohol) 103, 262
poly(vinyl chloride) 263
  article 105
  based

devices 195
material 185, 188
membrane 271
polyacrylic acid 201
polyamide 246
crystallinity 246
fiber 246
hydrogen bonding 246
polyamide-6 107, 109, 246
polycarbonate 248
lenses 338
membrane 249
moisture absorption 249
plasma treatment 249
polychromed sculpture 99
polycondensation 247
polyester 109, 246
polyurethane 259, 260
polyether PU 261
polyetheretherketone 250
polyethylene 250, 251
high density 251
tubing 194
/wood composite 288
low density oxidized 251
polyhexamethylene biguanide 280
polyimide 105, 254
film 254
polymeric biocides 68
polymers 297
polymethylmethacrylate 255
polyoxymethylene 255
polypropylene 107, 109, 256
surface colonization 256
wood composite 270
polysaccharide 33
polystyrene 257
ozone treated 258
polysulfone 258
polytetrafluoroethylene 259
polyurethane 189, 259, 261
devices 195
formulation 259
polyvinylpyridine 77
Pontiac fever 223
pore
size distribution 83
surface 38, 276
porins 41, 46
porosity 38
porphirin biosynthesis 45
positive pressure 340
post-mortal processes 117
potassium
bicarbonate 178
fluoride 178
povidone iodine 46
Prague Castle 177
prehistoric wood 284
preservative 5
pressure drop 340
primer 95
printed circuit 104, 105
process change 336
procollagen 119
producers 32
product preservation 1
progelatin 152
projectile shape 335
Prokaryota 8
prokaryotic 99
proliferation 33
propyl paraben 46
protease 108, 126, 133, 152, 173, 267
protective
barrier 336
clothing 336, 337
device 335
equipment 335
layer 338
protein 246
biosynthesis 45
residue 132
Proteobacteria 17
proteolitic
activity 126
capacity 99
enzyme 118
fungi 142
Protista 7
proton translocation 45
protonophores 44
protonophoric uncoupling 45
protoporphyrinogene oxidase 45

Protozoa 8, 29, 42
protozoal 42
  cysts 42
  oocysts 42
pseudoconidia 151
Pseudomonas aeruginosa 35
pseudo-tanning 146
pulmonary system 221
pulp 266
puncture resistance 337
pure components 338
purine biosynthesis 45
putrefactive bacteria 128
pyrazolidone 218
pyridine 69
  derivatives 69
  -triphenylborane concentration 292
pyrithione 44
pyrogallic acid 129

Q

quaternary ammonium
  compound 44, 46, 70, 170, 237
  salt 247
Qumran 148

R

radiation intensities 335
radical-mediated reaction 43
radiolabelled plastics 297, 298
radiolabelling 301
Raman spectroscopy 301
raphe 187
rawhide 110
  production 111
Recommended Exposure Limits 312
recyclers 24
red decomposition 144
"red heat" 121, 145, 162
redox catalyst 45
reductase 45
reduction reaction 198
refinery 296
regulations 309
relative humidity 342
release rate 76, 292
resistance to biocides 299
resistivity 254
respirable air 339
respirator 339
respiratory
  electron transport 45
  protection 335, 339
  tract 314
restoration composites 102
reticulins 119
reverse osmosis membrane 185, 188, 247
rhamnolipids 43
rheological properties 91
rhizines penetration 39
ribosomal
  DNA sequences 300
  genes 300
  protection 46
  sequence 300
ribosomes 45
risk 309
  assessment 310
rock caverns 207, 213
Roman catacombs 275
roofing 227, 269
  felt 227
  granules 270
  slate 270
root canal
  core material 102, 103
rosin 189
rotting 132
  bacteria 118
rough side 120
roughness 35, 89
rubber 271, 272
  oxidation 272
  straps in Russian space station 272
  wood 284
rulemaking procedures 311

S

safeguarding children's health 310
salicylic acid 2
salt
  stains 122, 123
  water 186
salted
  hides 120
  skin 122
salting 120, 153, 170

saltwater immersion 243
sanitation 87, 88, 91, 174
sanitizer 5
Santa Cruz la Real 39
saprophytic 174
  fungi 143
Scots pine 288
scratch resistance 338
scrolls 148
sealants 93, 238, 239, 266, 273, 291
seawater 319
  cooling 283
  desalination 283
secondary infections 221
sediment 318
selection 73
self-polishing coating 190
SEM 301
sequence of events 108
sericin 107
severity of exposure 336
sewage treatment plant 330
sewer pipes 84
shampoo 236
sheep 120, 122
sheepskin 113
shelf life 236
shell 190
ship 185, 214, 291
  coating 188
  hulls 186
shock treatment 280
shoe
  industry 111
  materials 173
Short-Term Exposure Limit 313
shower gel 236
shrink-wrap film 252
siding 82, 83, 84, 86, 263
silane 267
silicon gel 164
silicone 189, 265
silk 107
  fiber 107
silver 86, 110, 175
  coins 1
  ion 174
site-specific microcosms 293
skatole 143
skin 190
  antisepts 235, 236
  care 236
  chemical treatment 161
  damage 114, 122
  dehydration 161
  drying 153
  histological picture 116
  infection 115
  lession 111
  microflora 111, 292
  pH 114, 117
  preservation 153
  structure 113
slate 270
slaughterhouse 118
slime 267
  control 297
  producing bacteria 198
slimicides 269
sliminess 119, 120
soaking 126, 170
soaps 236
sodium
  hypochlorite 109, 247, 281
  pyrithione 167
  sulfide 127
software 342
soiling 38
Solar Boat 287
solid-phase micro extraction 294
solubility 99
sorbic acid 46
sorptive behavior 319
Southern pine 284, 288
spa 279
space shuttle 1
Spain 276
species 7
spill response 296
spiral wound membrane 283
splash
  protection 335
  resistance 339
spore 120

bacteria 127
germination 249
sporicidal activity 293
spring water 282
staining 297
stainless steel 86, 197, 198, 243
surface 199
stains 108, 268
standard classification system 309
standards 291
starch 251, 268
steady state permeation rate 336
steel pipe 200
sterigmatocystin 85
steroid biosynthesis 45
stone 1, 275, 278
consolidants 298
monuments 32
protective coating 265
storage 87
conditions 176
structural decomposition 108
stucco surface 36
subcutaneous tissue 112
sublimation 157
subsoles 136
sugars 132
sugi wood 288
suits 335
sulfate-reducing bacteria 203, 211, 212, 213, 219, 225, 229
sulfitolysis 108
sulfur content 203
sulfuric acid 84, 144
sunlight irradiation 325
super-chlorination 280
superhydrophilic 242
surface
active agents 70
changes 108
charge 35, 197, 198
degradation 105
energy 266
erosion 76, 260
wettability 198
surfactant 214, 294
surgical implants 298
survival rate 141
sweat 135
swimming pool 279
synergism 47
syntane 128
synthase 45
systemic toxic effects 335

T

talc 214
tank 204
tannase 129, 133
tannin
ferrous complex 143
hydrolysis 134
penetration 128
tanning 128, 170, 179
agent 143
biocides 168
industry 111, 132
process 128, 131
steps 168
tapestries 99
targets 43
tastes 267
taxonomic system 7
TCMTB 164, 165, 166
TEM 301
temperature 73, 258
range 203
regulator 214
tensile strength 260
teratogenicity 310
term 47
tetrahydrothiadiazinthione 46
textiles 2, 98, 100, 106, 108, 195
texture 338
thallus 278
thermal
insulation 82, 83, 85
protection 337
stability 203
Thermodesulfobacteria 23
Thermotogae 23
thickening agents 89
thin films 257
Threshold Limit Value 312
time scale 31

Time-Weighted Average 313
tire rubber 272
tissue engineered medical products 298
titanium dioxide 103, 193
toiletries 235
Tollens' reaction 247
tolylfluanid 292
toothpaste 236
Toxic Substances Control Act 309
toxicity 75, 170
  bioassay 322
toxin 221
  formation 90
traffic paint 95
transformer
  oil 13, 202, 217
  tank 104
transportation cost 187
transposon 46
triclosan 77, 175, 237
trifluoromethyl dichlorocarbanilide 46
tropocollagen 119
trypsin 246
tufa 276
turbidity 211
turbine oils 217
Turkey 276

U

ultrafiltration membrane 283
undersole 174
United States Congress 309
urease 86
ureolytic 86
UV
  degradation 75
  photopolymerization 240
  radiation 32

V

valine 119
van der Waals forces 33
vapor
  desorption 340
  pressure 341
varnish 98, 99
vascular catheter 195
vegetable
  tanned leather 178
  tannin 142
  tanning 128
vellum 148
Venetian turpentine 99
ventilation 336
Verrucomicrobiae 12
vinyl 189
viral penetration 337
virus 42
viscosity 203, 227
  change 90
  modifier 214
voids 35

W

wallpaper 82, 83, 85, 86
warning labels 335
water 281
  absorption 246
  control 269
  protection 1
  resistance 132
  systems 299
waterlogged wood 287
waxes 107
waxy cell walls 46
weight loss 245
wet-blue 164, 165, 166, 172
white
  biofilm 36
  leather 147, 179
  -rot fungi 285
  stains 149
WHO recommended classification 331
window 82, 83, 85
  profile 84
wire 105
  pit 199
wood 283, 287
  boxes 296
  cell wall 285, 286
  components 287
  decay 13
    fungi 284
  flooring 288
  flour 251, 252, 256, 263
  microfibril 285
  preservation 2

pulp 99
stain 92
wool 28, 107, 108
work intensity 340
workplace exposure limits 309
wound 110, 195
dressing 236

X

xerophylic fungi 83
X-ray diffraction 302

Y

yeast fungi 135
yellowing 99

Z

zeolite 242
zinc 86
chloride 2
ethylene-bis (dithiocarbamate) 292
pyrithione 188, 237
zineb 292
zirconium 128

Acknowlegment

The authors and publisher would like to thank Wydawnictwo Politechniki Lodzkiej in Lodz, Poland for giving permission to use two chapters previously published in Microbiologia Materialow, Ed., Bronislaw Zyska and Zofia Zakowska, Lodz 2005. These chapters were used for preparation of two new, updated chapters in this book:
Chapter 4.7 Leather and leather products by Michalina Falkiewicz-Dulík
Chapter 4.12 Petroleum products by Katarzyna Janda

The contents of the chapters were changed to account for current information and to prepare material for translation to English language.

LaVergne, TN USA
29 April 2010
180935LV00003B/2/P

9 781895 198447